T0137408

CRM Series in Mathematical Physics

The Centre de recherches mathématiques (CRM) was created in 1968 by the Université de Montréal to promote research in the mathematical sciences. It is now a national institute that hosts several groups and hold special theme years, summer schools, workshops, and a postdoctoral program. The focus of its scientific activities ranges from pure to applied mathematics and includes statistics, theoretical computer science, mathematical methods in biology and life sciences, and mathematical and theoretical physics. The CRM also promotes collaboration between mathematicians and industry. It is subsidized by the Natural Sciences and Engineering Research Council in Canada, the Fonds FCAR of the Province of Québec, and the Canadian Institute for Advanced Research and has private endowments. Current activities, fellowships, and annual reports can be found on the CRM Web page at www.CRM.UMontreal.CA.

The CRM Series in Mathematical Physics includes monographs, lecture notes, and proceedings based on research pursued and events held at the Centre de recherches mathématiques.

More information about this series at http://www.springer.com/series/3872

Şengül Kuru • Javier Negro • Luis M. Nieto
Editors

Integrability, Supersymmetry and Coherent States

A Volume in Honour of Professor Véronique Hussin

Springer

Editors
Şengül Kuru
Department of Physics
Ankara University
Ankara, Turkey

Javier Negro
Department of Theoretical Physics,
Atomic Physics and Optics
University of Valladolid
Valladolid, Spain

Luis M. Nieto
Department of Theoretical Physics,
Atomic Physics and Optics
University of Valladolid
Valladolid, Spain

CRM Series in Mathematical Physics
ISBN 978-3-030-20089-3 ISBN 978-3-030-20087-9 (eBook)
https://doi.org/10.1007/978-3-030-20087-9

This Springer imprint is published by the registered company Springer Nature Switzerland AG.
The registered company address is: Gewerbestrasse 11, 6330 Cham, Switzerland

Preface

The *6th International Workshop on New Challenges in Quantum Mechanics: Integrability and Supersymmetry* was organized in Valladolid (Spain) on June 27–30, 2017, in honor of Prof. Véronique Hussin. The conference was very successful; it attracted more than 60 researchers from different countries, and about 45 talks were delivered.

Prof. Hussin is an internationally recognized expert in many branches of mathematical physics. Her career started in her native city of Liège (Belgium); afterward she moved to Montréal, where she has stayed ever since at Université de Montréal and at CRM (Centre de Recherches Mathématiques). However, Véronique Hussin has had numerous stays in different research centers collaborating with many colleagues and friends. In particular, we want to mention her regular visits to Durham University (England), CINVESTAV (Mexico), and the University of Valladolid (Spain) where, besides her scientific activity, due to her open character, she made many friends. She has made remarkable contributions to several areas, like coherent and squeezed states, supersymmetry, sigma models, nonlinear equations, and other topics that show her wide range of interests.

The workshop program paid special attention to the topics in which Prof. Hussin has been working. Based on this conference, but open to scientists in these fields, we considered a good opportunity to edit a contributed book published by Springer Nature, under the title "Integrability, Supersymmetry and Coherent States". The book is devoted in part to review papers on the three main topics, whose authors are well-known experts in their fields and where the objective is to take a personal approach on some attractive aspects of these subjects. A second part is made up of

Special thanks are due to the University of Valladolid, where the meeting in honor of Prof. Véronique Hussin *6th International Workshop on New Challenges in Quantum Mechanics: Integrability and Supersymmetry* was held.

Partial financial support from Junta de Castilla y León and FEDER (Projects VA057U16, VA137G18, and BU229P18) and Ministerio de Economía y Competitividad of Spain (Project MTM2014-57129-C2-1-P) is acknowledged.

a list of contributions related to these topics. All the papers have satisfied a strict refereeing process by anonymous specialists before publishing.

It is a pleasure, for all of us, to dedicate this volume to Véronique Hussin on the occasion of her 60th birthday.

Ankara, Turkey
Valladolid, Spain
Valladolid, Spain
March 31, 2019

Şengül Kuru
Javier Negro
Luis M. Nieto

Contents

Curvature as an Integrable Deformation 1
Ángel Ballesteros, Alfonso Blasco, and Francisco J. Herranz

Trends in Supersymmetric Quantum Mechanics 37
David J. Fernández C.

Coherent States in Quantum Optics: An Oriented Overview 69
Jean-Pierre Gazeau

Higher Order Quantum Superintegrability: A New "Painlevé Conjecture" 103
Ian Marquette and Pavel Winternitz

Supersymmetries in Schrödinger–Pauli Equations and in Schrödinger Equations with Position Dependent Mass 133
Anatoly G. Nikitin

Nonlinear Supersymmetry as a Hidden Symmetry 163
Mikhail S. Plyushchay

Coherent and Squeezed States: Introductory Review of Basic Notions, Properties, and Generalizations 187
Oscar Rosas-Ortiz

Trace Formulas Applied to the Riemann ζ-Function 231
Mark S. Ashbaugh, Fritz Gesztesy, Lotfi Hermi, Klaus Kirsten, Lance Littlejohn, and Hagop Tossounian

Real- and Complex-Energy Non-conserving Particle Number Pairing Solution 255
Rodolfo M. Id Betan

Jacobi Polynomials as $su(2, 2)$ Unitary Irreducible Representation 267
Enrico Celeghini, Mariano A. del Olmo, and Miguel A. Velasco

Infinite Square-Well, Trigonometric Pöschl-Teller and Other Potential Wells with a Moving Barrier 285
Alonso Contreras-Astorga and Véronique Hussin

Variational Method Applied to Schrödinger-Like Equation 301
Elso Drigo Filho, Regina M. Ricotta, and Natália F. Ribeiro

The Lippmann–Schwinger Formula and One Dimensional Models with Dirac Delta Interactions 309
Fatih Erman, Manuel Gadella, and Haydar Uncu

Hermite Coherent States for Quadratic Refractive Index Optical Media 323
Zulema Gress and Sara Cruz y Cruz

Analysis of $\mathbb{C}P^{N-1}$ Sigma Models via Soliton Surfaces 341
Piotr P. Goldstein and Alfred M. Grundland

On the Equivalence Between Type I Liouville Dynamical Systems in the Plane and the Sphere 359
Miguel A. González León, Juan Mateos Guilarte, and Marina de la Torre Mayado

Construction of Partial Differential Equations with Conditional Symmetries 375
Decio Levi, Miguel A. Rodríguez, and Zora Thomova

An Integro-Differential Equation of the Fractional Form: Cauchy Problem and Solution 387
Fernando Olivar-Romero and Oscar Rosas-Ortiz

Quasi-Integrability and Some Aspects of $SU(3)$ Toda Field Theory 395
Wojtek Zakrzewski

On Some Aspects of Unitary Evolution Generated by Non-Hermitian Hamiltonians 411
Miloslav Znojil

Index 427

Contributors

Mark S. Ashbaugh Department of Mathematics, University of Missouri, Columbia, MO, USA

Ángel Ballesteros Departamento de Física, Facultad de Ciencias, Universidad de Burgos, Burgos, Spain

Alfonso Blasco Departamento de Física, Escuela Politécnica Superior, Universidad de Burgos, Burgos, Spain

Enrico Celeghini Dpto di Fisica, Università di Firenze and INFN–Sezione di Firenze, Firenze, Italy

Dpto de Física Teórica and IMUVA, Univ. de Valladolid, Valladolid, Spain

Alonso Contreras-Astorga Cátedras CONACYT—Departamento de Física, Cinvestav, Ciudad de México, Mexico

Department of Physics, Indiana University Northwest, Gary, IN, USA

Sara Cruz y Cruz Instituto Politécnico Nacional, UPIITA, Ciudad de México, Mexico

Marina de la Torre Mayado Departamento de Física Fundamental and IUFFyM, Universidad de Salamanca, Salamanca, Spain

Mariano A. del Olmo Dpto de Física Teórica and IMUVA, Univ. de Valladolid, Valladolid, Spain

Elso Drigo Filho Instituto de Biociências, Letras e Ciências Exatas, IBILCE-UNESP, São José do Rio Preto, SP, Brazil

Fatih Erman Department of Mathematics, Izmir Institute of Technology, Urla, Izmir, Turkey

David J. Fernández C. Departamento de Física, Cinvestav, Ciudad de México, México

Manuel Gadella Departamento de Física Teórica, Atómica y Óptica and IMUVA, Universidad de Valladolid, Valladolid, Spain

Jean-Pierre Gazeau APC, UMR 7164, Univ Paris Diderot, Sorbonne Paris Cité, Paris, France

Centro Brasileiro de Pesquisas Físicas, Rio de Janeiro, RJ, Brazil

Fritz Gesztesy Department of Mathematics, Baylor University, Waco, TX, USA

Piotr P. Goldstein Theoretical Physics Division, National Centre for Nuclear Research, Warsaw, Poland

Miguel A. González León Departamento de Matemática Aplicada and IUFFyM, Universidad de Salamanca, Salamanca, Spain

Zulema Gress Universidad Autónoma del Estado de Hidalgo, Ciudad del Conocimiento, Hidalgo, México

Alfred M. Grundland Centre de Recherches Mathématiques, Université de Montréal, Montréal, QC, Canada

Département de Mathématiques et Informatique, Université du Québec à Trois-Rivières, Trois-Rivières, QC, Canada

Lotfi Hermi Department of Mathematics and Statistics, Florida International University, Miami, FL, USA

Francisco J. Herranz Departamento de Física, Escuela Politécnica Superior, Universidad de Burgos, Burgos, Spain

Véronique Hussin Centre de Recherches Mathématiques & Département de Mathématiques et de Statistique, Université de Montréal, Montréal, QC, Canada

Rodolfo M. Id Betan Instituto de Física Rosario (CONICET-UNR), Facultad de Ciencias Exactas, Ingeniería y Agrimensura (UNR), Santa Fe, Argentina

Instituto de Estudios Nucleares y Radiaciones Ionizantes (UNR), Ocampo y Esmeralda, Rosario, Santa Fe, Argentina

Klaus Kirsten GCAP-CASPER, Department of Mathematics, Baylor University, Waco, TX, USA

Decio Levi INFN, Sezione Roma Tre, Roma, Italy

Lance Littlejohn Department of Mathematics, Baylor University, Waco, TX, USA

Ian Marquette School of Mathematics and Physics, The University of Queensland, Brisbane, St-Lucia, QLD, Australia

Juan Mateos Guilarte Departamento de Fìsica Fundamental and IUFFyM, Universidad de Salamanca, Salamanca, Spain

Anatoly G. Nikitin Institute of Mathematics, National Academy of Sciences of Ukraine, Kyiv, Ukraine

Fernando Olivar-Romero Physics Department, Cinvestav, Mexico City, Mexico

Mikhail S. Plyushchay Departamento de Física, Universidad de Santiago de Chile, Casilla, Santiago, Chile

Natália F. Ribeiro Centro Universitário do Norte Paulista, UNORP, São José do Rio Preto, SP, Brazil

Regina M. Ricotta Faculdade de Tecnologia de São Paulo, FATEC/SP-CEETEPS, São Paulo, SP, Brazil

Miguel A. Rodríguez Dept. de Física Teórica, Universidad Complutense de Madrid, Madrid, Spain

Oscar Rosas-Ortiz Physics Department, Cinvestav, Mexico City, Mexico

Zora Thomova SUNY Polytechnic Institute, Utica, NY, USA

Hagop Tossounian Department of Mathematics, Baylor University, Waco, TX, USA

Haydar Uncu Department of Physics, Adnan Menderes University, Aydın, Turkey

Miguel A. Velasco Departamento de Física Teórica, Atómica y Óptica, Universidad de Valladolid, Valladolid, Spain

Pavel Winternitz Centre de recherches mathématiques et Département de Mathématiques et de Statistique, Université de Montréal, Montréal, QC, Canada

Wojtek Zakrzewski Durham University, Durham, UK

Miloslav Znojil NPI ASCR, Řež, Czech Republic

Curvature as an Integrable Deformation

Ángel Ballesteros, Alfonso Blasco, and Francisco J. Herranz

Abstract The generalization of (super)integrable Euclidean classical Hamiltonian systems to the two-dimensional sphere and the hyperbolic space by preserving their (super)integrability properties is reviewed. The constant Gaussian curvature of the underlying spaces is introduced as an explicit deformation parameter, thus allowing the construction of new integrable Hamiltonians in a unified geometric setting in which the Euclidean systems are obtained in the vanishing curvature limit. In particular, the constant curvature analogue of the generic anisotropic oscillator Hamiltonian is presented, and its superintegrability for commensurate frequencies is shown. As a second example, an integrable version of the Hénon–Heiles system on the sphere and the hyperbolic plane is introduced. Projective Beltrami coordinates are shown to be helpful in this construction, and further applications of this approach are sketched.

Keywords Integrable systems · Curvature · Sphere · Hyperbolic plane · Integrable perturbations · Oscillator potential · Hénon–Heiles

1 Introduction

The aim of this contribution is to review some new recent results related to a seemingly elementary issue in the theory of finite-dimensional integrable systems [1–5], whose solution presents quite a number of interesting features. The problem can explicitly be stated as follows.

Á. Ballesteros
Departamento de Física, Facultad de Ciencias, Universidad de Burgos, Burgos, Spain
e-mail: angelb@ubu.es

A. Blasco · F. J. Herranz (✉)
Departamento de Física, Escuela Politécnica Superior, Universidad de Burgos, Burgos, Spain
e-mail: ablasco@ubu.es; fjherranz@ubu.es

Ş. Kuru et al. (eds.), *Integrability, Supersymmetry and Coherent States*, CRM Series in Mathematical Physics, https://doi.org/10.1007/978-3-030-20087-9_1

Let us consider a certain Liouville integrable natural Hamiltonian system for a particle with unit mass moving on the two-dimensional (2D) Euclidean space endowed with the standard bracket $\{q_i, p_j\} = \delta_{ij}$ in terms of canonical coordinates and momenta, namely

$$\mathcal{H} = \mathcal{T} + \mathcal{V} = \frac{1}{2}(p_1^2 + p_2^2) + \mathcal{V}(q_1, q_2), \tag{1}$$

where $\mathcal{T}$ is the kinetic energy and $\mathcal{V}$ is the potential. The Liouville integrability of this system will be provided by a constant of the motion given by a globally defined function $\mathcal{I}(p_1, p_2, q_1, q_2)$ such that $\{\mathcal{H}, \mathcal{I}\} = 0$.

The proposed problem consists in finding a one-parameter integrable deformation of $\mathcal{H}$ of the form

$$\mathcal{H}_\kappa = \mathcal{T}_\kappa(p_1, p_2, q_1, q_2) + \mathcal{V}_\kappa(q_1, q_2), \qquad \kappa \in \mathbb{R},$$

with integral of the motion given by the smooth and globally defined function $\mathcal{I}_\kappa(p_1, p_2, q_1, q_2)$ (therefore $\{\mathcal{H}_\kappa, \mathcal{I}_\kappa\} = 0$), and such that the following two conditions hold:

1. The smooth function $\mathcal{T}_\kappa$ is the kinetic energy of a particle on a 2D space whose constant curvature is given by the parameter κ, i.e. the 2D sphere $\mathbf{S}^2$ will arise in the case $\kappa > 0$ and the hyperbolic plane $\mathbf{H}^2$ when $\kappa < 0$.
2. The Euclidean system $\mathcal{H}$ given by (1) has to be smoothly recovered in the zero-curvature limit $\kappa \to 0$, namely

$$\mathcal{H} = \lim_{\kappa\to 0} \mathcal{H}_\kappa, \qquad \mathcal{I} = \lim_{\kappa\to 0} \mathcal{I}_\kappa.$$

If these two conditions are fulfilled, we will say that $\mathcal{H}_\kappa$ is an *integrable curved version of* $\mathcal{H}$ on the sphere and the hyperbolic space. We stress that within this framework the Gaussian curvature κ of the space enters as a deformation parameter, and the curved system $\mathcal{H}_\kappa$ can be thought of as smooth integrable perturbation of the flat one $\mathcal{H}$ in terms of the curvature parameter. Therefore, integrable Hamiltonian systems on $\mathbf{S}^2$ $(\kappa > 0)$, $\mathbf{H}^2$ $(\kappa < 0)$ and $\mathbf{E}^2$ $(\kappa = 0)$ will be simultaneously constructed and analysed.

Moreover, it could happen that the initial Hamiltonian $\mathcal{H}$ is not only integrable but superintegrable, i.e. another globally defined and functionally independent integral of the motion $\mathcal{K}(p_1, p_2, q_1, q_2)$ does exist such that

$$\{\mathcal{H}, \mathcal{I}\} = \{\mathcal{H}, \mathcal{K}\} = 0, \qquad \{\mathcal{I}, \mathcal{K}\} \neq 0.$$

In that case we could further impose the existence of the curved (and functionally independent) analogue $\mathcal{K}_\kappa$ of the second integral such that

$$\mathcal{K} = \lim_{\kappa\to 0} \mathcal{K}_\kappa.$$

If we succeed in finding such second integral fulfilling

$$\{\mathcal{H}_\kappa, \mathcal{I}_\kappa\} = \{\mathcal{H}_\kappa, \mathcal{K}_\kappa\} = 0, \qquad \{\mathcal{I}_\kappa, \mathcal{K}_\kappa\} \neq 0,$$

we will say that we have obtained a *superintegrable curved generalization* of the Euclidean superintegrable Hamiltonian $\mathcal{H}$.

The explicit curvature-dependent description of $\mathbf{S}^2$ and $\mathbf{H}^2$ is well known in the literature and can be found, for instance, in [6–28] (see also the references therein) where it has been mainly considered in the classification and description of superintegrable systems on these two spaces. In this contribution we will present several recent works in which this geometric framework has been applied for non-superintegrable systems where the lack of additional symmetries forces to make use of a purely integrable perturbation approach. Moreover, this perturbative viewpoint shows that the uniqueness of this construction is not guaranteed, since in general different $\mathcal{V}_\kappa$ integrable potentials (and their associated $\mathcal{I}_\kappa$ integrals) having the same $\kappa \to 0$ limit could exist and be found. As an outstanding example of this plurality, we will present the construction of different integrable curved analogues on $\mathbf{S}^2$ ($\kappa > 0$) and $\mathbf{H}^2$ ($\kappa < 0$) of some anisotropic oscillators.

The second novel technical aspect to be emphasized in the results here presented is that in some cases projective coordinates turn out to be helpful in order to construct the (super)integrable deformations $\mathcal{H}_\kappa$, since when these coordinates are considered on $\mathbf{S}^2$ and $\mathbf{H}^2$ then the curved kinetic energy $\mathcal{T}_\kappa$ is expressed as a polynomial in the canonical variables describing the projective phase space. Therefore, some of the examples here presented can be thought of as instances of integrable projective dynamics, in the sense of [29, 30].

The structure of the paper is as follows. In the next section we review the description of the geodesic dynamics on the sphere and the hyperboloid by making use of the above-mentioned curvature-dependent formalism. In particular, ambient space coordinates as well as geodesic parallel and geodesic polar coordinates for $\mathbf{S}^2$ and $\mathbf{H}^2$ will be introduced. In Sect. 3 the projective dynamics on the sphere and the hyperboloid in terms of Beltrami coordinates will also be summarized, thus providing a complete set of geometric possibilities for the description of dynamical systems on these curved spaces. In Sect. 4 we recall the (super)integrability properties of the 2D anisotropic oscillator with arbitrary frequencies and also with commensurate ones, and in Sect. 5 the explicit construction of the $\mathcal{H}_\kappa$ Hamiltonian defining its curved analogue will be presented. Section 6 will be devoted to recall the three integrable versions of the well-known (non-integrable) Hénon–Heiles Hamiltonian. In Sect. 7 the construction of the curved version on $\mathbf{S}^2$ and $\mathbf{H}^2$ of an integrable Hénon–Heiles system related to the KdV hierarchy will be constructed, thus exemplifying the usefulness of the approach here presented for the obtention of new integrable systems on curved spaces. Furthermore, the full Ramani–Dorizzi–Grammaticos series of integrable polynomial potentials will also be generalized to the curved case. Finally, a section including some remarks and open problems under investigation closes the paper.

2 Geodesic Dynamics on the Sphere and the Hyperboloid

Let us consider the one-parametric family of 3D real Lie algebras $\mathfrak{so}_\kappa(3) = \operatorname{span}\{J_{01}, J_{02}, J_{12}\}$ with commutation relations given by (in the sequel we follow the curvature-dependent formalism as presented in [31, 32]):

$$[J_{12}, J_{01}] = J_{02}, \qquad [J_{12}, J_{02}] = -J_{01}, \qquad [J_{01}, J_{02}] = \kappa J_{12}, \tag{2}$$

where κ is a real parameter. The Casimir invariant, coming from the Killing–Cartan form, reads

$$\mathcal{C} = J_{01}^2 + J_{02}^2 + \kappa J_{12}^2. \tag{3}$$

The family $\mathfrak{so}_\kappa(3)$ comprises three specific Lie algebras: $\mathfrak{so}(3)$ for $\kappa > 0$, $\mathfrak{so}(2,1) \simeq \mathfrak{sl}_2(\mathbb{R})$ for $\kappa < 0$, and $\mathfrak{iso}(2) \equiv \mathfrak{e}(2) = \mathfrak{so}(2) \oplus_S \mathbb{R}^2$ for $\kappa = 0$. Note that the value of κ can be reduced to $\{+1, 0, -1\}$ through a rescaling of the Lie algebra generators; therefore setting $\kappa = 0$ in (2) can be shown to be equivalent to applying an Inönü–Wigner contraction [33].

The involutive automorphism defined by

$$\Theta(J_{01}, J_{02}, J_{12}) = (-J_{01}, -J_{02}, J_{12}),$$

generates a $\mathbb{Z}_2$-grading of $\mathfrak{so}_\kappa(3)$ in such a manner that κ is a graded contraction parameter [34], and Θ gives rise to the following Cartan decomposition of the Lie algebra:

$$\mathfrak{so}_\kappa(3) = \mathfrak{h} \oplus \mathfrak{p}, \qquad \mathfrak{h} = \operatorname{span}\{J_{12}\} = \mathfrak{so}(2), \qquad \mathfrak{p} = \operatorname{span}\{J_{01}, J_{02}\}.$$

We denote $\mathrm{SO}_\kappa(3)$ and H the Lie groups with Lie algebras $\mathfrak{so}_\kappa(3)$ and $\mathfrak{h}$, respectively, and we consider the 2D symmetrical homogeneous space defined by

$$\mathbf{S}_\kappa^2 = \mathrm{SO}_\kappa(3)/H, \qquad H = \mathrm{SO}(2) = \langle J_{12} \rangle. \tag{4}$$

This coset space has constant Gaussian curvature equal to κ and is endowed with a metric having positive definite signature. The generator J_{12} leaves a point O invariant, the origin, so generating rotations around O, while J_{01} and J_{02} generate translations which move O along two basic orthogonal geodesics l_1 and l_2.

Therefore $\mathbf{S}_\kappa^2$ (4) covers the three classical 2D Riemannian spaces of constant curvature:

$\mathbf{S}_+^2$: Sphere	$\mathbf{S}_0^2$: Euclidean plane	$\mathbf{S}_-^2$: Hyperbolic space
$\mathbf{S}^2 = \mathrm{SO}(3)/\mathrm{SO}(2)$	$\mathbf{E}^2 = \mathrm{ISO}(2)/\mathrm{SO}(2)$	$\mathbf{H}^2 = \mathrm{SO}(2,1)/\mathrm{SO}(2)$

We recall that these three spaces (and their motion groups $\mathrm{SO}_\kappa(3)$) are contained within the family of the so-called 2D orthogonal Cayley–Klein geometries [6, 35, 36], which are parametrized in terms of two graded contraction parameters $\kappa \equiv \kappa_1$ and κ_2 [34].

In what follows we describe the metric structure and the geodesic motion on the above spaces in terms of several sets of coordinates that will be used throughout the paper. We stress that all the resulting expressions will have always a smooth and well-defined flat limit (contraction) $\kappa \to 0$ reducing to the corresponding Euclidean ones.

2.1 Ambient Space Coordinates

The *vector representation* of $\mathfrak{so}_\kappa(3)$ is provided by the following faithful matrix representation $\rho : \mathfrak{so}_\kappa(3) \to \mathrm{End}(\mathbb{R}^3)$ [8, 9]

$$\rho(J_{01}) = \begin{pmatrix} 0 & -\kappa & 0 \\ 1 & 0 & 0 \\ 0 & 0 & 0 \end{pmatrix}, \quad \rho(J_{02}) = \begin{pmatrix} 0 & 0 & -\kappa \\ 0 & 0 & 0 \\ 1 & 0 & 0 \end{pmatrix}, \quad \rho(J_{12}) = \begin{pmatrix} 0 & 0 & 0 \\ 0 & 0 & -1 \\ 0 & 1 & 0 \end{pmatrix}, \tag{5}$$

which satisfies

$$\rho(J_{ij})^T \mathbb{I}_\kappa + \mathbb{I}_\kappa \rho(J_{ij}) = 0, \qquad \mathbb{I}_\kappa = \mathrm{diag}(1, \kappa, \kappa). \tag{6}$$

The matrix exponentiation of (5) leads to the following one-parametric subgroups of $\mathrm{SO}_\kappa(3)$:

$$\begin{aligned} \mathrm{e}^{\alpha\rho(J_{01})} &= \begin{pmatrix} \mathrm{C}_\kappa(\alpha) & -\kappa\,\mathrm{S}_\kappa(\alpha) & 0 \\ \mathrm{S}_\kappa(\alpha) & \mathrm{C}_\kappa(\alpha) & 0 \\ 0 & 0 & 1 \end{pmatrix}, \quad \mathrm{e}^{\gamma\rho(J_{12})} = \begin{pmatrix} 1 & 0 & 0 \\ 0 & \cos\gamma & -\sin\gamma \\ 0 & \sin\gamma & \cos\gamma \end{pmatrix}, \\ \mathrm{e}^{\beta\rho(J_{02})} &= \begin{pmatrix} \mathrm{C}_\kappa(\beta) & 0 & -\kappa\,\mathrm{S}_\kappa(\beta) \\ 0 & 1 & 0 \\ \mathrm{S}_\kappa(\beta) & 0 & \mathrm{C}_\kappa(\beta) \end{pmatrix}, \end{aligned} \tag{7}$$

where we have introduced the κ-dependent cosine and sine functions [6, 8]

$$\mathrm{C}_\kappa(x) := \sum_{l=0}^{\infty} (-\kappa)^l \frac{x^{2l}}{(2l)!} = \begin{cases} \cos\sqrt{\kappa}\, x & \kappa > 0 \\ 1 & \kappa = 0 \\ \cosh\sqrt{-\kappa}\, x & \kappa < 0 \end{cases}$$

$$\mathrm{S}_\kappa(x) := \sum_{l=0}^{\infty}(-\kappa)^l \frac{x^{2l+1}}{(2l+1)!} = \begin{cases} \frac{1}{\sqrt{\kappa}} \sin \sqrt{\kappa}\, x & \kappa > 0 \\ x & \kappa = 0 \\ \frac{1}{\sqrt{-\kappa}} \sinh \sqrt{-\kappa}\, x & \kappa < 0 \end{cases}.$$

The κ-tangent function is defined as

$$\mathrm{T}_\kappa(x) := \frac{\mathrm{S}_\kappa(x)}{\mathrm{C}_\kappa(x)}.$$

These curvature-dependent trigonometric functions coincide with the circular and hyperbolic ones for $\kappa = \pm 1$, while under the contraction $\kappa = 0$ they reduce to the parabolic functions: $\mathrm{C}_0(x) = 1$ and $\mathrm{S}_0(x) = \mathrm{T}_0(x) = x$. Some trigonometric relations read [8]

$$\mathrm{C}_\kappa^2(x) + \kappa\, \mathrm{S}_\kappa^2(x) = 1, \quad \mathrm{C}_\kappa(2x) = \mathrm{C}_\kappa^2(x) - \kappa\, \mathrm{S}_\kappa^2(x), \quad \mathrm{S}_\kappa(2x) = 2\mathrm{S}_\kappa(x)\mathrm{C}_\kappa(x)$$

and their derivatives are given by [9]

$$\frac{\mathrm{d}}{\mathrm{d}x}\mathrm{C}_\kappa(x) = -\kappa\, \mathrm{S}_\kappa(x), \qquad \frac{\mathrm{d}}{\mathrm{d}x}\mathrm{S}_\kappa(x) = \mathrm{C}_\kappa(x), \qquad \frac{\mathrm{d}}{\mathrm{d}x}\mathrm{T}_\kappa(x) = \frac{1}{\mathrm{C}_\kappa^2(x)}.$$

Therefore, under the matrix realization (7), the Lie group $\mathrm{SO}_\kappa(3)$ becomes a group of isometries of the bilinear form $\mathbb{I}_\kappa$ (6),

$$g^T \mathbb{I}_\kappa\, g = \mathbb{I}_\kappa, \quad \forall g \in \mathrm{SO}_\kappa(3),$$

acting on a 3D linear ambient space $\mathbb{R}^3 = (x_0, x_1, x_2)$ through matrix multiplication. The subgroup $\mathrm{e}^{\gamma\rho(J_{12})}$ (7) is the isotropy subgroup of the point $O = (1, 0, 0)$, which is taken as the *origin* in the homogeneous space $\mathbf{S}_\kappa^2$ (4). The orbit of O is contained in the "κ-sphere" determined by $\mathbb{I}_\kappa$ (6):

$$\Sigma_\kappa \,:\, x_0^2 + \kappa\left(x_1^2 + x_2^2\right) = 1. \tag{8}$$

The connected component of Σ_κ is identified with the space $\mathbf{S}_\kappa^2$ and the action of $\mathrm{SO}_\kappa(3)$ is transitive on it. The coordinates (x_0, x_1, x_2) satisfying the constraint (8) are called *ambient space* or *Weierstrass coordinates*. Notice that for $\kappa > 0$ we recover the sphere, if $\kappa < 0$, we find the two-sheeted hyperboloid, and in the flat case with $\kappa = 0$ we get two Euclidean planes $x_0 = \pm 1$ with Cartesian coordinates (x_1, x_2). Since $O = (1, 0, 0)$, we identify the hyperbolic space $\mathbf{H}^2$ with the connected component corresponding to the sheet of the hyperboloid with $x_0 \geq 1$, and the Euclidean space $\mathbf{E}^2$ with the plane $x_0 = +1$.

The metric on $\mathbf{S}^2_\kappa$ comes from the flat ambient metric in $\mathbb{R}^3$ divided by the curvature κ and restricted to Σ_κ:

$$(\mathrm{d}s)^2_\kappa = \frac{1}{\kappa}\left(\mathrm{d}x_0^2 + \kappa\left(\mathrm{d}x_1^2 + \mathrm{d}x_2^2\right)\right)\Big|_{\Sigma_\kappa} = \frac{\kappa\,(x_1\mathrm{d}x_1 + x_2\mathrm{d}x_2)^2}{1-\kappa\left(x_1^2+x_2^2\right)} + \mathrm{d}x_1^2 + \mathrm{d}x_2^2 . \quad (9)$$

Isometry vector fields in ambient coordinates for $\mathfrak{so}_\kappa(3)$, fulfilling (2), are directly obtained from the vector representation (5):

$$J_{01} = \kappa\, x_1\partial_0 - x_0\partial_1, \qquad J_{02} = \kappa\, x_2\partial_0 - x_0\partial_2, \qquad J_{12} = x_2\partial_1 - x_1\partial_2, \quad (10)$$

where $\partial_\mu = \partial/\partial x_\mu$ ($\mu = 0, 1, 2$).

Now we consider the ambient momenta π_μ conjugate to x_μ fulfilling the canonical Poisson bracket $\{x_\mu, \pi_\nu\} = \delta_{\mu\nu}$ subjected to the constraint (8). The vector fields (10) give rise to a symplectic realization of $\mathfrak{so}_\kappa(3)$ in terms of ambient variables by setting $\partial_\mu \to -\pi_\mu$:

$$J_{01} = x_0\pi_1 - \kappa\, x_1\pi_0, \qquad J_{02} = x_0\pi_2 - \kappa\, x_2\pi_0, \qquad J_{12} = x_1\pi_2 - x_2\pi_1, \quad (11)$$

which close the Poisson brackets defining the Lie–Poisson algebra $\mathfrak{so}_\kappa(3)$

$$\{J_{12}, J_{01}\} = J_{02}, \qquad \{J_{12}, J_{02}\} = -J_{01}, \qquad \{J_{01}, J_{02}\} = \kappa\, J_{12}.$$

The metric (9) provides the free Lagrangian $\mathcal{L}_\kappa$ with ambient velocities $\dot{x}_\mu$ for a particle with unit mass, so determining geodesic motion on $\mathbf{S}^2_\kappa$:

$$\mathcal{L}_\kappa = \frac{1}{2\kappa}\left(\dot{x}_0^2 + \kappa\left(\dot{x}_1^2 + \dot{x}_2^2\right)\right)\Big|_{\Sigma_\kappa} = \frac{\kappa\,(x_1\dot{x}_1 + x_2\dot{x}_2)^2}{2\left(1-\kappa\left(x_1^2+x_2^2\right)\right)} + \frac{1}{2}\left(\dot{x}_1^2 + \dot{x}_2^2\right). \quad (12)$$

Thus the corresponding momenta $\pi_\mu = \partial\mathcal{L}_\kappa/\partial\dot{x}_\mu$ read

$$\pi_0 = \dot{x}_0/\kappa, \qquad \pi_1 = \dot{x}_1, \qquad \pi_2 = \dot{x}_2. \quad (13)$$

The time derivative of the constraint (8) provides the relation

$$\Sigma_\kappa\ :\ x_0\pi_0 + x_1\pi_1 + x_2\pi_2 = 0.$$

Finally, by introducing (13) in (12) we obtain that the kinetic energy $\mathcal{T}_\kappa$ in ambient variables is given by

$$\mathcal{T}_\kappa = \frac{1}{2}\left(\kappa\,\pi_0^2 + \pi_1^2 + \pi_2^2\right)\Big|_{\Sigma_\kappa} = \frac{\kappa\,(x_1\pi_1 + x_2\pi_2)^2}{2\left(1-\kappa\left(x_1^2+x_2^2\right)\right)} + \frac{1}{2}(\pi_1^2 + \pi_2^2). \quad (14)$$

Notice that the contraction $\kappa = 0$ is well defined in the r.h.s. of Eqs. (9), (12), and (14) yielding the Euclidean expressions

$$(\mathrm{d}s)_0^2 = \mathrm{d}x_1^2 + \mathrm{d}x_2^2, \qquad \mathcal{L}_0 = \tfrac{1}{2}(\dot{x}_1^2 + \dot{x}_2^2), \qquad \mathcal{T}_0 = \tfrac{1}{2}(\pi_1^2 + \pi_2^2).$$

2.2 Geodesic Parallel and Polar Coordinates

The ambient coordinates (8) can also be parametrized in terms of two intrinsic variables of geodesic type. For our purposes let us consider the so-called *geodesic parallel* (x, y) and *geodesic polar* (r, ϕ) coordinates of a point $Q = (x_0, x_1, x_2) \in \mathbf{S}_\kappa^2$ [7, 9], which are defined through the following action of the one-parametric subgroups (7) on the origin $O = (1, 0, 0)$:

$$\begin{aligned}(x_0, x_1, x_2)^T &= \exp(x\rho(J_{01}))\exp(y\rho(J_{02}))O^T \\ &= \exp(\phi\rho(J_{12}))\exp(r\rho(J_{01}))O^T,\end{aligned}$$

which gives

$$\begin{aligned}x_0 &= \mathrm{C}_\kappa(x)\mathrm{C}_\kappa(y) = \mathrm{C}_\kappa(r), \\ x_1 &= \mathrm{S}_\kappa(x)\mathrm{C}_\kappa(y) = \mathrm{S}_\kappa(r)\cos\phi, \\ x_2 &= \mathrm{S}_\kappa(y) = \mathrm{S}_\kappa(r)\sin\phi.\end{aligned} \tag{15}$$

In this construction, the variable r is the distance between the origin O and the point Q measured along the geodesic l that joins both points, while ϕ is the angle of l with respect to a base geodesic l_1 (associated with the translation generator J_{01}). Let Q_1 be the intersection point of l_1 with its orthogonal geodesic l_2' through Q. Then x is the geodesic distance between O and Q_1 measured along l_1 and y is the geodesic distance between Q_1 and Q measured along l_2'. On $\mathbf{E}^2$ with $\kappa = 0$, the relations (15) lead to $x_0 = 1$ and $(x_1, x_2) = (x, y) = (r\cos\phi, r\sin\phi)$ so reducing to Cartesian and polar coordinates.

These coordinates are shown in Fig. 1 for $\mathbf{S}^2$ and $\mathbf{H}^2$. In these pictures, l_2 is the base geodesic orthogonal to l_1 through O, so related to J_{02}, and Q_2 is the intersection point of l_2 with its orthogonal geodesic l_1' through Q.

We substitute (15) in the ambient metric (9) and in the free Lagrangian (12), finding that

$$\begin{aligned}(\mathrm{d}s)_\kappa^2 &= \mathrm{C}_\kappa^2(y)\mathrm{d}x^2 + \mathrm{d}y^2 = \mathrm{d}r^2 + \mathrm{S}_\kappa^2(r)\mathrm{d}\phi^2, \\ \mathcal{L}_\kappa &= \tfrac{1}{2}\big(\mathrm{C}_\kappa^2(y)\dot{x}^2 + \dot{y}^2\big) = \tfrac{1}{2}\big(\dot{r}^2 + \mathrm{S}_\kappa^2(r)\dot{\phi}^2\big).\end{aligned}$$

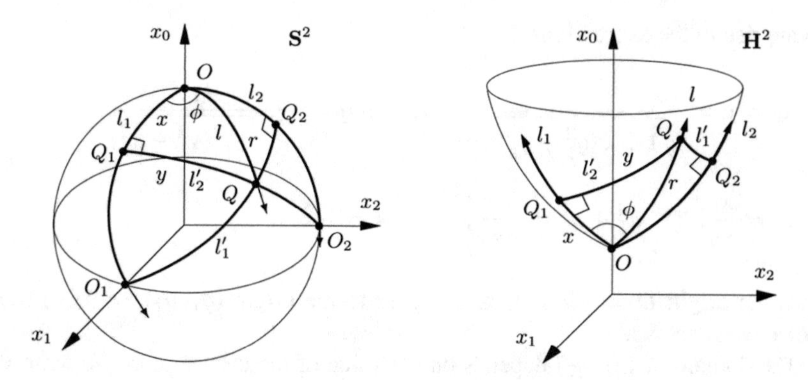

Fig. 1 Ambient (x_0, x_1, x_2), geodesic parallel (x, y) and geodesic polar (r, ϕ) coordinates of a point Q on the sphere $\mathbf{S}^2$ with $\kappa = +1$ and on the hyperbolic space $\mathbf{H}^2$ with $\kappa = -1$ and $x_0 \geq 1$. The origin of the space in ambient coordinates is $O = (1, 0, 0)$. Note that on $\mathbf{S}^2$, $O_1 = (0, 1, 0)$ and $O_2 = (0, 0, 1)$

Now, we denote (p_x, p_y) and (p_r, p_ϕ) the conjugate momenta of the coordinates (x, y) and (r, ϕ), respectively, and the free Hamiltonian (kinetic energy) turns out to be

$$\mathcal{T}_\kappa = \frac{1}{2}\left(\frac{p_x^2}{\mathrm{C}_\kappa^2(y)} + p_y^2\right) = \frac{1}{2}\left(p_r^2 + \frac{p_\phi^2}{\mathrm{S}_\kappa^2(r)}\right). \tag{16}$$

According to (15) and avoiding singularities in (16), we find that the domain of the geodesic coordinates on $\mathbf{S}^2$ and $\mathbf{H}^2$ reads (always $\phi \in [0, 2\pi)$)

$$\begin{aligned} &\mathbf{S}^2\ (\kappa > 0):\ -\frac{\pi}{\sqrt{\kappa}} < x \leq \frac{\pi}{\sqrt{\kappa}}, \quad -\frac{\pi}{2\sqrt{\kappa}} < y < \frac{\pi}{2\sqrt{\kappa}}, \quad 0 < r < \frac{\pi}{\sqrt{\kappa}}. \\ &\mathbf{H}^2\ (\kappa < 0):\ -\infty < x < \infty, \quad -\infty < y < \infty, \quad 0 < r < \infty. \end{aligned} \tag{17}$$

3 Beltrami Coordinates and Projective Dynamics

The quotients $(x_1/x_0, x_2/x_0) \equiv (q_1, q_2)$ of the ambient coordinates (8) are just the *Beltrami coordinates* of projective geometry for the sphere and the hyperbolic plane. They are obtained by applying the central stereographic projection with pole $(0, 0, 0) \in \mathbb{R}^3$ of a point $Q = (x_0, x_1, x_2)$ onto the projective plane with $x_0 = 1$ and coordinates (q_1, q_2):

$$(x_0, x_1, x_2) \in \Sigma_\kappa \ \rightarrow\ (0, 0, 0) + \mu\,(1, q_1, q_2) \in \Sigma_\kappa,$$

giving rise to the expressions

$$
\begin{aligned}
&x_0 = \mu = \frac{1}{\sqrt{1+\kappa(q_1^2+q_2^2)}}, \qquad x_i = \mu\, q_i = \frac{q_i}{\sqrt{1+\kappa(q_1^2+q_2^2)}}, \\
&q_i = \frac{x_i}{x_0}, \qquad q_1^2+q_2^2 = \frac{1-x_0^2}{\kappa x_0^2}, \qquad i=1,2.
\end{aligned}
\tag{18}
$$

Thus the origin $O = (1,0,0) \in \Sigma_\kappa$ goes to the origin $(q_1,q_2) = (0,0)$ in the projective space $\mathbf{S}^2_\kappa$.

The domain of (q_1,q_2) depends on the value of the curvature κ. We write κ in terms of the radius R of the space as $\kappa = \pm 1/R^2$ and we find that in the sphere $\mathbf{S}^2$ with $\kappa = 1/R^2 > 0$, $q_i \in (-\infty,+\infty)$. The points in the equator in Σ_κ with $x_0 = 0$ $(x_1^2+x_2^2 = R^2)$ go to infinity, so that the projection (18) is well defined for the hemisphere with $x_0 > 0$. In the hyperbolic or Lobachevski space $\mathbf{H}^2$ with $\kappa = -1/R^2 < 0$ and $x_0 \geq 1$ it is satisfied that

$$
q_1^2+q_2^2 = \frac{x_0^2-1}{|\kappa| x_0^2} < R^2,
$$

which is the *Poincaré disk* in Beltrami coordinates and

$$
q_i \in \big(-1/\sqrt{|\kappa|}, +1/\sqrt{|\kappa|}\big) = (-R,+R)\,.
$$

The points at the infinity in $\mathbf{H}^2$ $(x_0 \to \infty)$ are mapped onto to the circle $q_1^2 + q_2^2 = R^2$. Finally, in the Euclidean plane $\mathbf{E}^2$, with $\kappa = 0$ $(R \to \infty)$, the Beltrami coordinates are just the Cartesian ones $x_i = q_i \in (-\infty,+\infty)$.

By introducing (18) in the ambient metric (9) and in the free Lagrangian (12) we obtain that

$$
(\mathrm{d}s)^2_\kappa = \frac{(1+\kappa\,\mathbf{q}^2)\mathrm{d}\mathbf{q}^2 - \kappa(\mathbf{q}\cdot\mathrm{d}\mathbf{q})^2}{(1+\kappa\,\mathbf{q}^2)^2}, \qquad \mathcal{L}_\kappa = \frac{(1+\kappa\,\mathbf{q}^2)\dot{\mathbf{q}}^2 - \kappa(\mathbf{q}\cdot\dot{\mathbf{q}})^2}{2(1+\kappa\,\mathbf{q}^2)^2}, \tag{19}
$$

where $\mathbf{q} = (q_1,q_2)$ and hereafter we shall use the following notation for any 2-vectors $\mathbf{a} = (a_1,a_2)$ and $\mathbf{b} = (b_1,b_2)$:

$$
\mathbf{a}^2 = a_1^2 + a_2^2, \qquad \mathbf{a}\cdot\mathbf{b} = a_1 b_1 + a_2 b_2.
$$

The Beltrami momenta $\mathbf{p} = (p_1,p_2)$ conjugate to the coordinates $\mathbf{q}$, such that $\{q_i,p_j\} = \delta_{ij}$, come from $p_i = \partial\mathcal{L}_\kappa/\partial\dot{q}_i$

$$
p_i = \frac{(1+\kappa\,\mathbf{q}^2)\dot{q}_i - \kappa(\mathbf{q}\cdot\dot{\mathbf{q}})q_i}{(1+\kappa\,\mathbf{q}^2)^2}, \qquad \dot{q}_i = (1+\kappa\,\mathbf{q}^2)\big(p_i + \kappa(\mathbf{q}\cdot\mathbf{p})q_i\big). \tag{20}
$$

Table 1 Expressions for the ambient variables (x_μ, π_μ), free Hamiltonian $\mathcal{T}_\kappa$ and symplectic realization of the Lie–Poisson generators $J_{\mu\nu}$ of $\mathfrak{so}_\kappa(3)$ in terms of Beltrami, geodesic parallel and geodesic polar canonical variables

	Beltrami $(\mathbf{q}, \mathbf{p})$	Geodesic parallel (x, y, p_x, p_y)	Geodesic polar (r, ϕ, p_r, p_ϕ)
x_0	$\dfrac{1}{(1+\kappa\mathbf{q}^2)^{1/2}}$	$\mathrm{C}_\kappa(x)\mathrm{C}_\kappa(y)$	$\mathrm{C}_\kappa(r)$
x_1	$\dfrac{q_1}{(1+\kappa\mathbf{q}^2)^{1/2}}$	$\mathrm{S}_\kappa(x)\mathrm{C}_\kappa(y)$	$\mathrm{S}_\kappa(r)\cos\phi$
x_2	$\dfrac{q_2}{(1+\kappa\mathbf{q}^2)^{1/2}}$	$\mathrm{S}_\kappa(y)$	$\mathrm{S}_\kappa(r)\sin\phi$
π_0	$-\sqrt{1+\kappa\mathbf{q}^2}\,(\mathbf{q}\cdot\mathbf{p})$	$-\dfrac{\mathrm{S}_\kappa(x)}{\mathrm{C}_\kappa(y)}\,p_x - \mathrm{C}_\kappa(x)\mathrm{S}_\kappa(y)p_y$	$-\mathrm{S}_\kappa(r)\,p_r$
π_1	$\sqrt{1+\kappa\mathbf{q}^2}\,p_1$	$\dfrac{\mathrm{C}_\kappa(x)}{\mathrm{C}_\kappa(y)}\,p_x - \kappa\,\mathrm{S}_\kappa(x)\mathrm{S}_\kappa(y)p_y$	$\mathrm{C}_\kappa(r)\cos\phi\,p_r - \dfrac{\sin\phi}{\mathrm{S}_\kappa(r)}\,p_\phi$
π_2	$\sqrt{1+\kappa\mathbf{q}^2}\,p_2$	$\mathrm{C}_\kappa(y)p_y$	$\mathrm{C}_\kappa(r)\sin\phi\,p_r + \dfrac{\cos\phi}{\mathrm{S}_\kappa(r)}\,p_\phi$
$\mathcal{T}_\kappa$	$\dfrac{1}{2}(1+\kappa\mathbf{q}^2)(\mathbf{p}^2+\kappa(\mathbf{q}\cdot\mathbf{p})^2)$	$\dfrac{1}{2}\left(\dfrac{p_x^2}{\mathrm{C}_\kappa^2(y)} + p_y^2\right)$	$\dfrac{1}{2}\left(p_r^2 + \dfrac{p_\phi^2}{\mathrm{S}_\kappa^2(r)}\right)$
J_{01}	$p_1 + \kappa(\mathbf{q}\cdot\mathbf{p})q_1$	p_x	$\cos\phi\,p_r - \dfrac{\sin\phi}{\mathrm{T}_\kappa(r)}\,p_\phi$
J_{02}	$p_2 + \kappa(\mathbf{q}\cdot\mathbf{p})q_2$	$\mathrm{C}_\kappa(x)p_y + \kappa\,\mathrm{S}_\kappa(x)\mathrm{T}_\kappa(y)p_x$	$\sin\phi\,p_r + \dfrac{\cos\phi}{\mathrm{T}_\kappa(r)}\,p_\phi$
J_{12}	$q_1p_2 - q_2p_1$	$\mathrm{S}_\kappa(x)p_y - \mathrm{C}_\kappa(x)\mathrm{T}_\kappa(y)p_x$	p_ϕ

The specific expressions for $\mathbf{S}^2$, $\mathbf{H}^2$ and $\mathbf{E}^2$ correspond to set $\kappa > 0$, $\kappa < 0$ and $\kappa = 0$, respectively

And by inserting these expressions into $\mathcal{L}_\kappa$ (19) we get the free Hamiltonian

$$\mathcal{T}_\kappa = \tfrac{1}{2}(1+\kappa\,\mathbf{q}^2)\big(\mathbf{p}^2 + \kappa(\mathbf{q}\cdot\mathbf{p})^2\big). \tag{21}$$

By introducing (18) and (20) in (13) we obtain the ambient momenta written in terms of the Beltrami variables, $\pi_\mu(\mathbf{q}, \mathbf{p})$, and from this result a symplectic realization of the Lie–Poisson generators (11) in these variables is directly found. These expressions are displayed in Table 1. Notice that the kinetic energy (21) can also be recovered by computing the symplectic realization of the Casimir (3) of $\mathfrak{so}_\kappa(3)$ in Beltrami variables as $\mathcal{T}_\kappa \equiv \frac{1}{2}\mathcal{C}$. Likewise the ambient momenta π_μ and symplectic realization of the Lie–Poisson generators $J_{\mu\nu}$ can be computed in the geodesic variables introduced in Sect. 2.2, and these are also presented in Table 1.

We recall that a similar procedure can be performed with Poincaré coordinates [37] which come from the stereographic projection with pole $(-1, 0, 0)$. The resulting expressions can be found in [32].

4 Anisotropic Oscillators on the Euclidean Plane

To start with, let us consider the Hamiltonian determining the anisotropic oscillator with unit mass and frequencies ω_x and ω_y on the Euclidean plane in Cartesian coordinates $(x, y) \in \mathbb{R}^2$ and conjugate momenta (p_x, p_y):

$$H = \frac{1}{2}(p_x^2 + p_y^2) + \frac{1}{2}(\omega_x^2 x^2 + \omega_y^2 y^2). \tag{22}$$

Clearly, this Hamiltonian is always integrable due to its separability in Cartesian coordinates so that it Poisson commutes with the (quadratic in the momenta) integrals of motion

$$I_x = \frac{1}{2}p_x^2 + \frac{1}{2}\omega_x^2 x^2, \qquad I_y = \frac{1}{2}p_y^2 + \frac{1}{2}\omega_y^2 y^2,$$

which are not independent since

$$H = I_x + I_y.$$

Furthermore, it is also well known that for commensurate frequencies $\omega_x : \omega_y$ the Hamiltonian (22) provides a superintegrable system [38–40], in such a manner that an "additional" (in general higher order in the momenta) integral of motion does exist.

The (super)integrability properties of the commensurate oscillator will be sketched by following the approach given in [41, 42], which is based on a classical factorization formalism (see [43–47] and the references therein). If we denote

$$\omega_x = \gamma\omega_y, \qquad \omega_y = \omega, \qquad \gamma \in \mathbb{R}^+/\{0\}, \tag{23}$$

then H (22) can be written in terms of the parameter γ and frequency ω as

$$H = \frac{1}{2}(p_x^2 + p_y^2) + \frac{\omega^2}{2}\left((\gamma x)^2 + y^2\right). \tag{24}$$

Next we introduce new canonical variables

$$\xi = \gamma x, \qquad p_\xi = p_x/\gamma, \qquad \xi \in \mathbb{R}, \tag{25}$$

giving rise to

$$H = \frac{1}{2}p_y^2 + \frac{\omega^2}{2}y^2 + \gamma^2\left(\frac{1}{2}p_\xi^2 + \frac{\omega^2}{2\gamma^2}\xi^2\right). \tag{26}$$

Therefore we obtain two 1D Hamiltonians H^ξ and H^y given by

$$H^\xi = \frac{1}{2}\, p_\xi^2 + \frac{\omega^2}{2\gamma^2}\xi^2, \qquad H^y = \frac{1}{2}\, p_y^2 + \frac{\omega^2}{2} y^2, \qquad H = H^y + \gamma^2 H^\xi, \tag{27}$$

which are two integrals of the motion for H. The 1D Hamiltonian H^ξ (27) can then be factorized in terms of "ladder functions" $B^\pm$ as

$$H^\xi = B^+ B^-, \qquad B^\pm = \mp \frac{i}{\sqrt{2}}\, p_\xi + \frac{1}{\sqrt{2}} \frac{\omega}{\gamma}\, \xi\,, \tag{28}$$

fulfilling

$$\{H^\xi, B^\pm\} = \mp i\, \frac{\omega}{\gamma}\, B^\pm, \qquad \{B^-, B^+\} = -i\, \frac{\omega}{\gamma}.$$

The remaining 1D Hamiltonian H^y (27) can also be factorized through "shift functions" $A^\pm$ in the form

$$H^y = A^+ A^-, \qquad A^\pm = \mp \frac{i}{\sqrt{2}}\, p_y - \frac{\omega}{\sqrt{2}}\, y\,, \tag{29}$$

so that

$$\{H^y, A^\pm\} = \pm i\omega A^\pm, \qquad \{A^-, A^+\} = i\omega.$$

Notice that the sets of functions $(H^\xi, B^\pm, 1)$ and $(H^y, A^\pm, 1)$ span a Poisson–Lie algebra isomorphic to the harmonic oscillator Lie algebra $\mathfrak{h}_4$. Hence, the 2D Hamiltonian (26) can finally be expressed in terms of the above ladder and shift functions as

$$H = A^+ A^- + \gamma^2 B^+ B^-, \qquad \{H, B^\pm\} = \mp i\gamma\omega B^\pm, \qquad \{H, A^\pm\} = \pm i\omega A^\pm\,.$$

The remarkable fact now is that if we consider a rational value for γ,

$$\gamma = \frac{\omega_x}{\omega_y} = \frac{m}{n}, \qquad m, n \in \mathbb{N}^*, \tag{30}$$

we obtain two additional complex constants of the motion $X^\pm$ for H (26)

$$X^\pm = (B^\pm)^n (A^\pm)^m\,, \qquad \bar{X}^+ = X^-, \tag{31}$$

which are of $(m+n)$th-order in the momenta. Real-valued integrals of the motion can be defined through the expressions

$$X = \frac{1}{2}(X^+ + X^-), \qquad Y = \frac{1}{2i}(X^+ - X^-). \tag{32}$$

The final result can be summarized as follows [41, 42].

1. The Hamiltonian H (26) is always integrable for any value of the real parameter γ, since it is endowed with a quadratic constant of the motion given by either H^{ξ} or H^{y} (27).
2. When $\gamma = m/n$ is a rational parameter (30), the Hamiltonian (26) defines a superintegrable anisotropic oscillator with commensurate frequencies $\omega_x : \omega_y$ and the additional constant of the motion is given by either X or Y in (32). The sets (H, H^{ξ}, X) and (H, H^{ξ}, Y) are formed by three functionally independent functions.
3. When $(m + n)$ is even, the highest order constant of the motion in the momenta is X being of $(m + n)$-degree, while Y is of lower order $(m + n - 1)$. When $(m+n)$ is odd, the highest $(m+n)$-degree integral is Y while X is of lower order $(m + n - 1)$.

It is worth recalling that the Hamiltonian (22) can be enlarged by adding two Rosochatius (or Smorodinsky–Winternitz) terms as

$$H_{\lambda} = \frac{1}{2}(p_x^2 + p_y^2) + \frac{1}{2}(\omega_x^2 x^2 + \omega_y^2 y^2) + \frac{\lambda_1}{x^2} + \frac{\lambda_2}{y^2}, \tag{33}$$

where λ_1 and λ_2 are real parameters, which provide centrifugal barriers when both constants are positive ones. In the 3D case, the resulting Hamiltonian, called "caged anisotropic oscillator", has been solved in [48] (for both classical and quantum systems), and the general ND case has been fully studied in [40]. Despite the introduction of the λ_i-potentials, the Hamiltonian (33) is again (maximally) superintegrable for commensurate frequencies (in any dimension).

We also remark that any $m : n$ oscillator (labelled by γ) is equivalent to the $n : m$ one (with $1/\gamma$) via the interchanges $x \leftrightarrow y$ and $\gamma\omega \leftrightarrow \omega$. Consequently, according to the above statement the only anisotropic oscillators which are quadratically superintegrable correspond to the cases with $\gamma = 1$ and $\gamma = 2$ (so also $\gamma = 1/2$), in agreement with the classifications on superintegrable Euclidean systems given in [7, 49, 50]. In the sequel, we will illustrate the previous general results by working out these two particular cases.

4.1 The $\gamma = 1$ or 1:1 (Isotropic) Oscillator

We set $m = n = 1$ so that the relations (23) and (25) simply give $\omega_x = \omega_y = \omega$ and $\xi = x$ and $p_{\xi} = p_x$. Thus we recover the isotropic oscillator

$$H^{1:1} = \frac{1}{2}(p_x^2 + p_y^2) + \frac{\omega^2}{2}(x^2 + y^2), \tag{34}$$

and the constants of the motion (32) reduce to

$$X = -\tfrac{1}{2}(p_x p_y + \omega^2 xy), \qquad Y = -\tfrac{1}{2}\omega(xp_y - yp_x). \tag{35}$$

Since $(m+n) = 2$ is even, we find a quadratic integral X, which is one of the components of the Demkov–Fradkin tensor [51, 52], and a first-order one Y which is proportional to the angular momentum

$$\mathcal{J} = xp_y - yp_x. \tag{36}$$

We recall that if we add the two "centrifugal" λ_i-terms (33), then we get the 2D version of the so-called Smorodinsky–Winternitz system [53] which has been widely studied (see, e.g., [12, 15, 50, 54–56] and the references therein).

4.2 *The $\gamma = 2$ or 2:1 Oscillator*

If we take $m = 2$ and $n = 1$, then $\omega_x = 2\omega_y = 2\omega$, $\xi = 2x$ and $p_\xi = p_x/2$. The Hamiltonian (26) and the integrals (32) turn out to be

$$\begin{aligned} H^{2:1} &= \frac{1}{2}p_y^2 + \frac{\omega^2}{2}y^2 + 4\left(\frac{1}{2}p_\xi^2 + \frac{\omega^2}{8}\xi^2\right) = \frac{1}{2}(p_x^2 + p_y^2) + \frac{\omega^2}{2}\left(4x^2 + y^2\right), \\ X &= -\frac{\omega}{4\sqrt{2}}\left(p_y(\xi p_y - 4yp_\xi) - \omega^2 \xi y^2\right) = -\frac{\omega}{2\sqrt{2}}\left(p_y\mathcal{J} - \omega^2 xy^2\right), \\ Y &= \frac{1}{2\sqrt{2}}\left(p_\xi p_y^2 + \omega^2 y(\xi p_y - yp_\xi)\right) = \frac{1}{4\sqrt{2}}\left(p_x p_y^2 + \omega^2 y(4xp_y - yp_x)\right). \end{aligned} \tag{37}$$

In this case $(m+n) = 3$ is odd, so that we find a cubic integral Y and a quadratic one X, which involves the angular momentum $\mathcal{J}$ (36); the latter is the integral of the motion which is usually considered in the literature (see, e.g., [7, 57]) and shows that the 2:1 oscillator can be regarded as a superintegrable system with *quadratic* constants of the motion.

Notice that if we add a single Rosochatius–Winternitz potential λ_2/y^2 (by setting $\lambda_1 = 0$) the generalized system remains quadratically superintegrable [7], but if both λ_i-terms are introduced, the additional integral turns out to be of sixth-order in the momenta [40].

5 Anisotropic Oscillators on $\mathbf{S}^2$ and $\mathbf{H}^2$

Let us recall that the classification of all possible superintegrable systems on $\mathbf{S}^2$ and $\mathbf{H}^2$ with quadratic integrals of the motion was performed in [7], where two curved superintegrable oscillator potentials were found:

- The isotropic Higgs oscillator [58], whose Euclidean limit is the 1:1 isotropic oscillator (34).
- The curved version of the superintegrable Euclidean 2:1 oscillator (37).

The aim of this section is to present the construction of the constant-curvature Hamiltonian analogue H_κ of the anisotropic Euclidean Hamiltonian H (24) with arbitrary commensurate frequencies. The idea is to introduce appropriately the curvature parameter κ by requiring to keep the same (super)integrability properties as given in the previous section for (24) (or (26)) and, simultaneously, allowing for a smooth and well-defined flat limit $\kappa \to 0$ of the curved Hamiltonian and its constants of the motion.

This result has been achieved in [41] by using the geodesic parallel variables described in Sect. 2.2, so with kinetic term $\mathcal{T}_\kappa$ given by (16). Explicitly, the curved Hamiltonian H_κ has been shown to be of the form

$$H_\kappa = \mathcal{T}_\kappa + U_\kappa^\gamma = \frac{1}{2}\left(\frac{p_x^2}{\mathrm{C}_\kappa^2(y)} + p_y^2\right) + \frac{\omega^2}{2}\left(\frac{\mathrm{T}_\kappa^2(\gamma x)}{\mathrm{C}_\kappa^2(y)} + \mathrm{T}_\kappa^2(y)\right). \quad (38)$$

By taking into account (17) and the presence of the κ-tangent $\mathrm{T}_\kappa(\gamma x)$ in the potential U_κ^γ, we find that the domain of the geodesic parallel coordinates (x, y) is restricted to

$$\mathbf{S}^2\ (\kappa > 0): \ -\frac{\pi}{2\sqrt{\kappa}} < \gamma x < \frac{\pi}{2\sqrt{\kappa}}, \quad -\frac{\pi}{2\sqrt{\kappa}} < y < \frac{\pi}{2\sqrt{\kappa}}, \quad \gamma \geq \frac{1}{2}.$$

$$\mathbf{H}^2\ (\kappa < 0): \ x, y \in \mathbb{R}, \quad \gamma \in \mathbb{R}^+/\{0\}.$$

Now we assume that $\kappa \neq 0$ and since $1 + \kappa\,\mathrm{T}_\kappa^2(u) = 1/\mathrm{C}_\kappa^2(u)$ we rewrite the Hamiltonian H_κ (38) as

$$H_\kappa = \frac{p_y^2}{2} + \frac{1}{\mathrm{C}_\kappa^2(y)}\left(\frac{p_x^2}{2} + \frac{\omega^2}{2\kappa\,\mathrm{C}_\kappa^2(\gamma x)}\right) - \frac{\omega^2}{2\kappa}, \qquad \kappa \neq 0. \quad (39)$$

Next we introduce the canonical variables (ξ, p_ξ) (25) finding that

$$H_\kappa = \frac{p_y^2}{2} + \frac{\gamma^2}{\mathrm{C}_\kappa^2(y)}\left(\frac{p_\xi^2}{2} + \frac{\omega^2}{2\kappa\gamma^2\mathrm{C}_\kappa^2(\xi)}\right) - \frac{\omega^2}{2\kappa}, \qquad \kappa \neq 0.$$

And the curved Hamiltonian (39) is finally expressed as

$$\mathcal{H}_\kappa = \frac{p_y^2}{2} + \frac{\gamma^2 \mathcal{H}_\kappa^\xi}{\mathrm{C}_\kappa^2(y)} - \frac{\omega^2}{2\kappa}, \qquad \mathcal{H}_\kappa^\xi = \frac{p_\xi^2}{2} + \frac{\omega^2}{2\kappa\gamma^2 \mathrm{C}_\kappa^2(\xi)}, \qquad \kappa \neq 0, \tag{40}$$

where $\mathcal{H}_\kappa^\xi$ is a constant of the motion. Notice that in this form, the 1D Hamiltonians $\mathcal{H}_\kappa^\xi$ and $\mathcal{H}_\kappa$ correspond to Pöschl–Teller systems [44]. Consequently, $\mathcal{H}_\kappa$ determines an integrable system for any value of ω and γ.

In the sequel, we factorize the 1D Hamiltonians $\mathcal{H}_\kappa^\xi$ and $\mathcal{H}_\kappa$ (40) (see [41, 42, 44] for details). On the one hand, the Hamiltonian $\mathcal{H}_\kappa^\xi$ is factorized in terms of ladder functions as

$$\mathcal{H}_\kappa^\xi = B_\kappa^+ B_\kappa^- + \frac{\omega^2}{2\kappa\gamma^2}, \qquad B_\kappa^\pm = \mp\frac{i}{\sqrt{2}} \mathrm{C}_\kappa(\xi)\, p_\xi + \frac{\mathcal{E}_\kappa}{\sqrt{2}} \mathrm{S}_\kappa(\xi), \tag{41}$$

where $\mathcal{E}_\kappa$ is a constant of the motion defined by

$$\mathcal{E}_\kappa(p_\xi, \xi) := \sqrt{2\kappa\, \mathcal{H}_\kappa^\xi}\,. \tag{42}$$

Thus we get the Poisson algebra

$$\{\mathcal{H}_\kappa^\xi, B_\kappa^\pm\} = \mp i\, \mathcal{E}_\kappa\, B_\kappa^\pm, \qquad \{B_\kappa^-, B_\kappa^+\} = -i\, \mathcal{E}_\kappa\,,$$

so that

$$\{\mathcal{H}_\kappa, B_\kappa^\pm\} = \mp i\, \frac{\gamma^2 \mathcal{E}_\kappa}{\mathrm{C}_\kappa^2(y)}\, B_\kappa^\pm.$$

On the other hand, $\mathcal{H}_\kappa$ is factorized by means of shift functions in the form

$$\mathcal{H}_\kappa = A_\kappa^+ A_\kappa^- + \frac{1}{2\kappa}\left(\gamma^2 \mathcal{E}_\kappa^2 - \omega^2\right), \qquad A_\kappa^\pm = \mp\frac{i}{\sqrt{2}}\, p_y - \frac{\gamma\mathcal{E}_\kappa}{\sqrt{2}} \mathrm{T}_\kappa(y), \tag{43}$$

closing on the Poisson algebra

$$\{\mathcal{H}_\kappa, A_\kappa^\pm\} = \pm i\, \frac{\gamma \mathcal{E}_\kappa}{\mathrm{C}_\kappa^2(y)}\, A_\kappa^\pm, \qquad \{A_\kappa^-, A_\kappa^+\} = i\, \frac{\gamma \mathcal{E}_\kappa}{\mathrm{C}_\kappa^2(y)}.$$

As in the Euclidean case, when γ takes a rational value (30), two additional complex integrals of motion arise for $\mathcal{H}_\kappa$ (38) (under the change of variable (25)), namely

$$\mathcal{X}_\kappa^\pm = (B_\kappa^\pm)^n (A_\kappa^\pm)^m, \qquad \bar{\mathcal{X}}_\kappa^+ = \mathcal{X}_\kappa^-. \tag{44}$$

Nevertheless, in order to obtain real constants of the motion, X_κ and Y_κ, we are now led to distinguish between two situations [47] (due to the presence of powers of $\mathcal{E}_\kappa$ (42) in (44)):

$$\begin{aligned} &\text{When } m+n \text{ is even:} \quad X_\kappa^{\pm} = \pm i\, \mathcal{E}_\kappa Y_\kappa + X_\kappa . \\ &\text{When } m+n \text{ is odd:} \quad X_\kappa^{\pm} = \mathcal{E}_\kappa X_\kappa \pm i\, Y_\kappa . \end{aligned} \tag{45}$$

Summing up, the generalization to $\mathbf{S}^2$ and $\mathbf{H}^2$ of the anisotropic oscillator can be stated as follows [41].

1. For any value of γ, the Hamiltonian H_κ (38) always determines an integrable anisotropic curved oscillator on $\mathbf{S}^2$ and $\mathbf{H}^2$, with quadratic constant of motion given by H_κ^ξ (40).
2. When γ is a rational parameter (30), the Hamiltonian H_κ defines a superintegrable anisotropic curved oscillator and the additional constant of the motion is given by either X_κ or Y_κ in (45). The sets $(H_\kappa, H_\kappa^\xi, X_\kappa)$ and $(H_\kappa, H_\kappa^\xi, Y_\kappa)$ are formed by three functionally independent functions.
3. The integrals X_κ and Y_κ are polynomial in the momenta, whose degrees are $(m+n)$ and $(m+n-1)$ when $(m+n)$ is even, and $(m+n-1)$ and $(m+n)$ when $(m+n)$ is odd, respectively.

Some remarks are in order. Firstly, although the (flat) Euclidean limit $\kappa \to 0$ is precluded for H_κ written in the forms (39) and (40), it is actually well defined in all the remaining expressions. To perform the contractions one has to take into account the following flat limit of the integrals H_κ^ξ (40) and $\mathcal{E}_\kappa$ (42)

$$\lim_{\kappa\to 0} \kappa H_\kappa^\xi = \frac{\omega^2}{2\gamma^2}, \qquad \lim_{\kappa\to 0} \mathcal{E}_\kappa = \frac{\omega}{\gamma}. \tag{46}$$

Therefore, it can be easily checked that when $\kappa \to 0$, the curved Hamiltonian H_κ (38) reduces to H (24), the curved ladder functions $B_\kappa^{\pm}$ (41) to $B^{\pm}$ (28), the curved shift functions $A_\kappa^{\pm}$ (43) to $A^{\pm}$ (29) and so the curved integrals $X_\kappa^{\pm}$ (44) to $X^{\pm}$ (31).

Secondly, as in the Euclidean system (33), the curved Hamiltonian H_κ (38) can be generalized by adding two curved Rosochatius–Winternitz potentials which in ambient coordinates (15) adopt a very simple expression [12, 15]. Explicitly, the corresponding potential reads

$$U_{\kappa,\lambda}^{\gamma} = U_\kappa^\gamma + \frac{\lambda_1}{x_1^2} + \frac{\lambda_2}{x_2^2} = \left(\frac{\mathrm{T}_\kappa^2(\gamma x)}{\mathrm{C}_\kappa^2(y)} + \mathrm{T}_\kappa^2(y) \right) + \frac{\lambda_1}{\mathrm{S}_\kappa^2(x)\mathrm{C}_\kappa^2(y)} + \frac{\lambda_2}{\mathrm{S}_\kappa^2(y)}. \tag{47}$$

Then the corresponding Hamiltonian $H_{\kappa,\lambda} = \mathcal{T}_\kappa + U_{\kappa,\lambda}^{\gamma}$ can be written as (with $\kappa \neq 0$)

$$H_{\kappa,\lambda} = \frac{p_y^2}{2} + \frac{\lambda_2}{\mathrm{S}_\kappa^2(y)} + \frac{1}{\mathrm{C}_\kappa^2(y)} \left(\frac{p_x^2}{2} + \frac{\omega^2}{2\kappa\, \mathrm{C}_\kappa^2(\gamma x)} + \frac{\lambda_1}{\mathrm{S}_\kappa^2(x)} \right) - \frac{\omega^2}{2\kappa},$$

to be compared with (39). Consequently, $H_{\kappa,\lambda}$ defines an integrable system for any value of ω, γ, λ_1 and λ_2.

Now it could be expected that if γ is a rational number, $H_{\kappa,\lambda}$ should be again superintegrable but, to the best of our knowledge, this property has not been proven in general (except for $\gamma = 1$). We also point out that each λ_i-term gives rise to a centrifugal barrier on $\mathbf{H}^2$ when $\lambda_i > 0$, as in the Euclidean system but, surprisingly enough, both λ_1- and λ_2-potentials can be interpreted as noncentral 1D curved oscillators on $\mathbf{S}^2$ with centres at the points $O_1 = (0, 1, 0)$ and $O_2 = (0, 0, 1)$, respectively [10–13, 31, 59] (see Fig. 1).

Thirdly, it can be seen from H_κ (38) that, in general, U_κ^γ and $U_\kappa^{1/\gamma}$ determine two different systems, in contradistinction with the Euclidean case (recall that the equivalence was provided by the interchanges $x \leftrightarrow y$ and $\gamma\omega \leftrightarrow \omega$). However when $\kappa = 0$ both potentials reduce to equivalent Euclidean potentials. This clearly illustrates the fact that given a flat Hamiltonian system there could be not a single but several curved generalizations (or curvature integrable deformations) which would be non-equivalent in the sense that no canonical change of variables exists between them.

Fourthly, according to the results previously presented, the only anisotropic curved oscillators which are quadratically superintegrable correspond to the same values of γ as in the Euclidean system [7, 60, 61]: $\gamma = 1$, $\gamma = 2$ and (now the non-equivalent) $\gamma = 1/2$. In what follows we shall present the corresponding results for these three cases.

And finally, we recall that other integrable anisotropic oscillators on the spheres and hyperbolic spaces can be found in [62–65] (see also the references therein).

5.1 *The $\gamma = 1$ or 1:1 Curved (Isotropic) Oscillator*

This case is also known as the Higgs oscillator [58, 66] and it has been widely studied in the literature (see [7, 16, 31, 67–70] and the references therein).

We set $\gamma = m = n = 1$ so that $\xi = x$ and $p_\xi = p_x$. The Hamiltonian H_κ (38) reduces to

$$H_\kappa^{1:1} = \frac{1}{2}\left(\frac{p_x^2}{\mathrm{C}_\kappa^2(y)} + p_y^2\right) + \frac{\omega^2}{2}\left(\frac{\mathrm{T}_\kappa^2(x)}{\mathrm{C}_\kappa^2(y)} + \mathrm{T}_\kappa^2(y)\right) = \frac{p_y^2}{2} + \frac{H_\kappa^x}{\mathrm{C}_\kappa^2(y)} - \frac{\omega^2}{2\kappa},$$

where the quadratic integral $H_\kappa^x \equiv H_\kappa^\xi$ (40) is given by

$$H_\kappa^x = \frac{p_x^2}{2} + \frac{\omega^2}{2\kappa\,\mathrm{C}_\kappa^2(x)}.$$

Since $(m+n)=2$ is even the constants of the motion (45) read

$$X_\kappa = -\frac{1}{2}\left(\mathrm{C}_\kappa(x)p_x p_y + \mathcal{E}_\kappa^2 \mathrm{S}_\kappa(x)\mathrm{T}_\kappa(y)\right),$$

$$Y_\kappa = -\frac{1}{2}\left(\mathrm{S}_\kappa(x)p_y - \mathrm{C}_\kappa(x)\mathrm{T}_\kappa(y)p_x\right).$$

The integral Y_κ is proportional to the (curved) angular momentum $\mathcal{J}_\kappa$ which in geodesic parallel and polar variables is given by [7, 9]

$$\mathcal{J}_\kappa = \mathrm{S}_\kappa(x)p_y - \mathrm{C}_\kappa(x)\mathrm{T}_\kappa(y)p_x = p_\phi.$$

The flat limit $\kappa \to 0$ of all the above expressions leads to the results of the Euclidean isotropic oscillator given in Sect. 4.1. In particular, provided that $\mathcal{E}_\kappa \to \omega$ (46), the integrals X_κ, $\mathcal{E}_\kappa Y_\kappa$ and the curved angular momentum $\mathcal{J}_\kappa$ reduce to (35) and (36).

Notice that the potential of $H_\kappa^{1:1}$ is expressed in terms of ambient and geodesic polar coordinates (15) as

$$U_\kappa^{1:1} = \frac{\omega^2}{2}\left(\frac{x_1^2+x_2^2}{x_0^2}\right) = \frac{\omega^2}{2}\mathrm{T}_\kappa^2(r).$$

If we consider the two λ_i-potentials (47), we recover the curved Smorodinsky–Winternitz system

$$U_{\kappa,\lambda}^{1:1} = \frac{\omega^2}{2}\left(\frac{x_1^2+x_2^2}{x_0^2}\right) + \frac{\lambda_1}{x_1^2} + \frac{\lambda_2}{x_2^2},$$

which is known to be quadratically superintegrable [7, 12, 13, 15, 31, 71].

5.2 *The $\gamma = 2$ or 2:1 Curved Oscillator*

We set $\gamma = m = 2$ and $n = 1$ so that $\xi = 2x$ and $p_\xi = p_x/2$. The Hamiltonian H_κ (38) reads

$$H_\kappa^{2:1} = \frac{1}{2}\left(\frac{p_x^2}{\mathrm{C}_\kappa^2(y)} + p_y^2\right) + \frac{\omega^2}{2}\left(\frac{\mathrm{T}_\kappa^2(2x)}{\mathrm{C}_\kappa^2(y)} + \mathrm{T}_\kappa^2(y)\right) = \frac{p_y^2}{2} + \frac{4H_\kappa^\xi}{\mathrm{C}_\kappa^2(y)} - \frac{\omega^2}{2\kappa},$$

where

$$H_\kappa^\xi = \frac{p_\xi^2}{2} + \frac{\omega^2}{8\kappa\,\mathrm{C}_\kappa^2(\xi)} = \frac{p_x^2}{8} + \frac{\omega^2}{8\kappa\,\mathrm{C}_\kappa^2(2x)}.$$

Now $(m+n)=3$ is odd so that the constants of the motion (45) turn out to be

$$X_\kappa = -\frac{1}{2\sqrt{2}}\left(\left[\mathrm{S}_\kappa(2x)p_y - 2\mathrm{C}_\kappa(2x)\mathrm{T}_\kappa(y)p_x\right]p_y - 4\mathcal{E}_\kappa^2\mathrm{S}_\kappa(2x)\mathrm{T}_\kappa^2(y)\right),$$

$$Y_\kappa = \frac{1}{4\sqrt{2}}\left(\mathrm{C}_\kappa(2x)p_x p_y^2 + 4\mathcal{E}_\kappa^2\mathrm{T}_\kappa(y)\left[2\mathrm{S}_\kappa(2x)p_y - \mathrm{C}_\kappa(2x)\mathrm{T}_\kappa(y)p_x\right]\right),$$

that is X_κ is quadratic in the momenta, while Y_κ is cubic; this means that $H_\kappa^{2:1}$ is a quadratically superintegrable system. The limit $\kappa \to 0$ (46) gives $\mathcal{E}_\kappa \to \omega/2$, hence the Hamiltonian $H_\kappa^{2:1}$ and the integrals $\mathcal{E}_\kappa X_\kappa$ and Y_κ reduce to (37), thus reproducing the results of Sect. 4.2.

In terms of ambient and geodesic polar coordinates (15), the potential of $H_\kappa^{2:1}$ adopts the (cumbersome) expressions

$$U_\kappa^{2:1} = \frac{\omega^2}{2}\left(\frac{4x_0^2x_1^2}{(x_0^2+\kappa x_1^2)(x_0^2-\kappa x_1^2)^2} + \frac{x_2^2}{(1-\kappa x_2^2)}\right)$$

$$= \frac{\omega^2}{2}\left(\frac{4\mathrm{T}_\kappa^2(r)\cos^2\phi}{\left(1-\kappa\,\mathrm{S}_\kappa^2(r)\sin^2\phi\right)\left(1-\kappa\,\mathrm{T}_\kappa^2(r)\cos^2\phi\right)^2} + \frac{\mathrm{S}_\kappa^2(r)\sin^2\phi}{1-\kappa\,\mathrm{S}_\kappa^2(r)\sin^2\phi}\right)$$

the latter is the one formerly introduced in [7].

In this case it is only possible to add a single λ_i-potential (47) keeping the quadratic superintegrability of the system [7]

$$U_{\kappa,\lambda}^{2:1} = U_\kappa^{2:1} + \frac{\lambda_2}{x_2^2},$$

which has been studied in detail in [31, 32]. In this respect, we remark that an equivalent superintegrable system can be obtained by interchanging the ambient coordinates $x_1 \leftrightarrow x_2$ (so the role of the geodesics $l_1 \leftrightarrow l_2$ in Sect. 2.2) which means that the geodesic parallel coordinates are mapped as $(x, y) \to (y', x')$, that is,

$$x_0 = \mathrm{C}_\kappa(x')\mathrm{C}_\kappa(y'), \qquad x_1 = \mathrm{S}_\kappa(x'), \qquad x_2 = \mathrm{C}_\kappa(x')\mathrm{S}_\kappa(y').$$

In fact, the coordinates (x', y') are just the so-called geodesic parallel coordinates of type II [9]. These transformations provide the equivalent system

$$H'^{2:1}_\kappa = \mathcal{T}_\kappa + \frac{\omega^2}{2}\left(\frac{x_1^2}{(1-\kappa x_1^2)} + \frac{4x_0^2x_2^2}{(x_0^2+\kappa x_2^2)(x_0^2-\kappa x_2^2)^2}\right)$$

$$= \frac{1}{2}\left(p_{x'}^2 + \frac{p_{y'}^2}{\mathrm{C}_\kappa^2(x')}\right) + \frac{\omega^2}{2}\left(\mathrm{T}_\kappa^2(x') + \frac{\mathrm{T}_\kappa^2(2y')}{\mathrm{C}_\kappa^2(x')}\right). \tag{48}$$

This is exactly the expression for the 2:1 curved oscillator considered in [31, 32].

5.3 The $\frac{1}{2}$:1 Curved Oscillator

We set $\gamma = 1/2$, $m = 1$ and $n = 2$, so that $\xi = x/2$, $p_\xi = 2p_x$. Thus the Hamiltonian H_κ (38) is

$$H_\kappa^{\frac{1}{2}:1} = \frac{1}{2}\left(\frac{p_x^2}{\mathrm{C}_\kappa^2(y)} + p_y^2\right) + \frac{\omega^2}{2}\left(\frac{\mathrm{T}_\kappa^2(\frac{x}{2})}{\mathrm{C}_\kappa^2(y)} + \mathrm{T}_\kappa^2(y)\right) = \frac{p_y^2}{2} + \frac{H_\kappa^\xi}{4\mathrm{C}_\kappa^2(y)} - \frac{\omega^2}{2\kappa},$$

where

$$H_\kappa^\xi = \frac{p_\xi^2}{2} + \frac{2\omega^2}{\kappa\,\mathrm{C}_\kappa^2(\xi)} = 2p_x^2 + \frac{2\omega^2}{\kappa\,\mathrm{C}_\kappa^2(\frac{x}{2})}.$$

The sum $(m+n) = 3$ is again odd, and the additional integrals (45) read

$$X_\kappa = -\frac{1}{4\sqrt{2}}\left(4\left[\mathrm{S}_\kappa(x)p_y - \mathrm{C}_\kappa^2(\tfrac{x}{2})\mathrm{T}_\kappa(y)p_x\right]p_x + \mathcal{E}_\kappa^2\,\mathrm{S}_\kappa^2(\tfrac{x}{2})\mathrm{T}_\kappa(y)\right),$$

$$Y_\kappa = \frac{1}{2\sqrt{2}}\left(4\mathrm{C}_\kappa^2(\tfrac{x}{2})p_x^2 p_y - \mathcal{E}_\kappa^2\left[\mathrm{S}_\kappa^2(\tfrac{x}{2})p_y - \mathrm{S}_\kappa(x)\mathrm{T}_\kappa(y)p_x\right]\right),$$

which shows that $H_\kappa^{\frac{1}{2}:1}$ is again a quadratically superintegrable system.

As a consequence, when both Hamiltonians $H_\kappa^{2:1}$ and $H_\kappa^{\frac{1}{2}:1}$ are considered altogether, one finds a particular issue where the curvature-deformation approach gives rise to two non-equivalent systems starting from the common "seed" given by the Euclidean system $H^{2:1} \simeq H^{\frac{1}{2}:1}$ (37) described in Sect. 4.2. Therefore, the plurality of possible integrable curved generalizations of a given Euclidean system becomes evident, and a deeper analysis of the curved system $H_\kappa^{\frac{1}{2}:1}$ seems to be needed, since—to the best of our knowledge—it has not been appropriately considered in the literature so far.

6 Integrable Hénon–Heiles Systems

By making use of the results described in the previous sections, our aim now will be to present the generalization to the 2D sphere and the hyperbolic space of the integrable Hénon–Heiles Hamiltonian given by

$$\mathcal{H} = \frac{1}{2}(p_1^2 + p_2^2) + \Omega\left(q_1^2 + 4q_2^2\right) + \alpha\left(q_1^2 q_2 + 2q_2^3\right), \tag{49}$$

where Ω and α are real constants. Such curved Hénon–Heiles Hamiltonian will be constructed by considering it as an integrable cubic perturbation of the 1:2 anisotropic oscillator that we have introduced in the previous Sect. 5.2, in the form (48), although in this case projective Beltrami coordinates of Sect. 3 will be the ones that are naturally adapted to the construction of the curved system.

We recall that the original (non-integrable) Hénon–Heiles system

$$H = \frac{1}{2}(p_1^2 + p_2^2) + \frac{1}{2}(q_1^2 + q_2^2) + \lambda \left(q_1^2 q_2 - \frac{1}{3} q_2^3 \right),$$

was introduced in [72] in order to model a Newtonian axially symmetric galactic system. When the following generalization containing adjustable parameters was studied

$$\mathcal{H} = \frac{1}{2}(p_1^2 + p_2^2) + \Omega_1 q_1^2 + \Omega_2 q_2^2 + \alpha \left(q_1^2 q_2 + \beta q_2^3 \right),$$

it was found that the only Liouville-integrable members of this family of generalized Hénon–Heiles Hamiltonians were given by *three* specific choices of the real parameters Ω_1, Ω_2, α and β (see [73–85]):

- The Sawada–Kotera system, given by $\beta = 1/3$ and $\Omega_1 = \Omega_2 = \Omega$:

$$\mathcal{H} = \frac{1}{2}(p_1^2 + p_2^2) + \Omega \left(q_1^2 + q_2^2 \right) + \alpha \left(q_1^2 q_2 + \frac{1}{3} q_2^3 \right). \tag{50}$$

  This system is separable in rotated Euclidean coordinates, and therefore its integral of the motion is quadratic in the momenta.
- The Korteweg–de Vries (KdV) system, with $\beta = 2$ and (Ω_1, Ω_2) arbitrary parameters:

$$\mathcal{H} = \frac{1}{2}(p_1^2 + p_2^2) + \Omega_1 q_1^2 + \Omega_2 q_2^2 + \alpha \left(q_1^2 q_2 + 2 q_2^3 \right), \tag{51}$$

  which is separable in parabolic coordinates and has also a quadratic integral of the motion.

- The Kaup–Kupershdmit system, with $\beta = 16/3$ and $\Omega_2 = 16\Omega_1 = 16\Omega$:

$$\mathcal{H} = \frac{1}{2}(p_1^2 + p_2^2) + \Omega \left(q_1^2 + 16 q_2^2 \right) + \alpha \left(q_1^2 q_2 + \frac{16}{3} q_2^3 \right), \tag{52}$$

  whose integral is quartic in the momenta.

Hence the particular KdV case (51) arising when $\Omega_2 = 4\Omega_1$ gives the Hamiltonian (49) and this is connected to the so-called Ramani–Dorizzi–Grammaticos (RDG) series of integrable potentials [86, 87], which are just the polynomial potentials on the Euclidean plane that can be separated in parabolic coordinates and can freely be superposed by preserving integrability [78, 88]. Moreover, such separability in parabolic coordinates explains why a large collection of integrable rational perturbations can be added to the RDG potentials (see [88–92] and the references therein).

In the sequel we review the main results concerning the flat KdV Hénon–Heiles Hamiltonian (51) with $\Omega_2 = 4\Omega_1$ along with its associated RDG potentials. And in the next Sect. 7 we will sketch its integrable curved analogue on the 2D sphere $\mathbf{S}^2$ and the hyperbolic (or Lobachevski) space $\mathbf{H}^2$ which was constructed in [93], together with the full curved counterpart of the integrable RDG series of potentials. The corresponding integrable perturbations of the curved KdV system can be found in [85].

6.1 An Integrable KdV Hénon–Heiles System on the Euclidean Plane

Le us consider the integrable (albeit non-superintegrable) Hamiltonian system (49) defined on $\mathbf{E}^2$ whose constant of motion is quadratic in the momenta and given by

$$\mathcal{I} = p_1(q_1 p_2 - q_2 p_1) + q_1^2\left(2\Omega q_2 + \frac{\alpha}{4}(q_1^2 + 4q_2^2)\right). \tag{53}$$

This system can be regarded as an integrable cubic perturbation of the 1:2 oscillator with frequencies $(\omega, 2\omega)$ once the identification $\omega^2 = 2\Omega$ is performed (see Sect. 4.2).

The potential functions included in both the Hamiltonian (49) and its invariant (53) are directly connected to the so-called RDG series of integrable potentials, which consists of the homogeneous polynomial potentials of degree n given by [86, 87]

$$\mathcal{V}_n(q_1, q_2) = \sum_{i=0}^{[\frac{n}{2}]} 2^{n-2i}\binom{n-i}{i} q_1^{2i} q_2^{n-2i}, \qquad n = 1, 2, \ldots$$

Namely, the four members of this family read

$$\begin{aligned}
\mathcal{V}_1(q_1, q_2) &= 2q_2,\\
\mathcal{V}_2(q_1, q_2) &= q_1^2 + 4q_2^2,\\
\mathcal{V}_3(q_1, q_2) &= 4q_1^2 q_2 + 8q_2^3,\\
\mathcal{V}_4(q_1, q_2) &= q_1^4 + 12q_1^2 q_2^2 + 16q_2^4.
\end{aligned}$$

It is straightforward to realize that the quadratic and cubic potentials in the Hamiltonian (49) are just the second- and the third-order RDG potentials $\mathcal{V}_2$ and $\mathcal{V}_3$, respectively. Moreover, the integral $\mathcal{I}$ (53) contains the linear $\mathcal{V}_1$ and the quadratic $\mathcal{V}_2$ RDG potentials. Therefore, the integrable system (49) is constructed through the building block functions $\mathcal{V}_1$, $\mathcal{V}_2$ and $\mathcal{V}_3$.

In fact, it can be straightforwardly proven that a Hamiltonian $\mathcal{H}_n$ containing the RDG potential $\mathcal{V}_n$, namely,

$$\mathcal{H}_n = \frac{1}{2}\left(p_1^2 + p_2^2\right) + \alpha_n \mathcal{V}_n,$$

is always Liouville integrable, with integral of the motion $\mathcal{L}_n$ involving the $\mathcal{V}_{n-1}$ potential in the form

$$\mathcal{L}_n = p_1(q_1 p_2 - q_2 p_1) + \alpha_n q_1^2 \mathcal{V}_{n-1}, \qquad \{\mathcal{H}_n, \mathcal{L}_n\} = 0. \tag{54}$$

Note that formula (54) holds provided that the 0-th order RDG potential is defined as the constant $\mathcal{V}_0 := 1$, and the first integrable Hamiltonian system within the RDG series reads

$$\mathcal{H}_1 = \frac{1}{2}(p_1^2 + p_2^2) + \alpha_1(2q_2), \qquad \mathcal{L}_1 = p_1(q_1 p_2 - q_2 p_1) + \alpha_1 q_1^2.$$

Furthermore, all RDG potentials can freely be superposed by preserving integrability [87, 91, 92]. More explicitly, the Hamiltonian

$$\begin{aligned}\mathcal{H}_{(M)} &= \frac{1}{2}\left(p_1^2 + p_2^2\right) + \sum_{n=1}^{M} \alpha_n \mathcal{V}_n \\ &= \frac{1}{2}\left(p_1^2 + p_2^2\right) + \sum_{n=1}^{M} \sum_{i=0}^{[\frac{n}{2}]} \alpha_n 2^{n-2i} \binom{n-i}{i} q_1^{2i} q_2^{n-2i},\end{aligned} \tag{55}$$

where $M = 1, 2, \ldots$ and α_n are arbitrary real constants, has the following integral of the motion:

$$\begin{aligned}\mathcal{L}_{(M)} &= p_1(q_1 p_2 - q_2 p_1) + q_1^2 \sum_{n=1}^{M} \alpha_n \mathcal{V}_{n-1} \\ &= p_1(q_1 p_2 - q_2 p_1) + q_1^2 \left(\sum_{n=1}^{M} \sum_{i=0}^{[\frac{n-1}{2}]} \alpha_n 2^{n-1-2i} \binom{n-1-i}{i} q_1^{2i} q_2^{n-1-2i} \right).\end{aligned} \tag{56}$$

Therefore, the KdV Hénon–Heiles Hamiltonian $\mathcal{H}$ (49) and its integral $\mathcal{I}$ (53) can be thought of as the Hamiltonian $\mathcal{H}_{(M)}$ (55) and the integral $\mathcal{L}_{(M)}$ (56) by setting

$$M = 3, \qquad \alpha_1 = 0, \qquad \alpha_2 = \Omega, \qquad \alpha_3 = \alpha/4, \tag{57}$$

since in that case we obtain that

$$\mathcal{H}_{(3)} = \frac{1}{2}(p_1^2 + p_2^2) + \alpha_2 \mathcal{V}_2 + \alpha_3 \mathcal{V}_3,$$

$$\mathcal{L}_{(3)} = p_1(q_1 p_2 - q_2 p_1) + q_1^2 \left(\alpha_2 \mathcal{V}_1 + \alpha_3 \mathcal{V}_2\right).$$

As we will see in the sequel, this integrability structure associated with the RDG potentials can be fully generalized after introducing the integrable deformation generated by the curvature parameter.

7 An Integrable KdV Hénon–Heiles System on $\mathbf{S}^2$ and $\mathbf{H}^2$

The curved counterpart of the KdV Hénon–Heiles system (49) was constructed in [93] by making use of the approach we advocate in this paper, which can be summarized as follows. Given an integrable Euclidean Hénon–Heiles system

$$\mathcal{H} = \mathcal{T} + \mathcal{V} = \frac{1}{2}(p_1^2 + p_2^2) + \mathcal{V}_2(q_1, q_2) + \mathcal{V}_3(q_1, q_2),$$

an integrable generalization of this system to $\mathbf{S}^2$ and $\mathbf{H}^2$ of the form

$$\mathcal{H}_\kappa = \mathcal{T}_\kappa(p_1, p_2, q_1, q_2) + \mathcal{V}_{\kappa,2}(q_1, q_2) + \mathcal{V}_{\kappa,3}(q_1, q_2), \tag{58}$$

is constructed through the following steps:

1. Use the projective coordinates presented in Sect. 3 in order to describe the free motion on $\mathbf{S}^2$ and $\mathbf{H}^2$ (so such kinetic energy term $\mathcal{T}_\kappa$ is known and given by (21)).
2. Take the integrable curved anisotropic 1:2 oscillator and its integral of the motion given in Sect. 5.2 in the form (48) as the initial data in order to construct the curved family of RDG potentials.
3. Construct the full family of integrable curved RDG potentials on $\mathbf{S}^2$ and $\mathbf{H}^2$ (that we shall denote as $\mathcal{V}_{\kappa,n}$) through a recurrence procedure.
4. Show that the curved RDG potentials can be superposed by preserving integrability.
5. Obtain the curved 1:2 KdV Hénon–Heiles system as the particular case (58) of the latter curved RDG system.

Two important comments concerning this approach have to be pointed out: firstly, that projective coordinates will be the suitable ones in order to construct the curved RDG potentials and, secondly, that the integrability properties of $\mathcal{V}_{\kappa,2}$ will be our "initial conditions" that will guide the construction of the full integrability structure.

By following this procedure (see [93] for details), the RDG potentials on the sphere $\mathbf{S}^2$ and the hyperbolic space $\mathbf{H}^2$ can be defined in terms of projective Beltrami coordinates (q_1, q_2) as

$$\begin{aligned}\mathcal{V}_{\kappa,n} = &\left(\frac{1+\kappa\mathbf{q}^2}{1-\kappa q_2^2}\right)^2 \times \sum_{i=0}^{[\frac{n}{2}]} 2^{n-2i}\binom{n-i}{i}\left(\frac{q_1}{\sqrt{1+\kappa\mathbf{q}^2}}\right)^{2i} \\ &\times\left(1-\frac{i}{n-i}\left[\frac{\kappa q_1^2}{1+\kappa\mathbf{q}^2}\right]\right)\left(\frac{q_2}{1+\kappa\mathbf{q}^2}\right)^{n-2i}\end{aligned}$$

with $n = 1, 2, \ldots$. It is straightforward to prove that each curved RDG Hamiltonian

$$\mathcal{H}_{\kappa,n} = \mathcal{T}_\kappa + \alpha_n \mathcal{V}_{\kappa,n},$$

is integrable, with integral of motion $\mathcal{L}_{\kappa,n}$ being quadratic in the momenta and given by

$$\mathcal{L}_{\kappa,n} = J_{01}J_{12} + \alpha_n \frac{q_1^2}{1+\kappa\mathbf{q}^2}\mathcal{V}_{\kappa,n-1}, \qquad \{\mathcal{H}_{\kappa,n}, \mathcal{L}_{\kappa,n}\} = 0,$$

where $\mathcal{T}_\kappa$ is the kinetic energy (21) and J_{01}, J_{12} are the functions given in Table 1 in Beltrami variables. We stress that in order to get a suitable recurrence relation, the 0-term $\mathcal{V}_{\kappa,0}$ in the curved RGD series of potentials is by no means a constant and it has to be defined as the function

$$\mathcal{V}_{\kappa,0} := \frac{(1+\kappa q_2^2)(1+\kappa\mathbf{q}^2)}{(1-\kappa q_2^2)^2}.$$

Note that the quadratic curved RDG Hamiltonian, $\mathcal{H}_{\kappa,2} = \mathcal{T}_\kappa + \alpha_2\mathcal{V}_{\kappa,2}$, is just the superintegrable curved 1:2 oscillator (48), formerly introduced in [7] and further studied in [31, 32].

It is convenient to recall that in terms of the ambient coordinates (x_0, x_1, x_2), subjected to the constraint (8), the first curved RDG potentials turn out to be

$$\mathcal{V}_{\kappa,0} = \frac{1-\kappa x_1^2}{(x_0^2-\kappa x_2^2)^2},$$

$$V_{\kappa,1} = \frac{2x_0x_2}{(x_0^2-\kappa x_2^2)^2},$$

$$V_{\kappa,2} = \frac{x_1^2(1-\kappa x_1^2) + 4x_0^2 x_2^2}{(x_0^2 - \kappa x_2^2)^2},$$

$$V_{\kappa,3} = \frac{4x_0 x_1^2 x_2(1-\frac{1}{2}\kappa x_1^2) + 8x_0^3 x_2^3}{(x_0^2 - \kappa x_2^2)^2},$$

and the general formula for the curved RGD potentials is given by

$$\mathcal{V}_{\kappa,n} = \frac{1}{(x_0^2 - \kappa x_2^2)^2} \sum_{i=0}^{[\frac{n}{2}]} 2^{n-2i} \binom{n-i}{i} x_1^{2i} \left(1 - \frac{i}{n-i}\kappa x_1^2\right) (x_0 x_2)^{n-2i} .$$

Obviously, from these expressions these potentials can be written in any other coordinate system. Notice also that $V_{\kappa,2}$ is exactly the potential written in (48) due to the relation (8).

As in the Euclidean case the curved RDG potentials can be superposed and therefore expressions (55) and (56) can be generalized to the curved case [93]. In this way, it can straightforwardly be shown that the Hamiltonian

$$\mathcal{H}_{\kappa,(M)} = \mathcal{T}_\kappa + \sum_{n=1}^{M} \alpha_n \mathcal{V}_{\kappa,n} , \qquad M = 1, 2, \ldots$$

Poisson commutes with the function

$$\mathcal{L}_{\kappa,(M)} = J_{01} J_{12} + \frac{q_1^2}{1+\kappa \mathbf{q}^2} \sum_{n=1}^{M} \alpha_n \mathcal{V}_{\kappa,n-1} ,$$

where J_{01}, J_{12} and $\mathcal{T}_\kappa$ are again given in Table 1.

Finally, the integrable curved counterpart of the Hénon–Heiles KdV Hamiltonian (49) on $\mathbf{S}^2$ and $\mathbf{H}^2$ arises as a straightforward corollary of the previous result as the particular case $\mathcal{H}_{\kappa,(3)}$ and by considering (57). Explicitly,

$$\mathcal{H}_\kappa = \mathcal{T}_\kappa + \mathcal{V}_\kappa = \mathcal{T}_\kappa + \Omega\, \mathcal{V}_{\kappa,2} + \frac{\alpha}{4} \mathcal{V}_{\kappa,3} ,$$

and the curved analogue of the Hénon–Heiles KdV potential is so given by

$$\mathcal{V}_\kappa = \Omega \frac{q_1^2(1+\kappa q_2^2) + 4q_2^2}{(1-\kappa q_2^2)^2} + \alpha \frac{q_1^2 q_2(1+\kappa \mathbf{q}^2 - \frac{1}{2}\kappa q_1^2) + 2q_2^3}{(1-\kappa q_2^2)^2(1+\kappa \mathbf{q}^2)} .$$

The associated integral of the motion comes from $\mathcal{L}_{\kappa,(3)}$ and reads

$$\begin{aligned}\mathcal{I}_\kappa &= J_{01}J_{12} + \frac{q_1^2}{1+\kappa\mathbf{q}^2}\left(\Omega\,\mathcal{V}_{\kappa,1} + \frac{\alpha}{4}\mathcal{V}_{\kappa,2}\right)\\ &= (p_1 + \kappa(\mathbf{q}\cdot\mathbf{p})q_1)\,(q_1p_2 - q_2p_1)\\ &\quad + \frac{q_1^2}{1+\kappa\mathbf{q}^2}\left(\Omega\,\frac{2q_2(1+\kappa\mathbf{q}^2)}{(1-\kappa q_2^2)^2} + \alpha\,\frac{q_1^2(1+\kappa q_2^2)+4q_2^2}{4(1-\kappa q_2^2)^2}\right).\end{aligned}$$

We stress that, by construction, the $\kappa \to 0$ limit of all these expressions leads smoothly to their Euclidean counterparts (49) and (53) we started with.

8 Remarks and Open Problems

In this contribution we have intended to provide a summary of recent results concerning the construction of new (super)integrable systems on 2D spaces of constant curvature as (super)integrable deformations of the corresponding Euclidean systems, where the Gaussian curvature of the space plays the role of the parameter for an integrable deformation theory.

This approach can be developed in different coordinate systems, and we have stressed the fact that projective Beltrami coordinates are computationally very useful from the viewpoint of algebraic integrability, since in these coordinates the curved kinetic energy is just a polynomial in the canonical projective variables and the curved integrable potentials so obtained can be expressed as rational functions. As a summarizing example illustrating this fact we recall that the Higgs oscillator Hamiltonian $\mathcal{H}_\kappa^{1:1}$ [58, 94] (this is just the 1:1 oscillator on $\mathbf{S}^2$ and $\mathbf{H}^2$ presented in Sect. 5.1) is expressed, respectively, in terms of ambient, geodesic polar and Beltrami canonical variables as follows:

$$\begin{aligned}\mathcal{H}_\kappa^{1:1} &= \frac{\kappa\,(x_1\pi_1 + x_2\pi_2)^2}{2\left(1-\kappa\left(x_1^2+x_2^2\right)\right)} + \frac{1}{2}(\pi_1^2+\pi_2^2) + \delta\,\frac{\mathbf{x}^2}{(1-\kappa\mathbf{x}^2)}\,,\\ &= \frac{1}{2}\left(p_r^2 + \frac{p_\phi^2}{\mathrm{S}_\kappa^2(r)}\right) + \delta\,\mathrm{T}_\kappa^2(r)\,,\\ &= \tfrac{1}{2}(1+\kappa\,\mathbf{q}^2)(\mathbf{p}^2 + \kappa(\mathbf{q}\cdot\mathbf{p})^2) + \delta\,\mathbf{q}^2\,.\end{aligned}$$

The computational advantages of the projective dynamics approach become evident from these expressions, specially for the search of curved analogues of non-superintegrable systems (like Hénon–Heiles ones) where the lack of additional symmetries implies the need of making use of a purely computational approach. We also recall that in terms of Beltrami coordinates the superintegrable Kepler–

Coulomb potential on $\mathbf{S}^2$ and $\mathbf{H}^2$ is given by $\mathcal{V}^{\mathrm{KC}} = k/\sqrt{\mathbf{q}^2}$ (see [15, 19]), where again the potential in projective coordinates coincides formally with its corresponding Euclidean expression, and all the dynamical modifications arising from a non-vanishing curvature are concentrated in the kinetic energy term.

It should also be stressed that both anisotropic Euclidean oscillators and the integrable Hénon–Heiles Hamiltonian here considered preserve their integrability under the addition of some centrifugal terms, and the curved analogues of these "centrifugally perturbed" Hamiltonians can also be constructed. On the other hand, the wide applicability of the method here presented is currently being used in order to construct the curved analogue of the KdV Hénon–Heiles system (51) for arbitrary Ω_1 and Ω_2 parameters, as well as the curved analogue of the Sawada–Kotera case (50) as an integrable curvature perturbation of the Higgs 1:1 oscillator. Also, the construction of the curved Kaup–Kupershdmit Hénon–Heiles Hamiltonian (52) should be based on the constant curvature analogue of the superintegrable 1:4 curved oscillator, and is currently under investigation.

Finally, two further generalizations of the approach here presented should be mentioned. The first of them is the construction of integrable curved analogues of Minkowskian (instead of Euclidean) integrable systems, which could be addressed by following the same curvature-deformation approach, but considering the corresponding relativistic geometries with constant curvature (see [6, 8, 9, 14, 95] and the references therein). The second one deals with the construction of integrable systems on spaces with non-constant curvature, which in some cases can also be considered as (quantum) deformations of known (super)integrable systems on the Euclidean space. In these cases, a quite similar approach based on integrable perturbations in terms of a parameter related with the curvature has led to the obtention of new superintegrable oscillator and Kepler–Coulomb potentials on Darboux III and Taub-NUT spaces (see [96–105] for further details and references on integrability on spaces with non-constant curvature).

Acknowledgements This work has been partially supported by Ministerio de Ciencia, Innovación y Universidades (Spain) under grant MTM2016-79639-P (AEI/FEDER, UE) and by Junta de Castilla y León (Spain) under grant BU229P18.

References

1. A.M. Perelomov, *Integrable Systems of Classical Mechanics and Lie Algebras* (Birkhäuser, Berlin, 1990)
2. A. Goriely, *Integrability and Nonintegrability of Dynamical Systems* (World Scientific, Singapore, 2001)
3. T.G. Vozmischeva, *Integrable Problems of Celestial Mechanics in Spaces of Constant Curvature*. Astrophysics and Space Science Library, vol. 295 (Kluwer, Dordrecht, 2003)
4. D. Boccaletti, G. Pucacco, *Theory of Orbits* (Springer, Berlin, 2004)

5. W. Miller, Jr., S. Post, P. Winternitz, Classical and quantum superintegrability with applications. J. Phys. A Math. Theor. **46**, 423001 (2013). https://doi.org/10.1088/1751-8113/46/42/423001
6. A. Ballesteros, F.J. Herranz, M.A. del Olmo, M. Santander, Quantum structure of the motion groups of the two-dimensional Cayley–Klein geometries. J. Phys. A: Math. Gen. **26**, 5801–5823 (1993). https://doi.org/10.1088/0305-4470/26/21/019
7. M.F. Rañada, M. Santander, Superintegrable systems on the two-dimensional sphere S^2 and the hyperbolic plane H^2. J. Math. Phys. **40**, 5026–5057 (1999). https://doi.org/10.1063/1.533014
8. F.J. Herranz, R. Ortega, M. Santander, Trigonometry of spacetimes: a new self-dual approach to a curvature/signature (in)dependent trigonometry. J. Phys. A: Math. Gen. **33**, 4525–4551 (2000). https://doi.org/10.1088/0305-4470/33/24/309
9. F.J. Herranz, M. Santander, Conformal symmetries of spacetimes. J. Phys. A: Math. Gen. **35**, 6601–6618 (2002). https://doi.org/10.1088/0305-4470/35/31/306
10. M.F. Rañada, M. Santander, On some properties of harmonic oscillator on spaces of constant curvature. Rep. Math. Phys. **49**, 335–343 (2002). https://doi.org/10.1016/S0034-4877(02)80031-3
11. M.F. Rañada, M. Santander, On harmonic oscillators on the two-dimensional sphere S^2 and the hyperbolic plane H^2. J. Math. Phys. **43**, 431–451 (2002). https://doi.org/10.1063/1.1423402
12. A. Ballesteros, F.J. Herranz, M. Santander, T. Sanz-Gil, Maximal superintegrability on N-dimensional curved spaces. J. Phys. A: Math. Gen. **36**, L93–L99 (2003). https://doi.org/10.1088/0305-4470/36/7/101
13. F.J. Herranz, A. Ballesteros, M. Santander, T. Sanz-Gil, Maximally superintegrable Smorodinsky–Winternitz systems on the N-dimensional sphere and hyperbolic spaces, in *Superintegrability in Classical and Quantum Systems*, ed. by P. Tempesta et al. CRM Proceedings and Lecture Notes, vol. 37 (American Mathematical Society, Providence, 2004), pp. 75–89. https://doi.org/10.1090/crmp/037
14. F.J. Herranz, A. Ballesteros, Superintegrability on three-dimensional Riemannian and relativistic spaces of constant curvature. Symmetry Integrability Geom. Methods Appl. **2**, 010 (2006). https://doi.org/10.3842/SIGMA.2006.010
15. A. Ballesteros, F.J. Herranz, Universal integrals for superintegrable systems on N-dimensional spaces of constant curvature. J. Phys. A: Math. Theor. **40**, F51–F59 (2007). https://doi.org/10.1088/1751-8113/40/2/F01
16. J.F. Cariñena, M.F. Rañada, M. Santander, The quantum harmonic oscillator on the sphere and the hyperbolic plane. Ann. Phys. **322**, 2249–2278 (2007). https://doi.org/10.1016/j.aop.2006.10.010
17. J.F. Cariñena, M.F. Rañada, M. Santander, Superintegrability on curved spaces, orbits and momentum hodographs: revisiting a classical result by Hamilton. J. Phys. A: Math. Theor. **40**, 13645–13666 (2007). https://doi.org/10.1088/1751-8113/40/45/010
18. J.F. Cariñena, M.F. Rañada, M. Santander, The Kepler problem and the Laplace-Runge-Lenz vector on spaces of constant curvature and arbitrary signature. Qual. Theory Dyn. Syst. **7**, 87–99 (2008). https://doi.org/10.1007/s12346-008-0004-3
19. A. Ballesteros, F.J. Herranz, Maximal superintegrability of the generalized Kepler–Coulomb system on N-dimensional curved spaces. J. Phys. A: Math. Theor. **42**, 245203 (2009). https://doi.org/10.1088/1751-8113/42/24/245203
20. F. Diacu, E. Pérez-Chavela, Homographic solutions of the curved 3-body problem. J. Differ. Equ. **250**, 340–366 (2011). https://doi.org/10.1016/j.jde.2010.08.011
21. F. Diacu, E. Pérez-Chavela, M. Santoprete, The n-body problem in spaces of constant curvature. Part I: relative equilibria. J. Nonlinear Sci. **22**, 247–266 (2012). https://doi.org/10.1007/s00332-011-9116-z
22. F. Diacu, E. Pérez-Chavela, M. Santoprete, The n-body problem in spaces of constant curvature. Part II: singularities equilibria. J. Nonlinear Sci. **22**, 267–275 (2012). https://doi.org/10.1007/s00332-011-9117-y

23. F. Diacu, Relative equilibria in the 3-dimensional curved n-body problem. Memoirs Am. Math. Soc. **228**, 1071 (2014). http://dx.doi.org/10.1090/memo/1071
24. C. Gonera, M. Kaszubska, Superintegrable systems on spaces of constant curvature. Ann. Phys. **364**, 91–102 (2014). https://doi.org/10.1016/j.aop.2014.04.005
25. M.F. Rañada, The Tremblay-Turbiner-Winternitz system on spherical and hyperbolic spaces: superintegrability, curvature-dependent formalism and complex factorization. J. Phys. A: Math. Theor. **47**, 165203 (2014). https://doi.org/10.1088/1751-8113/47/16/165203
26. Rañada, M.F.: The Post-Winternitz system on spherical and hyperbolic spaces: a proof of the superintegrability making use of complex functions and a curvature-dependent formalism. Phys. Lett. A **379**, 2267–2271 (2015). https://doi.org/10.1016/j.physleta.2015.07.043
27. M.F. Rañada, Superintegrable deformations of superintegrable systems: quadratic superintegrability and higher-order superintegrability. J. Math. Phys. **56**, 042703 (2015). https://doi.org/10.1063/1.4918611
28. C.M. Chanu, L. Degiovanni, G. Rastelli, Warped product of Hamiltonians and extensions of Hamiltonian systems. J. Phys.: Conf. Ser. **597**, 012024 (2015). https://doi.org/10.1088/1742-6596/597/1/012024
29. A. Albouy, There is a projective dynamics. Eur. Math. Soc. Newsl. **89**, 37–43 (2013). http://www.ems-ph.org/journals/newsletter/pdf/2013-09-89.pdf
30. A. Albouy, Projective dynamics and first integrals. Regul. Chaot. Dyn. **20**, 247–276 (2015). https://doi.org/10.1134/S1560354715030041
31. A. Ballesteros, F.J. Herranz, F. Musso, The anisotropic oscillator on the 2D sphere and the hyperbolic plane. Nonlinearity **26**, 971–990 (2013). http://iopscience.iop.org/article/10.1088/0951-7715/26/4/971/pdf
32. A. Ballesteros, A. Blasco, F.J. Herranz, F. Musso, A new integrable anisotropic oscillator on the two-dimensional sphere and the hyperbolic plane. J. Phys. A: Math. Theor. **47**, 345204 (2014). http://iopscience.iop.org/article/10.1088/1751-8113/47/34/345204/pdf
33. E. Inönü, E.P. Wigner, On the contractions of groups and their representations. Proc. Natl. Acad. Sci. U.S.A. **39**, 510–524 (1953). https://doi.org/10.1073/pnas.39.6.510
34. F.J. Herranz, M. de Montigny, M.A. del Olmo, M. Santander, Cayley–Klein algebras as graded contractions of $so(N+1)$. J. Phys. A: Math. Gen. **27**, 2515–2526 (1994). https://doi.org/10.1088/0305-4470/27/7/027
35. Yaglom, I.M.: *A Simple Non-Euclidean Geometry and Its Physical Basis* (Springer, New York, 1979)
36. N.A. Gromov, V.I. Man'ko, The Jordan–Schwinger representations of Cayley–Klein groups. I. The orthogonal groups. J. Math. Phys. **31**, 1047–1053 (1990). https://doi.org/10.1063/1.528781
37. B. Doubrovine, S. Novikov, A. Fomenko, *Géométrie Contemporaine, Méthodes et Applications First Part* (MIR, Moscow, 1982)
38. J.M. Jauch, E.L. Hill, On the problem of degeneracy in quantum mechanics. Phys. Rev. **57**, 641–645 (1940). https://doi.org/10.1103/PhysRev.57.641
39. J.P. Amiet, S. Weigert, Commensurate harmonic oscillators: classical symmetries. J. Math. Phys. **43**, 4110–4126 (2002). https://doi.org/10.1063/1.1488672
40. M.A. Rodríguez, P. Tempesta, P. Winternitz, Reduction of superintegrable systems: the anisotropic harmonic oscillator. Phys. Rev. E **78**, 046608 (2008). https://doi.org/10.1103/PhysRevE.78.046608
41. A. Ballesteros, F.J. Herranz, S. Kuru, J. Negro, The anisotropic oscillator on curved spaces: a new exactly solvable model. Ann. Phys. **373**, 399–423 (2016). https://doi.org/10.1016/j.aop.2016.07.006
42. A. Ballesteros, F.J. Herranz, S. Kuru, J. Negro, Factorization approach to superintegrable systems: formalism and applications. Phys. Atom. Nuclei **80**, 389–396 (2017). https://doi.org/10.1134/S1063778817020053
43. C.D.J. Fernández, J. Negro, M.A. del Olmo, Group approach to the factorization of the radial oscillator equation. Ann. Phys. **252**, 386–412 (1996). https://doi.org/10.1006/aphy.1996.0138

44. S. Kuru, J. Negro, Factorizations of one-dimensional classical systems. Ann. Phys. **323**, 413–431 (2008). https://doi.org/10.1016/j.aop.2007.10.004
45. J.A. Calzada, S. Kuru, J. Negro, M.A. del Olmo, Dynamical algebras of general two-parametric Pöschl–Teller Hamiltonian. Ann. Phys. **327**, 808–822 (2012). https://doi.org/10.1016/j.aop.2011.12.014
46. E. Celeghini, S. Kuru, J. Negro, M.A. del Olmo, A unified approach to quantum and classical TTW systems based on factorizations. Ann. Phys. **332**, 27–37 (2013). https://doi.org/10.1016/j.aop.2013.01.008
47. J.A. Calzada, S. Kuru, J. Negro, Superintegrable Lissajous systems on the sphere. Eur. Phys. J. Plus **129**, 129–164 (2014). https://doi.org/10.1140/epjp/i2014-14164-5
48. N.W. Evans, P.E. Verrier, Superintegrability of the caged anisotropic oscillator. J. Math. Phys. **49**, 092902 (2008). https://doi.org/10.1063/1.2988133
49. N.W. Evans, Superintegrability in classical mechanics. Phys. Rev. A **41**, 5666–5676 (1990). https://doi.org/10.1103/PhysRevA.41.5666
50. E.G. Kalnins, G.C. Williams, W. Miller, G.S. Pogosyan, Superintegrability in the three–dimensional Euclidean space. J. Math. Phys. **40**, 708–725 (1999). https://doi.org/10.1063/1.532699
51. Y.N. Demkov, Symmetry group of the isotropic oscillator. Soviet Phys. JETP **36**, 63–66 (1959). http://www.jetp.ac.ru/cgi-bin/dn/e_009_01_0063.pdf
52. D.M. Fradkin, Three-dimensional isotropic harmonic oscillator and SU_3. Am. J. Phys. **33**, 207–211 (1965). https://doi.org/10.1119/1.1971373
53. T.I. Fris, V. Mandrosov, Y.A. Smorodinsky, M. Uhlir, P. Winternitz, On higher symmetries in quantum mechanics. Phys. Lett. **16**, 354–356 (1965). https://doi.org/10.1016/0031-9163(65)90885-1
54. N.W. Evans, Super-integrability of the Winternitz system. Phys. Lett. A **147**, 483–486 (1990). https://doi.org/10.1016/0375-9601(90)90611-Q
55. N.W. Evans, Group theory of the Smorodinsky-Winternitz system. J. Math. Phys. **32**, 3369–3375 (1991). https://doi.org/10.1063/1.529449
56. C. Grosche, G.S. Pogosyan, A.N. Sissakian, Path integral discussion for Smorodinsky–Winternitz potentials I. Two- and three dimensional Euclidean spaces. Fortschr. Phys. **43**, 453–521 (1995). https://doi.org/10.1002/prop.2190430602
57. K.B. Wolf, C.P. Boyer, The 2:1 anisotropic oscillator, separation of variables and symmetry group in Bargmann space. J. Math. Phys. **16**, 2215–2223 (1975). https://doi.org/10.1063/1.522471
58. P.W. Higgs, Dynamical symmetries in a spherical geometry I. J. Phys. A: Math. Gen. **12**, 309–323 (1979). https://doi.org/10.1088/0305-4470/12/3/006
59. M.F. Rañada, M. Santander, On harmonic oscillators on the two-dimensional sphere S^2 and the hyperbolic plane H^2 II. J. Math. Phys. **44**, 2149–2167 (2003). https://doi.org/10.1063/1.1560552
60. E.G. Kalnins, G.S. Pogosyan, W. Miller, Jr., Completeness of multiseparable superintegrability on the complex 2-sphere. J. Phys. A: Math. Gen. **33**, 6791–6806 (2000). https://doi.org/10.1088/0305-4470/33/38/310
61. E.G. Kalnins, J.M. Kress, G.S. Pogosyan, W. Miller, Jr., Completeness of superintegrability in two-dimensional constant-curvature spaces. J. Phys. A: Math. Gen. **34**, 4705–4720 (2001). https://doi.org/10.1088/0305-4470/34/22/311
62. E.G. Kalnins, S. Benenti, W. Miller, Jr., Integrability, Stäckel spaces, and rational potentials. J. Math. Phys. **38**, 2345–2365 (1997). https://doi.org/10.1063/1.531977
63. P. Saksida, Integrable anharmonic oscillators on spheres and hyperbolic spaces. Nonlinearity **14**, 977–994 (2001). https://doi.org/10.1088/0951-7715/14/5/304
64. A. Nerssesian, V. Yeghikyan, Anisotropic inharmonic Higgs oscillator and related (MICZ-)Kepler-like systems. J. Phys. A: Math. Theor. **41**, 155203 (2008). https://doi.org/10.1088/1751-8113/41/15/155203
65. I. Marquette, Generalized MICZ-Kepler system, duality, polynomial, and deformed oscillator algebras. J. Math. Phys. **51**, 102105 (2010). https://doi.org/10.1063/1.3496900

66. H.I. Leemon, Dynamical symmetries in a spherical geometry II. J. Phys. A: Math. Gen. **12**, 489–501 (1979). https://doi.org/10.1088/0305-4470/12/4/009
67. Y.M. Hakobyan, G.S. Pogosyan, A.N. Sissakian, S.I. Vinitsky, Isotropic oscillator in a space of constant positive curvature: interbasis expansions. Phys. Atom. Nucl. **62**, 623–637 (1999). https://arxiv.org/abs/quant-ph/9710045
68. A. Nersessian, G. Pogosyan, Relation of the oscillator and Coulomb systems on spheres and pseudospheres. Phys. Rev. A **63**, 020103 (2001). https://doi.org/10.1103/PhysRevA.63.020103
69. J.F. Cariñena, M.F. Rañada, M. Santander, M. Senthilvelan, A non-linear oscillator with quasi-harmonic behaviour: two-and n-dimensional oscillators. Nonlinearity **17**, 1941–1963 (2004). https://doi.org/10.1088/0951-7715/17/5/019
70. A. Ballesteros, A. Enciso, F.J. Herranz, O. Ragnisco, Superintegrability on N-dimensional curved spaces: central potentials, centrifugal terms and monopoles. Ann. Phys. **324**, 1219–1233 (2009). https://doi.org/10.1016/j.aop.2009.03.001
71. C. Grosche, G.S. Pogosyan, A.N. Sissakian, Path integral discussion for Smorodinsky–Winternitz potentials II. The two- and three-dimensional sphere. Fortschr. Phys. **43**, 523–563 (1995). https://doi.org/10.1002/prop.2190430603
72. M. Hénon, C. Heiles, The applicability of the third integral of motion: some numerical experiments. Astron. J. **69**, 73–79 (1964). https://doi.org/10.1086/109234
73. T. Bountis, H. Segur, F. Vivaldi, Integrable Hamiltonian systems and the Painlevé property. Phys. Rev. A **25**, 1257–1264 (1982). https://doi.org/10.1103/PhysRevA.25.1257
74. Y.F. Chang, M. Tabor, J. Weiss, Analytic structure of the Hénon–Heiles Hamiltonian in integrable and nonintegrable regimes. J. Math. Phys. **23**, 531–538 (1982). https://doi.org/10.1063/1.525389
75. B. Grammaticos, B. Dorizzi, R. Padjen, Painlevé property and integrals of motion for the Hénon–Heiles system. Phys. Lett. A **89**, 111–113 (1982). https://doi.org/10.1016/0375-9601(82)90868-4
76. J. Hietarinta, Integrable families of Hénon–Heiles-type Hamiltonians and a new duality. Phys. Rev. A **28**, 3670–3672 (1983). https://doi.org/10.1103/PhysRevA.28.3670
77. A.P. Fordy, Hamiltonian symmetries of the Hénon–Heiles system. Phys. Lett. A **97**, 21–23 (1983). https://doi.org/10.1016/0375-9601(83)90091-9
78. S. Wojciechowski, Separability of an integrable case of the Hénon–Heiles system. Phys. Lett. A **100**, 277–278 (1984). https://doi.org/10.1016/0375-9601(84)90535-8
79. R. Sahadevan, M. Lakshmanan, Invariance and integrability: Hénon–Heiles and two coupled quartic anharmonic oscillator systems. J. Phys. A: Math. Gen. **19**, L949–L954 (1986). https://doi.org/10.1088/0305-4470/19/16/001
80. A.P. Fordy, The Hénon–Heiles system revisited. Phys. D **52**, 204–210 (1991). https://doi.org/10.1016/0167-2789(91)90122-P
81. W. Sarlet, New aspects of integrability of generalized Hénon–Heiles systems. J. Phys. A: Math. Gen. **24**, 5245–5251 (1991). https://doi.org/10.1088/0305-4470/24/22/008
82. V. Ravoson, L. Gavrilov, R. Caboz, Separability and Lax pairs for Hénon–Heiles system. J. Math. Phys. **34**, 2385–2393 (1993). https://doi.org/10.1063/1.530123
83. G. Tondo, On the integrability of stationary and restricted flows of the KdV hierarchy. J. Phys. A: Math. Gen. **28**, 5097–5115 (1995). https://doi.org/10.1088/0305-4470/28/17/034
84. R. Conte, M. Musette, C. Verhoeven, Completeness of the cubic and quartic Hénon–Heiles Hamiltonians. Theor. Math. Phys. **144**, 888–898 (2005). https://doi.org/10.1007/s11232-005-0115-9
85. A. Ballesteros, A. Blasco, F.J. Herranz, A curved Hénon-Heiles system and its integrable perturbations. J. Phys.: Conf. Ser. **597**, 012013 (2015). https://doi.org/10.1088/1742-6596/597/1/012013
86. A. Ramani, B. Dorizzi, B. Grammaticos, Painlevé conjecture revisited. Phys. Rev. Lett. **49**, 1539–1541 (1982). https://doi.org/10.1103/PhysRevLett.49.1539
87. J. Hietarinta, Direct method for the search of the second invariant. Phys. Rep. **147**, 87–154 (1987). https://doi.org/10.1016/0370-1573(87)90089-5

88. E.V. Ferapontov, A.P. Fordy, Separable Hamiltonians and integrable systems of hydrodynamic type. J. Geom. Phys. **21**, 169–182 (1997). https://doi.org/10.1016/S0393-0440(96)00013-7
89. A.N.W. Hone, V. Novikov, C. Verhoeven, An integrable hierarchy with a perturbed Hénon–Heiles system. Inv. Probl. **22**, 2001–2020 (2006). https://doi.org/10.1088/0266-5611/22/6/006
90. A.N.W. Hone, V. Novikov, C. Verhoeven, An extended Hénon–Heiles system. Phys. Lett. A **372**, 1440–1444 (2008). https://doi.org/10.1016/j.physleta.2007.09.063
91. A. Blasco, Integrability of non-linear Hamiltonian systems with N degrees of freedom. Ph.D. Thesis, Burgos University, Burgos, 2009. http://riubu.ubu.es/bitstream/10259/106/4/Blasco_Sanz.pdf
92. A. Ballesteros, A. Blasco, Integrable Hénon–Heiles Hamiltonians: a Poisson algebra approach. Ann. Phys. **325**, 2787–2799 (2010). https://doi.org/10.1016/j.aop.2010.08.002
93. A. Ballesteros, A. Blasco, F.J. Herranz, F. Musso, An integrable Hénon–Heiles system on the sphere and the hyperbolic plane. Nonlinearity **28**, 3789–3801 (2015). https://doi.org/10.1088/0951-7715/28/11/3789
94. P. Serret, Théorie nouvelle géométrique et mécanique des lignes à double courbure. Paris, Mallet-Bachelier (1859)
95. D.R. Petrosyan, G.S. Pogosyan, Harmonic oscillator on the SO(2,2) hyperboloid. SIGMA Symmetry Integrability Geom. Methods Appl. **11**, 096 (2015). https://doi.org/10.3842/SIGMA.2015.096
96. E.G. Kalnins, J.M. Kress, P. Winternitz, Superintegrability in a two-dimensional space of nonconstant curvature. J. Math. Phys. **43**, 970–983 (2002). https://doi.org/10.1063/1.1429322
97. A. Ballesteros, F.J. Herranz, O. Ragnisco, Integrable potentials on spaces with curvature from quantum groups. J. Phys. A: Math. Theor. **38**, 7129–7144 (2005). https://doi.org/10.1088/0305-4470/38/32/004
98. A. Ballesteros, A. Enciso, F.J. Herranz, O. Ragnisco, A maximally superintegrable system on an n-dimensional space of nonconstant curvature. Phys. D **237**, 505–509 (2008). https://doi.org/10.1016/j.physd.2007.09.021
99. A. Ballesteros, A. Enciso, F.J. Herranz, O. Ragnisco, Bertrand spacetimes as Kepler/oscillator potentials. Class. Quant. Grav. **25**, 165005 (2008). https://doi.org/10.1088/0264-9381/25/16/165005
100. A. Ballesteros, A. Enciso, F.J. Herranz, O. Ragnisco, Hamiltonian systems admitting a Runge-Lenz vector and an optimal extension of Bertrand's theorem to curved manifolds. Commun. Math. Phys. **290**, 1033–1049 (2009). https://doi.org/10.1007/s00220-009-0793-5
101. A. Ballesteros, A. Blasco, F.J. Herranz, F. Musso, O. Ragnisco, (Super)integrability from coalgebra symmetry: formalism and applications. J. Phys.: Conf. Ser. **175**, 012004 (2009). https://doi.org/10.1088/1742-6596/175/1/012004
102. O. Ragnisco, D. Riglioni, A family of exactly solvable radial quantum systems on space of non-constant curvature with accidental degeneracy in the spectrum. SIGMA Symmetry Integrability Geom. Methods Appl. **6**, 097 (2010). https://doi.org/10.3842/SIGMA.2010.097
103. A. Ballesteros, A. Enciso, F.J. Herranz, O. Ragnisco, O. Riglioni, Superintegrable oscillator and Kepler systems on spaces of nonconstant curvature via the Säckel Transform. SIGMA Symmetry Integrability Geom. Methods Appl. **7**, 048 (2011). https://doi.org/10.3842/SIGMA.2011.048
104. A. Ballesteros, A. Enciso, F.J. Herranz, O. Ragnisco, O. Riglioni, Quantum mechanics on spaces of nonconstant curvature: the oscillator problem and superintegrability. Ann. Phys. **326**, 2053–2073 (2011). https://doi.org/10.1016/j.aop.2011.03.002
105. A. Ballesteros, A. Enciso, F.J. Herranz, O. Ragnisco, D. Riglioni, An exactly solvable deformation of the Coulomb problem associated with the Taub-NUT metric. Ann. Phys. **351**, 540–577 (2014). https://doi.org/10.1016/j.aop.2014.09.013

Trends in Supersymmetric Quantum Mechanics

David J. Fernández C.

Abstract Along the years, supersymmetric quantum mechanics (SUSY QM) has been used for studying solvable quantum potentials. It is the simplest method to build Hamiltonians with prescribed spectra in the spectral design. The key is to pair two Hamiltonians through a finite order differential operator. Some related subjects can be simply analyzed, as the algebras ruling both Hamiltonians and the associated coherent states. The technique has been applied also to periodic potentials, where the spectra consist of allowed and forbidden energy bands. In addition, a link with non-linear second-order differential equations, and the possibility of generating some solutions, can be explored. Recent applications concern the study of Dirac electrons in graphene placed either in electric or magnetic fields, and the analysis of optical systems whose relevant equations are the same as those of SUSY QM. These issues will be reviewed briefly in this paper, trying to identify the most important subjects explored currently in the literature.

Keywords Supersymmetric quantum mechanics · Coherent states · Painlevé equations · Painlevé transcendents · Polynomial Heisenberg algebras · Factorization method · Exact solutions · Spectral design · Graphene

1 Introduction

The birth of supersymmetric quantum mechanics (SUSY QM) in 1981, as a toy model to illustrate the properties that systems involving both bosons and fermions have, was a breakthrough in the study of solvable quantum mechanical models [1]. One of the reasons is that SUSY QM is tightly related to other approaches used

---

Dedicated to my dear friend and colleague Véronique Hussin.

D. J. Fernández C. (✉)
Departamento de Física, Cinvestav, Ciudad de México, Mexico
e-mail: david@fis.cinvestav.mx

Ş. Kuru et al. (eds.), *Integrability, Supersymmetry and Coherent States*, CRM Series in Mathematical Physics, https://doi.org/10.1007/978-3-030-20087-9_2

in the past to address this kind of systems, e.g., the factorization method, Darboux transformation, and intertwining technique [2–27].

On the other hand, it is well known that the factorization method was introduced by Dirac in 1935, to derive algebraically the spectrum of the harmonic oscillator [28]. The next important advance was done by Schrödinger in 1940, who realized that the procedure can be also applied to the Coulomb potential [29, 30]. Later on, Infeld and his collaborators push forward the technique [31, 32], supplying a general classification scheme including most of the exactly solvable Schrödinger Hamiltonians known up to that time [2]. As a consequence, the idea that the factorization method was essentially exhausted started to spread among the scientific community.

However, in 1984 Mielnik proved that this belief was wrong, by generalizing simply the Infeld–Hull factorization method when he was seeking the most general first-order differential operators which factorize the harmonic oscillator Hamiltonian in a certain given order [33]. The key point of his approach was that if the ordering of the generalized factorization operators is interchanged, then a new Hamiltonian is obtained which is intertwined with the oscillator one.

It is worth to stress that Mielnik's work represented the next breakthrough in the development of the factorization method, since it opened the way to look for new solvable quantum potentials. In particular, this generalization was immediately applied to the Coulomb problem [34]. Meanwhile, Andrianov's group [35, 36] and Nieto [37] identified the links of the factorization method with Darboux transformation and supersymmetric quantum mechanics, respectively. In addition, Sukumar indicated the way to apply Mielnik's approach to arbitrary potentials and factorization energies [38, 39], setting up the general framework where the factorization method would develop for the next decade [33, 34, 40–64].

Let us mention that up to the year 1993 the factorization operators, which at the same time are intertwining operators in this case, were first-order differential ones. A natural generalization, pursued by Andrianov and collaborators [65, 66], consists in taking the intertwining operators of order greater than one. This proposal was important, since it helped to circumvent the restriction of the first-order method, that only the energy of the initial ground state can be modified. Moreover, it made clear that the key of the generalization is the analysis of the intertwining relation rather than the factorized expressions. Let us note also that in 1995 Bagrov and Samsonov explored the same technique in a different but complementary way [67].

Our group got back to the subject in 1997 [68–72], although some works related with the method had been done previously [73]. In particular, several physically interesting potentials were addressed through this technique, as the standard harmonic oscillator [33, 69, 70], the radial oscillator, and Coulomb potentials [34, 73, 74], among others [75–77]. In addition, the coherent states associated to the SUSY partners of the harmonic oscillator were explored [78–81], and similar works dealing with more general one-dimensional Hamiltonians were done [82, 83]. Another important contribution has to do with the determination of the general systems ruled by polynomial Heisenberg algebras and the study of particular realizations based on the SUSY partners of the oscillator [80, 84–87]. The complex SUSY transformations involving either real or complex factorization "energies" were implemented as well [88–92]. In addition, the analysis of the confluent algorithm, the degenerate case in which all the factorization energies tend to a single one, was also elaborated [74, 93–101]. The SUSY techniques for exactly solvable periodic potentials, as the Lamé and associated Lamé potentials, have been explored as well[102–109].

Some other groups have addressed the same subjects through different viewpoints, e.g., the N-fold supersymmetry by Tanaka and collaborators [110–114], the hidden non-linear supersymmetry by Plyushchay et al. [115–119], among others.

Especially important is the connection of SUSY QM with non-linear second-order ordinary differential equations, as KdV and Painlevé IV and V equations, as well as the possibility of designing algorithms to generate some of their solutions [81, 84–86, 90, 92, 120–135].

Another relevant subject related to SUSY QM is the so-called exceptional orthogonal polynomials (EOP) [136–150]. In fact, it seems that most of these new polynomials appear quite naturally when the seed solutions which are employed reduce to polynomial solutions of the initial stationary Schrödinger equation [144].

Recently, the SUSY methods started to be used also in the study of Dirac electrons in graphene and some of its allotropes, when external electric or magnetic fields are applied [151–161]. It is worth to mention as well some systems in optics, since there is a well-known correspondence between Schrödinger equation and Maxwell equations in the paraxial approximation, which makes that the SUSY methods can be applied directly in some areas of optics [162–169].

As we can see, the number of physical systems which are related with supersymmetric quantum mechanics is large enough to justify the writing of a new review paper, in which we will present the recent advances in the subject. If the reader is looking for books and previous review papers addressing SUSY QM from an inductive viewpoint, we recommend Refs. [5–27].

2 Supersymmetric Quantum Mechanics

In this section we shall present axiomatically the supersymmetric quantum mechanics, as a tool for generating solvable potentials $\widetilde{V}(x)$ departing from a given initial one $V(x)$.

The supersymmetry algebra with two generators introduced by Witten in 1981 [1]

$$[Q_i, H_{\rm ss}] = 0, \quad \{Q_i, Q_j\} = \delta_{ij} H_{\rm ss}, \quad i, j = 1, 2, \tag{1}$$

when realized in the following way:

$$Q_1 = \frac{Q^+ + Q}{\sqrt{2}}, \qquad Q_2 = \frac{Q^+ - Q}{i\sqrt{2}}, \tag{2}$$

$$Q = \begin{pmatrix} 0 & 0 \\ B & 0 \end{pmatrix}, \quad Q^+ = \begin{pmatrix} 0 & B^+ \\ 0 & 0 \end{pmatrix}, \tag{3}$$

$$H_{\rm ss} = \{Q, Q^+\} = \begin{pmatrix} B^+B & 0 \\ 0 & BB^+ \end{pmatrix} \tag{4}$$

is called supersymmetric quantum mechanics, where $H_{\rm ss}$ is the supersymmetric Hamiltonian, while $Q_1,\ Q_2$ are the supercharges. The kth order differential operators $B,\ B^+$ intertwine two Schrödinger Hamiltonians

$$\widetilde{H} = -\frac{1}{2}\frac{d^2}{dx^2} + \widetilde{V}(x), \qquad H = -\frac{1}{2}\frac{d^2}{dx^2} + V(x), \tag{5}$$

in the way

$$\widetilde{H}B^+ = B^+H, \qquad HB = B\widetilde{H}. \tag{6}$$

There is a natural link with the *factorization method*, since the following relations are fulfilled:

$$B^+B = \prod_{j=1}^{k}(\widetilde{H} - \epsilon_j), \qquad BB^+ = \prod_{j=1}^{k}(H - \epsilon_j), \tag{7}$$

where $\epsilon_j,\ j = 1, \ldots, k$ are k *factorization energies* associated to k *seed solutions* required to implement the intertwining (see Eqs. (5) and (6) and Sects. 2.1 and 2.2). Taking into account these expressions, it turns out that the supersymmetric Hamiltonian $H_{\rm ss}$ is a polynomial of degree kth in the diagonal matrix operator $H_{\rm p}$ which involves the two Schrödinger Hamiltonians H and $\widetilde{H}$ as follows:

$$H_{\rm ss} = \prod_{j=1}^{k}(H_{\rm p} - \epsilon_j), \qquad H_{\rm p} = \begin{pmatrix} \widetilde{H} & 0 \\ 0 & H \end{pmatrix}. \tag{8}$$

In particular, if $k = 1$ the standard (first-order) supersymmetric quantum mechanics is recovered, for which $H_{\rm ss}$ is a first degree polynomial in $H_{\rm p}$, $H_{\rm ss} = H_{\rm p} - \epsilon_1$. For $k > 1$, however, we will arrive to the so-called higher-order supersymmetric quantum mechanics, in which $H_{\rm ss}$ is a polynomial of degree greater than one in $H_{\rm p}$ (see, for example, [23]).

2.1 Standard SUSY Transformations

Let us suppose now that we select k solutions u_j of the initial stationary Schrödinger equation for k *different* factorization energies ϵ_j, $j = 1, \ldots, k$,

$$Hu_j = \epsilon_j u_j, \tag{9}$$

which are called *seed solutions*. From them we implement the intertwining transformation of Eq. (6), leading to a new potential $\widetilde{V}(x)$ which is expressed in terms of the initial potential and the seed solutions as follows:

$$\widetilde{V}(x) = V(x) - [\log W(u_1, \ldots, u_k)]'', \tag{10}$$

where $W(u_1, \ldots, u_k)$ denotes the Wronskian of u_j, $j = 1, \ldots, k$. The eigenfunctions $\widetilde{\psi}_n$ and eigenvalues E_n of $\widetilde{H}$ are obtained from the corresponding ones of H, ψ_n, and E_n, as follows:

$$\widetilde{\psi}_n = \frac{B^+\psi_n}{\sqrt{(E_n - \epsilon_1)\cdots(E_n - \epsilon_k)}} \propto \frac{W(u_1, \ldots, u_k, \psi_n)}{W(u_1, \ldots, u_k)}. \tag{11}$$

Moreover, $\widetilde{H}$ could have additional eigenfunctions $\widetilde{\psi}_{\epsilon_j}$ for some of the factorization energies ϵ_j (at most k, depending on either they fulfill or not the required boundary conditions) which are given by:

$$\widetilde{\psi}_{\epsilon_j} \propto \frac{W(u_1, \ldots, u_{j-1}, u_{j+1}, \ldots, u_k)}{W(u_1, \ldots, u_k)}. \tag{12}$$

We can conclude that, given the initial potential $V(x)$, its eigenfunctions ψ_n, eigenvalues E_n, and the k chosen seed solutions u_j, $j = 1, \ldots, k$, it is possible to generate algorithmically its kth order SUSY partner potential $\widetilde{V}(x)$ as well as the associated eigenfunctions and eigenvalues through expressions (10)–(12).

It is important to stress that the seed solutions must be carefully chosen in order that the new potential will not have singularities additional to those of the initial potential $V(x)$. When this happens, we say that the transformation is *non-singular*. If the initial potential is real, and we require the same for the final potential, then there are some criteria for choosing the real seed solutions u_j according to their number of nodes, which also depend on the values taken by the associated factorization energies ϵ_j (see, for example, [23]). Although non-exhaustive, let us report next a list of some important criteria, which will make the final potential $\widetilde{V}(x)$ to be real and without any extra singularity with respect to $V(x)$.

- If $k = 1$ (first-order SUSY QM), the factorization energy ϵ_1 must belong to the infinite energy gap $\epsilon_1 < E_0$ in order that u_1 could be nodeless inside the x-domain of the problem, where E_0 is the ground state energy of H. Moreover, since in this ϵ_1-domain the seed solution u_1 could have either one node or none, then we additionally require to identify the right nodeless solution. With these conditions, the transformation will be non-singular and the spectrum of the new Hamiltonian $\widetilde{H}$ will have an extra level ϵ_1 with respect to H (creation of a new level). Note that also it is possible to select the seed solution with a node at one of the edges of the x-domain; thus, the SUSY transformation will be still non-singular but the factorization energy ϵ_1 will not belong to the spectrum of $\widetilde{H}$ (isospectral transformation).

- If $k = 1, \epsilon_1 = E_0$, and $u_1 = \psi_0$ (the seed solution is the ground state, which has one node at each edge of the x-domain), then the SUSY transformation will be non-singular and the spectrum of the new Hamiltonian will not have the level E_0 (deletion of one level).
- If $k = 2$ (standard second-order SUSY QM), first of all both ϵ_1 and ϵ_2 must belong to the same energy gap, either to the infinite one below E_0 or to a finite gap defined by two neighbor energy levels (E_m, E_{m+1}). Let us order the two factorization energies in the way $\epsilon_2 < \epsilon_1$. In order that the Wronskian of u_1 and u_2 would be nodeless, the seed solution u_2 associated to the lower factorization energy ϵ_2 should have one extra node with respect to the solution u_1 associated to the higher factorization energy ϵ_1 [23]. In particular, in the infinite gap u_2 should have one node and u_1 should be nodeless. On the other hand, when both factorization energies are in the finite gap (E_m, E_{m+1}) the seed solutions u_2 and u_1 should have $m + 2$ and $m + 1$ nodes, respectively. In both cases the spectrum of the new Hamiltonian will contain two extra eigenvalues ϵ_1, ϵ_2 (creation of two levels). Moreover, the seed solutions can be chosen such that the transformation is still non-singular but either ϵ_1, ϵ_2 or both will not belong to the spectrum of $\widetilde{H}$ (either creation of one new level or isospectral transformation).
- If $k = 2, \epsilon_2 = E_m, u_2 = \psi_m, \epsilon_1 = E_{m+1}, u_1 = \psi_{m+1}$, then the SUSY transformation will be non-singular and the spectrum of the new Hamiltonian will not have the two levels E_m, E_{m+1} (deletion of two levels).
- If $k > 2$, the corresponding non-singular SUSY transformation can be expressed as the product of a certain number of first and second-order SUSY transformations, each one having to be consistent with any of the previous criteria to be non-singular.

2.2 *Confluent SUSY Transformations*

An important degenerate case of the SUSY transformation for $k \geq 2$ appears when all the factorization energies $\epsilon_j, j = 1, \ldots, k$ tend to a fixed single value ϵ_1 [74, 93–101, 170–172]. Let us note that the expression for the new potential of Eq. (10) is still valid, but the seed solutions have to be changed if non-trivial modifications in the new potential are going to appear. In fact, the seed solutions $u_j, j = 1, \ldots, k$ instead of being just normal eigenfunctions of H should generate a Jordan chain of generalized eigenfunctions for H and ϵ_1 as follows:

$$(H - \epsilon_1)u_1 = 0, \tag{13}$$

$$(H - \epsilon_1)u_2 = u_1, \tag{14}$$

$$\vdots$$

$$(H - \epsilon_1)u_k = u_{k-1}. \tag{15}$$

First let us assume that the seed solution u_1 satisfying Eq. (13) is given, then we need to find the general solution for u_j, $j = 2, \ldots, k$ (precisely in that order!) in terms of u_1. There are two methods essentially different to determine such a general solution: the first one is known as integral method, in which through the technique of variation of parameters one simplifies each inhomogeneous equation in the chain and when integrating the resulting equation every solution u_j is found. In fact, by applying this procedure the solution to the inhomogeneous equations

$$(H - \epsilon_1)u_j = u_{j-1}, \quad j = 2, \ldots, k, \tag{16}$$

is given by

$$u_j(x) = -2\, u_1(x)\, v_j(x), \tag{17}$$

$$v_j(x) = v_j(x_0) + \int_{x_0}^{x} \frac{w_j(y)}{u_1^2(y)} dy, \tag{18}$$

$$w_j(x) = w_j(x_0) + \int_{x_0}^{x} u_1(z)\, u_{j-1}(z) dz, \tag{19}$$

where x_0 is a point in the initial domain of the problem. Thus, Eq. (19) with $j = 2$ determines w_2, by inserting then this result in Eq. (18) with $j = 2$ we find v_2 which in turn fixes u_2 through Eq. (17) [74]. By using then this expression for u_2 it is found w_3 through Eq. (19) and then v_3 and u_3 by means of Eqs. (18) and (17), respectively [95]. We continue this process to find at the end the expression for u_k, and then we insert all the u_j, $j = 1, \ldots, k$ in Eq. (10) in order to obtain the new potential [170].

An alternative is the so-called differential method, in which one identifies in a clever way (through parametric differentiation with respect to the factorization energy ϵ_1) one particular solution for each inhomogeneous equation of the chain [96, 100]. It is straightforward then to find the general solution for each u_j, $j = 2, \ldots, k$. Instead of supplying the resulting formulas for arbitrary $k > 1$, let us derive the results just for the simplest case with $k = 2$.

2.2.1 Confluent Second-Order SUSY QM

For $k = 2$ we just need to solve the following system of equations:

$$(H - \epsilon_1)u_1 = 0, \tag{20}$$

$$(H - \epsilon_1)u_2 = u_1. \tag{21}$$

The result for the integral method in this case is achieved by making $k = 2$ in Eqs. (17)–(19), which leads to [74]:

$$u_2(x) = -2\, u_1(x)\, v_2(x), \tag{22}$$

$$v_2(x) = v_2(x_0) + \int_{x_0}^{x} \frac{w_2(y)}{u_1^2(y)} dy, \tag{23}$$

$$w_2(x) = w_2(x_0) + \int_{x_0}^{x} u_1^2(y) dy. \tag{24}$$

Thus we obtain:

$$W(u_1, u_2) = -2\, w_2(x). \tag{25}$$

Up to a constant factor, this is the well-known formula generated for the first time in [94], which will induce non-trivial modifications in the new potential $\widetilde{V}(x)$ (see Eq. (10)).

Let us solve now the system of Eqs. (20)–(21) through the differential method [96]. If we derive Eq. (20) with respect to ϵ_1, assuming that the Hamiltonian H does not depend explicitly on ϵ_1, we obtain a particular solution of the inhomogeneous Eq. (21), namely

$$(H - \epsilon_1)\frac{\partial u_1}{\partial \epsilon_1} = u_1. \tag{26}$$

Thus, the general solution for u_2 we were looking for becomes:

$$u_2(x) = c_2\, u_1 + d_2\, u_1 \int_{x_0}^{x} \frac{dy}{u_1^2(y)} + \frac{\partial u_1}{\partial \epsilon_1}. \tag{27}$$

Hence:

$$W(u_1, u_2) = d_2 + W\left(u_1, \frac{\partial u_1}{\partial \epsilon_1}\right). \tag{28}$$

Let us note that both methods have advantages and disadvantages, as compared with each other. For instance, in the integral method often it is hard to find explicit analytic solutions for the involved integrals, then in such cases we can try to use the differential method. However, for numerical calculation of the new potential it is simple and straightforward to use the integral formulas. On the other hand, there are not many potentials for which we can calculate in a simple way the corresponding derivative with respect to the factorization energy. At the end both methods turn out to be complementary to each other. A final remark has to be done: the family of new potentials generated through both algorithms (the integral and differential one) is the same, but if we want to generate a specific member of the family through both methods we need to be sure that we are using the same pair of seed solutions u_1, u_2. In practice, given u_1, u_2, with u_2 generated, for example, through the integral method (which means that we have fixed the constants $v_2(x_0)$ and $w_2(x_0)$ of Eqs. (23), (24)) we have to look for the appropriate coefficients c_2 and d_2 of Eq. (27) in order to guarantee that the same seed solution u_2 is going to be used for the differential algorithm (see the discussion in [99]).

As in the non-confluent SUSY approach, once again we have to choose carefully the seed solution u_1 in order that the new potential will not have extra singularities with respect to $V(x)$. In the case of the second-order confluent algorithm, the way of selecting such a seed solution is the following [94]:

- In the first place u_1 must vanish at one of the two edges of the x-domain. If this happens, then there will be some domain of the parameter $w_2(x_0)$ for which the key function w_2 of Eq. (24) will not have any node.
- The above requirement can be satisfied, in principle, by seed solutions u_1 associated to any real factorization energy; thus, we can create an energy level at any place on the energy axis.
- In particular, any eigenfunction of H satisfies the conditions to produce non-singular confluent second-order SUSY transformations, and the corresponding energy eigenvalue can be also kept in the spectrum of the new Hamiltonian (isospectral transformations).
- When an eigenfunction of H is used, a zero for w_2 could appear at one of the edges of the x domain. In such a case, the SUSY transformation stays non-singular, but the corresponding eigenvalue will disappear from the spectrum of $\widetilde{H}$ (deletion of one level).

3 SUSY QM and Exactly Solvable Potentials

The methods discussed previously can be used to generate, from an exactly solvable potential, plenty of new exactly solvable Hamiltonians with spectra quite similar to the initial one. In this section we will employ the harmonic oscillator to illustrate the technique. Although in this case the spectrum consists of an infinite number of non-degenerate discrete energy levels, the method works as well for Hamiltonians with mixed spectrum (discrete and continuous) or even when there is just a continuous one (see, e.g., [173]). This is what happens for periodic potentials [102–109], where the spectrum consists of allowed energy bands separated by forbidden gaps.

Moreover, the technique has been applied also to a very special system whose spectrum is the full real line, with each level being doubly degenerate: the so-called repulsive oscillator [128].

3.1 Harmonic Oscillator

The harmonic oscillator potential is given by:

$$V(x) = \frac{x^2}{2}. \tag{29}$$

In order to apply the SUSY methods, it is required to find the general solution $u(x)$ of the stationary Schrödinger equation for an arbitrary factorization energy ϵ:

$$-\frac{1}{2}u''(x) + \frac{x^2}{2}u(x) = \epsilon\, u(x). \tag{30}$$

Up to a constant factor, the general solution to this equation is a linear combination (characterized by the parameter ν) of an even and odd linearly independent solutions, given by [80]:

$$\begin{aligned} u(x) &= e^{-\frac{x^2}{2}}\left[{}_1F_1\left(\frac{1-2\epsilon}{4},\frac{1}{2};x^2\right) + 2\nu\frac{\Gamma(\frac{3-2\epsilon}{4})}{\Gamma(\frac{1-2\epsilon}{4})}\,x\,{}_1F_1\left(\frac{3-2\epsilon}{4},\frac{3}{2};x^2\right)\right] \\ &= e^{\frac{x^2}{2}}\left[{}_1F_1\left(\frac{1+2\epsilon}{4},\frac{1}{2};-x^2\right) + 2\nu\frac{\Gamma(\frac{3-2\epsilon}{4})}{\Gamma(\frac{1-2\epsilon}{4})}\,x\,{}_1F_1\left(\frac{3+2\epsilon}{4},\frac{3}{2};-x^2\right)\right]. \end{aligned} \tag{31}$$

In order to produce non-singular SUSY transformations we need to know the number of nodes that u has, according to the position of the parameter ϵ on the energy axis. Let us note first of all that, if ϵ is any real number, u will have an even number of nodes for $|\nu| < 1$, while this number will be odd for $|\nu| > 1$. This implies that, when ϵ is in the infinite energy gap $\epsilon < E_0$, this solution will have one node for $|\nu| > 1$ and it will be nodeless for $|\nu| < 1$. On the other hand, if $E_m < \epsilon < E_{m+1}$ with m even, then u will have $m+2$ nodes for $|\nu| < 1$ and it will have $m+1$ nodes for $|\nu| > 1$, while for odd m it will have $m+2$ and $m+1$ nodes for $|\nu| > 1$ and $|\nu| < 1$, respectively.

Now, although the SUSY methods can supply an infinity of new exactly solvable potentials, their expressions become in general too long to be explicitly reported. The simplest formulas appear when the factorization energies become either some of the eigenvalues $E_n = n + \frac{1}{2}, n = 0, 1, \ldots$ of H or some other special values, defined by the sequence $\mathcal{E}_m = -(m + \frac{1}{2}), m = 0, 1, \ldots$ In both cases it is possible to reduce the Schrödinger solution u to the product of one exponential factor $e^{\pm x^2/2}$

times a Hermite polynomial, either of a real variable when one of the E_n is taken or of an imaginary one when any of the $\mathcal{E}_m$ is chosen [80]. We supply next some explicit expressions for exactly solvable potentials, generated through the SUSY methods for such special values of the factorization energies. Let us note that we have sticked strictly to the criteria pointed out at Sect. 1 for producing non-singular SUSY transformations on the full real line. It is just for the first-order transformation that we have employed one general solution to show explicitly the simplest family of exactly solvable potential generated through SUSY QM.

3.1.1 First-Order SUSY Partners of the Oscillator

For $k = 1$, $\epsilon_1 = -\frac{1}{2}$, $|\nu_1| < 1$ it is obtained (see also [33]):

$$\widetilde{V}(x) = \frac{x^2}{2} - \left(\frac{2\,\nu_1\, e^{-x^2}}{\sqrt{\pi}\,[1 + \nu_1\,\text{erf}(x)]}\right)' - 1, \tag{32}$$

where $\text{erf}(x)$ is the error function.

For $k = 1$, $\epsilon_1 = -\frac{5}{2}$, $\nu_1 = 0$ we get:

$$\widetilde{V}(x) = \frac{x^2}{2} - \left(\frac{4x}{2x^2 + 1}\right)' - 1. \tag{33}$$

For $k = 1$, $\epsilon_1 = -\frac{9}{2}$, $\nu_1 = 0$ it is obtained:

$$\widetilde{V}(x) = \frac{x^2}{2} - \left[\frac{8x(2x^2 + 3)}{4x^4 + 12x^2 + 3}\right]' - 1. \tag{34}$$

Let us note that in all these three cases the spectrum of the new Hamiltonian $\widetilde{H}$, besides having the eigenvalues of H, will contain also a new energy level at ϵ_1.

3.1.2 Second-Order SUSY Partners of the Oscillator

For $k = 2$, $\epsilon_1 = -\frac{5}{2}$, $\nu_1 = 0$, $\epsilon_2 = -\frac{7}{2}$, $\nu_2 \to \infty$ it is obtained:

$$\widetilde{V}(x) = \frac{x^2}{2} - \left(\frac{16x^3}{4x^4 + 3}\right)' - 2. \tag{35}$$

For $k = 2$, $\epsilon_1 = -\frac{9}{2}$, $\nu_1 = 0$, $\epsilon_2 = -\frac{11}{2}$, $\nu_2 \to \infty$ we get:

$$\widetilde{V}(x) = \frac{x^2}{2} - \left[\frac{32x^3(4x^4 + 12x^2 + 15)}{16x^8 + 64x^6 + 120x^4 + 45}\right]' - 2. \tag{36}$$

For $k = 2$, $\epsilon_1 = -\frac{5}{2}$, $\nu_1 = 0$, $\epsilon_2 = -\frac{11}{2}$, $\nu_2 \to \infty$ it is obtained:

$$\widetilde{V}(x) = \frac{x^2}{2} - \left[\frac{4x(12x^4 + 20x^2 + 5)}{8x^6 + 20x^4 + 10x^2 + 5}\right]' - 2. \tag{37}$$

Once again, in all these cases the spectrum of the new Hamiltonian $\widetilde{H}$ will have two new levels at ϵ_1, ϵ_2, besides the eigenvalues E_n of H.

On the other hand, when deleting two neighbor energy levels of H in order to create $\widetilde{H}$ we could obtain again some of the potentials reported above, up to an energy shift to align the corresponding energy levels (see, e.g., [174]). For instance, if we delete the first and second excited states of H we recover the potential given in Eq. (33), if we delete the second and third excited states we get again the potential in Eq. (35). Let us generate now a new potential by deleting the third and fourth excited states, which leads to:

$$\widetilde{V}(x) = \frac{x^2}{2} - \left[\frac{12x(4x^4 - 4x^2 + 3)}{8x^6 - 12x^4 + 18x^2 + 9}\right]' + 2. \tag{38}$$

Note that the corresponding Hamiltonian $\widetilde{H}$ will not have the levels $E_3 = 7/2$, $E_4 = 9/2$.

In order to present some potentials obtained through the confluent second-order SUSY QM, let us use once again the eigenstates of H. If the ground state is taken to implement the transformation, it is generated the same family of potentials of Eq. (32). However, if the first excited state is employed, the following one-parameter family of potentials isospectral to the oscillator is gotten (see Eqs. (10), (24), (25)):

$$\widetilde{V}(x) = \frac{x^2}{2} - \left[\frac{4x^2}{\sqrt{\pi}(2b_2 + 1)e^{x^2} + \sqrt{\pi}e^{x^2}\text{erf}(x) - 2x}\right]', \tag{39}$$

where $b_2 \equiv w_2(-\infty)$. For $b_2 > 0$ the new Hamiltonian $\widetilde{H}$ is isospectral to H. However, if $b_2 = 0$ the level E_1 will disappear from the spectrum of $\widetilde{H}$.

Let us note that if a general eigenfunction $\psi_n(x)$ of H is used to perform the confluent second-order transformation, an explicit expression for the key function $w_2(x)$ has been obtained, which will induce non-trivial modifications in the new potential [94].

4 Algebraic Structures of H, $\widetilde{H}$, and Coherent States

In this section we are going to analyze the kind of algebra that the SUSY partner Hamiltonian $\widetilde{H}$ will inherit from the initial one H. We are going to suppose that H has an algebraic structure general enough to include the most important one-dimensional Hamiltonians appearing currently in the literature, as the harmonic oscillator [82].

4.1 Algebraic Structure of H

Let us suppose that the initial Schrödinger Hamiltonian H has an infinite discrete spectrum whose non-degenerate energy levels E_n, $n = 0, 1, \ldots$ are ordered as usual, $E_n < E_{n+1}$. Moreover, there is an explicit functional dependence between the eigenvalues E_n and the index n, i.e., $E_n = E(n)$, where $E(n)$ is well defined on the non-negative integers. For example, for the harmonic oscillator it turns out that $E(n) = n + \frac{1}{2}$. In this section we will use Dirac notation, so that the eigenstates and eigenvalues satisfy:

$$H|\psi_n\rangle = E_n|\psi_n\rangle, \quad n = 0, 1, \ldots \tag{40}$$

The number operator N is now introduced as

$$N|\psi_n\rangle = n|\psi_n\rangle. \tag{41}$$

It can be defined now a pair of ladder operators of the system through

$$a^-|\psi_n\rangle = r(n)|\psi_{n-1}\rangle, \tag{42}$$

$$a^+|\psi_n\rangle = r^*(n+1)|\psi_{n+1}\rangle, \tag{43}$$

$$r(n) = e^{i\tau(E_n - E_{n-1})}\sqrt{E_n - E_0}, \quad \tau \in \mathbb{R}, \tag{44}$$

where $r^*(n)$ denotes the complex conjugate of $r(n)$. Thus, the *intrinsic algebra* of the system is defined by:

$$[N, a^\pm] = \pm a^\pm, \tag{45}$$

$$a^+a^- = E(N) - E_0, \tag{46}$$

$$a^-a^+ = E(N+1) - E_0, \tag{47}$$

$$[a^-, a^+] = E(N+1) - E(N) \equiv f(N), \tag{48}$$

$$[H, a^\pm] = \pm f(N - 1/2 \mp 1/2)a^\pm. \tag{49}$$

Let us note that, depending on the key function $E(n)$ associated to the initial Hamiltonian, the system could be ruled by a Lie algebra, in case that $E(n)$ is either linear or quadratic in n. However, it could be also ruled by non-Lie algebras, when $E(n)$ has a more involved dependence with n.

Once we have characterized the algebra for the initial Hamiltonian, it is possible to analyze the corresponding structure for its SUSY partner Hamiltonians $\widetilde{H}$.

4.2 Algebraic Structure of $\widetilde{H}$

The most important properties of $\widetilde{H}$ come from its connection with the initial Hamiltonian H through the intertwining operators (see Eq. (6)). In fact, from these expressions it is simple to identify the *natural* ladder operators for $\widetilde{H}$ as follows [33, 78, 80, 82]:

$$\widetilde{a}^{\pm} = B^{+} a^{\pm} B. \tag{50}$$

Its action on the eigenstates of $\widetilde{H}$ can be straightforwardly calculated, leading to:

$$\widetilde{a}^{\pm}|\widetilde{\psi}_{\epsilon_j}\rangle = 0, \tag{51}$$

$$\widetilde{a}^{-}|\widetilde{\psi}_{n}\rangle = \widetilde{r}(n)|\widetilde{\psi}_{n-1}\rangle, \tag{52}$$

$$\widetilde{a}^{+}|\widetilde{\psi}_{n}\rangle = \widetilde{r}^{*}(n+1)\, |\widetilde{\psi}_{n+1}\rangle, \tag{53}$$

$$\widetilde{r}(n) = \left[\prod_{i=1}^{k}[E(n) - \epsilon_i][E(n-1) - \epsilon_i]\right]^{\frac{1}{2}} r(n). \tag{54}$$

In order to simplify the discussion, from now on we will assume that none of the $\epsilon_j,\ j = 1, \ldots, k$ coincide with some eigenvalue of H, and that k new energy levels are created for $\widetilde{H}$ at $\epsilon_j,\ j = 1, \ldots, k$. It is important as well to define the number operator $\widetilde{N}$ for the system ruled by $\widetilde{H}$, through its action on the corresponding energy eigenstates:

$$\widetilde{N}|\widetilde{\psi}_{\epsilon_j}\rangle = 0, \tag{55}$$

$$\widetilde{N}|\widetilde{\psi}_{n}\rangle = n|\widetilde{\psi}_{n}\rangle. \tag{56}$$

The *natural algebra* of the system is now defined by:

$$[\widetilde{N}, \widetilde{a}^{\pm}] = \pm\widetilde{a}^{\pm}, \tag{57}$$

$$[\widetilde{a}^{-}, \widetilde{a}^{+}] = \left[\widetilde{r}^{*}(\widetilde{N}+1)\,\widetilde{r}(\widetilde{N}+1) - \widetilde{r}^{*}(\widetilde{N})\,\widetilde{r}(\widetilde{N})\right] \sum_{n=0}^{\infty} |\widetilde{\psi}_{n}\rangle\, \langle\widetilde{\psi}_{n}|, \tag{58}$$

where $\widetilde{r}(n)$ is given by Eqs. (54), (44).

4.3 Coherent States of H and $\widetilde{H}$

We have just identified the annihilation and creation operators for the SUSY partner Hamiltonians H and $\widetilde{H}$. The coherent states for such systems can be looked for as eigenstates of the annihilation operator with complex eigenvalues z, namely:

$$a^-|z,\tau\rangle = z|z,\tau\rangle, \tag{59}$$

$$\widetilde{a}^-|\widetilde{z,\tau}\rangle = z|\widetilde{z,\tau}\rangle. \tag{60}$$

If we expand the coherent states in the basis of energy eigenstates, substitute them in Eqs. (59), (60) to obtain a recurrence relation for the coefficients of the expansion, and express such coefficients in terms of the first one and normalize them, we arrive at the following expressions:

$$|z,\tau\rangle = \left(\sum_{m=0}^{\infty}\frac{|z|^{2m}}{\rho_m}\right)^{-\frac{1}{2}}\sum_{m=0}^{\infty} e^{-i\tau(E_m-E_0)}\frac{z^m}{\sqrt{\rho_m}}|\psi_m\rangle, \tag{61}$$

$$\rho_m = \begin{cases} 1 & \text{if } m=0 \\ (E_m-E_0)\cdots(E_1-E_0) & \text{if } m>0 \end{cases} \tag{62}$$

and

$$|\widetilde{z,\tau}\rangle = \left(\sum_{m=0}^{\infty}\frac{|z|^{2m}}{\widetilde{\rho}_m}\right)^{-\frac{1}{2}}\sum_{m=0}^{\infty} e^{-i\tau(E_m-E_0)}\frac{z^m}{\sqrt{\widetilde{\rho}_m}}|\widetilde{\psi}_m\rangle, \tag{63}$$

$$\widetilde{\rho}_m = \begin{cases} 1 & \text{if } m=0 \\ \rho_m\prod_{i=1}^{k}(E_m-\epsilon_i)(E_{m-1}-\epsilon_i)^2\ldots(E_1-\epsilon_i)^2(E_0-\epsilon_i) & \text{if } m>0. \end{cases} \tag{64}$$

It is important to ensure that our coherent states fulfill a completeness relation, in order that an arbitrary state can be decomposed in terms of them. In our case the two completeness relations are:

$$\int |z,\tau\rangle\langle z,\tau|d\mu(z) = 1, \tag{65}$$

$$d\mu(z) = \frac{1}{\pi}\left(\sum_{m=0}^{\infty}\frac{|z|^{2m}}{\rho_m}\right)\rho(|z|^2)\,d^2z, \tag{66}$$

and

$$\sum_{i=1}^{k}|\widetilde{\psi}_{\epsilon_i}\rangle\langle\widetilde{\psi}_{\epsilon_i}| + \int |\widetilde{z,\tau}\rangle\langle\widetilde{z,\tau}|\,d\widetilde{\mu}(z) = 1, \tag{67}$$

$$d\widetilde{\mu}(z) = \frac{1}{\pi}\left(\sum_{m=0}^{\infty}\frac{|z|^{2m}}{\widetilde{\rho}_m}\right)\widetilde{\rho}(|z|^2)\,d^2z. \tag{68}$$

They will be fulfilled if we would find two measure functions $\rho(y)$ and $\widetilde{\rho}(y)$ solving the following moment problems [78, 80, 175–177]:

$$\int_0^\infty y^m \rho(y)\, dy = \rho_m, \tag{69}$$

$$\int_0^\infty y^m \widetilde{\rho}(y)\, dy = \widetilde{\rho}_m, \qquad m = 0, 1, \ldots. \tag{70}$$

The fact that two coherent states of a given family in general are not orthogonal is contained in the so-called reproducing kernel, which turns out to be:

$$\langle z_1, \tau | z_2, \tau\rangle = \left(\sum_{m=0}^\infty \frac{|z_1|^{2m}}{\rho_m}\right)^{-\frac{1}{2}} \left(\sum_{m=0}^\infty \frac{|z_2|^{2m}}{\rho_m}\right)^{-\frac{1}{2}} \left(\sum_{m=0}^\infty \frac{(\bar{z}_1 z_2)^m}{\rho_m}\right), \tag{71}$$

$$\langle \widetilde{z_1, \tau} | \widetilde{z_2, \tau}\rangle = \left(\sum_{m=0}^\infty \frac{|z_1|^{2m}}{\widetilde{\rho}_m}\right)^{-\frac{1}{2}} \left(\sum_{m=0}^\infty \frac{|z_2|^{2m}}{\widetilde{\rho}_m}\right)^{-\frac{1}{2}} \left(\sum_{m=0}^\infty \frac{(\bar{z}_1 z_2)^m}{\widetilde{\rho}_m}\right). \tag{72}$$

Concerning dynamics, the coherent states evolve as follows:

$$U(t)|z, \tau\rangle = \exp(-itH)|z, \tau\rangle = e^{-itE_0}|z, \tau + t\rangle, \tag{73}$$

$$\widetilde{U}(t)|\widetilde{z, \tau}\rangle = \exp(-it\widetilde{H})|\widetilde{z, \tau}\rangle = e^{-itE_0}|\widetilde{z, \tau + t}\rangle. \tag{74}$$

Let us note that, while the eigenvalue $z = 0$ of a^- is non-degenerate (if $z = 0$ is made in Eq. (61) the ground state of H is achieved), for $\widetilde{a}^-$ this eigenvalue is $(k+1)$th degenerate, since all states $\widetilde{\psi}_{\epsilon_i}$, $i = 1, \ldots, k$ are annihilated by $\widetilde{a}^-$ and for $z = 0$ Eq. (63) reduces to the eigenstate $|\widetilde{\psi}_0\rangle$ of $\widetilde{H}$ associated to E_0.

4.4 Example: Harmonic Oscillator

The simplest system available to illustrate the previous treatment is the harmonic oscillator. In this case there is a linear relation between the number operator and the Hamiltonian H, $H = E(N) = N + 1/2$. In addition, the function characterizing the action of $a^\pm$ onto the eigenstates of H becomes:

$$r(n) = \sqrt{E_n - E_0} = \sqrt{n}, \tag{75}$$

where, since the phase factors of Eq. (44) are independent of n, we have fixed them by taking $\tau = 0$. The function characterizing the commutator between the annihilation and creation operators is now (see Eq. (48)):

$$f(N) = E(N + 1) - E(N) = 1. \tag{76}$$

Thus, the commutation relations for the intrinsic algebra of the oscillator become:

$$[N, a^{\pm}] = \pm a^{\pm}, \tag{77}$$

$$[a^{-}, a^{+}] = 1, \tag{78}$$

which is the well-known Heisenberg–Weyl algebra.

On the other hand, for the SUSY partner Hamiltonian $\widetilde{H}$ we have that:

$$\widetilde{r}(n) = \left[\prod_{i=1}^{k} (E_n - \epsilon_i - 1)(E_n - \epsilon_i)\right]^{\frac{1}{2}} r(n). \tag{79}$$

If we insert this expression in Eq. (58) it is obtained a polynomial Heisenberg algebra, since in this case the commutator of $\widetilde{a}^{-}$ and $\widetilde{a}^{+}$ is a polynomial of degree $2k$ either in $\widetilde{H}$ or in $\widetilde{N}$.

Concerning coherent states, in the first place the coefficients ρ_m and $\widetilde{\rho}_m$, which are also the moments arising in Eqs. (69), (70), become:

$$\rho_m = m!, \tag{80}$$

$$\widetilde{\rho}_m = m! \prod_{i=1}^{k} \left(\frac{1}{2} - \epsilon_i\right)_m \left(\frac{3}{2} - \epsilon_i\right)_m, \tag{81}$$

where $(c)_m = \Gamma(c+m)/\Gamma(c)$ is a Pochhammer's symbol. It is straightforward to find now the explicit expressions for the coherent states:

$$|z\rangle = e^{-\frac{|z|^2}{2}} \sum_{m=0}^{\infty} \frac{z^m}{\sqrt{m!}} |\psi_m\rangle, \tag{82}$$

$$|\widetilde{z}\rangle = \sum_{m=0}^{\infty} \frac{z^m |\widetilde{\psi}_m\rangle}{\sqrt{{}_0F_{2k}\left(\frac{1}{2}-\epsilon_1, \frac{3}{2}-\epsilon_1, \ldots, \frac{1}{2}-\epsilon_k, \frac{3}{2}-\epsilon_k; |z|^2\right) m! \prod_{i=1}^{k} \left(\frac{1}{2}-\epsilon_i\right)_m \left(\frac{3}{2}-\epsilon_i\right)_m}}. \tag{83}$$

The solutions to the moment problems of Eqs. (69), (70) are given by:

$$\rho(y) = \exp(-y), \tag{84}$$

$$\widetilde{\rho}(y) = \frac{G^{2k+1\ 0}_{0\ 2k+1}\left(y|0, -\epsilon_1 - \frac{1}{2}, \ldots, -\epsilon_k - \frac{1}{2}, \frac{1}{2} - \epsilon_1, \ldots, \frac{1}{2} - \epsilon_k\right)}{\prod_{i=1}^{k} \Gamma\left(\frac{1}{2} - \epsilon_i\right) \Gamma\left(\frac{3}{2} - \epsilon_i\right)}, \tag{85}$$

where G is a Meijer G-function [80]. The reproducing kernel in both cases turns out to be:

$$\langle z_1|z_2\rangle = \exp\left[-\tfrac{1}{2}(|z_1|^2 + |z_2|^2 - 2z_1^* z_2)\right], \tag{86}$$

$$\langle\widetilde{z_1}|\widetilde{z_2}\rangle = \frac{{}_0F_{2k}\left(\frac{1}{2}-\epsilon_1,\frac{3}{2}-\epsilon_1,\ldots,\frac{1}{2}-\epsilon_k,\frac{3}{2}-\epsilon_k;z_1^* z_2\right)}{\sqrt{{}_0F_{2k}\left(\frac{1}{2}-\epsilon_1,\frac{3}{2}-\epsilon_1,\ldots,\frac{1}{2}-\epsilon_k,\frac{3}{2}-\epsilon_k;|z_1|^2\right){}_0F_{2k}\left(\frac{1}{2}-\epsilon_1,\frac{3}{2}-\epsilon_1,\ldots,\frac{1}{2}-\epsilon_k,\frac{3}{2}-\epsilon_k;|z_2|^2\right)}}. \tag{87}$$

As we can see, the coherent states for the initial Hamiltonian H are the standard ones, which minimize the Heisenberg uncertainty relation, namely $(\Delta X)(\Delta P) = 1/2$. It would be important to know if the coherent states associated to $\widetilde{H}$ have also this property. However, the calculation of $(\Delta X)(\Delta P)$ for general SUSY transformations, with arbitrary factorization energies and associated constants $\epsilon_j,\ \nu_j,\ j = 1,\ldots,k$ involved in the Schrödinger solution of Eq. (31), is difficult. Such an uncertainty can be analytically calculated in the harmonic oscillator limit for an arbitrary k. In particular, for $k = 1,\ \epsilon_1 = -\frac{1}{2},\ \nu_1 = 0$ it is obtained [78] $(r = |z|)$:

$$(\Delta X)(\Delta P) = \sqrt{\left\{\tfrac{3}{2} - [\mathrm{Re}(z)]^2\xi_1(r)\right\}\left\{\tfrac{3}{2} - [\mathrm{Im}(z)]^2\xi_1(r)\right\}}, \tag{88}$$

$$\xi_1(r) = 2\left[\frac{{}_0F_2(2,2;r^2)}{{}_0F_2(1,2;r^2)}\right]^2 - \left[\frac{{}_0F_2(2,3;r^2)}{{}_0F_2(1,2;r^2)}\right], \tag{89}$$

while for $k = 2,\ (\epsilon_1,\epsilon_2) = (-\frac{1}{2},-\frac{3}{2}),\ (\nu_1,\nu_2) = (0,\infty)$ we arrive at [80]:

$$(\Delta X)(\Delta P) = \sqrt{\left\{\tfrac{5}{2} - [\mathrm{Re}(z)]^2\xi_2(r)\right\}\left\{\tfrac{5}{2} - [\mathrm{Im}(z)]^2\xi_2(r)\right\}}, \tag{90}$$

$$\xi_2(r) = \frac{1}{2}\left[\frac{{}_0F_4(2,2,3,3;r^2)}{{}_0F_4(1,2,2,3;r^2)}\right]^2 - \frac{1}{6}\left[\frac{{}_0F_4(2,3,3,4;r^2)}{{}_0F_4(1,2,2,3;r^2)}\right]. \tag{91}$$

Plots of the Heisenberg uncertainty relations of Eqs. (88) and (90) as functions of z are shown in Figs. 1 and 2, respectively. It is seen that these coherent states are no longer minimum uncertainty states. However, for $k = 1$ there are some directions in the complex plane for which the minimum value $(\Delta X)(\Delta P) = 1/2$ is achieved when $|z| \to \infty$ (see Fig. 1).

5 SUSY QM and Painlevé Equations

In a general context, the polynomial Heisenberg algebras (PHA) of degree m are deformations of the Heisenberg–Weyl algebra for which the commutators of the Hamiltonian H (of form given in Eq. (5)) with $(m+1)$th order differential ladder operators $L^\pm$ are standard, while the commutator between L^- and L^+ is a

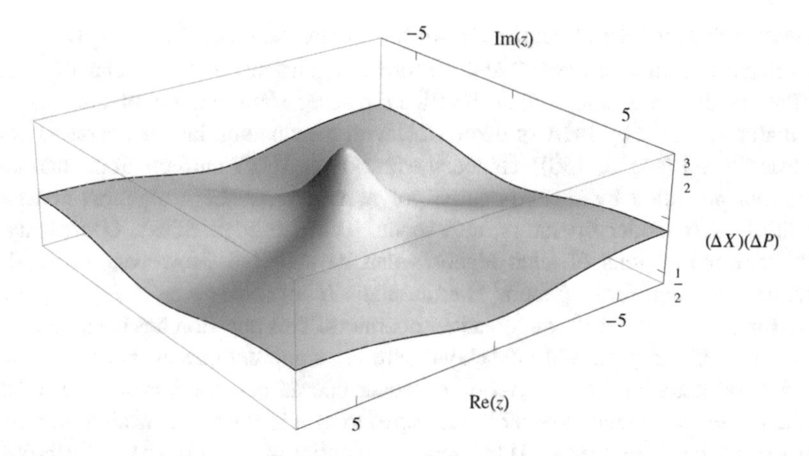

Fig. 1 Uncertainty relation $(\Delta X)(\Delta P)$ for the coherent states $|\widetilde{z}\rangle$ with $k = 1$ in the harmonic oscillator limit, when $\epsilon_1 = -\frac{1}{2}$, $\nu_1 = 0$

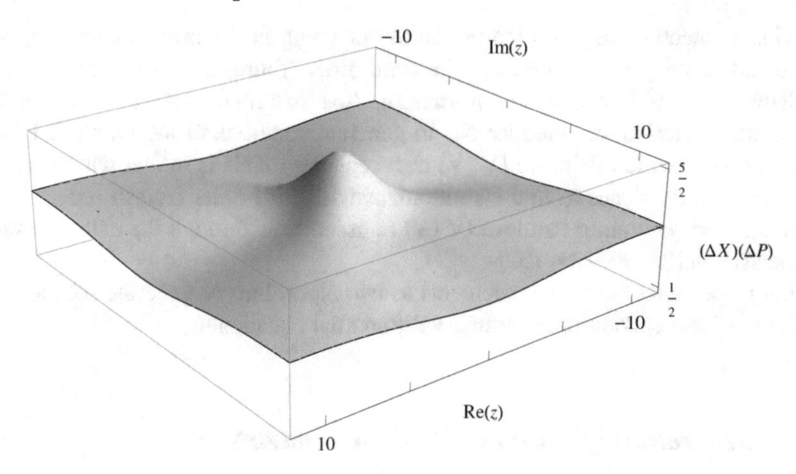

Fig. 2 Uncertainty relation $(\Delta X)(\Delta P)$ for the coherent states $|\widetilde{z}\rangle$ with $k = 2$ in the harmonic oscillator limit, when $(\epsilon_1, \epsilon_2) = (-\frac{1}{2}, -\frac{3}{2})$, $(\nu_1, \nu_2) = (0, \infty)$

polynomial of degree mth in H [85], i.e.,

$$[H, L^{\pm}] = \pm L^{\pm}, \tag{92}$$

$$[L^{-}, L^{+}] = q_{m+1}(H+1) - q_{m+1}(H) = p_m(H), \tag{93}$$

$$L^{+}L^{-} = q_{m+1}(H) = \prod_{j=1}^{m+1}(H - \mathcal{E}_j), \tag{94}$$

$$L^{-}L^{+} = q_{m+1}(H+1) = \prod_{j=1}^{m+1}(H - \mathcal{E}_j + 1). \tag{95}$$

Systems ruled by PHA of degree m have $m+1$ extremal states $\psi_{\mathcal{E}_j}$, $j = 1, \ldots, m+1$, which are annihilated by L^- and are formal eigenstates of H associated to $\mathcal{E}_j$.

Previously it was shown that the SUSY partner Hamiltonians of the harmonic oscillator are ruled by PHA of degree $2k$, with their natural ladder operators being of order $2k+1$ (see Eq. (50)). Hence, the first-order SUSY partners of the harmonic oscillator are ruled by second-degree polynomial Heisenberg algebras generated by third-order ladder operators, and so on. Thus, through SUSY QM plenty of particular realizations of such algebras can be supplied. However, it would be important to identify the general Hamiltonians H, of form given in Eq. (5), which have $(m+1)$th order differential ladder operators. This question has been addressed recurrently in the past, and nowadays there are some definite answers: if $m = 0$ the general potential having first-order ladder operators is the harmonic oscillator, while for $m = 1$ (second-order ladder operators) it is the radial oscillator. On the other hand, for $m = 2$ ($m = 3$) the general potential with third-order (fourth-order) ladder operators is expressed in terms of a function which satisfies the Painlevé IV (V) equation [85].

This connection suggests the possibility of going in the inverse direction, so if we could identify a Hamiltonian with third-order (fourth-order) ladder operators, perhaps we could use some information (the extremal state expressions and associated factorization energies $\mathcal{E}_j$) to generate solutions to the Painlevé IV (V) equation (also called Painlevé IV (V) transcendents). This is in fact what happens; thus, the game reduces to find Hamiltonians with third-order (fourth-order) ladder operators for generating Painlevé IV (V) transcendents through the extremal states of the system [85, 90, 129, 132].

Let us present next these statements as two algorithms to generate solutions for such non-linear second-order ordinary differential equations.

5.1 Generation of Painlevé IV Transcendents

Let us suppose that we have identified a Hamiltonian of the form given in Eq. (5), which has third-order differential ladder operators $L^\pm$ satisfying Eqs. (92)–(95) with $m = 2$, as well as its three extremal states $\psi_{\mathcal{E}_j}$ and associated factorization energies $\mathcal{E}_j$, $j = 1, 2, 3$. Thus, a solution to the Painlevé IV (PIV) equation

$$g'' = \frac{g'^2}{2g} + \frac{3}{2}g^3 + 4xg^2 + 2(x^2 - \alpha)g + \frac{\beta}{g} \tag{96}$$

is given by

$$g(x) = -x - \{\ln[\psi_{\mathcal{E}_3}(x)]\}', \tag{97}$$

where the parameters α, β of the PIV equation are related with $\mathcal{E}_1$, $\mathcal{E}_2$, $\mathcal{E}_3$ in the way

$$\alpha = \mathcal{E}_1 + \mathcal{E}_2 - 2\mathcal{E}_3 - 1, \qquad \beta = -2(\mathcal{E}_1 - \mathcal{E}_2)^2. \tag{98}$$

Let us note that, if the indices assigned to the extremal states are permuted cyclically, we will obtain three PIV transcendents, one for each extremal state when it is labeled as $\psi_{\mathcal{E}_3}$.

Summarizing, our task has been reduced to identify systems ruled by second-degree PHA and the corresponding extremal states [85, 90]. The harmonic oscillator supplies several such possibilities, for instance, the two operator pairs $\{a^3, (a^+)^3\}$, $\{a^+a^2, (a^+)^2a\}$ are third-order ladder operators satisfying Eqs. (92)–(95) (the level spacing has to be adjusted in the first case), and it is simple to identify the corresponding extremal states. On the other hand, the first-order SUSY partners of the oscillator also have natural third-order ladder operators, and well-identified extremal states. For the SUSY partners of the oscillator with $k \geq 2$ the natural ladder operators are not of third order (they are in general of order $2k + 1$). However, it is possible to induce a reduction process, by choosing connected seed solutions $u_{j+1} = au_j$, $\epsilon_{j+1} = \epsilon_j - 1$, $j = 1, \ldots, k - 1$ instead of general seed solutions, so that the $(2k + 1)$th order ladder operators reduce to third-order ones.

Some examples of real PIV transcendents associated to real PIV parameters α, β, which are generated through this algorithm, are presented next.

5.1.1 Harmonic Oscillator

If we take the ladder operators $L^- = a^3$, $L^+ = (a^+)^3$ for the harmonic oscillator Hamiltonian we get the PIV transcendents reported in Table 1 [178]. Note that in order that the level spacing induced by this pair of ladder operators coincides with the standard one ($\Delta E = 1$) of Eqs. (92)–(95), we need to change variables $y = \sqrt{3}x$ and scale the factorization energies (dividing by 3). Remember also that $\psi_j(x)$ are the eigenfunctions of the harmonic oscillator associated to the first three energy levels $E_j = j + 1/2$, $j = 0, 1, 2$.

Table 1 PIV transcendents generated from the harmonic oscillator Hamiltonian with $L^- = a^3$, $L^+ = (a^+)^3$

$\psi_{\mathcal{E}_3}$	$\psi_0(x)$	$\psi_1(x)$	$\psi_2(x)$
$\mathcal{E}_3$	$\frac{1}{2}$	$\frac{3}{2}$	$\frac{5}{2}$
$g(y)$	$-\frac{2y}{3}$	$-\frac{2y}{3} - \frac{1}{y}$	$-\frac{2y}{3} - \frac{4y}{2y^2-3}$
α	0	-1	-2
β	$-\frac{2}{9}$	$-\frac{8}{9}$	$-\frac{2}{9}$

Table 2 PIV transcendents generated from the first-order SUSY partner Hamiltonian $\widetilde{H}$ with $L^- = B^+aB,\ L^+ = B^+a^+B$

$\psi_{\mathcal{E}_3}$	$\frac{1}{u_1}$	$B^+\psi_0$	$B^+a^+u_1$
$\mathcal{E}_3$	$-\frac{5}{2}$	$\frac{1}{2}$	$-\frac{3}{2}$
$g(x)$	$\frac{4x}{1+2x^2}$	$-\frac{4x^4+3}{4x^5+8x^3+3x}$	$\frac{8x^5+6x}{1-4x^4}$
α	3	-6	0
β	-8	-2	-18

Table 3 PIV transcendents generated from the second-order SUSY partner Hamiltonian $\widetilde{H}$ and the third-order ladder operators obtained by reducing $L^- = B^+aB,\ L^+ = B^+a^+B$

$\psi_{\mathcal{E}_3}$	$\frac{u_1}{W(u_1,u_2)}$	$B^+\psi_0$	$B^+a^+u_1$
$\mathcal{E}_3$	$-\frac{7}{2}$	$\frac{1}{2}$	$-\frac{3}{2}$
$g(x)$	$\frac{4x(4x^4+4x^2-3)}{8x^6+4x^4+6x^2+3}$	$-\frac{4x(16x^8+72x^2+27)}{32x^{10}+48x^8+96x^6+54x^2-27}$	$\frac{-16x^8+32x^6-48x^4+9}{x(2x^2-3)(4x^4+3)}$
α	5	-7	-1
β	-8	-8	-32

5.1.2 First-Order SUSY Partner of the Harmonic Oscillator

For $\epsilon_1 = -\frac{5}{2},\ \nu_1 = 0$, and the third-order ladder operators $L^- = B^+aB,\ L^+ = B^+a^+B$ of $\widetilde{H}$, we get the PIV transcendents reported in Table 2. The seed solution employed is $u_1(x) = e^{\frac{x^2}{2}}(1 + 2x^2)$.

5.1.3 Second-Order SUSY Partner of the Harmonic Oscillator

For $\epsilon_1 = -\frac{5}{2},\ \nu_1 = 0$, and the third-order ladder operators of $\widetilde{H}$ obtained from the reduction of the fifth-order ones $L^- = B^+aB,\ L^+ = B^+a^+B$, we get the PIV transcendents reported in Table 3. Once again, the seed solution u_1 employed is $u_1(x) = e^{\frac{x^2}{2}}(1 + 2x^2)$ and $u_2 = au_1$.

5.2 Generation of Painlevé V Transcendents

Let us suppose now that the Hamiltonian H we have identified has fourth-order ladder operators and satisfy Eqs. (92)–(95) with $m = 3$. We know also its four extremal states $\psi_{\mathcal{E}_j}$ and associated factorization energies $\mathcal{E}_j,\ j = 1, 2, 3, 4$. Thus, one solution to the Painlevé V (PV) equation

$$w'' = \left(\frac{1}{2w} - \frac{1}{w-1}\right)(w')^2 - \frac{w'}{z} + \frac{(w-1)^2}{z^2}\left(\alpha\, w + \frac{\beta}{w}\right) + \gamma\,\frac{w}{z} + \delta\,\frac{w(w+1)}{w-1} \quad (99)$$

is given by

$$w(z) = 1 + \frac{\sqrt{z}}{g(\sqrt{z})}, \tag{100}$$

$$g(x) = -x - \frac{d}{dx}\left\{\ln\left[W(\psi_{\mathcal{E}_3}(x), \psi_{\mathcal{E}_4}(x))\right]\right\}, \tag{101}$$

where the prime in Eq. (99) means derivative with respect to z, and the PV parameters $\alpha,\ \beta,\ \gamma,\ \delta$ are related with $\mathcal{E}_1,\ \mathcal{E}_2,\ \mathcal{E}_3,\ \mathcal{E}_4$ through

$$\alpha = \frac{(\mathcal{E}_1-\mathcal{E}_2)^2}{2}, \quad \beta = -\frac{(\mathcal{E}_3-\mathcal{E}_4)^2}{2}, \quad \gamma = \frac{\mathcal{E}_1+\mathcal{E}_2}{2} - \frac{\mathcal{E}_3+\mathcal{E}_4+1}{2}, \quad \delta = -\frac{1}{8}. \tag{102}$$

Note that if the indices of the extremal states are permuted, we will obtain at the end six PV transcendents (in principle different), one for each pair of extremal states when they are labeled as $\psi_{\mathcal{E}_3},\ \psi_{\mathcal{E}_4}$ [132].

Once again, now we require just to identify systems ruled by third degree PHA and their four extremal states. The harmonic oscillator also supplies some possibilities, the simplest one through the fourth order ladder operators $\{L^- = a^4, L^+ = (a^+)^4\}$, which satisfy Eqs. (92)–(95) if we change variables and adjust the levels spacing, with the extremal states being the eigenstates associated to the four lowest energy levels of the oscillator. Another system closely related to PV equation is the radial oscillator, for which its ladder operators $b^\pm$ are of second order [132]. Thus, the second powers of such operators are also fourth order ladder operators that will give place to PV transcendents. Concerning SUSY partners, those of the radial oscillator give place to PHA of degree $2k+1$, with natural ladder operators of order $2k+2$. Thus, the first-order SUSY partners of the radial oscillator have natural fourth-order ladder operators and well-identified extremal states. For $k \geq 2$, it is possible to produce again a reduction process, by connecting the seed solutions in the way $u_{j+1} = b^- u_j,\ \epsilon_{j+1} = \epsilon_j - 1,\ j = 1, \ldots, k-1$, so that the $(2k+2)$th order natural ladder operators reduce to fourth-order ones [132]. Remember that the first-order SUSY partners of the harmonic oscillator also have fourth-order ladder operators, given by $L^- = B^+ a^2 B,\ L^+ = B^+ (a^+)^2 B$, but we will have to change variables and adjust the level spacing to stick to the standard convention $\Delta E = 1$.

Some examples of real PV transcendents associated to real parameters $\alpha,\ \beta,\ \gamma,\ \delta$, generated through this algorithm, are now presented.

5.2.1 Harmonic Oscillator

If we take $L^- = a^4, L^+ = (a^+)^4$ as ladder operators, we generate the PV transcendents reported in Table 4. Note that here $z = 4x^2$ and $\psi_j(x),\ j = 0, 1, 2, 3$ are the eigenfunctions for the four lowest eigenvalues of the harmonic oscillator. We

Table 4 PV transcendents generated from the harmonic oscillator Hamiltonian and $L^- = a^4$, $L^+ = (a^+)^4$

Permutation	α	β	γ	$w(z)$
1234	$\frac{1}{32}$	$-\frac{1}{32}$	0	-1
4231	$\frac{1}{8}$	$-\frac{1}{8}$	$-\frac{1}{4}$	$\frac{2-z}{z+2}$
1432	$\frac{1}{32}$	$-\frac{9}{32}$	$-\frac{1}{2}$	$\frac{6-z}{z+2}$
3241	$\frac{9}{32}$	$-\frac{1}{32}$	$-\frac{1}{2}$	$\frac{2-z}{z+6}$
3142	$\frac{1}{8}$	$-\frac{1}{8}$	$-\frac{3}{4}$	$\frac{6-z}{z+6}$
3412	$\frac{1}{32}$	$-\frac{1}{32}$	-1	$-\frac{(z-6)(z-2)}{(z+2)(z+6)}$

Table 5 PV transcendents generated from the first-order SUSY partner Hamiltonian $\widetilde{H}$ of the oscillator and $L^- = B^+a^2B$, $L^+ = B^+(a^+)^2B$

Permutation	α	β	γ	$w(z)$
1234	$\frac{1}{8}$	$-\frac{1}{2}$	$\frac{3}{4}$	$-\frac{2}{z-1}$
4231	$\frac{1}{2}$	$-\frac{9}{8}$	$\frac{1}{4}$	$\frac{z+3}{2}$
1432	$\frac{1}{8}$	-2	$-\frac{1}{4}$	$\frac{z^2+2z-1}{z-1}$
3241	2	$-\frac{1}{8}$	$-\frac{3}{4}$	$\frac{z+3}{z^2+2z+3}$
3142	$\frac{9}{8}$	$-\frac{1}{2}$	$-\frac{5}{4}$	$\frac{2(z^2+2z-1)}{z^3+z^2+z-3}$
3412	$\frac{1}{2}$	$-\frac{1}{8}$	$-\frac{7}{4}$	$-\frac{z^3+5z^2+5z-3}{2(z^2+2z+3)}$

initially order the extremal states as

$$\psi_{\mathcal{E}_1}(x) = \psi_2(x), \qquad \mathcal{E}_1 = \tfrac{5}{2}, \tag{103}$$

$$\psi_{\mathcal{E}_2}(x) = \psi_3(x), \qquad \mathcal{E}_2 = \tfrac{7}{2}, \tag{104}$$

$$\psi_{\mathcal{E}_3}(x) = \psi_0(x), \qquad \mathcal{E}_3 = \tfrac{1}{2}, \tag{105}$$

$$\psi_{\mathcal{E}_4}(x) = \psi_1(x), \qquad \mathcal{E}_4 = \tfrac{3}{2}, \tag{106}$$

and this permutation will be denoted as 1234. We do not include the parameter δ in this table since it is constant ($\delta = -\frac{1}{8}$).

5.2.2 First-Order SUSY Partner of the Harmonic Oscillator

For $\epsilon_1 = -\frac{5}{2}$, $\nu_1 = 0$, and the fourth-order ladder operators $L^- = B^+a^2B$, $L^+ = B^+(a^+)^2B$ of $\widetilde{H}$, we will get the PV transcendents reported in Table 5, where $z = 2x^2$. The seed solution employed is $u_1(x) = e^{\frac{x^2}{2}}(1 + 2x^2)$. The initial order for the extremal states, denoted as 1234 in the table, is

$$\psi_{\mathcal{E}_1}(x) = \frac{W(u_1, \psi_0)}{u_1}, \qquad \mathcal{E}_1 = \frac{1}{2}, \tag{107}$$

$$\psi_{\mathcal{E}_2}(x) = \frac{W(u_1, \psi_1)}{u_1}, \qquad \mathcal{E}_2 = \frac{3}{2}, \tag{108}$$

$$\psi_{\mathcal{E}_3}(x) = \frac{1}{u_1}, \qquad \mathcal{E}_3 = -\frac{5}{2}, \tag{109}$$

$$\psi_{\mathcal{E}_4}(x) = B^+(a^+)^2 u_1, \qquad \mathcal{E}_4 = -\frac{1}{2}. \tag{110}$$

We conclude this section by stating that an infinity of PIV and PV transcendents can be derived through the techniques described here. It is an open question to determine if any exact solution to such equations that exists in the literature can be derived through these methods. However, the algorithms are so simple and direct that we felt it was the right time to try to make them known to a wider and diversified community, not just to people working on solutions to non-linear differential equations.

6 Recent Applications of SUSY QM

Some recent interesting applications of SUSY QM are worth of some discussion. We would like to mention in the first place the motion of electrons in graphene, a single layer of carbon atoms arranged in a hexagonal honeycomb lattice. Since close to the Dirac points in the Brillouin zone there is a gapless linear dispersion relation, obtained in the low energy regime through a tight binding model, one ends up with an electron description in terms of the massless Dirac–Weyl equation, with Fermi velocity $v_F \approx c/300$ instead of the speed of light c. If the graphene layer is subject to external magnetic fields orthogonal to its surface (the $x - y$ plane), the Dirac–Weyl equation reads:

$$\mathbf{H}\Psi(x, y) = v_F \boldsymbol{\sigma} \cdot \left[\mathbf{p} + \frac{e\mathbf{A}}{c}\right] \Psi(x, y) = E\Psi(x, y), \tag{111}$$

where $v_F \sim 8 \times 10^5$ m/s is the Fermi velocity, $\boldsymbol{\sigma} = (\sigma_x, \sigma_y)$ are the Pauli matrices, $\mathbf{p} = -i\hbar(\partial_x, \partial_y)^T$ is the momentum operator in the $x - y$ plane, $-e$ is the electron charge, and $\mathbf{A}$ is the vector potential leading to the magnetic field through $\mathbf{B} = \nabla \times \mathbf{A}$. For magnetic fields which change just along x-direction, $\mathbf{B} = \mathcal{B}(x)\hat{e}_z$, in the Landau gauge we have that $\mathbf{A} = \mathcal{A}(x)\hat{e}_y$, $\mathcal{B}(x) = \mathcal{A}'(x)$. Since there is a translational invariance along y axis, we can propose

$$\Psi(x, y) = e^{iky} \begin{bmatrix} \psi^+(x) \\ i\psi^-(x) \end{bmatrix}, \tag{112}$$

where k is the wave number in the y direction and $\psi^\pm(x)$ describe the electron amplitudes on two adjacent sites in the unit cell of graphene. Thus we arrive to:

$$\left(\pm\frac{d}{dx} + \frac{e}{c\hbar}\mathcal{A} + k\right)\psi^\mp(x) = \frac{E}{\hbar v_F}\psi^\pm(x). \tag{113}$$

By decoupling these set of equations it is obtained:

$$H^{\pm}\psi^{\pm}(x) = \mathcal{E}\psi^{\pm}(x), \qquad \mathcal{E} = \frac{E^2}{\hbar^2 v_F^2}, \tag{114}$$

$$H^{\pm} = -\frac{d^2}{dx^2} + V^{\pm} = -\frac{d^2}{dx^2} + \left(\frac{e\mathcal{A}}{c\hbar} + k\right)^2 \pm \frac{e}{c\hbar}\frac{d\mathcal{A}}{dx}. \tag{115}$$

Let us note that these expressions are characteristic of the first-order SUSY QM. In fact, through the identification[1]:

$$B^{\pm} = \mp\frac{d}{dx} + \mathcal{W}(x), \tag{116}$$

where

$$\mathcal{W}(x) = \frac{e\mathcal{A}(x)}{c\hbar} + k \tag{117}$$

is the superpotential, it turns out that

$$B^{\mp}\psi^{\mp}(x) = \sqrt{\mathcal{E}}\psi^{\pm}(x). \tag{118}$$

The SUSY partner Hamiltonians $H^{\pm}$ thus satisfy:

$$H^{\pm} = B^{\mp}B^{\pm}, \quad V^{\pm}(x) = \mathcal{W}^2 \pm \mathcal{W}', \tag{119}$$

$$H^{\pm}B^{\mp} = B^{\mp}H^{\mp}. \tag{120}$$

By comparing these expressions with the formalism of Sect. 2, one realizes that $H^{\pm}$ can be identified with any of the two SUSY partner Hamiltonians H and $\widetilde{H}$ (up to a constant factor), depending on which one will be taken as the departure Hamiltonian. Moreover, by deriving the superpotential with respect to x it is obtained:

$$\mathcal{B}(x) = \frac{c\hbar}{e}\frac{d\mathcal{W}}{dx}. \tag{121}$$

This formula suggests a method to proceed further: the magnetic field $\mathcal{B}(x)$ has to be chosen cleverly, in order to arrive to a pair of exactly solvable potentials $V^{\pm}$. In particular, it has been chosen in several different ways but taking care that $V^{\pm}$ are shape invariant potentials [151]. An important case of this type appears for constant homogeneous magnetic fields: in such a situation both $V^{\pm}$ become

[1] We choose here a notation consistent with Sect. 2. Please do not confuse the intertwining operators of Eq. (116) with the magnetic field **B**, its magnitude $\mathcal{B}(x)$, or any of its components.

harmonic oscillator potentials. It is worth to mention also that the shape invariance condition has been generalized, thus supplying a method for generating magnetic fields which are deformed with respect to the chosen initial one, but leading once again to an exactly solvable problem [157].

Let us note that the SUSY methods have been applied also to other carbon allotropes, as the carbon nanotubes, and it has been successfully implemented when electrostatic fields are applied, with or without static magnetic fields. In addition, the coherent state methods have been started to be applied recently to graphene subject to static homogeneous magnetic fields [179]. As can be seen, the SUSY methods applied to Dirac materials is a very active field which surely will continue its development in the near future [151–161].

At this point, it is worth to mention also the applications of SUSY QM to optical system, since there is a well-known correspondence between Schrödinger equation and Maxwell equations in the paraxial approximation. Thus, it seems natural to think that many techniques successfully used to deal with quantum mechanical problems can be directly applied to optical systems in an appropriate approximation. In a way, we are dealing with the optical analogues of quantum phenomena, which have been realized, for example, in waveguide arrays, optimization of quantum cascade lasers, among others. In particular, the optical analogues of SUSY QM is an emergent field which could supply a lot of interesting physical information [162–169].

7 Conclusions

It has been shown that supersymmetric quantum mechanics is a simple powerful tool for generating potentials with known spectra departing from a given initial solvable one. Since the spectrum of the new Hamiltonian differs slightly from the initial one, the method can be used to implement the spectral design in quantum mechanics.

In this direction, let us note that here we have discussed real SUSY transformations, by employing just real seed solutions which will produce at the end real SUSY partner potentials $\widetilde{V}(x)$. However, most of these formulas can be used without any change for implementing complex SUSY transformations. If we would introduce this procedure gradually, in the first place we could use complex seed solutions associated to real factorization energies in order to generate complex potentials with real spectrum [90, 180]. This offers immediately new possibilities of spectral design which were not available for real SUSY transformations, for example, through a complex first-order SUSY transformation with real factorization energies a new energy level can be created at any position on the real energy axis. In a second step of this approach, one can use complex seed solutions associated to complex factorization energies for an initial potential which is real [88], thus generating new levels at arbitrary positions in the complex energy plane. The third step for making complex the SUSY transformation is to apply the method to initial potentials which are complex from the very beginning [92]. In all these steps we will get at the end new potentials which are complex, but the spectrum will depend on the initial potential as well as of the kind of seed solutions employed.

We want to finish this paper by noting that the factorization method and intertwining techniques have been also applied with success to some discrete versions of the stationary Schrödinger equation [181–186]. The connections that could be established between such problems and well-known finite difference equations [187, 188] could contribute to the effort of classifying the known solutions and generate new ones, as it has happened in the continuous case for more than 80 years.

As it was pointed out previously, one of our aims when writing this article was to make a short review of the most recent advances of SUSY QM, either on purely theoretical or applied directions. We hope to have succeeded; perhaps the reader will find interesting and/or useful the ideas here presented.

Acknowledgements The author acknowledges the financial support of the Spanish MINECO (project MTM2014-57129-C2-1-P) and Junta de Castilla y León (VA057U16).

References

1. E. Witten, Nucl. Phys. B **185**, 513–554 (1981)
2. L. Infeld, T.E. Hull, Rev. Mod. Phys. **23**, 21–68 (1951)
3. W. Miller, *Lie Theory and Special Functions* (Academic, New York, 1968)
4. A. Lahiri, P.K. Roy, B. Bagchi, Int. J. Mod. Phys. A **5**, 1383–1456 (1990)
5. B. Roy, P. Roy, R. Roychoudhury, Fortschr. Phys. **39**, 211–258 (1991)
6. O.L. de Lange, R.E. Raab, *Operator Methods in Quantum Mechanics* (Clarendon Press, Oxford, 1991)
7. F. Cooper, A. Khare, U. Sukhatme, Phys. Rep. **251**, 267–385 (1995)
8. G. Junker, *Supersymmetric Methods in Quantum and Statistical Mechanics* (Springer, Berlin, 1996)
9. B.N. Zakhariev, V.M. Chabanov, Inv. Prob. **13**, R47–R79 (1997)
10. A. Ramos, J.F. Cariñena, in *Symmetries in Quantum Mechanics and Quantum Optics*, ed. by A. Ballesteros et al. (Universidad de Burgos, Burgos, 1999), pp. 259–269
11. H.C. Rosu, in *Symmetries in Quantum Mechanics and Quantum Optics*, ed. by A. Ballesteros et al. (Universidad de Burgos, Burgos, 1999), pp. 301–315
12. B. Bagchi, *Supersymmetry in Quantum and Classical Mechanics* (Chapman & Hall, Boca Raton, 2001)
13. F. Cooper, A. Khare, U. Sukhatme, *Supersymmetry in Quantum Mechanics* (World Scientific, Singapore, 2001)
14. R. de Lima Rodrigues (2002). arXiv:hep-th/0205017v6
15. I. Aref'eva, D.J. Fernández, V. Hussin, J. Negro, L.M. Nieto, B.F. Samsonov (eds.) *Special issue dedicated to the subject of the International Conference on Progress in Supersymmetric Quantum Mechanics*. J. Phys. A: Math. Gen. **37**(43) (2004)
16. B. Mielnik, O. Rosas-Ortiz, J. Phys. A: Math. Gen. **37**, 10007–10035 (2004)
17. A. Khare, U. Sukhatme, J. Phys. A: Math. Gen. **37**, 10037–10055 (2004)
18. C.M. Bender, J. Brod, A. Refig, M.E. Reuter, J. Phys. A: Math. Gen. **37**, 10139–10165 (2004)
19. D. Baye, J.M. Sparenberg, J. Phys. A: Math. Gen. **37**, 10223–10249 (2004)
20. A.A. Andrianov, F. Cannata, J. Phys. A: Math. Gen. **37**, 10297–10321 (2004)

21. A. Khare, AIP Conf. Proc. **744**, 133–165 (2005)
22. C.V. Sukumar, AIP Conf. Proc. **744**, 166–235 (2005)
23. D.J. Fernández, N. Fernández-García, AIP Conf. Proc. **744**, 236–273 (2005)
24. S.H. Dong, *Factorization Method in Quantum Mechanics* Springer, Dordrecht (2007)
25. D.J. Fernández, AIP Conf. Proc. **1287**, 3–36 (2010)
26. A.A. Andrianov, M.V. Ioffe, J. Phys. A: Math. Theor. **45**, 503001 (2012)
27. D. Baye, J.M. Sparenberg, A.M. Pupasov-Maksimov, B.F. Samsonov, J. Phys. A: Math. Theor. **47**, 243001 (2014)
28. P.A.M. Dirac, *The Principles of Quantum Mechanics*, 2nd edn. (Clarendon, Oxford, 1935)
29. E. Schrödinger, Proc. R. Irish Acad. **46A**, 9–16 (1940)
30. E. Schrödinger, Proc. R. Irish Acad. **46A**, 183–206 (1941)
31. L. Infeld, Phys. Rev. **59**, 737–747 (1941)
32. T.E. Hull, L. Infeld, Phys. Rev. **74**, 905–909 (1948)
33. B. Mielnik, J. Math. Phys. **25**, 3387–3389 (1984)
34. D.J. Fernández, Lett. Math. Phys. **8**, 337–343 (1984)
35. A.A. Andrianov, N.V. Borisov, M.V. Ioffe, Phys. Lett. A **105**, 19–22 (1984)
36. A.A. Andrianov, N.V. Borisov, M.V. Ioffe, Theor. Math. Phys. **61**, 1078–1088 (1984)
37. M.M. Nieto, Phys. Lett. B **145**, 208–210 (1984)
38. C.V. Sukumar, J. Phys. A: Math. Gen. **18**, 2917–2936 (1985)
39. C.V. Sukumar, J. Phys. A: Math. Gen. **18**, 2937–2955 (1985)
40. M. Luban, D.L. Pursey, Phys. Rev. D **33**, 431–436 (1986)
41. J. Beckers, V. Hussin, Phys. Lett. A **118**, 319–321 (1986)
42. J. Beckers, D. Dehin, V. Hussin, J. Phys. A: Math. Gen. **20**, 1137–1154 (1987)
43. N.A. Alves, E. Drigo-Filho, J. Phys. A: Math. Gen. **21**, 3215–3225 (1988)
44. A. Arai, Lett. Math. Phys. **19**, 217–227 (1990)
45. R. Adhikari, R. Dutt, Y.P. Varshni, J. Math. Phys. **32**, 447–456 (1991)
46. J. Beckers, N. Debergh, A.G. Nikitin, J. Math. Phys. **33**, 152–160 (1992)
47. J. Casahorrán, Physica A **217**, 429–439 (1995)
48. H.C. Fu, R. Sasaki, J. Phys. A: Math. Gen. **29**, 2129–2138 (1996)
49. G. Junker, P. Roy, Phys. Lett. A **232**, 155–161 (1997)
50. N. Aizawa, H.T. Sato, Prog. Theor. Phys. **98**, 707–718 (1997)
51. E. Drigo-Filho, J.R. Ruggiero, Phys. Rev. E **56**, 4486–4488 (1997)
52. L.J. Boya, H.C. Rosu, A.J. Segui-Santonja, F.J. Vila, Nuovo Cim. B **113**, 409–414 (1998)
53. I.F. Márquez, J. Negro, L.M. Nieto, J. Phys. A: Math. Gen. **31**, 4115–4125 (1998)
54. J.F. Cariñena, G. Marmo, A.M. Perelomov, M.F. Rañada, Int. J. Mod. Phys. A **13**, 4913–4929 (1998)
55. C. Quesne, N. Vansteenkiste, Phys. Lett. A **240**, 21–28 (1998)
56. C. Quesne, N. Vansteenkiste, Helv. Phys. Acta **72**, 71–92 (1999)
57. V.M. Eleonsky, V.G. Korolev, J. Math. Phys. **40**, 1977–1992 (1999)
58. J.F. Cariñena, A. Ramos, Mod. Phys. Lett. A **15**, 1079–1088 (2000)
59. B. Bagchi, S. Mallik, C. Quesne, Int. J. Mod. Phys. A **16**, 2859–2872 (2001)
60. C. Quesne, Mod. Phys. Lett. A **18**, 515–526 (2003)
61. H.C. Rosu, J.M. Morán-Mirabal, O. Cornejo, Phys. Lett. A **310**, 353–356 (2003)
62. M.V. Ioffe, S. Kuru, J. Negro, L.M. Nieto, J. Phys. A: Math. Theor. **39**, 6987–7001 (2006)
63. S. Kuru, J. Negro, Ann. Phys. **323**, 413–431 (2008)
64. C. Quesne, J. Phys. A: Math. Theor. **41**, 244002 (2008)
65. A.A. Andrianov, M.V. Ioffe, V.P. Spiridonov, Phys. Lett. A **174**, 273–279 (1993)
66. A.A. Andrianov, M.V. Ioffe, F. Cannata, J.P. Dedonder, Int. J. Mod. Phys. A **10**, 2683–2701 (1995)
67. V.G.Bagrov, B.F. Samsonov, Theor. Math. Phys. **104**, 1051–1060 (1995)
68. D.J. Fernández, Int. J. Mod. Phys. A **12**, 171–176 (1997)
69. D.J. Fernández, M.L. Glasser, L.M. Nieto, Phys. Lett. A **240**, 15–20 (1998)
70. D.J. Fernández, V. Hussin, B. Mielnik, Phys. Lett. A **244**, 309–316 (1998)
71. D.J. Fernández, H.C. Rosu, Phys. Scripta **64**, 177–183 (2001)

72. J.F. Cariñena, A. Ramos, D.J. Fernández, Ann. Phys. **292**, 42–66 (2001)
73. D.J. Fernández, J. Negro, M.A. del Olmo, Ann. Phys. **252**, 386–412 (1996)
74. D.J. Fernández, E. Salinas-Hernández, Phys. Lett. A **338**, 13–18 (2005)
75. A. Contreras-Astorga, D.J. Fernández, AIP Conf. Proc. **960**, 55–60 (2007)
76. A. Contreras-Astorga, D.J. Fernández, J. Phys. A: Math. Theor. **41**, 475303 (2008)
77. D.J. Fernández, V.S. Morales-Salgado, Ann. Phys. **388**, 122–134 (2018)
78. D.J. Fernández, V. Hussin, L.M. Nieto, J. Phys. A: Math. Gen. **27**, 3547–3564 (1994)
79. D.J. Fernández, L.M. Nieto, O. Rosas-Ortiz, J. Phys. A: Math. Gen. **28**, 2693–2708 (1995)
80. D.J. Fernández, V. Hussin, J. Phys. A: Math. Gen. **32**, 3603–3619 (1999)
81. D. Bermudez, A. Contreras-Astorga, D.J. Fernández, Ann. Phys. **350**, 615–634 (2014)
82. D.J. Fernández, V. Hussin, O. Rosas-Ortiz, J. Phys. A: Math. Theor. **40**, 6491–6511 (2007)
83. D.J. Fernández, V. Hussin, V.S. Morales-Salgado, Eur. Phys. J. Plus **134**, 18 (2019)
84. D.J. Fernández, J. Negro, L.M. Nieto, Phys. Lett. A **324**, 139–144 (2004)
85. J.M. Carballo, D.J. Fernández, J. Negro, L.M. Nieto, J. Phys. A: Math. Gen. **37**, 10349–10362 (2004)
86. J. Mateo, J. Negro, J. Phys. A: Math. Theor. **41**, 045204 (2008)
87. M. Castillo-Celeita, E. Díaz-Bautista, D.J. Fernández, Phys. Scripta **94**, 045203 (2019)
88. D.J. Fernández, R. Muñoz, A. Ramos, Phys. Lett. A **308**, 11–16 (2003)
89. N. Fernández-García, O. Rosas-Ortiz, Ann. Phys. **323**, 1397–1414 (2008)
90. D. Bermudez, D.J. Fernández, Phys. Lett. A **375**, 2974–2978 (2011)
91. D. Bermudez, SIGMA **8**, 069 (2012)
92. D.J. Fernández, J.C. González, Ann. Phys. **359**, 213–229 (2015)
93. B. Mielnik, L.M. Nieto, O. Rosas-Ortiz, Phys. Lett. A **269**, 70–78 (2000)
94. D.J. Fernández, E. Salinas-Hernández, J. Phys. A: Math. Gen. **36**, 2537–2543 (2003)
95. D.J. Fernández, E. Salinas-Hernández, J. Phys. A: Math. Theor. **44**, 365302 (2011)
96. D. Bermudez, D.J. Fernández, N. Fernández-García, Phys. Lett. A **376**, 692–696 (2012)
97. A. Contreras-Astorga, A. Schulze-Halberg, J. Math. Phys. **55**, 103506 (2014)
98. A. Contreras-Astorga, A. Schulze-Halberg, Ann. Phys. **354**, 353–364 (2015)
99. A. Contreras-Astorga, A. Schulze-Halberg, J. Phys. A: Math. Theor. **48**, 315202 (2015)
100. D. Bermudez, Ann. Phys. **364**, 35–52 (2016)
101. A. Contreras-Astorga, A. Schulze-Halberg, J. Phys. A: Math. Theor. **50**, 105301 (2017)
102. D.J. Fernández, J. Negro, L.M. Nieto, Phys. Lett. A **275**, 338–349 (2000)
103. J. Negro, L.M. Nieto, D.J. Fernández, Czech. J. Phys. **50**, 1303–1308 (2000)
104. D.J. Fernández, B. Mielnik, O. Rosas-Ortiz, B.F. Samsonov, Phys. Lett. A **294**, 168–174 (2002)
105. D.J. Fernández, B. Mielnik, O. Rosas-Ortiz, B.F. Samsonov, J. Phys. A: Math. Gen. **35**, 4279–4291 (2002)
106. D.J. Fernández, A. Ganguly, Phys. Lett. A **338**, 203–208 (2005)
107. A. Ganguly, M. Ioffe, L.M. Nieto, J. Phys. A: Math. Gen. **39**, 14659–14680 (2006)
108. D.J. Fernández, A. Ganguly, Ann. Phys. **322**, 1143–1161 (2007)
109. D.J. Fernández, A. Ganguly, Phys. Atom. Nucl. **73**, 288–294 (2010)
110. H. Aoyama, M. Sato, T. Tanaka, Nucl. Phys. B **619**, 105–127 (2001)
111. M. Sato, T. Tanaka, J. Math. Phys. **43**, 3484–3510 (2002)
112. T. Tanaka, Nucl. Phys. B **662**, 413–446 (2003)
113. A. González-López, T. Tanaka, Phys. Lett. B **586**, 117–124 (2004)
114. B. Bagchi, T. Tanaka, Ann. Phys. **324**, 2438–2451 (2009)
115. C. Leiva, M.S. Plyushchay, J. High Energy Phys. **0310**, 069 (2003)
116. M. Plyushchay, J. Phys. A: Math. Gen. **37**, 10375–10384 (2004)
117. F. Correa, L.M. Nieto, M.S. Plyushchay, Phys. Lett. B **644**, 94–98 (2007)
118. F. Correa, V. Jakubsky, L.M. Nieto, M.S. Plyushchay, Phys. Rev. Lett. **101**, 030403 (2008)
119. F. Correa, V. Jakubsky, M.S. Plyushchay, J. Phys. A: Math. Theor. **41**, 485303 (2008)
120. S.Y. Dubov, V.M. Eleonsky, N.E. Kulagin, Sov. Phys. JETP **75** 446–451 (1992)
121. A.P. Veselov, A.B. Shabat, Funct. Anal. Appl. **27**, 81–96 (1993)
122. V.E. Adler, Physica D **73**, 335–351 (1994)

123. U.P. Sukhatme, C. Rasinariu, A. Khare, Phys. Lett. A **234**, 401–409 (1997)
124. A.A. Andrianov, F. Cannata, M. Ioffe, D. Nishnianidze, Phys. Lett. A **266**, 341–349 (2000)
125. I. Marquette, J. Math. Phys. **50**, 012101 (2009)
126. I. Marquette, J. Math. Phys. **50**, 095202 (2009)
127. D. Bermudez, D.J. Fernández, SIGMA **7**, 025 (2011)
128. D. Bermudez, D.J. Fernández, Ann. Phys. **333**, 290–306 (2013)
129. D. Bermudez, D.J. Fernández, AIP Conf. Proc. **1575**, 50–88 (2014)
130. D.J. Fernández, V.S. Morales-Salgado, J. Phys. A: Math. Theor. **47**, 035304 (2014)
131. D.J. Fernández, V.S. Morales-Salgado, J. Phys. A: Math. Theor. **49**, 195202 (2016)
132. D. Bermudez, D.J. Fernández, J. Negro, J. Phys. A: Math. Theor. **49**, 335203 (2016)
133. I. Marquette, C. Quesne, J. Math. Phys. **57**, 052101 (2016)
134. V.Y. Novokshenov, SIGMA **14**, 106 (2018)
135. P.A. Clarkson, D. Gómez-Ullate, Y. Grandati, R. Milson (2018). arXiv:1811.09274 [math-ph]
136. J.F. Cariñena, A.M Perelomov, M. Rañada, M. Santander, J. Phys. A: Math. Theor. **41**, 085301 (2008)
137. S. Odake, R. Sasaki, Phys. Lett. B **679**, 414–417 (2009)
138. S. Odake, R. Sasaki, Phys. Lett. B **682**, 130–136 (2009)
139. D. Gómez-Ullate, N. Kamran, R. Milson, J. Phys. A: Math. Theor. **43**, 434016 (2010)
140. S. Odake, R. Sasaki, Phys. Lett. B **702**, 164–170 (2011)
141. C. Quesne, Mod. Phys. Lett. A **26**, 1843–1852 (2011)
142. C. Quesne, Int. J. Mod. Phys. A **26**, 5337–5347 (2011)
143. D. Gómez-Ullate, N. Kamran, R. Milson, J. Math. Ann. App. **387**, 410–418 (2012)
144. D. Gómez-Ullate, N. Kamran, R. Milson, Found. Comp. Math. **13**, 615–666 (2013)
145. I. Marquette, C. Quesne, J. Math. Phys. **54**, 042102 (2013)
146. I. Marquette, C. Quesne, J. Math. Phys. **55**, 112103 (2014)
147. D. Gómez-Ullate, Y. Grandati, R. Milson, J. Phys. A: Math. Theor. **47**, 015203 (2014)
148. D. Gómez-Ullate, Y. Grandati, R. Milson, J. Math. Phys. **55**, 043510 (2014)
149. A. Schulze-Halberg, B. Roy, Ann. Phys. **349**, 159–170 (2014)
150. D. Gómez-Ullate, Y. Grandati, R. Milson, J. Phys. A: Math. Theor. **51**, 345201 (2018)
151. S. Kuru, J. Negro, L.M. Nieto, J. Phys.: Condens. Matter **21**, 455305 (2009)
152. E. Milpas, M. Torres, G. Murguia, J. Phys.: Condens. Matter **23**, 245304 (2011)
153. V. Jakubsky, M.S. Plyushchay, Phys. Rev. D **85**, 045035 (2012)
154. O. Panella, P. Roy, Phys. Lett. A **376**, 2580–2583 (2012)
155. V. Jakubsky, S. Kuru, J. Negro, S. Tristao, J. Phys.: Condens. Matter **25**, 165301 (2013)
156. F. Queisser, R. Schuetzhold, Phys. Rev. Lett. **111**, 046601 (2013)
157. B. Midya, D.J. Fernández, J. Phys. A: Math. Theor. **47**, 285302 (2014)
158. S. Kuru, J. Negro, S. Tristao, J. Phys.: Condens. Matter **27**, 285501 (2015)
159. A. Schulze-Halberg, P. Roy, J. Phys. A: Math. Theor. **50**, 365205 (2017)
160. Y. Concha, A. Huet, A. Raya, D. Valenzuela, Mat. Res. Express **5**, 065607 (2018)
161. D. Jahani, F. Shahbazi, M.R. Setare, Eur. Phys. J. Plus **133**, 328 (2018)
162. S.M. Chumakov, K.B. Wolf, Phys. Lett. A **193**, 51–53 (1994)
163. M.A. Miri, M. Heinrich, D.N. Christodoulides, Phys. Rev. A **87**, 043819 (2013)
164. A. Zúñiga-Segundo, B.M. Rodríguez-Lara, D.J. Fernández, H.M. Moya-Cessa, Opt. Express **22**, 987–994 (2014)
165. B. Midya, Phys. Rev. A **89**, 032116 (2014)
166. S. Dehdashti, R. Li, X. Liu, M. Raoofi, H. Chen, Laser Phys. **25**, 075201 (2015)
167. S. Cruz y Cruz, Z. Gress, Ann. Phys. **383**, 257–277 (2017)
168. B. Midya, W. Walasik, N.M. Litchinitser, L. Feng, Opt. Lett. **43**, 4927 (2018)
169. A. Contreras-Astorga, V. Jakubsky, Phys. Rev. A **99**, 053812 (2019)
170. A. Schulze-Halberg, Eur. Phys. J. Plus **128**, 68 (2013)
171. Y. Grandati, C. Quesne, SIGMA **11**, 061 (2015)
172. J.C. Krapez, Int. J. Heat Mass Transfer **99**, 485–503 (2016)
173. A.R.P. Rau, J. Phys. A: Math. Gen. **37**, 10421–10427 (2004)

174. M.I. Estrada-Delgado, D.J. Fernández, J. Phys.: Conf. Ser. **698**, 012027 (2016)
175. J.M. Sixdeniers, K.A. Penson, A.I. Solomon, J. Phys. A: Math. Gen. **32**, 7543–7563 (1999)
176. J.M. Sixdeniers, K.A. Penson, J. Phys. A: Math. Gen. **33**, 2907–2916 (2000)
177. J.R. Klauder, K.A. Penson, J.M. Sixdeniers, Phys. Rev. A **64**, 013817 (2001)
178. M. Castillo-Celeita, D.J. Fernández, in *Physical and Mathematical Aspects of Symmetries*, ed. by S. Duarte et al. (Springer International Publishing AG, Basel, 2017), pp. 111–117
179. E. Díaz-Bautista, D.J. Fernández, Eur. Phys. J. Plus **132**, 499 (2017)
180. A.A. Andrianov, M.V. Ioffe, F. Cannata, J.P. Dedonder, Int. J. Mod. Phys. A **14**, 2675–2688 (1999)
181. T. Goliński, A. Odzijewicz, J. Comput. Appl. Math. **176**, 331–355 (2005)
182. A. Dobrogowska, A. Odzijewicz, J. Comput. Appl. Math. **193**, 319–346 (2006)
183. A. Dobrogowska, A. Odzijewicz, J. Phys. A: Math. Theor. **40**, 2023–2036 (2007)
184. A. Dobrogowska, K. Janglajew, J. Differ. Equ. Appl. **13**, 1171–1177 (2007)
185. A. Dobrogowska, G. Jakimowicz, Appl. Math. Lett. **74**, 161–166 (2017)
186. A. Dobrogowska, D.J. Fernández (2018). arXiv:1807.06895 [math.DS]
187. A.F. Nikiforov, S.K. Suslov, V.B. Uvarov, *Classical Orthogonal Polynomials of a Discrete Variable*. Springer Series in Computational Physics (Springer, Berlin, 1991)
188. R. Alvarez-Nodarse, N.M. Atakishiyev, R.S. Costas-Santos, Electron. Trans. Numer. Anal. **27** 34–50 (2007)

Coherent States in Quantum Optics: An Oriented Overview

Jean-Pierre Gazeau

Abstract In this survey, various generalizations of Glauber–Sudarshan coherent states are described in a unified way, with their statistical properties and their possible role in non-standard quantizations of the classical electromagnetic field. Some statistical photon-counting aspects of Perelomov SU(2) and SU(1, 1) coherent states are emphasized.

Keywords Coherent states · Quantum optics · Quantization · Photon-counting statistics · Group theoretical approaches

1 Introduction

The aim of this contribution is to give a restricted review on coherent states in a wide sense (linear, non-linear, and various other types), and on their possible relevance to quantum optics, where they are generically denoted by $|\alpha\rangle$, for a complex parameter α, with $|\alpha| < R$, $R \in (0, \infty)$. Many important aspects of these states, understood here in a wide sense, will not be considered, like photon-added, intelligent, squeezed, dressed, "non-classical," all those cat superpositions of any type, involved into quantum entanglement and information, ... Of course, such a variety of features can be found in existing articles or reviews. A few of them [1–6] are included in the list of references in order to provide the reader with an extended palette of various other references.

We have attempted to give a minimal framework for all various families of $|\alpha\rangle$'s which are described in the present review. Throughout the paper we put

J.-P. Gazeau (✉)
APC, UMR 7164, Univ Paris Diderot, Sorbonne Paris Cité, Paris, France

Centro Brasileiro de Pesquisas Físicas, Rio de Janeiro, RJ, Brazil
e-mail: gazeau@apc.in2p3.fr

Ş. Kuru et al. (eds.), *Integrability, Supersymmetry and Coherent States*, CRM Series in Mathematical Physics, https://doi.org/10.1007/978-3-030-20087-9_3

$\hbar = 1 = c$, except if we need to make precise physical units. In Sect. 2 we recall the main characteristics of the Hilbertian framework (one-mode) Fock space with the underlying Weyl–Heisenberg algebra of its lowering and raising operators, and the basic statistical interpretation in terms of detection probability. In Sect. 3 we introduce coherent states in Fock space as superpositions of number states with coefficients depending on a complex number α. These "PHIN" states are requested to obey two fundamental properties, normalization and resolution of the identity in Fock space. The physical meaning of the parameter α is explained in terms of the number of photons, and may or not be interpreted in terms of classical optics quadratures. A first example is given in terms of holomorphic Hermite polynomials. We then define an important subclass AN in PHIN. Section 4 is devoted to the celebrated prototype of all CS in class AN, namely the Glauber–Sudarshan states. Their multiple properties are recalled, and their fundamental role in quantum optics is briefly described by following the seminal 1963 Glauber paper. We end the section with a description of the CS issued from unitary displacement of an arbitrary number eigenstate in place of the vacuum. The latter belong to the PHIN class, but not in the AN class. The so-called non-linear CS in the AN class are presented in Sect. 5, and an example of q-deformed CS illustrates this important extension of standard CS. In Sect. 6 we adapt the Gilmore–Perelomov spin or SU(2) CS to the quantum optics framework and we emphasize their statistical meaning in terms of photon counting. We extend them also these CS to those issued from an arbitrary number state. We follow a similar approach in Sect. 7 with Perelomov and Barut–Girardello SU(1, 1) CS. Section 8 is devoted to another type of AN CS, named Susskind–Glogower, which reveal to be quite attractive in the context of quantum optics. We end in Sect. 9 this list of various CS with a new type of non-linear CS based on deformed binomial distribution. In Sect. 10 we briefly review the statistical aspects of CS in quantum optics by focusing on their potential statistical properties, like sub- or super-Poissonian or just Poissonian. The content of Sect. 11 concerns the role of all these generalizations of CS belonging to the AN class in the quantization of classical solutions of the Maxwell equations and the corresponding quadrature portraits. Some promising features of this CS quantization are discussed in Sect. 12.

2 Fock Space

In their number or Fock representation, the eigenstates of the harmonic oscillator are simply denoted by kets $|n\rangle$, where $n = 0, 1, \dots,$ stands for the number of elementary quanta of energy, named photons when the model is applied to a quantized monochromatic electromagnetic wave. These kets form an orthonormal basis of the Fock Hilbert space $\mathcal{H}$. The latter is actually a physical model for all separable Hilbert spaces, namely the space $\ell^2(\mathbb{N})$ of square summable sequences. For such a basis (actually for any Hilbertian basis $\{e_n\,,\, n = 0, 1, \dots\}$), the *lowering* or *annihilation* operator a, and its adjoint $a^\dagger$, the *raising* or *creation* operator, are defined by

$$a|n\rangle = \sqrt{n}|n-1\rangle\,, \quad a^{\dagger}|n\rangle = \sqrt{n+1}|n+1\rangle\,, \tag{2.1}$$

together with the action of a on the ground or "vacuum" state $a|0\rangle = 0$. They obey the so-called canonical commutation rule (ccr) $[a, a^{\dagger}] = I$. In this context, the *number* operator $\hat{N} = a^{\dagger}a$ is diagonal in the basis $\{|n\rangle,\ n \in \mathbb{N}\}$, with spectrum $\mathbb{N}$: $\hat{N}|n\rangle = n|n\rangle$.

3 General Setting for Coherent States in a Wide Sense

3.1 The PHIN Class

A large class of one-mode optical coherent states can be written as the following normalized superposition of photon number states:

$$|\alpha\rangle = \sum_{n=0}^{\infty} \phi_n(\alpha)|n\rangle\,, \tag{3.1}$$

where the complex parameter α lies in some bounded or unbounded subset $\mathfrak{S}$ of $\mathbb{C}$. Its physical meaning will be discussed below in terms of detection probability. Note that the adjective "coherent" is used in a generic sense and should not be understood in the restrictive sense it was given originally by Glauber [7]. The complex-valued functions $\alpha \mapsto \phi_n(\alpha)$, from which the name "PHIN class," obey the two conditions

$$1 = \sum_{n=0}^{\infty} |\phi_n(\alpha)|^2\,, \quad \alpha \in \mathfrak{S}\,, \quad \text{(normalisation)} \tag{3.2}$$

$$\delta_{nn'} = \int_{\mathfrak{S}} \mathrm{d}^2\alpha\, \mathfrak{w}(\alpha)\, \overline{\phi_n(\alpha)}\, \phi_{n'}(\alpha)\,, \quad \text{(orthonormality)}\,, \tag{3.3}$$

where $\mathfrak{w}(\alpha)$ is a weight function, with support $\mathfrak{S}$ in $\mathbb{C}$. While Eq. (3.2) is necessary, Eq. (3.3) might be optional, except if we request resolution of the identity in the Fock Hilbert space spanned by the number states:

$$\int_{\mathfrak{S}} \mathrm{d}^2\alpha\, \mathfrak{w}(\alpha)\, |\alpha\rangle\langle\alpha| = I\,. \tag{3.4}$$

A finite sum in (3.1) due to $\phi_n = 0$ for all n larger than a certain $n_{\max}$ may be considered in this study.

If the orthonormality condition (3.3) is satisfied with a positive weight function, it allows us to interpret the map

$$\alpha \mapsto |\phi_n(\alpha)|^2 \equiv \varpi_n(\alpha) \tag{3.5}$$

as a probability distribution, with parameter n, on the support $\mathfrak{S}$ of $\mathfrak{w}$ in $\mathbb{C}$, equipped with the measure $\mathfrak{w}(\alpha)\, \mathrm{d}^2\alpha$.

On the other hand, the normalization condition (3.2) allows to interpret the discrete map

$$n \mapsto \varpi_n(\alpha) \tag{3.6}$$

as a probability distribution on $\mathbb{N}$, with parameter α, precisely the probability to detect n photons when the quantum light is in the coherent state $|\alpha\rangle$. The average value of the number operator

$$\bar{n} = \bar{n}(\alpha) := \langle\alpha|\hat{N}|\alpha\rangle = \sum_{n=0}^{\infty} n\, \varpi_n(\alpha) \tag{3.7}$$

can be viewed as the intensity (or energy up to a physical factor like $\hbar\omega$) of the state $|\alpha\rangle$ of the quantum monochromatic radiation under consideration. An optical phase space associated with this radiation may be defined as the image of the map

$$\mathfrak{S} \ni \alpha \mapsto \xi_\alpha = \sqrt{\bar{n}(\alpha)}\, e^{\mathrm{i} \arg\alpha} \in \mathbb{C}\,. \tag{3.8}$$

A statistical interpretation of the original set $\mathfrak{S}$ is made possible if one can invert the map (3.8). Two examples of such an inverse map will be given in Sects. 6 and 7.1, respectively, with interesting statistical interpretations.

3.2 A First Example of PHIN CS with Holomorphic Hermite Polynomials

These coherent states were introduced in [8]. Given a real number $0 < s < 1$, the functions $\phi_{n;s}$ are defined as

$$\phi_{n;s}(\alpha) := \frac{1}{\sqrt{b_n(s)\mathcal{N}_s(\alpha)}}\, e^{-\alpha^2/2}\, H_n(\alpha)\,, \quad \alpha \in \mathbb{C}\,. \tag{3.9}$$

The non-holomorphic part lies in the expression of $\mathcal{N}_s$

$$\mathcal{N}_s(\alpha) = \frac{s^{-1} - s}{2\pi}\, e^{-s\,X^2 + s^{-1}\,Y^2}\,, \; \alpha = X + \mathrm{i}Y\,.$$

The constant $b_n(s)$ is given by

$$b_n(s) = \frac{\pi\sqrt{s}}{1-s}\left(2\frac{1+s}{1-s}\right)^n n!\,.$$

The function $H_n(\alpha)$ is the usual Hermite polynomial of degree n [9], considered here as a holomorphic polynomial in the complex variable α. The corresponding normalized coherent states

$$|\alpha; s\rangle = \sum_{n=0}^{\infty} \phi_{n;s}(\alpha)|n\rangle \tag{3.10}$$

solve the identity in $\mathcal{H}$,

$$\frac{s^{-1}-s}{2\pi}\int_{\mathbb{C}} \mathrm{d}^2\alpha\, |\alpha; s\rangle\langle\alpha; s| = I\,. \tag{3.11}$$

Thus, in the present case we have the constant weight $\mathfrak{w}(\alpha) = \frac{s^{-1}-s}{2\pi}$. This resolution of the identity results from the orthogonality relations verified by the holomorphic Hermite polynomials in the complex plane:

$$\int_{\mathbb{C}} \mathrm{d}X\, \mathrm{d}Y\, \overline{H_n(X+\mathrm{i}Y)}\, H_{n'}(X+\mathrm{i}Y) \exp\left[-(1-s)X^2 - \left(\frac{1}{s}-1\right)Y^2\right] = b_n(s)\delta_{nn'}\,. \tag{3.12}$$

Note that the map $\alpha \mapsto \bar{n}(\alpha) = \sum_n n\left|e^{-\alpha^2/2}H_n(\alpha)\right|^2$ is not rotationally invariant.

3.3 The AN Class

Particularly convenient to manage and mostly encountered are coherent states $|\alpha\rangle$ for which the functions ϕ_n factorize as

$$\phi_n(\alpha) = \alpha^n\, h_n(|\alpha|^2)\,, \quad \sum_{n=0}^{\infty} |\alpha|^{2n}|h_n(\alpha)|^2 = 1\,, \quad |\alpha| < R\,, \tag{3.13}$$

where R can be finite or infinite. All coherent states of the above type lie in the so-called AN class (AN for "αn"). Then, due to Fourier angular integration in (3.3), the orthonormality condition holds if there exists an isotropic weight function w such that the h_n's solve the following kind of moment problem on the interval $[0, R^2]$:

$$\int_0^{R^2} \mathrm{d}u\, w(u)\, u^n |h_n(u)|^2 = 1\,, \quad n \in \mathbb{N}\,. \tag{3.14}$$

This w is related to the above $\mathfrak{w}$ through

$$\mathfrak{w}(\alpha) = \frac{w(|\alpha|^2)}{\pi} \,. \tag{3.15}$$

Note that the probability (3.6) to detect n photons when the quantum light is in such a AN coherent state $|\alpha\rangle$ is expressed as a function of $u = |\alpha|^2$ only

$$n \mapsto \varpi_n(\alpha) \equiv \mathsf{P}_n(u) = u^n \, (h_n(u))^2 \,. \tag{3.16}$$

Hence, the map $\alpha \mapsto \bar{n}$ is here rotationally invariant: $\bar{n} = \bar{n}(u)$. On the other hand, the probability distribution on the interval $[0, R^2]$, for a detected n, that CS $|\alpha\rangle$ have classical intensity u is given by

$$u \mapsto \varpi_n(\alpha) \equiv \mathsf{P}_n(u) \,. \tag{3.17}$$

4 Glauber–Sudarshan CS

4.1 Definition and Properties

They are the most popular, of course, among the AN families, and historically the first ones to appear in QED with Schwinger [10], and in quantum optics with the 1963 seminal papers by Glauber [7, 11, 12] and Sudarshan [13]. See also some key papers like [14–16] for further developments in quantum optics and quantum field theory. They were introduced in quantum mechanics by Schrödinger [17] and later by Klauder [18–20]. They correspond to the Gaussian

$$h_n(u) = \frac{e^{-u/2}}{\sqrt{n!}} \,, \tag{4.1}$$

and read

$$|\alpha\,\rangle = e^{-|\alpha|^2/2} \sum_{n=0}^{\infty} \frac{\alpha^n}{\sqrt{n!}} |n\rangle. \tag{4.2}$$

Here, the parameter, i.e., the *amplitude*, $\alpha = X + \mathrm{i}Y$ represents an element of the optical phase space. Its Cartesian components X and Y in the Euclidean plane are called quadratures. In complete analogy with the harmonic oscillator model, the quantity $u = |\alpha|^2$ is considered as the classical *intensity* or *energy* of the coherent state $|\alpha\rangle$. The corresponding detection distribution is the familiar Poisson distribution

$$n \mapsto \mathsf{P}_n(u) = e^{-u}\,\frac{u^n}{n!}\,, \tag{4.3}$$

and the average value of the number operator is just the intensity.

$$\bar{n}(\alpha) = |\alpha|^2 = u\,. \tag{4.4}$$

Hence, the detection distribution is written in terms of this average value as

$$\mathsf{P}_n(u) = e^{-\bar{n}}\,\frac{\bar{n}^n}{n!}\,. \tag{4.5}$$

From now on the states (4.2) will be called *standard coherent states*. They are called harmonic oscillator CS when we consider the $|n\rangle$'s as eigenstates of the corresponding quantum Hamiltonian $H_{\mathrm{osc}} = \left(P^2 + Q^2\right)/2 = \hat{N} + 1/2$ with $Q = \dfrac{a + a^\dagger}{\sqrt{2}}$ and $P = \dfrac{a - a^\dagger}{\mathrm{i}\sqrt{2}}$. They are exceptional in the sense that they obey the following long list of properties that give them, on their whole own, a strong status of uniqueness.

P$_0$ *The map $\mathbb{C} \ni \alpha \to |\alpha\rangle \in \mathcal{H}$ is continuous.*
P$_1$ *$|\alpha\rangle$ is eigenvector of annihilation operator: $a|\alpha\rangle = \alpha|\alpha\rangle$.*
P$_2$ *The CS family resolves the unity: $\int_{\mathbb{C}} \frac{\mathrm{d}^2\alpha}{\pi}\,|\alpha\rangle\langle\alpha| = I$.*
P$_3$ *The CS saturate the Heisenberg inequality : $\Delta X\,\Delta Y = \Delta Q\,\Delta P = 1/2$.*
P$_4$ *The CS family is temporally stable : $e^{-\mathrm{i}H_{\mathrm{osc}}t}|\alpha\rangle = e^{-\mathrm{i}t/2}|e^{-\mathrm{i}t}\alpha\rangle$.*
P$_5$ *The mean value (or "lower symbol") of the Hamiltonian H_{osc} mimics the classical relation energy-action: $\check{H}_{\mathrm{osc}}(\alpha) := \langle\alpha|H_{\mathrm{osc}}|\alpha\rangle = |\alpha|^2 + \frac{1}{2}$.*
P$_6$ *The CS family is the orbit of the ground state under the action of the Weyl displacement operator: $|\alpha\rangle = e^{(\alpha a^\dagger - \bar{\alpha} a)}|0\rangle \equiv D(\alpha)|0\rangle$.*
P$_7$ *The unitary Weyl–Heisenberg covariance follows from the above:*
$\mathcal{U}(s,\zeta)|\alpha\rangle = e^{\mathrm{i}(s+\mathrm{Im}(\zeta\bar{\alpha}))}|\alpha+\zeta\rangle$, *where* $\mathcal{U}(s,\zeta) := e^{\mathrm{i}s}\,D(\zeta)$.
P$_8$ *From P$_2$ the coherent states provide a straightforward quantization scheme:*
Function $f(\alpha) \to$ *Operator* $A_f = \int_{\mathbb{C}} \frac{\mathrm{d}^2\alpha}{\pi}\, f(\alpha)\,|\alpha\rangle\langle\alpha|$.

These properties cover a wide spectrum, starting from the "wave-packet" expression (4.2) together with Properties P$_3$ and P$_4$, through an algebraic side (P$_1$), a group representation side (P$_6$ and P$_7$), a functional analysis side (P$_2$) to end with the ubiquitous problematic of the relationship between classical and quantum models (P$_5$ and P$_8$). Starting from this exceptional palette of properties, the game over the past almost seven decades has been to build families of CS having some of these properties, if not all of them, as it can be attested by the huge literature, articles, proceedings, special issues, and author(s) or collective books, a few of them being [21–32].

4.2 *Why the Adjective* Coherent*? (Partially Extracted from [30])*

Let us compare the two equations :

$$a|\alpha\rangle = \alpha|\alpha\rangle\,, \qquad a|n\rangle = \sqrt{n}|n-1\rangle\,. \tag{4.6}$$

Hence, *an infinite superposition of number states $|n\rangle$, each of the latter describing a determinate number of elementary quanta, describes a state which is left unmodified (up to a factor) under the action of the operator annihilating an elementary quantum. The factor is equal to the parameter α labeling the considered coherent state.*

More generally, we have $f(a)|\alpha\rangle = f(\alpha)|\alpha\rangle$ for an analytic function f. This is precisely the idea developed by Glauber [7, 11, 12]. Indeed, an electromagnetic field in a box can be assimilated to a countably infinite assembly of harmonic oscillators. This results from a simple Fourier analysis of Maxwell equations. The (canonical) quantization of these classical harmonic oscillators yields the Fock space $\mathcal{F}$ spanned by all possible tensor products of number eigenstates $\bigotimes_k |n_k\rangle \equiv |n_1, n_2, \dots, n_k, \dots\rangle$, where "$k$" is a shortening for labeling the mode (including the photon polarization)

$$k \equiv \begin{cases} \quad\mathbf{k} & \text{wave vector,} \\ \omega_k = \|\mathbf{k}\|c & \text{frequency,} \\ \lambda = 1, 2 & \text{helicity,} \end{cases} \tag{4.7}$$

and n_k is the number of photons in the mode "k." The Fourier expansion of the quantum vector potential reads as

$$\vec{A}\,(\mathbf{r}, t) = c \sum_k \sqrt{\frac{\hbar}{2\omega_k}} \left(a_k \mathbf{u}_k(\mathbf{r}) e^{-\mathrm{i}\omega_k t} + a_k^\dagger \overline{\mathbf{u}_k(\mathbf{r})} e^{\mathrm{i}\omega_k t} \right). \tag{4.8}$$

As an operator, it acts (up to a gauge) on the Fock space $\mathcal{F}$ via a_k and $a_k^\dagger$ defined by

$$a_{k_0} \prod_k |n_k\rangle = \sqrt{n_{k_0}} |n_{k_0} - 1\rangle \prod_{k \neq k_0} |n_k\rangle\,, \tag{4.9}$$

and obeying the canonical commutation rules

$$[a_k, a_{k'}] = 0 = [a_k^\dagger, a_{k'}^\dagger]\,, \qquad [a_k, a_{k'}^\dagger] = \delta_{kk'}\, I\,. \tag{4.10}$$

Let us now give more insights on the modes, observables, and Hamiltonian. On the level of the mode functions $\mathbf{u}_k$ the Maxwell equations read as

$$\Delta \mathbf{u}_k(\mathbf{r}) + \frac{\omega_k^2}{c^2}\mathbf{u}_k(\mathbf{r}) = \mathbf{0}. \tag{4.11}$$

When confined to a cubic box C_L with size L, these functions form an orthonormal basis

$$\int_{C_L} \overline{\mathbf{u}_k(\mathbf{r})} \cdot \mathbf{u}_l(\mathbf{r})\, \mathrm{d}^3\mathbf{r} = \delta_{kl}\,,$$

with obvious discretization constraints on "k." By choosing the gauge $\nabla \cdot \mathbf{u}_k(\mathbf{r}) = 0$, their expression is

$$\mathbf{u}_k(\mathbf{r}) = L^{-3/2}\widehat{e}^{(\lambda)} e^{\mathrm{i}\mathbf{k}\cdot\mathbf{r}}\,, \quad \lambda = 1 \text{ or } 2\,, \quad \mathbf{k}\cdot\widehat{e}^{(\lambda)} = 0\,, \tag{4.12}$$

where the $\widehat{e}^{(\lambda)}$'s stand for polarization vectors. The respective expressions of the electric and magnetic field operators are derived from the vector potential:

$$\vec{E} = -\frac{1}{c}\frac{\partial \vec{A}}{\partial t}\,, \quad \vec{B} = \vec{\nabla} \times \vec{A}\,.$$

Finally, the electromagnetic field Hamiltonian is given by

$$H_{\text{e.m.}} = \frac{1}{2}\int \left(\| \vec{E} \|^2 + \| \vec{B} \|^2\right) \mathrm{d}^3\mathbf{r} = \frac{1}{2}\sum_k \hbar\omega_k \left(a_k^\dagger a_k + a_k a_k^\dagger\right).$$

Let us now decompose the electric field operator into positive and negative frequencies

$$\vec{E} = \vec{E}^{(+)} + \vec{E}^{(-)}, \; \vec{E}^{(-)} = \vec{E}^{(+)\dagger}\,,$$

$$\vec{E}^{(+)}(\mathbf{r},t) = \mathrm{i}\sum_k \sqrt{\frac{\hbar\omega_k}{2}}\, a_k \mathbf{u}_k(\mathbf{r}) e^{-\mathrm{i}\omega_k t}\,. \tag{4.13}$$

We then consider the field described by the density (matrix) operator :

$$\rho = \sum_{(n_k)} c_{(n_k)} \prod_k |n_k\rangle\langle n_k|\,, \quad c_{(n_k)} \geq 0\,, \quad \operatorname{tr}\rho = 1\,, \tag{4.14}$$

and the derived sequence of correlation functions $G^{(n)}$. The Euclidean tensor components for the simplest one read as

$$G_{ij}^{(1)}(\mathbf{r},t;\mathbf{r}',t') = \operatorname{tr}\left\{\rho E_i^{(-)}(\mathbf{r},t) E_j^{(+)}(\mathbf{r}',t')\right\}, \quad i,j = 1,2,3\,. \tag{4.15}$$

They measure the correlation of the field state at different space-time points. A *coherent state* or *coherent radiation* $|\text{c.r.}\rangle$ for the electromagnetic field is then defined by

$$|\text{c.r.}\rangle = \prod_k |\alpha_k\rangle\,, \tag{4.16}$$

where $|\alpha_k\rangle$ is precisely the standard coherent state for the "k" mode :

$$|\alpha_k\rangle = e^{-\frac{|\alpha_k|^2}{2}} \sum_{n_k} \frac{(\alpha_k)^{n_k}}{\sqrt{n_k!}} |n_k\rangle\,, \quad a_k|\alpha_k\rangle = \alpha_k|\alpha_k\rangle\,, \tag{4.17}$$

with $\alpha_k \in \mathbb{C}$. The particular status of the state $|\text{c.r.}\rangle$ is well understood through the action of the positive frequency electric field operator

$$\overrightarrow{E}^{(+)}(\mathbf{r}, t)|\text{c.r.}\rangle = \overrightarrow{\mathcal{E}}^{(+)}(\mathbf{r}, t)|\text{c.r.}\rangle\,. \tag{4.18}$$

The expression $\overrightarrow{\mathcal{E}}^{(+)}(\mathbf{r}, t)$ which shows up is precisely the classical field expression, solution to the Maxwell equations

$$\overrightarrow{\mathcal{E}}^{(+)}(\mathbf{r}, t) = \mathrm{i} \sum_k \sqrt{\frac{\hbar\omega_k}{2}} \alpha_k \mathbf{u}_k(\mathbf{r}) e^{-\mathrm{i}\omega_k t}\,. \tag{4.19}$$

Now, if the density operator is chosen as a pure coherent state, i.e.,

$$\rho = |\text{c.r.}\rangle\langle\text{c.r.}|\,, \tag{4.20}$$

then the components (4.15) of the first order correlation function factorize into independent terms :

$$G^{(1)}_{ij}(\mathbf{r}, t; \mathbf{r}', t') = \overline{\mathcal{E}^{(-)}_i(\mathbf{r}, t)\mathcal{E}^{(+)}_j}(\mathbf{r}', t')\,. \tag{4.21}$$

An electromagnetic field operator is said "fully coherent" in the Glauber sense if all of its correlation functions factorize like in (4.21). Nevertheless, one should notice that such a definition does not imply monochromaticity.

A last important point concerns the production of such states in quantum optics. They can be manufactured by adiabatically coupling the e.m. field to a classical source, for instance, a radiating current $\mathbf{j}(\mathbf{r}, t)$. The coupling is described by the Hamiltonian

$$H_{\text{coupling}} = -\frac{1}{c} \int \mathrm{d}\mathbf{r}\, \overrightarrow{j}(\mathbf{r}, t) \cdot \overrightarrow{A}(\mathbf{r}, t)\,. \tag{4.22}$$

From the Schrödinger equation, the time evolution of a field state supposed to be originally, say at t_0, the state $|\text{vacuum}\rangle$(no photons) is given by

$$|t\rangle = \exp\left[\frac{\mathsf{i}}{\hbar c}\int_{t_0}^{t} \mathrm{d}t' \int \mathrm{d}\mathbf{r}\, \overrightarrow{j}(\mathbf{r},t')\cdot \overrightarrow{A}(\mathbf{r},t') + \mathsf{i}\varphi(t)\right] |\text{vacuum}\rangle\,, \tag{4.23}$$

where $\varphi(t)$ is some phase factor, which cancels if one deals with the density operator $|t\rangle\langle t|$ and can be dropped. From the Fourier expansion (4.8) we easily express the above evolution operator in terms of the Weyl displacement operators corresponding to each mode

$$\exp\left[\frac{\mathsf{i}}{\hbar c}\int_{t_0}^{t} \mathrm{d}t' \int \mathrm{d}\mathbf{r}\, \overrightarrow{j}(\mathbf{r},t')\cdot \overrightarrow{A}(\mathbf{r},t')\right] = \prod_k D(\alpha_k(t))\,, \tag{4.24}$$

where the complex amplitudes are given by

$$\alpha_k(t) = \frac{\mathsf{i}}{\hbar c}\int_{t_0}^{t} \mathrm{d}t' \int \mathrm{d}\mathbf{r}\, \overrightarrow{j}(\mathbf{r},t')\cdot \overline{\mathbf{u}_k(\mathbf{r})} e^{\mathsf{i}\omega_k t'}\,. \tag{4.25}$$

Hence, we obtain the time-dependent e.m. CS

$$|t\rangle = \otimes_k |\alpha_k(t)\rangle\,. \tag{4.26}$$

4.3 Weyl–Heisenberg CS with Laguerre Polynomials

The construction of the standard CS is minimal from the point of view of the action of the Weyl unitary operator $D(\alpha)$ on the vacuum $|0\rangle$ (Property $\mathbf{P}_6$). More elaborate states are issued from the action of $D(\alpha)$ on other states $|s\rangle$, $s = 1, 2, \dots,$ of the Fock basis, which might be considered as initial states in the evolution described by (4.23). Hence, let us define the family of CS

$$|\alpha; s\rangle = D(\alpha)|s\rangle = \sum_{n=0}^{\infty} D_{ns}(\alpha)|n\rangle\,. \tag{4.27}$$

The coefficients in this Fock expansion are the matrix elements $D_{ns} = \langle n|D(\alpha)|s\rangle$ of the displacement operator. They are given in terms of the generalized Laguerre polynomials [9] as

$$\begin{aligned} D_{ns}(\alpha) &:= \sqrt{\frac{s!}{n!}}\, e^{-\frac{|\alpha|^2}{2}}\, \alpha^{n-s}\, L_s^{(n-s)}\left(|\alpha|^2\right) \quad \text{for} \quad s \le n\,, \\ &= \sqrt{\frac{n!}{s!}}\, e^{-\frac{|\alpha|^2}{2}}\, (-\bar{\alpha})^{s-n}\, L_n^{(s-n)}\left(|\alpha|^2\right) \quad \text{for} \quad s > n\,. \end{aligned} \tag{4.28}$$

As matrix elements of a projective square-integrable UIR of the Weyl–Heisenberg group they obey the orthogonality relations

$$\int_{\mathbb{C}} \frac{\mathrm{d}^2\alpha}{\pi} \overline{D_{ns}(\alpha)}\, D_{n's'}(\alpha) = \delta_{nn'}\, \delta_{ss'}\,. \tag{4.29}$$

Like for the general case presented in (3.3)–(3.4) this property validates the resolution of the identity

$$\int_{\mathbb{C}^2} \frac{\mathrm{d}^2\alpha}{\pi} |\alpha; s\rangle\langle\alpha; s| = I\,. \tag{4.30}$$

The corresponding detection distribution is the "Laguerre weighted" Poisson distribution

$$n \mapsto \mathsf{P}_n(u) = \begin{cases} e^{-u}\,\dfrac{u^{s-n}}{(s-n)!}\,\dfrac{\left(L_n^{(s-n)}(u)\right)^2}{\binom{s}{n}} & n \leq s \\ e^{-u}\,\dfrac{u^{n-s}}{(n-s)!}\,\dfrac{\left(L_s^{(n-s)}(u)\right)^2}{\binom{n}{s}} & n \geq s \end{cases}\,. \tag{4.31}$$

Of course, the optical phase space made of the complex $\sqrt{\bar{n}(\alpha)}e^{\mathrm{i}\arg\alpha}$ is here less immediate.

We notice that for $s > 0$, these CS $|\alpha; s\rangle$ do not pertain to the AN class, since we find in the expansion a finite number of terms in $\bar{\alpha}^n$ besides an infinite number of terms in α^n. On the other hand, there exist families of coherent states in the AN class (or their complex conjugate) which are related to the generalized Laguerre polynomials in a quasi-identical way [33, 34].

5 Non-linear CS

5.1 General

We define as non-linear CS those AN CS for which the functions $h_n(u)$ assume the simple form

$$h_n(u) = \frac{\lambda_n}{\sqrt{\mathcal{N}(u)}}\,, \quad \mathcal{N}(u) = \sum_{n=0}^{\infty} |\lambda_n|^2 u^n\,. \tag{5.1}$$

5.2 Deformed Poissonian CS

They are particular cases of the above. All λ_n form a strictly decreasing sequence of positive numbers tending to 0:

$$\lambda_0 = 1 > \lambda_1 > \cdots \lambda_n > \lambda_{n+1} > \cdots, \quad \lambda_n \to 0. \tag{5.2}$$

We now introduce the strictly increasing sequence

$$x_n = \left(\frac{\lambda_{n-1}}{\lambda_n}\right)^2, \quad x_0 = 0. \tag{5.3}$$

It is straightforward to check that

$$\lambda_n = \frac{1}{\sqrt{x_n!}}, \quad \text{with} \quad x_n! := x_1 x_2 \cdots x_n. \tag{5.4}$$

Then $\mathcal{N}(u)$ is the generalized exponential with convergence radius R^2

$$\mathcal{N}(u) = \sum_{n=0}^{\infty} \frac{u^n}{x_n!}, \tag{5.5}$$

and the corresponding CS take the form extending to the non-linear case the familiar Glauber–Sudarshan one

$$|\alpha\rangle = \frac{1}{\sqrt{\mathcal{N}(|\alpha)|^2)}} \sum_{n=0}^{\infty} \frac{\alpha^n}{\sqrt{x_n!}} |n\rangle. \tag{5.6}$$

The orthonormality condition (3.3) is completely fulfilled if there exists a weight $w(u)$ solving the moment problem for the sequence $(x_n!)_{n \in \mathbb{N}}$

$$x_n! = \int_0^{R^2} \mathrm{d}u \, \frac{w(u)}{\mathcal{N}(u)} u^n. \tag{5.7}$$

The detection probability distribution is the deformed Poisson distribution:

$$n \mapsto \mathsf{P}_n(u) = \frac{1}{\mathcal{N}(u)} \frac{u^n}{x_n!}. \tag{5.8}$$

The average value of the number operator $\bar{n}$ is given by

$$\bar{n}\left(|\alpha|^2\right)) = \langle \alpha | \hat{N} | \alpha \rangle = u \left. \frac{\mathrm{d} \log \mathcal{N}(u)}{\mathrm{d}u} \right|_{u=|\alpha|^2}. \tag{5.9}$$

5.3 Example with q Deformations of Integers

These coherent states have been studied by many authors, see [35], that we follow here, and the references therein. They are built from the symmetric or bosonic q-deformation of natural numbers:

$$x_n = {}^{[s]}[n]_q = \frac{q^n - q^{-n}}{q - q^{-1}} = {}^{[s]}[n]_{q^{-1}}\,, \quad q > 0\,. \tag{5.10}$$

$$|\alpha\rangle_q = \frac{1}{\sqrt{\mathcal{N}_q(|\alpha|^2)}} \sum_{n=0}^{\infty} \frac{\alpha^n}{\sqrt{{}^{[s]}[n]_q!}} |n\rangle\,, \tag{5.11}$$

where its associated exponential is one of the so-called q exponentials [36]

$$\mathcal{N}_q(u) = \mathfrak{e}_q(u) \equiv= \sum_{n=0}^{+\infty} \frac{u^n}{{}^{[s]}[n]_q!}\,. \tag{5.12}$$

This series defines the analytic entire function $\mathfrak{e}_q(z)$ in the complex plane for any positive q. The CS $|\alpha\rangle_q$ in the limit $q \to 1$ goes to the standard CS $|\alpha\rangle$. The solution to the moment problem (3.14) for $0 < q < 1$ is given by

$$\int_0^{\infty} \mathrm{d}u\, w_q(u)\, \frac{u^n}{\mathfrak{e}_q(u)\, {}^{[s]}[n]_q!} = 1$$

with positive density

$$w_q(t) = (q^{-1} - q) \sum_{j=0}^{\infty} g_q\left(t\, \frac{q^{-1} - q}{q^{2j}}\right) \mathfrak{E}_q\left(-\frac{q^{2j}}{q^{-1} - q}\right)\,.$$

The function g_q is given by

$$g_q(u) = \frac{1}{\sqrt{2\pi\, |\ln q|}} \exp\left[-\frac{\left[\ln\left(\frac{u}{\sqrt{q}}\right)\right]^2}{2|\ln q|}\right],$$

and a second q-exponential [36] appears here

$$\mathfrak{E}_q(u) := \sum_{n=0}^{\infty} q^{\frac{n(n+1)}{2}} \frac{u^n}{{}^{[s]}[n]_q!}\,.$$

Its radius of convergence is ∞ for $0 < q \le 1$ (it is equal to $1/(q - q^{-1})$ for $q > 1$). There results the resolution of the identity

$$\int_{\mathbb{C}} \mathrm{d}^2\alpha \, \mathfrak{w}_q(\alpha) \, |\alpha\rangle_{qq}\langle\alpha| = I \,, \quad \mathfrak{w}_q(\alpha) = \frac{w_q(|\alpha|^2)}{\pi} \,. \tag{5.13}$$

More exotic families of non-linear CS are, for instance, presented in [37].

6 Spin CS as Optical CS

These states are an adaptation to the quantum optical context of the well-known Gilmore or Perelomov SU(2)-CS, also called spin CS [22, 23]. The Fock space reduces to the finite-dimensional subspace $\mathcal{H}_j$, with dimension $n_j + 1 := 2j + 1$, for j positive integer or half-integer, consistently with the fact that the functions h_n, given here by

$$h_n(u) = \sqrt{\binom{n_j}{n}}(1+u)^{-\frac{n_j}{2}} \,, \quad \binom{n_j}{n} = \frac{n_j!}{n!(n_j - n)!} \,, \tag{6.1}$$

cancel for $n > n_j$. The corresponding spin CS read

$$|\alpha; n_j\rangle = \left(1 + |\alpha|^2\right)^{-\frac{n_j}{2}} \sum_{n=0}^{n_j} \sqrt{\binom{n_j}{n}} \alpha^n \, |n\rangle \,. \tag{6.2}$$

They resolve the unity in $\mathcal{H}_{n_j}$ in the following way:

$$\frac{n_j + 1}{\pi} \int_{\mathbb{C}} \frac{\mathrm{d}^2\alpha}{(1 + |\alpha|^2)^2} \, |\alpha; n_j\rangle\langle\alpha; n_j| = I \,. \tag{6.3}$$

The detection probability distribution is binomial:

$$n \mapsto \mathsf{P}_n(u) = (1 + u)^{-n_j} \binom{n_j}{n} u^n \,. \tag{6.4}$$

There results the average value of the number operator

$$\bar{n}(u) = n_j \frac{u}{1+u} \Leftrightarrow u = \frac{\bar{n}/n_j}{1 - \bar{n}/n_j} \,. \tag{6.5}$$

Thus the probability (6.4) is expressed in terms of the ratio $p := \bar{n}/n_j$ as

$$\mathsf{P}_n(u) \equiv \widetilde{\mathsf{P}}_n(p) = \binom{n_j}{n} (1-p)^{n_j - n}\, p^n\,, \tag{6.6}$$

which allows to define the optical phase space as the open disk of radius $\sqrt{n_j}$, $\mathcal{D}_{\sqrt{n_j}} = \left\{\xi_\alpha = \sqrt{\bar{n}\left(|\alpha|^2\right)} e^{\mathrm{i}\arg\alpha}\,,\ |\xi_\alpha| < \sqrt{n_j}\right\}$.

The interpretation of $\mathsf{P}_n(u)$ together with the number n_j in terms of photon statistics (see Sect. 10 for more details) is luminous if we consider a beam of perfectly coherent light with a constant intensity. If the beam is of finite length L and is subdivided into n_j segments of length L/n_j, then $\widetilde{\mathsf{P}}_n(p)$ is the probability of finding n subsegments containing one photon and $(n_j - n)$ containing no photons, in any possible order [38]. A more general statistical interpretation of (6.4) or (6.6) is discussed in [39].

Note that the standard coherent states are obtained from the above CS at the limit $n_j \to \infty$ through a contraction process. The latter is carried out through a scaling of the complex variable α, namely $\alpha \mapsto \sqrt{n_j}\,\alpha$. Then the binomial distribution $\widetilde{\mathsf{P}}_n(p)$ becomes the Poissonian (4.5), as expected.

Actually, these states are the simplest ones among a whole family issued from the Perelomov construction [22, 30, 40], and based on spin spherical harmonics. For our present purpose we modify their definition by including an extra phase factor and delete the factor $\sqrt{\frac{2j+1}{4\pi}}$. For $j \in \mathbb{N}/2$ and a given $-j \leq \sigma \leq j$, the spin spherical harmonics are the following functions on the unit sphere $\mathbb{S}^2$:

$$\begin{aligned} {}_\sigma\mathfrak{Y}_{j\mu}(\Omega) &:= (-1)^{(j-\mu)} \sqrt{\frac{(j-\mu)!(j+\mu)!}{(j-\sigma)!(j+\sigma)!}} \times \\ &\times \frac{1}{2^\mu}\,(1+\cos\theta)^{\frac{\mu+\sigma}{2}}\,(1-\cos\theta)^{\frac{\mu-\sigma}{2}}\,P_{j-\mu}^{(\mu-\sigma,\mu+\sigma)}(\cos\theta)\,e^{-\mathrm{i}(j-\mu)\varphi}\,, \end{aligned} \tag{6.7}$$

where $\Omega = (\theta, \varphi)$ (polar coordinates), $-j \leq \mu \leq j$, and the $P_n^{(a,b)}(x)$ are Jacobi polynomials [9] with $P_0^{(a,b)}(x) = 1$. Singularities of the factors at $\theta = 0$ (resp. $\theta = \pi$) for the power $\mu - \sigma < 0$ (resp. $\mu + \sigma < 0$) are just apparent. To remove them it is necessary to use alternate expressions of the Jacobi polynomials based on the relations:

$$P_n^{(-a,b)}(x) = \frac{\binom{n+b}{a}}{\binom{n}{a}} \left(\frac{x-1}{2}\right)^a P_{n-a}^{(a,b)}(x)\,. \tag{6.8}$$

The functions (6.7) obey the two conditions required in the construction of coherent states

$$\frac{2j+1}{4\pi}\int_{\mathbb{S}^2} \mathrm{d}\Omega\, \overline{{}_\sigma\mathfrak{Y}_{j\mu}(\Omega)}\, {}_\sigma\mathfrak{Y}_{j\mu'}(\Omega) = \delta_{\mu\mu'} \quad \text{(orthogonality)} \tag{6.9}$$

$$\sum_{\mu=-j}^{j} |{}_\sigma\mathfrak{Y}_{j\mu}(\Omega)|^2 = 1 \quad \text{(normalisation)}\,. \tag{6.10}$$

At $j = l$ integer and $\sigma = 0$, $\mu = m$ we recover the spherical harmonics $Y_{lm}(\Omega)$ (up to the factor $(-1)^l e^{-\mathrm{i}j\varphi}\sqrt{\frac{2l+1}{4\pi}}$). We now consider the parameter α in (6.2) as issued from the stereographic projection $\mathbb{S}^2 \ni \Omega \mapsto \alpha \in \mathbb{C}$:

$$\alpha = \tan\frac{\theta}{2}\, e^{-\mathrm{i}\varphi}\,, \quad \text{with} \quad \mathrm{d}\Omega = \sin\theta \mathrm{d}\theta \mathrm{d}\varphi = \frac{4\mathrm{d}^2\alpha}{(1+|\alpha|^2)^2}\,. \tag{6.11}$$

In this regard, the probability $p = \bar{n}/n_j$ is equal to $\sin\theta/2$, while $\varphi = \arg\alpha$. With the notations $n_j = 2j \in \mathbb{N}$, $n = j - \mu = 0, 1, 2, \ldots, n_j$, $0 \le s = j - \sigma \le n_j$, adapted to the content of the present paper, and from the expression of the Jacobi polynomials, we get the functions (6.7) in terms of $\alpha \in \mathbb{C}$:

$${}_\sigma\mathfrak{Y}_{j\mu}(\Omega) = \alpha^n\, h_{n;s}\left(|\alpha|^2\right)\,, \tag{6.12}$$

where

$$h_{n;s}(u) = \sqrt{\frac{n!(n_j-n)!}{s!(n_j-s)!}}\,(1+u)^{-\frac{n_j}{2}} \sum_{r=\max(0,n+s-n_j)}^{\min(n,s)} \binom{s}{r}\binom{n_j-s}{n-r}(-1)^r\, u^{s/2-r}\,. \tag{6.13}$$

The corresponding “Jacobi” CS are in the AN class and read

$$|\alpha; n_j; s\rangle = \sum_{n=0}^{n_j} \alpha^n\, h_{n;s}\left(|\alpha|^2\right)|n\rangle\,. \tag{6.14}$$

They solve the identity as

$$\frac{n_j+1}{\pi}\int_{\mathbb{C}} \frac{\mathrm{d}^2\alpha}{(1+|\alpha|^2)^2}\,|\alpha; n_j; s\rangle\langle\alpha; n_j; s| = I\,. \tag{6.15}$$

The states (6.2) are recovered for $s = 0$. Similarly to CS (4.27) states (6.14) can be also viewed as displaced occupied states. Indeed, they can be written in the

Perelomov way as

$$|\alpha; n_j; s\rangle = \mathcal{D}^{n_j/2}(\zeta_\alpha)\,|s\rangle\,, \tag{6.16}$$

where $\zeta_\alpha = \begin{pmatrix} \left(1+|\alpha|^2\right)^{-1/2} & \left(1+|\alpha|^2\right)^{-1/2}\alpha \\ -\left(1+|\alpha|^2\right)^{-1/2}\bar{\alpha} & \left(1+|\alpha|^2\right)^{-1/2} \end{pmatrix}$ is the element of SU(2) which brings 0 to α under the homographic action

$$\alpha \mapsto \begin{pmatrix} a & b \\ -\bar{b} & \bar{a} \end{pmatrix} \cdot \alpha := \frac{a\alpha + b}{-\bar{b}\alpha + \bar{a}}$$

of this group on the complex plane, and $\mathcal{D}^{n_j/2}$ is the corresponding $n_j + 1$-dimensional UIR of SU(2). Let us write $\mathcal{D}^{n_j/2}(\zeta_\alpha)$ as a displacement operator similar to the Weyl–Heisenberg one (propriety $\mathbf{P}_6$) and involving the usual angular momentum generators $J_\pm$ for the representation $\mathcal{D}^{n_j/2}$

$$\mathcal{D}^{n_j/2}(\zeta_\alpha) = e^{\varsigma_\alpha J_+ - \bar{\varsigma}_\alpha J_-} \equiv D_{n_j}(\varsigma_\alpha)\,, \quad \varsigma_\alpha = -\tan^{-1}|\alpha|\,e^{-\mathrm{i}\arg\alpha}\,. \tag{6.17}$$

Note that we could have adopted here the historical approaches by Jordan, Holstein, Primakoff, Schwinger [41–43] in transforming these angular momentum operators in terms of "bosonic" a and $a^\dagger$. Nevertheless this QFT artificial flavor is not really useful in the present context.

7 SU(1, 1)-CS as Optical CS

7.1 Perelomov CS

These states are also an adaptation to the quantum optical context of the Perelomov SU(1, 1)-CS [22, 23, 30, 44]. They are yielded through a SU(1, 1) unitary action on a number state. The Fock Hilbert space $\mathcal{H}$ is infinite-dimensional, while the complex number α is restricted to the open unit disk $\mathcal{D} := \{\alpha \in \mathbb{C}\,,\, |\alpha| < 1\}$. Let $\varkappa > 1/2$ and $s \in \mathbb{N}$. We then define the $(\varkappa; s)$-dependent CS family as the "SU(1, 1)-displaced s-th state"

$$|\alpha; \varkappa; s\rangle = U^{\varkappa}(p(\bar{\alpha}))|s\rangle = \sum_{n=0}^{\infty} U^{\varkappa}_{ns}(p(\bar{\alpha}))|n\rangle \equiv \sum_{n=0}^{\infty} \phi_{n;\varkappa;s}(\alpha)\,|n\rangle\,, \tag{7.1}$$

where the $U^{\varkappa}_{ns}(p(\bar{\alpha}))$'s are matrix elements of the UIR $U^{\varkappa}$ of SU(1, 1) in its discrete series and $p(\bar{\alpha})$ is the particular matrix

$$\begin{pmatrix} \left(1-|\alpha|^2\right)^{-1/2} & \left(1-|\alpha|^2\right)^{-1/2}\bar{\alpha} \\ \left(1-|\alpha|^2\right)^{-1/2}\alpha & \left(1-|\alpha|^2\right)^{-1/2} \end{pmatrix} \in \mathrm{SU}(1,1)\,. \tag{7.2}$$

They are given in terms of Jacobi polynomials as

$$U^{\varkappa}_{ns}(p(\bar{\alpha})) = \left(\frac{n_<!\,\Gamma(2\varkappa + n_>)}{n_>!\,\Gamma(2\varkappa + n_<)}\right)^{1/2} \left(1 - |\alpha|^2\right)^{\varkappa} (\mathrm{sgn}(n-s))^{n-s} \times$$
$$\times P^{(n_> - n_<,\, 2\varkappa - 1)}_{n_<}\left(1 - 2|\alpha|^2\right) \times \begin{cases} \alpha^{n-s} \text{ if } n_> = n \\ \bar{\alpha}^{s-n} \text{ if } n_> = s \end{cases} \tag{7.3}$$

with $n_{\gtrless} = \begin{cases} \max \\ \min \end{cases} (n, s) \geq 0$. The states (7.1) solve the identity:

$$\frac{2\varkappa - 1}{\pi} \int_{\mathcal{D}} \frac{d^2\alpha}{\left(1 - |\alpha|^2\right)^2} |\alpha; \varkappa; s\rangle\langle\alpha; \varkappa; s| = I\,. \tag{7.4}$$

The simplest case $s = 0$ pertains to the AN class

$$|\alpha; \varkappa; 0\rangle \equiv |\alpha; \varkappa\rangle = \sum_{n=0}^{\infty} \alpha^n\, h_{n;\varkappa}\left(|\alpha|^2\right) |n\rangle\,,\ h_{n;\varkappa}(u) := \sqrt{\binom{2\varkappa - 1 + n}{n}}\, (1-u)^{\varkappa}\,. \tag{7.5}$$

The corresponding detection probability distribution is negative binomial

$$n \mapsto \mathsf{P}_n(u) = (1-u)^{2\varkappa} \binom{2\varkappa - 1 + n}{n} u^n\,. \tag{7.6}$$

The average value of the number operator reads as

$$\bar{n}(u) = 2\varkappa \frac{u}{1-u} \Leftrightarrow u = \frac{\bar{n}/2\varkappa}{1 + \bar{n}/2\varkappa}\,. \tag{7.7}$$

By introducing the “efficiency” $\eta := 1/2\varkappa \in (0, 1)$ the probability (7.6) is expressed in terms of the corrected average value $\bar{N} := \eta\bar{n}$ as

$$\mathsf{P}_n(u) \equiv \widetilde{\mathsf{P}}_n(\bar{N}) = (1 + \bar{N})^{-1/\eta} \binom{1/\eta - 1 + n}{n} \left(\frac{\bar{N}}{1 + \bar{N}}\right)^n\,. \tag{7.8}$$

It is remarkable that such a distribution reduces to the celebrated Bose–Einstein one for the thermal light at the limit $\eta = 1$, i.e., at the lowest bound $\varkappa = 1/2$ of the discrete series of SU(1, 1). For $\eta < 1$, the difference might be understood from the fact that we consider the average photocount number $\bar{N}$ instead of the mean photon number $\bar{n}$ impinging on the detector in the same interval [38]. For a related interpretation within the framework of thermal equilibrium states of the oscillator see [45].

Note that the above CS, built from the negative binomial distribution, were also discussed in [39].

Like for CS (4.27), the CS $|\alpha; \varkappa; s\rangle$ in (7.1) do not pertain to the AN class for $s > 0$. In their expansion there are s terms in $\bar{\alpha}^{s-n}$, $s > n$, besides an infinite number of terms in α^{n-s}, $s \leq n$. Finally, like for the Weyl–Heisenberg and SU(2) cases, the representation operator $U^{\varkappa}(p(\bar{\alpha}))$ used in (7.1) to build the SU(1, 1) CS can be given the following form of a displacement operator involving the generators $K_{\pm}$ for the representation U^{κ} [23]:

$$U^{\kappa}(p(\bar{\alpha})) = e^{\varrho_\alpha K_+ - \bar{\varrho}_\alpha K_-} \equiv D_\kappa(\varrho_\alpha), \quad \varrho_\alpha = \tanh^{-1}|\alpha|\, e^{\mathrm{i}\arg\alpha}. \tag{7.9}$$

7.2 *Barut–Girardello CS*

These non-linear CS states [46, 47] pertain to the AN class. They are requested to be eigenstates of the SU(1, 1) lowering operator in its discrete series representation $U^{\varkappa}$, $\varkappa > 1/2$. The Fock Hilbert space $\mathcal{H}$ is infinite-dimensional, while the complex number α has no domain restriction in $\mathbb{C}$. With the notations of (5.6) they read

$$|\alpha; \varkappa\rangle_{\mathrm{BG}} = \frac{1}{\sqrt{\mathcal{N}_{\mathrm{BG}}(|\alpha|^2)}} \sum_{n=0}^{\infty} \frac{\alpha^n}{\sqrt{x_n!}} |n\rangle, \; x_n = n(2\varkappa + n - 1), \; x_n! = n! \frac{\Gamma(2\varkappa + n)}{\Gamma(2\varkappa)}, \tag{7.10}$$

with

$$\mathcal{N}_{\mathrm{BG}}(u) = \Gamma(2\varkappa) \sum_{n=0}^{\infty} \frac{u^n}{n!\Gamma(2\varkappa + n)} = \Gamma(2\varkappa)\, u^{-\varkappa}\, I_{2\varkappa-1}(2\sqrt{u}), \tag{7.11}$$

where I_ν is a modified Bessel function [9]. In the present case the moment problem (3.14) is solved as

$$\int_0^{\infty} du\, w_{\mathrm{BG}}(u) \frac{u^n}{\mathcal{N}_{\mathrm{BG}}(u)\, x_n!} = 1, \; w_{\mathrm{BG}}(u) = \mathcal{N}_{\mathrm{BG}}(u) \frac{2}{\Gamma(2\varkappa)} u^{\varkappa - 1/2} K_{2\varkappa-1}(2\sqrt{u}), \tag{7.12}$$

where K_ν is the second modified Bessel function. The resolution of the identity follows:

$$\int_{\mathbb{C}} \mathrm{d}^2\alpha\, \mathfrak{w}_{\mathrm{BG}}(\alpha)\, |\alpha; \varkappa\rangle_{\mathrm{BG}}{}_{\mathrm{BG}}\langle\alpha; \varkappa| = I, \quad \mathfrak{w}_{\mathrm{BG}}(u) = \frac{w_{\mathrm{BG}}(u)}{\pi}. \tag{7.13}$$

8 Adapted Susskind–Glogower CS

Let us examine the Susskind–Glogower CS [48] presented in [49]. These normalized states read for real $\alpha \equiv x \in \mathbb{R}$

$$|x\rangle_{\mathrm{SG}} = \sum_{n=0}^{\infty} (n+1)\, \frac{J_{n+1}(2x)}{x}\, |n\rangle\,, \tag{8.1}$$

where the Bessel function J_ν is given by

$$J_\nu(z) = \left(\frac{z}{2}\right)^\nu \sum_{m=0}^{\infty} \frac{(-1)^m \left(\frac{z}{2}\right)^{2m}}{m!\,\Gamma(\nu+m+1)}\,. \tag{8.2}$$

The normalization implies the interesting identity (E. Curado, private communication)

$$\sum_{n=1}^{\infty} n^2\, (J_n(2x))^2 = x^2\,. \tag{8.3}$$

The above expression allows us to extend the formula (8.1) in a non-analytic way to complex α as

$$(n+1)\frac{J_{n+1}(2x)}{x} \mapsto \alpha^n\, (n+1) \sum_{m=0}^{\infty} \frac{(-1)^m |\alpha|^{2m}}{m!\,\Gamma(n+m+2)} \equiv \alpha^n\, h_n^{\mathrm{SG}}(|\alpha|^2)\,, \tag{8.4}$$

i.e.,

$$h_n^{\mathrm{SG}}(u) = (n+1)\, \frac{1}{u^{\frac{n+1}{2}}}\, J_{n+1}(2\sqrt{u})\,, \tag{8.5}$$

and thus

$$|\alpha\rangle_{\mathrm{SG}} = \sum_{n=0}^{\infty} \alpha^n\, h_n^{\mathrm{SG}}(|\alpha|^2)\, |n\rangle\,. \tag{8.6}$$

The moment Eq. (3.14) reads here

$$\int_0^\infty \mathrm{d}u\, \frac{w(u)}{u}\, \left(J_n(2\sqrt{u})\right)^2 = 2\int_0^\infty \mathrm{d}t\, \frac{w(t^2)}{t}\, (J_n(2t))^2 = \frac{1}{n^2}\,. \tag{8.7}$$

Let us examine the following integral formula for Bessel functions [9]:

$$\int_0^{\infty} \frac{\mathrm{d}t}{t}\, (J_n(2t))^2 = \frac{1}{2n}\,. \tag{8.8}$$

This leads us to replace the SG-CS of (8.1) by the modified

$$|\alpha\rangle_{\mathrm{SGm}} = \sum_{n=0}^{\infty} \alpha^n\, h_n^{\mathrm{SGm}}(|\alpha|^2)\, |n\rangle\,, \quad h_n^{\mathrm{SGm}}(u) = \sqrt{\frac{n+1}{\mathcal{N}(u)}}\, \frac{1}{u^{\frac{n+1}{2}}}\, J_{n+1}(2\sqrt{u})\,, \tag{8.9}$$

with

$$\mathcal{N}(u) = \frac{1}{u}\sum_{n=1}^{\infty} n\, \left(J_n(2\sqrt{u})\right)^2\,. \tag{8.10}$$

Then the formula (8.8) allows us to prove that the resolution of the identity is fulfilled by these $|\alpha\rangle_{\mathrm{SGm}}$ with $w(u) = \mathcal{N}(u)$. More details, particularly those concerning statistical aspects, are given in [50].

9 CS from Symmetric Deformed Binomial Distributions (DFB)

In [51] (see also the related works [52–54]) was presented the following generalization of the binomial distribution:

$$\mathfrak{p}_k^{(n)}(\xi) = \frac{x_n!}{x_{n-k}!x_k!} q_k(\xi) q_{n-k}(1-\xi)\,, \tag{9.1}$$

where the $\{x_n\}$'s form a non-negative sequence and the $q_k(\xi)$ are polynomials of degree k, while ξ is a running parameter on the interval $[0, 1]$. The $\mathfrak{p}_k^{(n)}(\xi)$ are constrained by

(a) the normalization

$$\forall n \in \mathbb{N}, \quad \forall \xi \in [0,1], \quad \sum_{k=0}^{n} \mathfrak{p}_k^{(n)}(\xi) = 1, \tag{9.2}$$

(b) the non-negativeness condition (requested by statistical interpretation)

$$\forall n, k \in \mathbb{N}, \quad \forall \xi \in [0,1], \quad \mathfrak{p}_k^{(n)}(\xi) \geq 0. \tag{9.3}$$

These conditions imply that $q_0(\xi) = \pm 1$. With the choice $q_0(\xi) = 1$ one easily proves that the non-negativeness condition (9.3) is equivalent to the non-negativeness of the polynomials q_n on the interval [0, 1]. Hence the quantity $\mathfrak{p}_k^{(n)}(\xi)$ can be interpreted as the probability of having k wins and $n-k$ losses in a sequence of *correlated* n trials. Besides, as we recover the invariance under $k \to n-k$ and $\xi \to 1-\xi$ of the binomial distribution, no bias (in the case $\xi = 1/2$) can exist favoring either win or loss. The polynomials $q_n(\xi)$ are viewed here as *deformations* of ξ^n. We now suppose that the generating function for the polynomials q_n, defined as

$$F(\xi; t) := \sum_{n=0}^{\infty} \frac{q_n(\xi)}{x_n!} t^n , \tag{9.4}$$

can be expressed as

$$F(\xi; t) = e^{\sum_{n=1}^{\infty} a_n t^n} \quad \text{with} \quad a_1 = 1 \,, \; a_n = a_n(\xi) \geq 0 \,, \; \sum_{n=1}^{\infty} a_n < \infty \,. \tag{9.5}$$

It is proved in [51] that conditions of normalization (a) and non-negativeness (b) on $\mathfrak{p}_k^{(n)}(\xi)$ are satisfied. We now define

$$f_n = \int_0^{\infty} q_n(\xi)\, e^{-\xi}\, \mathrm{d}\xi \quad \text{and} \quad b_{m,n} = \int_0^1 q_m(\xi)\, q_n(1-\xi)\, \mathrm{d}\xi \,. \tag{9.6}$$

The f_n and $b_{m,n}$ are deformations of the usual factorial and beta function, respectively, deduced from their usual integral definitions through the substitution $\xi^n \mapsto q_n(\xi)$. The following properties are proven in [51]:

$$\begin{aligned} & q_n(\xi) \geq 0 \; \forall \xi \in \mathbb{R}^+ , \quad x_n! \leq f_n , \\ & \sum_{n=0}^{\infty} \frac{q_n(\xi)}{f_n} < \infty \; \forall \xi \in \mathbb{R}^+ , \quad \text{and} \quad b_{m,n} \geq \frac{x_m! x_n!}{(m+n+1)!} \,. \end{aligned} \tag{9.7}$$

Then let us introduce the function $\mathcal{N}(z)$ defined on $\mathbb{C}$ as

$$\forall z \in \mathbb{C} \quad \mathcal{N}(z) = \sum_{n=0}^{\infty} \frac{q_n(z)}{f_n} \,. \tag{9.8}$$

This definition makes sense since from Eq. (9.7)

$$\sum_{n=0}^{\infty} \left| \frac{q_n(z)}{f_n} \right| \leq \sum_{n=0}^{\infty} \frac{q_n(|z|)}{f_n} < \infty. \tag{9.9}$$

The above material allows us to present below two new generalizations of standard and spin coherent states.

9.1 DFB Coherent States on the Complex Plane

They are defined in the Fock space as

$$|\alpha\rangle_{\text{dfb}} = \frac{1}{\sqrt{\mathcal{N}(|\alpha|^2)}} \sum_{n=0}^{\infty} \frac{1}{\sqrt{f_n}} \sqrt{q_n(|\alpha|^2)}\, e^{\mathrm{i}\, n \arg(\alpha)} |n\rangle\,. \tag{9.10}$$

These states verify the following resolution of the unity:

$$\int_{\mathbb{C}} \frac{\mathrm{d}^2\alpha}{\pi} e^{-|\alpha|^2} \mathcal{N}(|\alpha|^2)\, |\alpha\rangle_{\text{dfb}\,\text{dfb}}\langle\alpha| = I\,. \tag{9.11}$$

They are a natural generalization of the standard coherent states that correspond to the special polynomials $q_n(\xi) = \xi^n$. The latter are associated to the generating function $F(t) = e^t$ that gives the usual binomial distribution.

9.2 DFB Spin Coherent States

These states can be considered as generalizing the spin coherent states (6.2)

$$|\alpha; n_j\rangle_{\text{dfb}} = \frac{1}{\sqrt{\mathcal{N}(|\alpha|^2)}} \sum_{n=0}^{n_j} \sqrt{\frac{q_n\left(\frac{1}{1+|\alpha|^2}\right) q_{n_j-n}\left(\frac{|\alpha|^2}{1+|\alpha|^2}\right)}{b_{n,n_j-n}}}\, e^{\mathrm{i}\arg(\alpha)} |n\rangle\,, \tag{9.12}$$

where the $b_{m,n}$ are defined in Eq. (9.6) and $\mathcal{N}(u)$ is given by

$$\mathcal{N}(u) = \sum_{n=0}^{n_j} \frac{q_n\left(\frac{1}{1+u}\right) q_{n_j-n}\left(\frac{u}{1+u}\right)}{b_{n,n_j-n}}\,. \tag{9.13}$$

The family of states (9.12) resolves the unity:

$$\int_{\mathbb{C}} \mathrm{d}^2\alpha\, \varpi(\alpha)\, |\alpha; n_j\rangle_{\text{dfb}\,\text{dfb}}\langle\alpha; n_j| = I\,, \quad \varpi(\alpha) = \frac{\mathcal{N}\left(|\alpha|^2\right)}{\pi\left(1+|\alpha|^2\right)^2}\,. \tag{9.14}$$

10 Photon Counting: Basic Statistical Aspects

In this section, we mainly follow the inspiring chapter 5 of Ref. [38] (see also the seminal papers [55–57] on the topic, the renowned [58], the pedagogical [59], and the more recent [60–62]). In quantum optics one views a beam of light as a stream of discrete energy packets named "photons" rather than a classical wave. With a photon counter the average count rate is determined by the intensity of the light beam, but the actual count rate fluctuates from measurement to measurement. Whence, one easily understands that two statistics are in competition here, on one hand the statistical nature of the photodetection process, and on the other hand, the intrinsic photon statistics of the light beam, e.g., the average $\bar{n}(\alpha)$ for a CS $|\alpha\rangle$. Photon-counting detectors are specified by their quantum efficiency η, which is defined as the ratio of the number of photocounts to the number of incident photons. For a perfectly coherent monochromatic beam of angular frequency ω, constant intensity I, and area A, and for a counting time T

$$\eta = \frac{N(T)}{\Phi T}\,, \tag{10.1}$$

where the photon flux is $\Phi = \dfrac{IA}{\hbar\omega} \equiv \dfrac{P}{\hbar\omega}$, P being the power. Thus the corresponding count rate is $\mathcal{R} = \dfrac{\eta P}{\hbar\omega}$ counts s^{-1}. Due to a "dead time" of $\sim 1\,\mu$s for the detector reaction, the count rate cannot be larger than $\sim 10^6$ counts s^{-1}, and due to weak values $\eta \sim 10\%$ for standard detectors, photon counters are only useful for analyzing properties of very faint beams with optical powers of $\sim 10^{-12}$W or less. The detection of light beams with higher powers requires other methods.

Although the average photon flux can have a well-defined value, the photon number on short time-scales fluctuates due to the discrete nature of the photons. These fluctuations are described by the photon statistics of the light.

One proves that the photon statistics for a coherent light wave with constant intensity (e.g., a light beam described by the electric field $\mathcal{E}(x,t) = \mathcal{E}_0 \sin(kx - \omega t + \phi)$ with constant angular frequency ω, phase ϕ, and intensity $\mathcal{E}_0$) is encoded by the Poisson distribution

$$n \mapsto \mathsf{P}_n(\bar{n}) = e^{-\bar{n}}\,\frac{(\bar{n})^n}{n!}\,, \tag{10.2}$$

This randomness of the count rate of a photon-counting system detecting individual photons from a light beam with constant intensity originates from chopping the continuous beam into discrete energy packets with an equal probability of finding the energy packet within any given time subinterval.

Let us introduce the variance as the quantity

$$\mathrm{Var}_n(\bar{n}) \equiv (\Delta n)^2 = \sum_{n=0}^{\infty} (n-\bar{n})^2 \mathsf{P}_n(\bar{n})\,.$$

Thus, for a Poissonian coherent beam, $\Delta n = \sqrt{\bar{n}}$. There results that three different types of photon statistics can occur: Poissonian, super-Poissonian, and sub-Poissonian. The two first ones are consistent as well with the classical theory of light, whereas sub-Poissonian statistics is not and constitutes direct confirmation of the photon nature of light. More precisely

(i) if the Poissonian statistics holds, e.g., for a perfectly coherent light beam with constant optical power P, we have

$$\Delta n = \sqrt{\bar{n}}\,, \tag{10.3}$$

(ii) if the super-Poissonian statistics, e.g., classical light beams with time-varying light intensities, like thermal light from a black-body source, or like partially coherent light from a discharge lamp, we have

$$\Delta n > \sqrt{\bar{n}}\,, \tag{10.4}$$

(iii) finally, the sub-Poissonian statistics is featured by a narrower distribution than the Poissonian case

$$\Delta n < \sqrt{\bar{n}}\,. \tag{10.5}$$

This light is "quieter" than the perfectly coherent light. Since a perfectly coherent beam is the most stable form of light that can be envisaged in classical optics, sub-Poissonian light has no classical counterpart.

In this context popular useful parameters are introduced to account for CS statistical properties, e.g., the Mandel parameter $Q = (\Delta n)^2/\bar{n} - 1$, where $(\Delta n)^2 = \overline{n^2} - \bar{n}^2$, which is <0 (resp. >0, $=0$) for sub-Poissonian (resp. super-Poissonian, Poissonian), the parameter $Q/\bar{n} + 1$ which is >1 for "bunching" CS and <1 for "anti-bunching" CS, etc.

The aim of the quantum theory of photodetection is to relate the photocount statistics observed in a particular experiment to those of the incoming photons, more precisely the average photocount number $\bar{N}$ to the mean photon number $\bar{n}$ incident on the detector in a same time interval. The quantum efficiency η of the detector, defined as $\eta = \bar{N}/\bar{n}$ is the critical parameter that determines the relationship between the photoelectron and photon statistics. Indeed, consider the relation between variances $(\Delta N)^2 = \eta^2\,(\Delta n)^2 + \eta\,(1-\eta)\,\bar{n}$.

- If $\eta = 1$, we have $\Delta N = \Delta n$: the photocount fluctuations faithfully reproduce the fluctuations of the incident photon stream.
- If the incident light has Poissonian statistics $\Delta n = \sqrt{\bar{n}}$, then $(\Delta N)^2 = \eta\,\bar{n}$ for all values of η: photocount is Poisson.
- If $\eta \ll 1$, the photocount fluctuations tend to the Poissonian result with $(\Delta N)^2 = \eta\,\bar{n} = \bar{N}$ irrespective of the underlying photon statistics.

Observing sub-Poissonian statistics in the laboratory is a delicate matter since it depends on the availability of single-photon detectors with high quantum efficiencies.

11 AN CS Quantization

11.1 The Quantization Map and Its Complementary

If the resolution of the identity (3.4) is valid for a given family of AN CS determined by the sequence of functions $\mathsf{h} := (h_n(u))$, it makes the quantization of functions (or distributions) $f(\alpha)$ possible along the linear map

$$f(\alpha) \mapsto A^{\mathsf{h}}_f = \int_{|\alpha|<R} \frac{d^2\alpha}{\pi}\, w(|\alpha|^2)\, f(\alpha)\, |\alpha\rangle\langle\alpha| \,, \tag{11.1}$$

together with its complementary map, likely to provide a "semi-classical" optical phase space portrait, or *lower symbol*, of A^{h}_f through the map (3.8)

$$\langle\alpha|A^{\mathsf{h}}_f|\alpha\rangle = \int_{|\beta|<R} \frac{d^2\beta}{\pi}\, w(|\beta|^2)\, f(\beta)\, |\langle\alpha|\beta\rangle|^2 \equiv \breve{f}^{\mathsf{h}}(\alpha) \,. \tag{11.2}$$

Since for fixed α the map $\beta \mapsto w(|\beta|^2)\, |\langle\alpha|\beta\rangle|^2$ is a probability distribution on the centered disk $\mathcal{D}_R$ of radius R, the map $f(\alpha) \mapsto \breve{f}^{\mathsf{h}}(\alpha)$ is a local, generally regularizing, averaging, of the original f.

The quantization map (11.1) can be extended to cases comprising geometric constraints in the optical phase portrait through the map (3.8), and encoded by distributions like Dirac or Heaviside functions.

11.2 AN CS Quantization of Simple Functions

When applied to the simplest functions α and $\bar{\alpha}$ weighted by a positive $\mathfrak{n}\left(|\alpha|^2\right)$, the quantization map (11.1) yields lowering and raising operators

$$\alpha \mapsto a^{\mathsf{h}} = \int_{|\alpha|<R} \frac{d^2\alpha}{\pi}\, \tilde{w}(|\alpha|^2)\, \alpha\, |\alpha\rangle\langle\alpha| = \sum_{n=1}^{\infty} a^{\mathsf{h}}_{n-1n} |n-1\rangle\langle n| \,, \tag{11.3}$$

$$\bar{\alpha} \mapsto \left(a^{\mathsf{h}}\right)^{\dagger} = \sum_{n=0}^{\infty} \overline{a^{\mathsf{h}}_{nn+1}} |n+1\rangle\langle n| \,, \tag{11.4}$$

where $\tilde{w}(u) := \mathfrak{n}(u) w(u)$. Their matrix elements are given by the integrals

$$a^{\mathsf{h}}_{n-1n} := \int_0^{R^2} du\, \tilde{w}(u)\, u^n\, h_{n-1}(u)\, \overline{h_n(u)} \,, \tag{11.5}$$

and $a^{\mathsf{h}}|0\rangle = 0$.

The lower symbol of a^{h} and its adjoint read, respectively:

$$\check{a^{\mathsf{h}}}(\alpha) = \langle\alpha|a^{\mathsf{h}}|\alpha\rangle = \alpha\,\tau\left(|\alpha|^2\right)\,, \quad \check{\left(a^{\mathsf{h}}\right)^{\dagger}}(\alpha) = \overline{\check{a^{\mathsf{h}}}(\alpha)}\,, \tag{11.6}$$

in which the “weighting” factor is given by $\tau(u) = \sum_{n\geq 0} a^{\mathsf{h}}_{nn+1}\, u^n\, \overline{h_n(u)}\, h_{n+1}(u)$.

In the above, as it was mentioned in Sect. 3 and, as it occurred in the spin case, the involved sums can be finite, and a finite number of matrix elements (11.5) are not zero. As a generalization of the number operator we get in the present case

$$a^{\mathsf{h}}\left(a^{\mathsf{h}}\right)^{\dagger} = \mathsf{X}^{\mathsf{h}}_{\hat{N}+I}\,, \quad \left(a^{\mathsf{h}}\right)^{\dagger} a = \mathsf{X}^{\mathsf{h}}_{\hat{N}}\,, \quad \left[a^{\mathsf{h}}, \left(a^{\mathsf{h}}\right)^{\dagger}\right] = \mathsf{X}^{\mathsf{h}}_{\hat{N}+I} - \mathsf{X}^{\mathsf{h}}_{\hat{N}}\,, \tag{11.7}$$

with the notations

$$\mathsf{X}^{\mathsf{h}}_n = |a^{\mathsf{h}}_{n-1n}|^2\,, \quad \mathsf{X}^{\mathsf{h}}_0 = 0\,, \quad \mathsf{X}^{\mathsf{h}}_{\hat{N}}|n\rangle = \mathsf{X}^{\mathsf{h}}_n|n\rangle\,, \quad \mathsf{X}^{\mathsf{h}}_{\hat{N}+I}|n\rangle = \mathsf{X}^{\mathsf{h}}_{n+1}|n\rangle\,. \tag{11.8}$$

When all the h_n's are real, the diagonal elements in (11.7) are given by the product of integrals

$$\begin{aligned} \mathsf{X}^{\mathsf{h}}_{n+1} - \mathsf{X}^{\mathsf{h}}_n = &\left[\int_0^{R^2} \mathrm{d}u\, \tilde{w}(u)\, u^n\, h_n(u)\, \left(u h_{n+1}(u) - h_{n-1}(u)\right)\right] \\ &\times \left[\int_0^{R^2} \mathrm{d}u\, \tilde{w}(u)\, u^n\, h_n(u)\, \left(u h_{n+1}(u) + h_{n-1}(u)\right)\right]. \end{aligned} \tag{11.9}$$

The quantum version of $u = |\alpha|^2$ and its lower symbol read as

$$\begin{aligned} A^{\mathsf{h}}_u &= \sum_n \langle u\rangle_n |n\rangle\langle n|\,, \quad \langle u\rangle_n := \int_0^{R^2} \mathrm{d}u\, \tilde{w}(u)\, u^{n+1}\, h_n(u) \\ \langle\alpha|A^{\mathsf{h}}_u|\alpha\rangle &= \langle\langle u\rangle_n\rangle_\alpha\,(u) := \sum_n \langle u\rangle_n\, u^n\, |h_n(u)|^2 = \sum_n \langle u\rangle_n\, \mathsf{P}^{\mathsf{h}}_n\,. \end{aligned} \tag{11.10}$$

We notice here an interesting duality between classical $(\langle\cdot\rangle_n)$ and quantum $(\langle\cdot\rangle_\alpha)$ statistical averages.

11.3 AN CS as a-Eigenstates

One crucial property of the Glauber–Sudarshan CS is that they are eigenstates of the lowering operator a. Imposing this property to AN CS leads to a supplementary condition on the functions h_n.

$$a^{\mathsf{h}}|\alpha\rangle = \alpha|\alpha\rangle \Rightarrow h_n(u) = h_{n+1}(u) \int_0^{R^2} \mathrm{d}t\, \tilde{w}(t)\, t^{n+1}\, h_n(t)\, \overline{h_{n+1}(t)}\,. \tag{11.11}$$

Let us examine the particular case of non-linear CS of the deformed Poissonian type (5.6). In this case, $\mathsf{X}_n = x_n$, and whence the construction formula

$$|\alpha\rangle = \frac{\mathcal{N}(\alpha a^{\mathsf{h}\dagger})}{\sqrt{\mathcal{N}(|\alpha|^2)}}|0\rangle\,. \tag{11.12}$$

Moreover (11.11) imposes that the sequence $x_n!$ derives from the following moment problem:

$$x_n! = \int_0^{R^2} \mathrm{d}u\, \frac{w(u)}{\mathcal{N}(u)}\, u^n\,. \tag{11.13}$$

Now, instead of starting from a known sequence (x_n), one can reverse the game by choosing a suitable function $f(u) = \dfrac{w(u)}{\mathcal{N}(u)}$ to calculate the corresponding $x_n!$ (from which we deduce the x_n's), the resulting generalized exponential $\mathcal{N}(u)$ (and checking the finiteness of the convergence radius), and eventually the weight function $w(u) = f(u)\,\mathcal{N}(u)$. There are an infinity of "manufactured" products in this non-linear CS factory!

11.4 AN CS from Displacement Operator

One can attempt to build (other?) AN CS by following the standard procedure involving the unitary "displacement" operator built from a^{h} and $a^{\mathsf{h}\dagger}$ and acting on the vacuum

$$|\breve{\alpha}\rangle_{\mathrm{disp}} := D_{\mathsf{h}}(\breve{\alpha})\,|0\rangle = \sum_{n=0}^{\infty} \breve{\alpha}^n\, h_n^{\mathrm{disp}}(|\breve{\alpha}|^2)\,|n\rangle\,, \quad D_{\mathsf{h}}(\breve{\alpha}) := e^{\breve{\alpha} a^{\mathsf{h}\dagger} - \bar{\breve{\alpha}} a^{\mathsf{h}}}\,, \tag{11.14}$$

where the notation $\breve{\alpha}$ is used to make the distinction from the original α. Of course, $D_{\mathsf{h}}{}^{\dagger}(\breve{\alpha}) = D_{\mathsf{h}}{}^{-1}(\breve{\alpha})$ is not equal in general to $D_{\mathsf{h}}(-\breve{\alpha})$. Besides the two examples (6.17) and (7.9) encountered in the SU(2) and SU(1, 1) CS constructions, for which the respective weights $\mathfrak{n}(u)$ can be given explicitly, another recent interesting example is given in [63].

So an appealing program is to establish the relation between the original h_n's and these (new?) h_n^{disp}'s, through a suitable choice of the weight $\mathfrak{n}(u)$, actually a big challenge in the general case! More interesting yet is the fact that these new CS's might be experimentally produced in the Glauber's way (4.23), once we accept that the a^{h} and $a^{\mathsf{h}\dagger}$ appearing in the quantum version (4.8) of the classical e.m. field are yielded by a CS quantization different from the historical Dirac (canonical) one [64]. Hence one introduces a kind of duality between two families of coherent states, the first one used in the quantization procedure $f(\alpha) \mapsto A^{\mathsf{h}}_f$, producing the operators $\mathfrak{n}(u)\alpha \mapsto a^{\mathsf{h}}$ and $\mathfrak{n}(u)\bar{\alpha} \mapsto a^{\mathsf{h}\dagger}$, and so the unitary displacement $D^{\mathsf{h}}(\breve{\alpha}) := e^{\breve{\alpha}a^{\mathsf{h}\dagger} - \bar{\breve{\alpha}}a^{\mathsf{h}}}$, while the other one uses this $D_{\mathsf{h}}(\breve{\alpha})$ to build potentially experimental CS yielded in the Glauber's way.

12 Conclusion

We have presented in this paper a unifying approach to build coherent states in a wide sense that are potentially relevant to quantum optics. Of course, for most of them, their experimental observation or production comes close to being impossible with the current experimental physics. Nevertheless, when one considers the way quantum optics has emerged from the golden 1920s of quantum mechanics, nothing prevents us to enlarge the Dirac quantization of the classical e.m. field in order to include all these deformations (non-linear or others) by adopting the consistent method exposed in the previous section.

Acknowledgements This research is supported in part by the Ministerio de Economía y Competitividad of Spain under grant MTM2014-57129-C2-1-P and the Junta de Castilla y León (grant VA137G18). The author is also indebted to the University of Valladolid. He thanks M. del Olmo (UVA) for helpful discussions about this review. He addresses special thanks to Y. Hassoumi (Rabat University) and to the Organizers of the Workshop QIQE'2018 in Al-Hoceima, Morocco, for valuable comments and questions which allowed to improve significantly the content of this review.

References

1. A.H. El Kinani, M. Daoud, Generalized intelligent states for an arbitrary quantum system. J. Phys. A Math. Gen. **34**, 5373–5387 (2001)
2. E.E. Hach III, P.M. Alsing, C.C. Gerry, Violations of a Bell inequality for entangled SU(1, 1) coherent states based on dichotomic observables. Phys. Rev. A **93**, 042104-1–042104-8 (2016)
3. S. Cruz y Cruz, Z. Gress, Group approach to the paraxial propagation of Hermite-Gaussian modes in a parabolic medium. Ann. Phys. **383**, 257–277 (2017)
4. S.E. Hoffmann, V. Hussin, I. Marquette, Y.-Z. Zhang, Non-classical behaviour of coherent states for systems constructed using exceptional orthogonal polynomials. J. Phys. A Math. Theor. **51**, 085202-1–085202-16 (2018)
5. K. Górska, A. Horzela, F.H. Szafraniec, Coherence, squeezing and entanglement: an example of peaceful coexistence, in J.-P. Antoine, F. Bagarello, J.P. Gazeau, eds. Coherent States and their applications: a contemporary panorama, in *Proceedings of the CIRM Workshop*, 13–18 Nov 2016. Springer Proceedings in Physics (SPPHY), vol. **205** (2018), pp. 89–117

6. E.E. Hach, R. Birrittella, P.M. Alsing, C.C. Gerry, SU(1, 1) parity and strong violations of a Bell inequality by entangled Barut-Girardello coherent states. J. Opt. Soc. Am. B **35**, 2433–2442 (2018)
7. R.J. Glauber, Photons correlations. Phys. Rev. Lett. **10**, 84–86 (1963)
8. J.-P. Gazeau, F.H. Szafraniec, Holomorphic Hermite polynomials and a non-commutative plane. J. Phys. A Math. Theor. **44**, 495201-1–495201-13 (2011)
9. W. Magnus, F. Oberhettinger, R.P. Soni, *Formulas and Theorems for the Special Functions of Mathematical Physics* (Springer, Berlin, 1966)
10. J. Schwinger, The theory of quantized fields. III. Phys. Rev. **91**, 728–740 (1953)
11. R.J. Glauber, The quantum theory of optical coherence. Phys. Rev. **130**, 2529–2539 (1963)
12. R.J. Glauber, Coherent and incoherent states of radiation field. Phys. Rev. **131**, 2766–2788 (1963)
13. E.C.G. Sudarshan, Equivalence of semiclassical and quantum mechanical descriptions of statistical light beams. Phys. Rev. Lett. **10**, 277–279 (1963)
14. L. Mandel, E. Wolf, Coherence properties of optical fields. Rev. Mod. Phys. **37**, 231–287 (1965)
15. K.E. Cahill, R.J. Glauber, Ordered expansions in Boson amplitude operators. Phys. Rev. **177**, 1857–1881 (1969)
16. B.S. Agarwal, E. Wolf, Calculus for functions of noncommuting operators and general phase-space methods in quantum mechanics. Phys. Rev. D **2**, 2161–2186 (I), 2187–2205 (II), 2206–2225 (III) (1970)
17. E. Schrödinger, Der stetige Übergang von der Mikro- zur Makromechanik. Naturwiss **14**, 664 (1926)
18. J.R. Klauder, The action option and the Feynman quantization of spinor fields in terms of ordinary c-numbers. Ann. Phys. **11**, 123 (1960)
19. J.R. Klauder, Continuous-representation theory I. Postulates of continuous-representation theory. J. Math. Phys. **4**, 1055–1058 (1963)
20. J.R. Klauder, Continuous-representation theory II. Generalized relation between quantum and classical dynamics. J. Math. Phys. **4**, 1058–1073 (1963)
21. J.R. Klauder, B.S. Skagerstam (ed.), *Coherent States. Applications in Physics and Mathematical Physics* (World Scientific, Singapore, 1985)
22. A.M. Perelomov, Coherent states for arbitrary lie group. Commun. Math. Phys. **26**, 222–236 (1972)
23. A.M. Perelomov, *Generalized Coherent States and Their Applications* (Springer, Berlin, 1986)
24. W.-M. Zhang, D.H. Feng, R. Gilmore, Coherent states: theory and some applications. Rev. Mod. Phys. **26**, 867–927 (1990)
25. D.H. Feng, J.R. Klauder, M. Strayer (ed.) Coherent States: Past, Present and Future, in *Proceedings of the 1993 Oak Ridge Conference* (World Scientific, Singapore, 1994)
26. S.T. Ali, J.-P Antoine, J.-P. Gazeau, *Coherent States, Wavelets and their Generalizations* (2000), 2d edn., Theoretical and Mathematical Physics (Springer, New York, 2014)
27. V.V. Dodonov, 'Nonclassical' states in quantum optics: a 'squeezed' review of the first 75 years. J. Opt. B Quantum Semiclass. Opt. **4**, R1 (2002)
28. V.V. Dodonov, V.I. Man'ko (ed.), *Theory of Nonclassical States of Light* (Taylor & Francis, London, 2003)
29. A. Vourdas, Analytic representations in quantum mechanics. J. Phys. A **39**, R65 (2006)
30. J.-P. Gazeau, *Coherent States in Quantum Physics* (Wiley-VCH, Berlin, 2009)
31. S.T. Ali, J.P. Antoine, F. Bagarello, J.P. Gazeau, Special issue on coherent states: mathematical and physical aspects. J. Phys. A Math. Theor. **45** (2012)
32. J.-P. Antoine, F. Bagarello, J.P. Gazeau, Coherent States and their applications: a contemporary panorama, in *Proceedings of the CIRM Workshop*, 13–18 Nov 2016. Springer Proceedings in Physics (SPPHY), vol. **205** (2018)
33. N. Cotfas, J.-P. Gazeau, K. Górska, Complex and real Hermite polynomials and related quantizations. J. Phys. A Math. Theor. **43**, 305304-1–305304-14 (2010)

34. S.T. Ali, F. Bagarello, J.-P. Gazeau, Quantizations from reproducing kernel spaces. Ann. Phys. **332**, 127–142 (2012)
35. J.-P. Gazeau, M.A. del Olmo, Pisot q-coherent states quantization of the harmonic oscillator. Ann. Phys. **330**, 220–245 (2013)
36. A. De Sole, V. Kac, On integral representations of q-gamma and q-beta functions. Rend. Mat. Acc. Lincei **9**, 11–29 (2005). ArXiv: math.QA/0302032
37. M. El Baz, R. Fresneda, J.-P. Gazeau, Y. Hassouni, Coherent state quantization of paragrassmann algebras. J. Phys. A Math. Theor. **43**, 385202-1–385202-15 (2010); Corrigendum J. Phys. A Math. Theor. **45**, 079501-1–079501-2 (2012)
38. M. Fox, *Quantum Optics: An Introduction* (Oxford University, New York, 2006)
39. S.T. Ali, J.-P. Gazeau, B. Heller, Coherent states and Bayesian duality. J. Phys. A Math. Theor. **41**, 365302-1–365302-22 (2008)
40. J.-P. Gazeau, E. Huguet, M. Lachièze-Rey, J. Renaud, Fuzzy spheres from inequivalent coherent states quantizations. J. Phys. A Math. Theor. **40**, 10225–10249 (2007)
41. P. Jordan, Der Zusammenhang der symmetrischen und linearen Gruppen und das Mehrkörperproblem". Z. Phys. **94**, 531–535 (1935)
42. T. Holstein, H. Primakoff, Phys. Rev. **58**, 1098–1113 (1940)
43. J. Schwinger, On Angular Momentum, Unpublished Report, Harvard University, Nuclear Development Associates, Inc., United States Department of Energy (through predecessor agency the Atomic Energy Commission), Report Number NYO-3071 (1952).
44. J.-P. Gazeau, M. del Olmo, Covariant integral quantization of the unit disk, submitted (2018). ArXiv:1810.10399 [math-ph]
45. Y. Aharonov, E.C. Lerner, H.W. Huang, J.M. Knight, Oscillator phase states, thermal equilibrium and group representations. J. Math. Phys. **14**, 746–755 (2011)
46. A. O. Barut, L. Girardello, New "Coherent" states associated with non-compact groups. Commun. Math. Phys. **21**, 41–55 (1971)
47. J.-P. Antoine, J.-P. Gazeau, J.R. Klauder, P. Monceau, K.A. Penson, J. Math. Phys. **42**, 2349–2387 (2001)
48. L. Susskind, J. Glogower, Quantum mechanical phase and time operator. Phys. Phys. Fiz. 1 **1**, 49–61 (1964)
49. H.M. Moya-Cessa, F. Soto-Eguibar, *Introduction to Quantum Optics* (Rinton, Paramus, 2011)
50. E.M.F. Curado, S. Faci, J.-P. Gazeau, D. Noguera, *in progress*.
51. H. Bergeron, E.M.F. Curado, J.-P. Gazeau, Ligia M.C.S. Rodrigues, Symmetric generalized binomial distributions. J. Math. Phys. **54**, 123301-1–123301-22 (2013)
52. E.M.F. Curado, J.-P. Gazeau, Ligia M.C.S. Rodrigues, Nonlinear coherent states for optimizing quantum information. Phys. Scr. **82**, 038108-1–038108-9 (2010)
53. E.M.F. Curado, J.-P. Gazeau, Ligia M.C.S. Rodrigues, On a generalization of the binomial distribution and its Poisson-like limit. J. Stat. Phys. **146**, 264–280 (2012)
54. H. Bergeron, E.M.F. Curado, J.-P. Gazeau, Ligia M.C.S. Rodrigues, Generating functions for generalized binomial distributions. J. Math. Phys. **53**, 103304-1–103304-22 (2012)
55. L. Mandel, Fluctuations of photons beams and their correlations. Proc. Phys. Soc. (London) **72**, 1037–1048 (1958); Fluctuations of photon beams: the distribution of photoelectrons. Proc. Phys. Soc. **74**, 233–243 (1959)
56. L. Mandel, E. Wolf, *Selected Papers on Coherence and Fluctuations of Light*, vols. 1, 2 (Dover, New York, 1970)
57. D.N. Klyshko, Observable signs of nonclassical light. Phys. Lett. A **213**, 7–15 (1996)
58. R. Loudon, *The Quantum Theory of Light*, 3rd edn. (Oxford University, Oxford 2000)
59. P. Koczyk, P. Wiewior, C. Radzewicz, Photon counting statistics - undergraduate experiment. Am. J. Phys. **64**(1996), 240–245 (1996)
60. C. Gerry, P. Knight, *Introductory Quantum Optics* (Cambridge University, Cambridge, 2004)
61. H.A. Bachor, T.C. Ralph, *A Guide to Experiments in Quantum Optics* (Wiley-VCH, Weinheim, 2004)

62. M.D. Eisaman, J. Fan, A. Migdall, S.V. Polyakov, Single-photon sources and detectors (Invited Review Article). Rev. Sci. Instrum. **82**, 071101-25 (2011)
63. C. Huerta Alderete, Liliana Villanueva Vergara, B.M. Rodríguez-Lara, Nonclassical and semiclassical para-Bose states. Phys. Rev. A **95**, 043835-1–043835-7 (2017)
64. P.A.M. Dirac, The quantum theory of emission and absorption of radiation. Proc. R. Soc. Lond. A **114**, 243–265 (1927)

Higher Order Quantum Superintegrability: A New "Painlevé Conjecture"

Higher Order Quantum Superintegrability

Ian Marquette and Pavel Winternitz

Abstract We review recent results on superintegrable quantum systems in a two-dimensional Euclidean space with the following properties. They are integrable because they allow the separation of variables in Cartesian coordinates and hence allow a specific integral of motion that is a second order polynomial in the momenta. Moreover, they are superintegrable because they allow an additional integral of order $N > 2$. Two types of such superintegrable potentials exist. The first type consists of "standard potentials" that satisfy linear differential equations. The second type consists of "exotic potentials" that satisfy nonlinear equations. For $N = 3$, 4, and 5 these equations have the Painlevé property. We conjecture that this is true for all $N \geq 3$. The two integrals X and Y commute with the Hamiltonian, but not with each other. Together they generate a polynomial algebra (for any N) of integrals of motion. We show how this algebra can be used to calculate the energy spectrum and the wave functions.

Keywords Superintegrable systems · Painlevé transcendents · Polynomial algebras · Exact solvability · Higher order integrals · Chazy class

PACS 03.65.Fd

I. Marquette (✉)
School of Mathematics and Physics, The University of Queensland, Brisbane, St-Lucia, QLD, Australia
e-mail: i.marquette@uq.edu.au

P. Winternitz
Centre de recherches mathématiques et Département de Mathématiques et de Statistique, Université de Montréal, Montréal, QC, Canada
e-mail: wintern@CRM.UMontreal.CA

Ş. Kuru et al. (eds.), *Integrability, Supersymmetry and Coherent States*, CRM Series in Mathematical Physics, https://doi.org/10.1007/978-3-030-20087-9_4

1 Introduction

Let us first consider a classical system in an n-dimensional Riemannian space with Hamiltonian

$$H = \sum_{\substack{i,k=1 \\ j,k\geq 0}}^{n} g_{ik}(\mathbf{x}) p_i p_k + V(\mathbf{x}) \,, \mathbf{x} \in \mathbb{R}^n . \tag{1}$$

The system is called *integrable* (or Liouville integrable) if it allows $n-1$ Poisson commuting integrals of motion (in addition to H)

$$\begin{aligned} X_a &= f_a(\mathbf{x}, \mathbf{p}) \,, \qquad a = 1, \ldots, n-1, \\ \frac{dX_a}{dt} &= \{H, X_a\}_p = 0 \,, \{X_a, X_b\}_p = 0, \end{aligned} \tag{2}$$

where $\{,\}_p$ is the Poisson bracket, and p_i are the momenta canonically conjugate to the coordinates x_i.

This system is *superintegrable* if it allows further integrals

$$\begin{aligned} Y_b &= f_b(\mathbf{x}, \mathbf{p}) \,, \qquad b = 1, \ldots, k \qquad 1 \leq k \leq n-1, \\ \frac{dY_b}{dt} &= \{H, Y_b\}_p = 0 \,. \end{aligned} \tag{3}$$

In addition, the integrals must satisfy the following requirements:

1. The integrals H, X_a, Y_b are well-defined functions on phase space, i.e., polynomials or convergent power series on phase space (or an open submanifold of phase space).
2. The integrals H, X_a are in involution, i.e., Poisson commute as indicated in (2). The integrals Y_b Poisson commute with H but not necessarily with each other nor with X_a.
3. The entire set of integrals is functionally independent, i.e., the Jacobian matrix satisfies

$$\text{rank} \frac{\partial(H, X_1, \ldots, X_{n-1}, Y_1, \ldots, Y_k)}{\partial(x_1, \ldots, x_n, p_1, \ldots, p_n)} = n + k. \tag{4}$$

In quantum mechanics we define integrability and superintegrability in the same way; however, in this case, H, X_a, and Y_b are operators. The condition on the integrals of motion must also be modified, e.g., as follows:

1. H, X_a, and Y_b are well-defined Hermitian operators in the enveloping algebra of the Heisenberg algebra $H_n \sim \{\mathbf{x}, \mathbf{p}, \hbar\}$ or some generalization thereof.
2. The integrals satisfy the Lie bracket relations

$$[H, X_a] = [H, Y_b] = 0 \,, [X_i, X_k] = 0. \tag{5}$$

3. No polynomial in the operators H, X_a, Y_b formed entirely using Lie anticommutators (i.e., Jordan polynomials) should vanish identically.

The two best known superintegrable systems are the Kepler-Coulomb system with potential $V(r) = \frac{\alpha}{r}$ and the isotropic harmonic oscillator $V(r) = \alpha r^2$ [10, 32, 75, 84]. In both cases the integrals X_a correspond to angular momentum, the additional integrals Y_a to the Laplace–Runge–Lenz vector for $V(r) = \frac{\alpha}{r}$ and to the quadrupole tensor $T_{ik} = p_i p_k + \alpha x_i x_k$, for $V(r) = \beta r^2$. No further ones were discovered until a 1940 paper by Jauch and Hill [49] on the rational anisotropic harmonic oscillator $V(\mathbf{x}) = \alpha \sum_{i=1}^{n} n_i x_i^2$, $n_i \in \mathbb{Z}$. A systematic search for superintegrable systems was started in 1965 [33, 58, 100] and a real proliferation of them was observed during the last few years [74]. This research program remains very active [8, 16, 17, 25, 27–31, 45–47, 54, 57, 59–61, 78, 79, 85, 87, 90]. The search has also been extended to systems with spin, magnetic fields, and monopoles. Many families of superintegrable systems have been constructed using combinations of approaches such as the co-algebra [6, 7, 91] and the recurrence method [9, 52, 64, 66, 74]. Let us just list some of the reasons why superintegrable systems are interesting both in classical and quantum physics.

1. In classical mechanics, superintegrability restricts trajectories to an $n-k$ dimensional subspace of phase space [76]. For $k = n-1$ (maximal superintegrability), this implies that all finite trajectories are closed and motion is periodic.
2. Moreover, at least in principle, the trajectories can be calculated without any calculus.
3. Bertrand's theorem states that the only spherically symmetric potentials $V(r)$ for which all bounded trajectories are closed are $\dfrac{\alpha}{r}$ and αr^2 [11, 38]; hence, no other maximally superintegrable systems are spherically symmetric.
4. The algebra of integrals of motion $\{H, X_a, Y_b\}$ is a non-Abelian and interesting one. Usually it is a finitely generated polynomial algebra, only exceptionally a finite dimensional Lie algebra. In the special case of quadratic superintegrability (all integrals of motion are at most quadratic polynomials in the moment), second order integrability is related to separation of variables in the Hamilton–Jacobi equation.

In quantum mechanics,

1. Superintegrability leads to an additional degeneracy of energy levels, sometimes called "accidental degeneracy." The term was coined by Fock [32] and used by Moshinsky and Smirnov [75], though the point of their studies was to show that this degeneracy is certainly no accident. Quadratic integrability is related to separation of variables to the corresponding Schrodinger equation. Quadratic superintegrability implies multiseparability of the Schrodinger equation.
2. A conjecture, born out by all known examples, is that all maximally superintegrable systems are exactly solvable [94]. If the conjecture is true, then the energy levels can be calculated algebraically. The wave functions are polynomials (in appropriately chosen variables) multiplied by some overall gauge factor.

3. The non-Abelian polynomial algebra of integrals of motion has been obtained for various models [12, 26–28, 35–37, 41, 45–47, 54, 56, 57]. In many cases rank polynomial algebras. They provide energy spectra and information on wave functions via Casimir operators and representation theory. Moreover, it has been demonstrated how Inonu–Wigner and more generally Bocher contractions of quadratic algebras play a role [28, 54] in connecting all quadratically superintegrable models in conformally flat spaces. Interesting relations exist between superintegrability and supersymmetry in quantum mechanics [64] and even more generally other types of operator algebras appear [71].
4. Relation to special function theory: multivariable orthogonal polynomials, new "nonclassical" orthogonal polynomials, Askey–Wilson classification [54, 98], and exceptional orthogonal polynomials [39, 40, 65, 89].

The theory of superintegrable systems has also been formulated in the context of Lie theory and generalized symmetries [77, 93]. As a comment, let us mention that superintegrability has also been called non-Abelian integrability. From this point of view, infinite dimensional integrable systems (soliton systems) described, e.g., by the Korteweg–de-Vries equation, the nonlinear Schrödinger equation, the Kadomtsev–Petviashvili equation, etc. are actually superintegrable [80–82]. Indeed, the generalized symmetries of these equations form infinite dimensional non-Abelian algebras (the Orlov–Shulman symmetries) with infinite dimensional Abelian subalgebras of commuting flows. There is another connection between superintegrable systems in quantum mechanics and soliton theory [1], namely the important role of the Painlevé property and Painlevé transcendents (of second and higher order) in both.

The paper is organized as follows: In Sect. 2, we present the case of second order superintegrable systems in two-dimensional Euclidean space. In Sect. 3, we present a summary of results for integrals of motion of order N in E_2. In Sect. 4, we review the case of $N = 4$ with exotic potentials and separable in Cartesian coordinates and present the connection with the Chazy class of equations. We present a summary of the classification of exotic potentials with fourth order integrals separable in Cartesian coordinates in Sect. 5. Section 6 is devoted to a discussion of the algebraic derivation of the spectrum using a cubic algebra. In Sect. 7 we discuss the connection with supersymmetric quantum mechanics.

2 Second Order Superintegrability

Let us consider the Hamiltonian (1) in the Euclidian space E_2 and search for second order integrals of motion [33, 74, 100]. We have

$$H = \frac{1}{2}\left(p_1^2 + p_2^2\right) + V(x_1, x_2), \qquad X = \sum_{j+k=0}^{2} \left\{ f_{jk}(x_1, x_2), p_1^j p_2^k \right\}, \quad (6)$$

where $j, k \in \mathbb{Z} \geq 0$ and $\{,\}$ is the anti-commutator. In the quantum case we have

$$p_j = -i\hbar \frac{\partial}{\partial x_j}, \qquad L_3 = x_1 p_2 - x_2 p_1. \tag{7}$$

The commutativity condition $[H, X] = 0$ implies that the even terms $j + k = 0, 2$ and odd terms $j + k = 1$ in X commute with H separately. Hence we can, with no loss of generality, set $f_{10} = f_{01} = 0$. Further we find that the leading (second order) term in X lies in the enveloping algebra of the Euclidian algebra $e(2)$. Thus we obtain

$$X = aL_3^2 + b_1(L_3 p_1 + p_1 L_3) + b_2(L_3 p_2 + p_2 L_3) + c_1(p_1^2 - p_2^2) \tag{8}$$
$$+2c_2 p_1 p_2 + \phi(x_1, x_2),$$

where a, b_i, c_i are constants. The terms $c_0(p_1^2 + p_2^2)$ have been removed by linear combinations with the Hamiltonian.

The function $\phi(x_1, x_2)$ must satisfy the determining equations

$$\phi_{x_1} = -2(ax_2^2 + 2b_1 x_2 + c_1)V_{x_1} + 2(ax_1 x_2 + b_1 x_1 - b_2 x_2 - c_2)V_{x_2}$$
$$\phi_{x_2} = -2(ax_1 x_2 + b_1 x_1 - b_2 x_2 - c_2)V_{x_1} + 2(-ax_1^2 + 2b_2 x_1 + c_1)V_{x_2}. \tag{9}$$

The compatibility condition $\phi_{x_1 x_2} = \phi_{x_2 x_1}$ implies

$$(-ax_1 x_2 - b_1 x_1 + b_2 x_2 + c_2)(V_{x_1 x_1} - V_{x_2 x_2})$$
$$-(a(x_1^2 + x_2^2) + 2b_1 x_1 + 2b_2 x_2 + 2c_1)V_{x_1 x_1}$$
$$-(ax_2 + b_1)V_{x_1} + 3(ax_1 - b_2)V_{x_2} = 0. \tag{10}$$

Equation (10) is exactly the same equation that we would have obtained if we had required that the potential should allow the separation of variables in the Schrödinger equation in one of the coordinate systems in which the Helmholtz equation allows separation ($V(x_1, x_2) = 0$ in (6)). Another important observation is that (9) and (10) do not involve the Planck constant. Indeed, if we consider the classical functions H and X in (6) and require that they Poisson commute, we arrive at exactly the same conclusions and to Eqs. (9) and (10). Thus for quadratic integrability (and superintegrability) the potentials and integrals of motion coincide in classical and quantum mechanics (up to a possible symmetrization of the integrals). The Hamiltonian (1) is form-invariant under Euclidian transformations, so we can classify the integrals X into equivalence classes under rotations, translations, and linear combinations with H. There are two invariants in the space of parameters a, b_i, c_i, namely

$$I_1 = a, \qquad I_2 = (2ac_1 - b_1^2 + b_2^2)^2 + 4(ac_2 - b_1 b_2)^2. \tag{11}$$

Solving (6) for different values of I_1 and I_2 we obtain

$$I_1 = I_2 = 0 \qquad V_C = f_1(x_1) + f_2(x_2)$$

$$I_1 = 1,\ I_2 = 0 \qquad V_R = f(r) + \frac{1}{r^2} g(\phi) \qquad x_1 = r\cos\phi,\ x_2 = r\sin\phi$$

$$I_1 = 0,\ I_2 = 1 \qquad V_P = \frac{f(\xi) + g(\eta)}{\xi^2 + \eta^2} \qquad x_1 = \frac{\xi^2 - \eta^2}{2},\ x_2 = \xi\eta$$

$$I_1 = 1,\ I_2 = l^2 \neq 0 \qquad V_E = \frac{f(\sigma) + g(\eta)}{\cos^2\sigma - \cosh^2\rho} \qquad \begin{array}{l} x_1 = l\cosh\rho\cos\sigma \\ x_2 = l\sinh\rho\sin\sigma \\ 0 < l < \infty. \end{array} \tag{12}$$

We see that V_C, V_R, V_P, and V_E correspond to separation of variables in Cartesian, polar, parabolic, and elliptic coordinates, respectively, and that second order integrability (in E_2) is equivalent to the separation of variables in the Hamilton–Jacobi and the Schrodinger equation. For second order superintegrability, two integrals of the form (9) exist and the Hamiltonian separates in at least two coordinate systems. Four three-parameter families of superintegrable systems exist, namely

$$V_I = \alpha(x^2 + y^2) + \frac{\beta}{x^2} + \frac{\gamma}{y^2},\ V_{II} = \alpha(x^2 + 4y^2) + \frac{\beta}{x^2} + \gamma y$$

$$V_{III} = \frac{\alpha}{r} + \frac{1}{r^2}\left(\frac{\beta}{\cos^2\frac{\phi}{2}} + \frac{\gamma}{\sin^2\frac{\phi}{2}}\right),\ V_{IV} = \frac{\alpha}{r} + \frac{1}{\sqrt{r}}\left(\beta\cos\frac{\phi}{2} + \gamma\sin\frac{\phi}{2}\right). \tag{13}$$

The classical trajectories, quantum energy levels, and wave functions for all of these systems are known. The potentials V_I and V_{II} are isospectral deformations of the isotropic and an anisotropic harmonic oscillator, respectively, whereas V_{III} and V_{IV} are isospectral deformations of the Kepler-Coulomb potential. In n-dimensional space E_n, a set of n commuting second order integrals corresponds to a separable coordinate system. All of the above results on quadratic superintegrability have been generalized to arbitrary dimensions, to spaces of constant curvature, and to other real and complex spaces [51, 53, 73, 74].

3 Summary of Results for Integrals of Motion of Order N in E_2

In quantum mechanics on two-dimensional Euclidean space E_2 the most general N-th order integral has the form

$$X = \frac{1}{2}\sum_{l=0}^{[\frac{N}{2}]}\sum_{j=0}^{N-2l}\{f_{j,2l}, p_1^j p_2^{N-j-2l}\}, \tag{14}$$

where $f_{j,2l}$ are real functions of x,y and we set $f_{j,2l} = 0$ for $j, l < 0$ or $j > N-2l$. The brackets $\{,\}$ denote a symmetrization. In classical mechanics the brackets are inessential. The determining equations following from the commutativity relation $[H, X] = 0$ were obtained in [88] for arbitrary $N \geq 2$, both in the classical and quantum cases. The equations are quite complicated but completely explicit.

A priori the Lie or Poisson commutator $[H, X]$ is a polynomial of order $N+1$ in the components of the momenta p_i. The terms of order $N+1$ are linear and do not involve the potential $V(x, y)$. All lower order terms are nonlinear since they involve products of the unknown potential and the unknown coefficients $f_{j,2l}$.

An analysis of the highest and second to highest order determining equations provides several important results.

1. Even and odd parity terms in X commute with H separately, so all terms in (14) have the same parity (this is already built into Eq. (14)).
2. The leading terms in X are polynomials of order N in the enveloping algebra of the Euclidean Lie algebra, i.e.,

$$X = X_L + l.o.t \tag{15}$$

$$X_L = \frac{1}{2} \sum_{0 \leq m+n \leq N} A_{N-m-n,m,n} \{L_3^{N-m-n}, p_1^m p_2^n\},$$

where the coefficient $A_{N-m-n,m,n}$ are real constants. Indeed the leading terms are obtained for $l = 0$ in (14) and are polynomials

$$f_{j0} = \sum_{n=0}^{N-j} \sum_{m=0}^{j} \binom{N-n-m}{j-m} A_{N-n-m,m,n} x^{N-j-m} (-y)^{j-m}. \tag{16}$$

3. The set of determining equations f_{j2} does involve the potential and is nonlinear. However, the equations are in general incompatible. A compatibility condition for arbitrary N is the linear PDE

$$\sum_{j=0}^{N-1} \partial_x^{N-1-j} \partial_y^j (-1)^j [(j+1) f_{j+1,0} \partial_x V + (N-j) f_{j0} \partial_y V] = 0. \tag{17}$$

This is a linear PDE for V alone, since the coefficients f_{j0} are already known in terms of the constants $A_{N-m-n,m,n}$. Other compatibility condition exists, but they are nonlinear PDEs for the potential $V(x, y)$ and are less useful than (17).

For $N = 2$ the condition (17) reduces to the condition (10) and provides the connection between second order integrability and the separation of variables.

For $N \geq 3$ Eq. (17) is also the starting point for all further studies. Right from the beginning we distinguish two types of integrable potentials:

1. Standard potentials. For these the linear compatibility condition LCC (17) is satisfied nontrivially. For $N = 2$ all integrable potentials are standard.

2. Exotic potentials. These exist for $N \geq 3$ and for them the LCC is satisfied trivially, i.e., all coefficients $A_{N-n-m,m,n}$ that figure in the LCC vanish identically. Surprisingly that does not imply that the integral X vanishes; it does however greatly simplify.

Solving the remaining nonlinear PDEs is still a formidable task for any $N \geq 3$, especially in quantum mechanics. Instead of attempting this task we turn to a simpler problem, namely construct superintegrable systems in E_2 with two independent integrals of motion X and Y, where X is of first or second order and Y is of the order N. The integrals X implies that $V(x, y)$ has one of the form given in (12). The potential in (12) depends on two arbitrary functions of one variable. Hence the LCC (17) is no longer a PDE but reduces to one or several ODEs. The most interesting cases occur when the potential has the form V_C and V_R of (12), i.e., allows separation in Cartesian [3, 42, 43, 62, 63, 68–70, 92, 99] or polar coordinates [29–31, 86, 95–97]. Let us now turn to the example of exotic potentials allowing the separation of variables in Cartesian coordinates and admitting an additional independent integral of order $N = 4$.

4 Fourth Order Superintegrability and Exotic Potentials

The article [70] is part of a general program, the aim of which is to derive, classify, and solve the equations of motion of superintegrable systems with integrals of motion that are polynomials of finite order N in the components of linear momentum. The search has been performed in two-dimensional Euclidean space. The study of Hamiltonians with integrals of motion of order $N = 3$ was started in [43] and a classification of Hamiltonians separable in Cartesian coordinates with an integrals of order $N = 3$ was performed [42]. The obtained classical and quantum Hamiltonian systems have been studied in [62, 63, 68, 69, 99]. In [70] the case $N = 4$ was considered and all exotic potentials have been classified. The connection with the Painlevé property and Chazy class of equations was also highlighted. Partial results which consist in classifying all doubly exotic potentials were performed for $N = 5$ [3]. The results are known for systems with integrals of arbitrary order N [92] and anisotropic oscillator complemented by Painlevé transcendents [64]. In this review we concentrate on superintegrable systems with Hamiltonians of the form

$$H = \frac{1}{2}(p_1^2 + p_2^2) + V(x, y), \tag{18}$$

in two-dimensional Euclidean space E_2. In classical mechanics, p_1 and p_2 are the canonical momenta conjugate to the Cartesian coordinates x and y. In quantum mechanics, we have p_i and L_i in Eq. (7).

The determining equations for fourth order classical and quantum integrals of motion were derived earlier and they are a special case of Nth order ones given in [88]. In the quantum case, the integral is $Y^{(4)} = Y$:

$$Y = \sum_{j+k+l=4} \frac{A_{jkl}}{2}\{L_3^j, p_1^k p_2^l\} + \frac{1}{2}(\{g_1(x,y), p_1^2\} + \{g_2(x,y), p_1 p_2\} + \{g_3(x,y), p_2^2\}) + l(x,y), \tag{19}$$

where A_{jkl} are real constants, the brackets $\{.,.\}$ denote anti-commutators, and the Hermitian operators p_1, p_2, and L_3 are given in (7). The functions $g_1(x,y), g_2(x,y), g_3(x,y)$, and $l(x,y)$ are real and the operator Y is self-adjoint. Equation (19) is also valid in classical mechanics where p_1, p_2 are the canonical momenta conjugate to x and y, respectively (and the symmetrization becomes irrelevant). The commutation relation $[H, Y] = 0$ with H in (18) provides the determining equations

$$g_{1,x} = 4 f_1 V_x + f_2 V_y, \tag{20a}$$

$$g_{2,x} + g_{1,y} = 3 f_2 V_x + 2 f_3 V_y, \tag{20b}$$

$$g_{3,x} + g_{2,y} = 2 f_3 V_x + 3 f_4 V_y, \tag{20c}$$

$$g_{3,y} = f_4 V_x + 4 f_5 V_y. \tag{20d}$$

These four equations are linear PDEs and involve four unknown functions g_1, g_2, g_3, V. Furthermore we have the following two further equations:

$$\begin{aligned}\ell_x =& 2 g_1 V_x + g_2 V_y + \frac{\hbar^2}{4}\Big((f_2 + f_4) V_{xxy} - 4(f_1 - f_5) V_{xyy} - (f_2 + f_4) V_{yyy} \\ &+ (3 f_{2,y} - f_{5,x}) V_{xx} - (13 f_{1,y} + f_{4,x}) V_{xy} - 4(f_{2,y} - f_{5,x}) V_{yy} \\ &- 2(6 A_{400} x^2 + 62 A_{400} y^2 + 3 A_{301} x - 29 A_{310} y + 9 A_{220} + 3 A_{202}) V_x \\ &+ 2(56 A_{400} xy - 13 A_{310} x + 13 A_{301} y - 3 A_{211}) V_y\Big),\end{aligned} \tag{21a}$$

$$\begin{aligned}\ell_y =& g_2 V_x + 2 g_3 V_y + \frac{\hbar^2}{4}\Big(-(f_2 + f_4) V_{xxx} + 4(f_1 - f_5) V_{xxy} + (f_2 + f_4) V_{xyy} \\ &+ 4(f_{1,y} - f_{4,x}) V_{xx} - (f_{2,y} + 13 f_{5,x}) V_{xy} - (f_{1,y} - 3 f_{4,x}) V_{yy} \\ &+ 2(56 A_{400} xy - 13 A_{310} x + 13 A_{301} y - 3 A_{211}) V_x \\ &- 2(62 A_{400} x^2 + 6 A_{400} y^2 + 29 A_{301} x - 3 A_{310} y + 9 A_{202} + 3 A_{220}) V_y\Big).\end{aligned} \tag{21b}$$

The quantities f_i, $i = 1, 2, .., 5$ are polynomials in x and y. They are obtained from the highest order terms in the condition $[H, Y] = 0$.

These two nonlinear PDEs for l, g_1, g_2, g_3, V will give nonlinear compatibility condition. Explicitly for these polynomials we have

$$\begin{aligned} f_1 &= A_{400}y^4 - A_{310}y^3 + A_{220}y^2 - A_{130}y + A_{040} \\ f_2 &= -4A_{400}xy^3 - A_{301}y^3 + 3A_{310}xy^2 + A_{211}y^2 - 2A_{220}xy - A_{121}y \\ &\quad + A_{130}x + A_{031} \\ f_3 &= 6A_{400}x^2y^2 + 3A_{301}xy^2 - 3A_{310}x^2y + A_{202}y^2 - 2A_{211}xy + A_{220}x^2 \\ &\quad - A_{112}y + +A_{121}x + A_{022} \\ f_4 &= -4A_{400}yx^3 + A_{310}x^3 - 3A_{301}x^2y + A_{211}x^2 - 2A_{202}xy + A_{112}x \\ &\quad - A_{103}y + A_{013} \\ f_5 &= A_{400}x^4 + A_{301}x^3 + A_{202}x^2 + A_{103}x + A_{004}, \end{aligned} \tag{22}$$

with 15 constants A_{jkl}. For a known potential the determining Eqs. (20) and (21) form a set of six linear PDEs for the functions g_1, g_2, g_3, and l. If V is not known, we have a system of six nonlinear PDEs for g_i, l, and V. In any case the four equations (20) are a priori incompatible. The compatibility equation is a fourth order linear PDE for the potential $V(x, y)$ alone, namely

$$\begin{aligned} \partial_{yyy}(4f_1V_x + f_2V_y) - \partial_{xyy}(3f_2V_x + 2f_3V_y) + \partial_{xxy}(2f_3V_x + 3f_4V_y) \\ -\partial_{xxx}(f_4V_x + 4f_5V_y) = 0. \end{aligned} \tag{23}$$

This is a special case of the Nth order linear compatibility equation (17). We see that Eq. (23) does not contain the Planck constant and is hence the same in quantum and classical mechanics (this is true for any N). The difference between classical and quantum mechanics manifests itself in the two equations (21). They greatly simplify in the classical limit $\hbar \to 0$. Further compatibility conditions on the potential $V(x, y)$ can be derived for the systems (20) and (21), they will however be nonlinear. We will not go further into the problem of the fourth order integrability of the Hamiltonian (18). Instead, we turn to the problem of superintegrability formulated in Sect. 1.

4.1 Potentials Separable in Cartesian Coordinates

We shall now assume that the potential in the Hamiltonian (18) has the form

$$V(x, y) = V_1(x) + V_2(y). \tag{24}$$

This is equivalent to saying that a second order integral exists which can be taken in the form

$$X = \frac{1}{2}(p_1^2 - p_2^2) + V_1(x) - V_2(y). \tag{25}$$

Equivalently, we have two one-dimensional Hamiltonians

$$H_1 = \frac{p_1^2}{2} + V_1(x), \quad H_2 = \frac{p_2^2}{2} + V_2(y). \tag{26}$$

We are looking for a third integral of the form (19) satisfying the determining equations (20) and (21). This means that we wish to find all potentials of the form (24) that satisfy the linear compatibility condition (23). Once (24) is substituted, (23) is no longer a PDE and will split into a set of ODEs which we will solve for $V_1(x)$ and $V_2(y)$.

The task thus is to determine and classify all potentials of the considered form that allow the existence of at least one fourth order integral of motion. As in every classification we must avoid triviality and redundancy. Since H_1 and H_2 of (26) are integrals, we immediately obtain 3 "trivial" fourth order integrals, namely H_1^2, H_2^2, and $H_1 H_2$. The fourth order integral Y of Eq. (19) can be simplified by taking linear combination with polynomials in the second order integrals H_1 and H_2 of (26):

$$Y \to Y' = Y + a_1 H_1^2 + a_2 H_2^2 + a_3 H_1 H_2 + b_1 H_1 + b_2 H_2 + b_0, \quad a_i, b_i \in \mathbb{R}. \tag{27}$$

Using the constants a_1, a_2, and a_3 we set

$$A_{004} = A_{040} = A_{022} = 0, \tag{28}$$

in the integral Y we are searching for. At a later stage we will use the constants b_0, b_1, and b_2 to eliminate certain terms in g_1, g_2, g_3, and l.

Substituting (24) into the compatibility condition (23), we obtain a linear condition, relating the functions $V_1(x)$ and $V_2(y)$

$$\begin{aligned}
&(-60A_{310} + 240yA_{400})V_1'(x) + (-20A_{211} + 60yA_{301} \\
&\quad - 60xA_{310} + 240xyA_{400})V_1''(x) + (-5A_{112} + 10yA_{202} - 10xA_{211} \\
&\quad + 30xyA_{301} - 15x^2A_{310} + 60x^2yA_{400})V_1^{(3)}(x) + (-A_{013} + yA_{103} \\
&\quad - xA_{112} + 2xyA_{202} - x^2A_{211} + 3x^2yA_{301} - x^3A_{310} + 4x^3yA_{400})V_1^{(4)}(x) \\
&\quad + (-60A_{301} - 2140xA_{400})V_2'(y) + (20A_{211} - 60yA_{301} + 60xA_{310} \\
&\quad - 240xyA_{400})V_2''(y) + (-5A_{121} + +10yA_{211} - 10xA_{220} - 15y^2A_{301} \\
&\quad + 30xyA_{310} - 60xy^2A_{400})V_2^{(3)}(y) + (A_{031} - yA_{121} + xA_{130} + y^2A_{211} \\
&\quad - 2xyA_{220} - y^3A_{301} + 3xy^2A_{310} - 4xy^3A_{400})V_2^{(4)}(y) = 0.
\end{aligned} \tag{29}$$

It should be stressed that this is no longer a PDE, since the unknown functions $V_1(x)$ and $V_2(y)$ both depend on one variable only.

We differentiate (29) twice with respect to x and thus eliminate $V_2(y)$ from the equation. The resulting equation for $V_1(x)$ splits into two linear ODEs (since the coefficients contain terms proportional to y^0, and y^1), namely

$$210A_{310}V_1^{(3)}(x) + 42(A_{211} + 3A_{310}x)V_1^{(4)}(x) + 7(A_{112} + 2A_{211}x + 3A_{310}x^2)V_1^{(5)}(x) + (A_{013} + A_{112}x + A_{211}x^2 + A_{310}x^3)V_1^{(6)}(x) = 0, \tag{30a}$$

$$840A_{400}V_1^{(3)}(x) + (126A_{301} + 504A_{400}x)V_1^{(4)}(x) + 14(A_{202} + 3A_{301}x + 6A_{400}x^2)V_1^{(5)}(x) + (A_{103} + 2A_{202}x + 3A_{301}x^2 + 4A_{400}x^3)V_1^{(6)}(x) = 0. \tag{30b}$$

Similarly, differentiating (29) with respect to y we obtain two linear ODEs for $V_2(y)$

$$210A_{301}V_2^{(3)}(y) - 42(A_{211} - 3A_{301}y)V_2^{(4)}(y) + 7(A_{121} - 2A_{211}y + 3A_{301}y^2)V_2^{(5)}(y) - (A_{031} - A_{121}y + A_{211}y^2 - A_{301}y^3)V_2^{(6)}(y) = 0, \tag{31a}$$

$$840A_{400}V_2^{(3)}(y) - (126A_{310} - 504A_{400}y)V_2^{(4)}(y) + 14(A_{220} - 3A_{310}y + 6A_{400}y^2)V_2^{(5)}(y) - (A_{130} - 2A_{220}y + 3A_{310}y^2 - 4A_{400}y^3)V_2^{(6)}(y) = 0. \tag{31b}$$

The compatibility condition $\ell_{xy} = \ell_{yx}$ for (21a) and (21b) implies

$$\begin{aligned}
&- g_2V_1''(x) + g_2V_2''(y) + (2g_{1y} - g_{2x})V_1'(x) + (g_{2y} - 2g_{3x})V_2'(y) \\
&+ \frac{\hbar^2}{4}\Big((f_2 + f_4)(V_1^{(4)} - V_2^{(4)}) + (f_{2x} - 4f_1'(y))V_1^{(3)} + (4f_5'(x) - 5f_{2y} - f_{4y})V_2^{(3)} \\
&+ (3f_{2yy} + 4f_{4xx} + 6A_{211} - 26A_{301}y + 26A_{310}x - 112A_{400}xy)V_1'' \\
&- (4f_{2yy} + 3f_{4xx} + 6A_{211} - 26A_{301}y + 26A_{310}x - 112A_{400}xy)V_2'' \\
&+ (84A_{310} - 360A_{400}y)V_1' + (84A_{310} + 360A_{400}y)V_2'\Big) = 0.
\end{aligned} \tag{32}$$

This equation, contrary to (30) and (31), is nonlinear since it still involves the unknown functions g_1, g_2, and g_3 (in addition to $V_1(x)$ and $V_2(y)$).

4.2 ODEs with the Painlevé Property

In order to study exotic potentials $V(x, y) = V_1(x) + V_2(y)$, allowing fourth order integrals of motion in quantum mechanics we must first recall some known results on Painlevé type equations [34, 48, 83]. Painlevé and Gambier showed that 50 classes of second order ODE exist that are single valued about their singular points. Six of them are "'irreducible," i.e., cannot be solved in terms of linear ODEs or elliptic functions, namely:

$$
\begin{aligned}
P_1''(z) &= 6P_1^2(z) + z,\\
P_2''(z) &= 2P_2(z)^3 + zP_2(z) + \alpha,\\
P_3(z)'' &= \frac{P_3'(z)^2}{P_3(z)} - \frac{P_3'(z)}{z} + \frac{\alpha P_3^2(z) + \beta}{z} + \gamma P_3^3(z) + \frac{\delta}{P_3(z)},\\
P_4(z)'' &= \frac{P_4'^2(z)}{2P_4(z)} + \frac{3}{2}P_4^3(z) + 4zP_4^2(z) + 2(z^2-\alpha)P_4(z) + \frac{\beta}{P_4(z)},\\
P_5''(z) &= \left(\frac{1}{2P_5(z)} + \frac{1}{P_5(z)-1}\right)P_5'(z)^2 - \frac{1}{z}P_5'(z)\\
&\quad + \frac{(P_5(z)-1)^2}{z^2}\left(\frac{\alpha P_5^2(z)+\beta}{P_5(z)}\right) + \frac{\gamma P_5(z)}{z} + \frac{\delta P_5(z)(P_5(z)+1)}{P_5(z)-1}\\
P_6''(z) &= \frac{1}{2}\left(\frac{1}{P_6(z)} + \frac{1}{P_6(z)-1} + \frac{1}{P_6(z)-z}\right)P_6'(z)^2\\
&\quad - \left(\frac{1}{z} + \frac{1}{z-1} + \frac{1}{P_6(z)-z}\right)P_6'(z) + \frac{P_6(z)(P_6(z)-1)(P_6(z)-z)}{z^2(z-1)^2}\\
&\quad \times\left(\gamma_1 + \frac{\gamma_2 z}{P_6(z)^2} + \frac{\gamma_3(z-1)}{(P_6(z)-1)^2} + \frac{\gamma_4 z(z-1)}{(P_6(z)-z)^2}\right).
\end{aligned}
\tag{33}
$$

An ODE has the Painlevé property if its general solution has no movable branch points (i.e., branch points whose location depends on one or more constants of integration). For a review and further developments see [19, 20, 44, 55]. Passing the so called Painlevé [2] test is a necessary condition for having the Painlevé property. We shall need it only for equations of the form

$$W^{(n)} = F(y, W, W', W'', \ldots, W^{(n-1)}), \tag{34}$$

where F is polynomial in $W, W', W'', \ldots, W^{(n-1)}$ and rational in y.

The general solution must have the form of a Laurent series with a finite number of negative power terms

$$W = \Sigma_{k=0}^{\infty} d_k (y - y_0)^{k+p}, \; d_0 \neq 0, \tag{35}$$

satisfying the requirements:

1. The constant p is a negative integer.
2. The coefficients d_k satisfy a recursion relation of the form

$$P(k) d_k = \phi_k(y_0, d_0, d_1, \ldots, d_{k-1}),$$

   where $P(k)$ is a polynomial that has $n - 1$ distinct nonnegative integer zeros. The values of k_j for which we have $P(k_j) = 0$ are called resonances and the values of d_k for $k = k_j$ are free parameters. Together with the position y_0 of the singularity we thus have n free parameters in the general solution (35) of the n-th order ODE (34).
3. A compatibility condition, also called the resonance condition,

$$\phi_k(y_0, d_0, d_1, \ldots, d_{k-1}) = 0,$$

   must be satisfied identically in y_0 and in the values of d_{k_j} for all k_j; $j = 1, 2, \ldots, n-1$.

This test is a generalization of the Frobenius method used to study fixed singularities of linear ODEs. Passing the Painlevé test is a necessary condition only. To make it sufficient one would have to prove that the series (35) has a nonzero radius of convergence and that the n free parameters can be used to satisfy arbitrary initial conditions. A more practical procedure that we shall adopt is the following. Once a nonlinear ODE passes the Painlevé test one can try to integrate it explicitly.

Let us first investigate the cases that may lead to "exotic potentials," that is, potentials which do not satisfy any linear differential equations. That means that either (30) or (31) (or both) must be satisfied trivially. The linear ODEs (30) are satisfied identically if we have

$$A_{400} = A_{310} = A_{301} = A_{211} = A_{202} = A_{112} = A_{103} = A_{013} = 0. \tag{36}$$

The linear ODEs (31) are satisfied identically if we have

$$A_{400} = A_{310} = A_{301} = A_{211} = A_{220} = A_{121} = A_{130} = A_{031} = 0. \tag{37}$$

If (36) and (37) both hold, then the only fourth order integrals are the trivial ones H_1^2, H_2^2, and $H_1 H_2$. Their existence does not imply superintegrability, it is simply a consequence of second order integrability. In other words, no fourth order superintegrable systems, satisfying (36) and (37) simultaneously, exist. This means that at most one of the functions $V_1(x)$ or $V_2(y)$ can be "exotic." The other one will be a solution of a linear ODE. For third order integrals both $V_1(x)$ and $V_2(y)$ can be exotic.

Let us consider the case when (37) is valid and (36) not. The leading-order term for the nontrivial fourth order integral has the form

$$Y_L = A_{202}\{L_3^2, p_2^2\} + A_{112}\{L_3, p_1 p_2^2\} + A_{103}\{L_3, p_2^3\} + 2A_{013} p_1 p_2^3. \tag{38}$$

We proceed in several steps.

1. Let us classify the integrals (38) under translations (they leave the form of the potential (24) invariant). The three classes are

$$\begin{aligned} &I. \quad A_{202} \neq 0,\, A_{112} = A_{103} = 0. \\ &II. \quad A_{202} = 0,\, A_{112}^2 + A_{103}^2 \neq 0,\, A_{013} = 0, \\ &\quad IIa.\ A_{103} \neq 0, \\ &\quad IIb.\ A_{103} = 0,\, A_{112} \neq 0. \\ &III. \quad A_{202} = A_{112} = A_{103} = 0,\, A_{013} \neq 0. \end{aligned} \tag{39}$$

2. Let us solve the linear ODE for $V(x)$
   The functions f_i in (22) reduce to

$$\begin{aligned} f_1 &= f_2 = 0, \\ f_3(y) &= A_{202}y^2 - A_{112}y, \\ f_4(x, y) &= -2A_{202}xy + A_{112}x - A_{103}y + A_{013}, \\ f_5(x) &= A_{202}x^2 + A_{103}x. \end{aligned} \tag{40}$$

   We obtain two equations for $V_1(x)$, namely

$$5A_{112}V_1^{(3)}(x) + (A_{013} + A_{112}x)V_1^{(4)}(x) = 0, \tag{41a}$$

$$10A_{202}V_1^{(3)}(x) + (A_{103} + 2A_{202}x)V_1^{(4)}(x) = 0. \tag{41b}$$

   (They replace Eq. (30)). These two equations imply $V_1^{(3)} = V_1^{(4)} = 0$ unless we have

$$A_{112}A_{103} - 2A_{202}A_{013} = 0. \tag{42}$$

   The result is that $V_1(x)$ can have one of the following forms: $V_1(x) = 0,\ ax,\ ax^2,\ \frac{a}{x^2} + bx + cx^2$.
3. Let us solve the nonlinear ODEs for $V_2(y)$. We first introduce an auxiliary function:

$$W(y) = \int V_2(y)\mathrm{d}y,$$

$$\widetilde{W} \Leftrightarrow W + \alpha y + \beta$$

Case I $A_{202} \neq 0, A_{112} = 0; Y_L = A_{202}\{L_3^2, p_2^2\}$.
Let $A_{202} = 1$. We obtain

$$\frac{1}{2}\hbar^2 y W^{(4)} + 2\hbar^2 W^{(3)} - 6yW'W'' - 2WW'' + \frac{8}{3}c_2 y^3 W'' - 8W'^2$$
$$+ 16c_2 y^2 W' + 16c_2 yW - \frac{16}{3}c_2^2 y^4 + k_1 = 0, \quad (43)$$

integrating once we get

$$\hbar^2 y^2 W^{(3)} + 2\hbar^2 yW'' - 6y^2 W'^2 - 4yWW' + \left(\frac{16}{3}c_2 y^4 - 2\hbar^2\right) W' + 2W^2$$
$$+ \frac{32}{3}c_2 y^3 W - \frac{16}{9}c_2^2 y^6 + k_1 y^2 + k_2 = 0. \quad (44)$$

Equation (44) passes the Painlevé test. Substituting the Laurent series (35) into (44), we find $p = -1$. The resonances are $r = 1$ and $r = 6$, and we obtain $d_0 = -\hbar^2$. The constants d_1 and d_6 are arbitrary, as they should be. We now proceed to integrate (44). Using the results of Chazy, Bureau, Cosgrove, and Scoufis [13, 14, 18, 21–24].

By the following transformation:

$$Y = y^2, \; U(Y) = -\frac{y}{2\hbar^2}W(y) + \frac{c_2}{6\hbar^2}y^4 + \frac{1}{16},$$

we transform (44) to

$$Y^2 U^{(3)} = -2(U'(3YU' - 2U) - \frac{c_2}{\hbar^2}Y(YU' - U) + k_3 Y + k_4) - YU'', \quad (45)$$

where $k_3 = \frac{-2k_1 - 12c_2\hbar^2}{64\hbar^4}$, $k_4 = \frac{-k_2}{32\hbar^4}$. Equation (45) is a special case of the Chazy class I equation. It admits the first integral

$$Y^2 U''^2 = -4(U'^2(YU' - U) - \frac{c_2}{2\hbar^2}(YU' - U)^2 + k_3(YU' - U) + k_4 U' + k_5), \quad (46)$$

where k_5 is the integration constant. The equation is of the Chazy canonical form SD-I .b.

When c_2 and k_3 are both nonzero the solution is

$$U = \frac{1}{4}\left(\frac{1}{P_5}\left(\frac{YP_5'}{P_5 - 1} - P_5\right)^2 - (1 - \sqrt{2\alpha})^2(P_5 - 1) - 2\beta\frac{P_5 - 1}{P_5} + \gamma Y\frac{P_5 + 1}{P_5 - 1}\right.$$
$$\left. + 2\delta\frac{Y^2 P_5}{(P_5 - 1)^2}\right),$$

$$U' = -\frac{Y}{4P_5(P_5-1)}\left(P_5' - \sqrt{2\alpha}\frac{P_5(P_5-1)}{Y}\right)^2 - \frac{\beta}{2Y}\frac{P_5-1}{P_5} - \frac{1}{2}\delta Y \frac{P_5}{P_5-1}$$
$$-\frac{1}{4}\gamma, \tag{47}$$

where $P_5 = P_5(Y)$; $Y = y^2$, satisfies the fifth Painlevé equation

$$P_5'' = \left(\frac{1}{2P_5} + \frac{1}{P_5-1}\right)P_5'^2 - \frac{1}{Y}P_5' + \frac{(P_5-1)^2}{Y^2}\left(\alpha P_5 + \frac{\beta}{P_5}\right) + \gamma\frac{P_5}{Y}$$
$$+\delta\frac{P_5(P_5+1)}{P_5-1},$$

with

$$c_2 = -\hbar^2\delta, \; k_3 = -\frac{1}{4}\left(\frac{1}{4}\gamma^2 + 2\beta\delta - \delta(1-\sqrt{2\alpha})^2\right),$$
$$k_4 = -\frac{1}{4}\left(\beta\gamma + \frac{1}{2}\gamma(1-\sqrt{2\alpha})^2\right),$$

$$k_5 = -\frac{1}{32}(\gamma^2((1-\sqrt{2\alpha})^2 - 2\beta) - \delta((1-\sqrt{2\alpha})^2 + 2\beta)^2).$$

The solution for the potential up to a constant is

$$V(x,y) = \frac{c_{-2}}{x^2} - \delta\hbar^2(x^2+y^2)$$
$$+\hbar^2\left(\frac{\gamma}{P_5-1} + \frac{1}{y^2}(P_5-1)\left(\sqrt{2\alpha} + \alpha(2P_5-1) + \frac{\beta}{P_5}\right)\right.$$
$$\left. + y^2\left(\frac{P_5'^2}{2P_5} + \delta P_5\right)\frac{(2P_5-1)}{(P_5-1)^2} - \frac{P_5'}{P_5-1} - 2\sqrt{2\alpha}P_5'\right) + \frac{3\hbar^2}{8y^2}. \tag{48}$$

The list of exotic superintegrable quantum potentials in quantum case that admit one second order Cartesian and one fourth order integral is given below. We also give their fourth order integrals by listing the leading terms Y_L and the functions $g_i(x, y)$; $i = 1, 2, 3$; and $l(x, y)$. Each of the exotic potentials has a nonexotic part that comes from $V_1(x)$. By construction $V_2(y)$ is exotic; however, in four cases a nonexotic part proportional to y^2 splits off from $V_2(y)$ and can be combined with an x^2 term in $V_1(x)$. We order the final list below in such a manner that the first two potentials are isotropic harmonic oscillators (possibly with an additional $\frac{1}{x^2}$ term) with an added exotic part. The next two are 2 : 1 anisotropic harmonic oscillators, plus an exotic part (in y).

Based on previous experience, we expect these harmonic terms to determine the bound state spectrum. The remaining eight cases have either $\frac{a}{x^2}$ or $c_1 x$ as their nonexotic terms and we expect the energy spectrum to be continuous.

These results also highlight how the study of higher order Painlevé equations plays a role in the classification of superintegrable systems with higher order integrals of motion. Classes of such equations of third, fourth, and fifth order have been studied by Chazy, Bureau, Cosgrove, and Scoufis [13, 14, 18, 21–24].

5 Summary of the Classification of Exotic Potentials with Fourth Order Integrals Separable in Cartesian Coordinates

In this section we give a list of some of these exotic potentials and their fourth order integrals. There are 12 cases that are divided into three types. We present one case among each of them.

1. Isotropic harmonic oscillator: $Q_1^1 : (Y_L = \{L_3^2, p_2^2\})$

$$\begin{aligned} V(x,y) &= -\delta\hbar^2(x^2+y^2) + \frac{a}{x^2} \\ &+ \hbar^2\left(\frac{\gamma}{P_5-1} + \frac{1}{y^2}(P_5-1)\left(\sqrt{2\alpha} + \alpha(2P_5-1) + \frac{\beta}{P_5}\right)\right. \\ &\left. + y^2\left(\frac{P_5'^2}{2P_5} + \delta P_5\right)\frac{(2P_5-1)}{(P_5-1)^2} - \frac{P_5'}{P_5-1} - 2\sqrt{2\alpha}P_5'\right) + \frac{3\hbar^2}{8y^2}. \end{aligned} \tag{49}$$

$$g_1(x,y) = 2y\left(yW' + W + \frac{1}{3}\hbar^2\delta y^3\right),$$

$$g_2(x,y) = -2x\left(3yW' + W + \frac{4}{3}\hbar^2\delta y^3\right)$$

$$\begin{aligned} l(x,y) &= \hbar^2x^2\left(\frac{1}{4}yW^{(4)} + W^{(3)}\right) - x^2(3yW'+W)W'' \\ &- \hbar^2 y\left(\frac{4}{3}\delta x^2y^2 + \frac{3}{2}\right)W'' + \left(4\left(\frac{a}{x^2} - \hbar^2\delta x^2\right)y^2 - 3\hbar^2\right)W' \\ &+ 4y\left(\frac{a}{x^2} - \hbar^2\delta x^2\right)W + \frac{4a}{3x^2}\hbar^2\delta y^4 - 2\hbar^2\delta x^2\left(\frac{2}{3}\hbar^2\delta y^4 - \hbar^2\right) \\ &- 2\hbar^4\delta y^2. \end{aligned}$$

$$W(y) = \frac{-\hbar^2}{2y}\left(\frac{1}{P_5}\left(\frac{YP_5'}{P_5-1} - P_5\right)^2 - (1-\sqrt{2\alpha})^2(P_5-1) - 2\beta\frac{P_5-1}{P_5}\right.$$
$$\left. +\gamma Y\frac{P_5+1}{P_5-1} + \frac{2\delta Y^2 P_5}{(P_5-1)^2}\right) + \frac{\hbar^2}{8y} - \frac{\delta\hbar^2}{3}y^3,$$

where $P_5 = P_5(Y)$; $Y = y^2$.

2. Anisotropic harmonic oscillator: $Q_2^1 : (Y_L = \{L_3, p_1 p_2^2\})$

$$V(x,y) = c_2(x^2+4y^2) + \frac{a}{x^2} - 4\sqrt[4]{2c_2^3\hbar^2}\,yP_4 + \sqrt{2c_2}\hbar(\epsilon P_4' + P_4^2) \tag{50}$$

$$g_1(x,y) = -2yW' - W + \frac{4}{3}c_2 y^3,\ g_2(x,y) = 3xW' - 4c_2xy^2,\ g_3(x,y)$$
$$= 2c_2x^2y - 2a\frac{y}{x^2},$$
$$l(x,y) = -\frac{1}{8}\hbar^2x^2W^{(4)} + \frac{3}{2}x^2W'W'' - \left(2c_2x^2y^2 - \frac{3}{4}\hbar^2\right)W''$$
$$-2\left(2a\frac{y}{x^2} + 2c_2x^2y\right)W' - 2\left(\frac{a}{x^2} + c_2x^2\right)W$$
$$+\frac{8}{3}c_2y^3\left(c_2x^2 + \frac{a}{x^2}\right) - 2c_2\hbar^2 y.$$
$$W(y) = \sqrt[4]{8c_2\hbar^6}\left(\frac{1}{8P_4}P_4'^2 - \frac{1}{8}P_4^3 - \frac{1}{2}YP_4^2 - \frac{1}{2}(Y^2-\alpha+\epsilon)P_4\right.$$
$$\left. +\frac{1}{3}(\alpha-\epsilon)Y + \frac{\beta}{4P_4}\right) + \frac{4c_2}{3}y^3,$$

where $P_4 = P_4(Y)$; $Y = -\sqrt[4]{\frac{8c_2}{\hbar^2}}\,y$.

3. Potentials with no confining (harmonic oscillator) term: eight cases occur involving P_1, P_2, P_3, or elliptic functions.

   For confining potentials the potentials involve P_4 and P_5. (P_6 appears in the case of separation in polar coordinates.)

   $Q_3^1 : (Y_L = \{L_3^2, p_2^2\})$

$$V(x,y) = \frac{a}{x^2} + \frac{\hbar^2}{2}\left(\sqrt{\alpha}P_3' + \frac{3}{4}\alpha(P_3)^2 + \frac{\delta}{4P_3^2} + \frac{\beta P_3}{2y} + \frac{\gamma}{2yP_3}\right.$$
$$\left. -\frac{P_3'}{2yP_3} + \frac{P_3'^2}{4P_3^2}\right). \tag{51}$$

$$g_1(x,y) = 2y^2W' + 2yW,\ g_2(x,y) = -6xyW' - 2xW,$$

$$g_3(x,y) = 4x^2W' + 2a\frac{y^2}{x^2},$$

$$l(x,y) = \hbar^2x^2\left(\frac{1}{4}yW^{(4)} + W^{(3)}\right) - x^2(3yW' + W)W'' - \frac{3}{2}\hbar^2yW''$$

$$+\left(4\frac{a}{x^2}y^2 - 3\hbar^2\right)W' + 4\frac{a}{x^2}yW.$$

$$W(y) = -\frac{\hbar^2}{2y}\left(\frac{1}{4}\left(y\frac{P_3'}{P_3} - 1\right)^2 - \frac{1}{16}\alpha y^2P_3^2 - \frac{1}{8}(\beta + 2\sqrt{\alpha})yP_3\right.$$

$$\left.+\frac{\gamma}{8P_3}y + \frac{\delta}{16P_3^2}y^2\right) + \frac{\hbar^2}{8y}.$$

The potentials Q_1^2, Q_3^6, and Q_3^7 are in the list of quantum potentials obtained by Gravel [42], respectively, Q_{18}, Q_{19}, Q_{21}. Among the integrals of motion we have $\{L_3^2, p_2^2\}$ and $\{L_3, p_2^3\}$. These cannot be obtained by commuting a third and a second order integral. Let us mention that the classical limit $\hbar \to 0$ cannot be taken in the expressions for the potentials like (49), (50), and (51). The limit is singular and must be taken in the original determining equations. In particular for $N = 4$ in Eq. (21) (the other determining equation (20) and their linear compatibility equation (23) do not contain $\hbar$). In the potential Q_1^1, the isotropic harmonic oscillator term appears with the coefficient $-\delta\hbar^2$. In Q_2^1 the coefficient of the anisotropic harmonic oscillator is c_2. We do not attach any importance to this fact since both c_2 and δ are arbitrary real constants ($\hbar^2$ could be absorbed into δ). Moreover, as stated above the limit $\hbar \to 0$ is not allowed in these formulas.

For a complete list of exotic potentials of the form $V(x,y) = V_1(x) + V_2(y)$ with fourth order integrals we refer to the original article [70].

The results can be summed up as follows:

1. For $N = 4$ one of the two $V_a (a = 1, 2)$ must be standard, i.e., satisfy a linear ODE. We choose $V_2(y)$ to be exotic.
2. The exotic part satisfies a nonlinear ODE that not only passes the Painlevé test but actually has the Painlevé property. Moreover $V_2(y)$ can always be expressed in terms of either elliptic functions or one of the original Painlevé-Gambier transcendents $P_1,\ldots,P_5$. The sixth transcendent does not occur. However the sixth Painlevé transcendent P_6 plays a crucial role when the potential allows separation in polar coordinates instead of Cartesian ones [28–30].
3. The exotic potentials may have a nonexotic part that makes them confining. For $N = 4$ this occurs in one of the three versions

$$V(x,y) = a(x^2 + y^2) + \frac{b}{x^2} + \frac{c}{y^2} + V_E(y)$$

$$V(x,y) = a(x^2 + 4y^2) + V_E(y)$$

$$V(x,y) = a(x^2 + y^2) + V_E(y),$$

where V_E is expressed in terms of P_4 or P_5. The nonexotic parts in other cases are nonconfining like

$$V = \frac{a}{x} + V_E(y), V = ax + V_E(y)$$

with $V_E(y)$ expressed in terms of P_1, P_2, P_3, or an elliptic function. We expect the confining potentials to correspond to a bound spectrum in quantum mechanics.

6 Example of Schrödinger Equation with Painlevé Potential

Let us consider the example of an exotic potential expressed in terms of P_4. The Hamiltonian and two integrals of motion in this case are [62, 63]

$$\begin{aligned} H &= \frac{1}{2}\left[p_1^2 + p_2^2 + \omega^2(x^2 + y^2)\right] + V_E(x) \\ A &= p_1^2 - p_2^2 + \omega^2(x^2 - y^2) + V_E(x) \\ B &= \frac{1}{2}\{L_3, p_1^2\} + \frac{1}{2}\left\{\frac{\omega^2}{2}x^2 y - 3xy - 3yV_E'), p_1\right\} \\ &\quad -\frac{1}{\omega^2}\left\{\frac{\hbar^2}{4}V_E''' + (-\omega^2 x^2 - 3V_E)(\omega x + V_E'), p_1\right\} \end{aligned} \tag{52}$$

with

$$\begin{aligned} V_E &= \epsilon\frac{\hbar\omega}{2}P_4'\left(\sqrt{\frac{\omega}{\hbar}}x\right) + \frac{\omega\hbar}{2}P_4^2\left(\sqrt{\frac{\omega}{\hbar}}x\right) \\ &\quad +\omega\sqrt{\hbar\omega}x P_4\left(\sqrt{\frac{\omega}{\hbar}}x\right) + \frac{\hbar\omega}{3}(-\alpha + \epsilon), \qquad \epsilon = \pm 1 \\ P_4 &= P_4\left(\sqrt{\frac{\omega}{\hbar}}x, \alpha, \beta\right). \end{aligned} \tag{53}$$

The integrals of motion form a polynomial (cubic) algebra, satisfying

$$\begin{aligned} &[A, B] = C \qquad [A, C] = 16\omega^2\hbar^2 B \\ &[B, C] = -2\hbar^2 A^3 - 6\hbar^2 H A^2 + 8\hbar^2 H^3 \\ &\quad + \frac{\omega^2\hbar^4}{3}(4\alpha^2 - 20 - 6\beta - 8\epsilon\alpha)A - 8\omega^2\hbar^4 H \\ &\quad + \frac{\hbar^5\omega^3}{27}(-8\alpha^3 - 24\alpha - 36\alpha\beta + 24\epsilon\alpha^2 + 8\epsilon + 36\epsilon\beta), \end{aligned} \tag{54}$$

$$K = -16\hbar^2 H^4 + \frac{4\hbar^4\omega^2}{3}(4\alpha^2 - 8\alpha + 4 - \alpha\beta)H^2$$
$$-\frac{4\hbar^5\omega^3}{27}(8\alpha^3 - 24\epsilon\alpha^2 + 24\alpha + 36\alpha\beta - 8\epsilon - 36\epsilon\beta)H$$
$$-\frac{4\hbar^6\omega^4}{3}(4\alpha - 8\epsilon\alpha - 8 - 6\beta). \tag{55}$$

The algebra has a Casimir operator that is a fourth order polynomial in the Hamiltonian H (with constant coefficients). The representation theory of the algebra (54) and its realization in terms of a deformed oscillator algebra is used to calculate the energy spectrum and wave functions of the system. A connection with "higher order supersymmetry" also gives the wave functions. One obtains three series of states with energies

$$E_1 = \hbar\omega\Big(p + \frac{\epsilon + 3}{3} - \frac{\alpha}{3}\Big)$$
$$E_2 = \hbar\omega\Big(p + \frac{-\epsilon + 6}{6} + \frac{\alpha}{6} + \sqrt{\frac{-\beta}{8}}\Big), \qquad \beta < 0$$
$$E_3 = \hbar\omega\Big(p + \frac{-\epsilon + 6}{6} + \frac{\alpha}{6} - \sqrt{\frac{-\beta}{8}}\Big), \qquad p = 0, 1, 2, 3, ... \tag{56}$$

and 3 "zero modes," all in terms of the Painlevé transcendent $\mathcal{P}_{IV}$.

It has been demonstrated that this construction may not provide the appropriate number of degeneracies via algebraic approaches and these case are associated with parameters of the fourth Painlevé transcendents related to exceptional orthogonal polynomials. The connection has been established via generalized Hermite and Okamoto polynomials [67]. Constructions involving other integrals and their higher order polynomial algebras have been presented elsewhere [66]. It has been shown how more complicated patterns of finite dimensional unitary representations can provide the degeneracies in these cases [66].

7 SUSYQM Construction and Wave Functions

The wave functions can be calculated using another approach that is also in essence algebraic. Supersymmetric quantum mechanics has been studied using many approaches and the intertwining of differential operators can be traced back to Darboux and Moutard [50]. Second order supersymmetric quantum mechanics has been introduced in [4] and has been exploited to generate ladder operators of third order [5, 15, 63, 64, 72].

Let us present a construction using first and second order supersymmetry given by the following intertwining relation:

$$H_1 q^\dagger = q^\dagger (H_2 + 2\lambda), \quad H_1 M^\dagger = M^\dagger H_2. \tag{57}$$

These relations correspond to a third order ladder operator

$$H_1 a^\dagger = a^\dagger (H_1 + 2\lambda), \tag{58}$$

where $a^\dagger$ and a are third order operators.

$$a^\dagger = q^\dagger M, a = M^\dagger q. \tag{59}$$

Similarly

$$H_2 a^\dagger = a^\dagger (H_2 + 2\lambda), \tag{60}$$

where $a^\dagger$ and a are third order operators.

$$a^\dagger = M q^\dagger, \quad a = q M^\dagger. \tag{61}$$

The explicit form is the following:

$$\begin{aligned}
H_i &= \frac{P_x^2}{2} + V_i(x),\\
q^\dagger &= \sqrt{\frac{\hbar}{2}}\partial + W_3(x),\\
q &= -\sqrt{\frac{\hbar}{2}}\partial + W_3(x),\\
M^\dagger &= \left(\sqrt{\frac{\hbar}{2}}\partial + W_1(x)\right)\left(\sqrt{\frac{\hbar}{2}}\partial + W_2(x)\right),\\
M &= \left(-\sqrt{\frac{\hbar}{2}}\partial + W_2(x)\right)\left(-\sqrt{\frac{\hbar}{2}}\partial + W_1(x)\right).
\end{aligned} \tag{62}$$

The potentials V_1 and V_2 correspond up to an additive constant the one given by (53) with $\epsilon = 1$ and $\epsilon = -1$. Moreover, the functions W_1, W_2, and W_3 that appear in the intertwining operators (or supercharges) are also expressed in terms of the fourth Painlevé transcendent

$$W_{1,2} = \sqrt{\frac{\omega}{8}} P_4\left(\sqrt{\frac{\omega}{\hbar}}x\right) \pm \sqrt{\frac{\hbar}{2}} P_4'\left(\sqrt{\frac{\omega}{\hbar}}x\right) - \frac{2\sqrt{-\beta}}{\omega},$$

$$W_3 = \sqrt{\frac{\omega}{2}} P_4\left(\sqrt{\frac{\omega}{\hbar}} x\right) - \frac{\omega}{2\hbar} x. \tag{63}$$

The spectrum is obtained for cases when normalizable zero modes of the annihilation operator exist

$$a\psi_k^{(0)} = 0.$$

The energy of the zero modes are for $\epsilon = 1$ associated with the three solutions of the cubic algebra

$$\begin{aligned}
\psi_a^0(x) &= e^{\int^{\sqrt{\frac{2}{\hbar}}x} \sqrt{\frac{2}{\hbar}} W_3(x')dx'},\\
\psi_b^0(x) &= \left(\sqrt{\frac{2}{\hbar}} W_2(x) - \sqrt{\frac{2}{\hbar}} W_3(x)\right) e^{-\int^{\sqrt{\frac{2}{\hbar}}x} \sqrt{\frac{2}{\hbar}} W_2(x')dx'},\\
\psi_c^0(x) &= \left(\frac{4\sqrt{-\beta}}{\omega} + \left(\sqrt{\frac{2}{\hbar}} W_2(x) - \sqrt{\frac{2}{\hbar}} W_3(x)\right)\right.\\
&\quad \left.\left(\sqrt{\frac{2}{\hbar}} W_1(x) + \sqrt{\frac{2}{\hbar}} W_2(x)\right)\right) e^{-\int^{\sqrt{\frac{2}{\hbar}}x} \sqrt{\frac{2}{\hbar}} W_1(x')dx'},
\end{aligned} \tag{64}$$

with the corresponding zero modes for $\epsilon = -1$

$$\begin{aligned}
\psi_a^0(x) &= \left(\frac{\omega}{\hbar}(\alpha - 1) - \frac{2\sqrt{-\beta}}{\omega} + \left(\sqrt{\frac{2}{\hbar}} W_1(x) + \sqrt{\frac{2}{\hbar}} W_2(x)\right),\right.\\
&\quad \left.\left(\sqrt{\frac{2}{\hbar}} W_1(x) - \sqrt{\frac{2}{\hbar}} W_3(x)\right)\right) e^{\int^{\sqrt{\frac{2}{\hbar}}x} \sqrt{\frac{2}{\hbar}} W_3(x')dx'},\\
\psi_b^0(x) &= e^{-\int^{\sqrt{\frac{2}{\hbar}}x} \sqrt{\frac{2}{\hbar}} W_2(x')dx'},\\
\psi_c^0(x) &= \left(\sqrt{\frac{2}{\hbar}} W_1(x) + \sqrt{\frac{2}{\hbar}} W_2(x)\right) e^{-\int^{\sqrt{\frac{2}{\hbar}}x} \sqrt{\frac{2}{\hbar}} W_1(x')dx'}.
\end{aligned} \tag{65}$$

In both cases $\epsilon = 1$ and $\epsilon = -1$ the complete spectrum is recovered by acting with the raising operators. In addition the raising ladder operators also admit zero modes. However due to conflicting asymptotics we can have in total three, two, or one infinite sequence of levels. When a potential allows only one infinite sequence of energies, this potential may also possess a singlet state or doublet states

$$a^+\psi(x) = a^-\psi(x) = 0, \quad (a^+)^2\psi(x) = a^-\psi(x) = 0. \tag{66}$$

8 Conclusion

This review is devoted to superintegrable quantum systems with Hamiltonians of the form (6) with a potential satisfying (24). They allow two integrals of motion $\{X, Y\}$ with X (of order 2) as in (25) and Y (of order N) as in (14) and (15) (for $N = 4$ see Eq. (19); for N arbitrary see Ref. [88]). So far the cases $N = 3, 4$, and 5 have been investigated in detail [3, 42, 43, 70, 71]. Some conclusions for general N can be drawn. The general situation can be summed up as follows:

1. The commutator $[H,Y]$ is a priori a linear differential operator of order $N + 1$. The coefficients of all powers must vanish simultaneously. From terms of order $N + 1$ we deduce that the terms of order N in Y are contained in the enveloping algebra of the Euclidean Lie algebra e(2). Moreover, all terms in Y have the same parity (after an appropriate symmetrization) [88].
2. Terms of order $N - 1$ in the commutator provide nonlinear determining equations for the potential $V(x, y) = V_1(x) + V_2(y)$. However, for any $N > 2$ a linear compatibility condition must be satisfied. It amounts to linear ODEs for $V_1(x)$ and $V_2(y)$. These may be satisfied trivially (all coefficients equal to zero). Then we obtain "exotic potentials." If the linear compatibility condition is satisfied nontrivially, we obtain "standard potentials." So far, for $N < 7$ all standard potentials are expressed in terms of elementary functions and all exotic ones pass the Painlevé test [2]. We conjecture that this is true for all N.
3. For a different approach to superintegrable systems in E_2 where such systems separating in Cartesian coordinates are obtained from operator algebras in one dimension we refer to [71].
4. For recent results on superintegrable systems in E_2 separable in polar coordinates we refer to the original articles [29–31].

Acknowledgements The research of I. M. was supported by the Australian Research Council through Discovery Early Career Researcher Award DE130101067 and Australian Research Council Discovery Project DP 160101376. The research of P.W. was partially supported by an NSERC discovery research grant.

References

1. M.J. Ablowitz, P.A. Clarkson, *Solitons, Nonlinear Evolution Equations and Inverse Scattering* (Cambridge University Press, Cambridge, 1991)
2. M.J. Ablowitz, A. Ramani, H. Segur, Non-linear evolution equations and ordinary differential-equations of Painlevé type. Lett. al Nuovo Cimento **23**, 333 (1978)

3. I. Abouamal, P. Winternitz, Fifth-order superintegrable quantum system separating in Cartesian coordinates. Doubly exotic potentials. J. Math. Phys. **59**, 022104 (2018)
4. A. Andrianov, M. Ioffe, V.P. Spiridonov, Higher-derivative supersymmetry and the Witten index. Phys. Lett. A **174**, 273 (1993)
5. A. Andrianov, F. Cannata, M. Ioffe, D. Nishnianidze, Systems with higher-order shape invariance: spectral and algebraic properties. Phys. Lett. A **266**, 341–349 (2000)
6. A. Ballesteros, O. Ragnisco, A systematic construction of completely integrable Hamiltonians from coalgebras. J. Phys. A Math. Gen. **31**, 3791 (1998)
7. A. Ballesteros, A. Blasco, F.J. Herranz, F. Musso, O. Ragnisco, (Super)integrability from coalgebra symmetry: formalism and applications. J. Phys. Conf. Ser. **175**, 012004 (2009)
8. A. Ballesteros, A. Enciso, F.J. Herranz, D. Latini, O. Ragnisco, D. Riglioni, The classical Darboux III oscillator: factorization, spectrum generating algebra and solution to the equations of motion. J. Phys. Conf. Ser. **670**, 012031 (2016)
9. A. Ballesteros, F.J. Herranz, S. Kuru, J. Negro, The anisotropic oscillator on curved spaces: a new exactly solvable model. Ann. Phys. **373**, 399 (2016)
10. V. Bargmann, Zur theorie des Wasserstoffatoms. Z. Phys. **99**, 576 (1936)
11. J.L.F. Bertrand, Théoreme relatif au mouvement d'un point attiré vers un centre fixe. C. R. Acad. Sci. **77**, 849 (1873)
12. D. Bonatsos, C. Daskaloyannis, K. Kokkotas, Quantum algebraic description of quantum superintegrable systems in 2 dimensions. Phys. Rev. A **48**(5), R23407–R3410 (1993)
13. F.J. Bureau, Differential equations with fixed critical points. Annali di Matematica **LXIV**, 229–364 (1964)
14. F.J. Bureau, Differential equations with fixed critical points. Annali di Matematica **LXVI**, 1–116 (1964)
15. J.M. Carballo, D.J. Fernandez C, J. Negro, L.M. Nieto, Polynomial Heisenberg algebras. J. Phys. A **37**, 10349, 25J (2004)
16. J.F. Carinena, F.J. Herranz, M.F. Ranada, Superintegrable systems on 3-dimensional curved spaces: Eisenhart formalism and separability. J. Math. Phys. **58**, 022701 (2017)
17. E. Celeghini, S. Kuru, J. Negro, M.A. del Olmo, A unified approach to quantum and classical TTW systems based on factorization. Ann. Phys. **332**, 27–37 (2013)
18. J. Chazy, Sur les équations différentielles du troisieme ordre et d'ordre supérieur dont l'intégrale générale a ses points critiques fixes. Acta Math. **34**, 317–385 (1911)
19. R. Conte, *The Painlevé Approach to Nonlinear Ordinary Differential Equations*. The Painlevé property, one century later, pp. 77–180 (Springer, New York, 1999)
20. R. Conte, M. Musette, *The Painlevé Handbook* (Springer, Berlin, 2008)
21. C.M. Cosgrove, Higher-order Painlevé equation in the polynomial class I: bureau Symbol P2. Stud. Appl. Math. **104**, 1–65 (2000)
22. C.M. Cosgrove, Chazy classes IX–XI of third-order differential equations. Stud. Appl. Math. **104**, 171–228 (2000)
23. C.M. Cosgrove, Higher-order Painlevé equation in the polynomial class II: bureau symbol P1. Stud. Appl. Math. **116**, 321–413 (2006)
24. C.M. Cosgrove, G. Scoufis, Painlevé classification of a class of differential equations of the second order and second degree. Stud. Appl. Math. **88**, 25–87 (1993)
25. E. D'Hoker, L. Vinet, Supersymmetry of the Pauli equation in the presence of a magnetic monopole. Phys. Lett. B **137**, 1, 72 (1984)
26. C. Daskaloyannis, Quadratic Poisson algebras of two-dimensional classical superintegrable systems and quadratic algebras of quantum superintegrable systems. J. Math. Phys. **42**, 1100–1119 (2001)
27. H. De Bie, V.X. Genest, J.-M. Lemay, L. Vinet, A superintegrable model with reflections on S^{n-1} and the higher rank Bannai-Ito algebra. J. Phys. A Math. Theor. **50**(19), 195202 (2017)
28. A.M. Escobar Ruiz, E.G. Kalnins, W. Miller Jr., E. Subag, Bocher and abstract contractions of 2nd order quadratic algebras. SIGMA **13**, 013, 38 pp. (2017)
29. A.M. Escobar-Ruiz, J.C. Lopez Vieyra, P. Winternitz. Fourth order superintegrable systems separating in Polar Coordinates. I. Exotic potentials. J. Phys. A **50**(49), 495206 (2017)

30. A.M. Escobar-Ruiz, J.C. Lopez Vieyra, P. Winternitz, I. Yurdusen. Fourth order superintegrable systems separating in Polar Coordinates. II. Standard potentials. J. Phys. A: Math. Theor. **51**, 455202 (2018)
31. A.M. Escobar-Ruiz, P. Winternitz, I. Yurdusen, General Nth order superintegrable systems separating in polar coordinates. J. Phys. A: Math. Theor. **51**, 40LT01 (2018)
32. V. Fock, Zur theorie des wasserstoffatoms. Z. Phys. A **98**, 145 (1935)
33. I. Fris, V. Mandrosov, J. Smorodinsky, M. Uhlíř, P. Winternitz, On higher symmetries in quantum mechanics. Phys. Lett. **16**, 354 (1965)
34. B. Gambier, Sur les équations différentielles du second ordre et du premier degré dont l'intégrale générale est à points critiques fixes. Acta Math. **33**, 1 (1910)
35. V. Genest, I. Mourad, The Dunkl oscillator in the plane: I. Superintegrability, separated wavefunctions and overlap coefficients. J. Phys. A: Math. Theor. **46**, 14, 145201 (2013)
36. V. Genest, L. Vinet, A. Zhedanov, Superintegrability in two dimensions and the Racah-Wilson algebra. Lett. Math. Phys. **104**, 931 (2011)
37. V.X. Genest, L. Vinet, A. Alexei, Superintegrability in two dimensions and the Racah-Wilson algebra. Lett. Math. Phys. **104**, 8, 931 (2014)
38. H. Goldstein, C.P. Poole, J.L. Safko, *Classical Mechanics* (Addison-Wesley, Reading, 2001)
39. D. Gomez-Ullate, N. Kamran, R. Milson, Exceptional orthogonal polynomials and the Darboux transformation. J. Phys. A **43**, 434016 (2010)
40. D. Gomez Ullate, Y. Grandati, R. Milson, Rational extensions of the quantum harmonic oscillator and exceptional Hermite polynomials. J. Phys. A Math. Theor. **47**, 015203 (2014)
41. Y. Granovskii, I. Lutzenko, A.Z. Zhedanov, Mutual integrability, quadratic algebras and dynamic symmetry. Ann. Phys. **217**, 1–20 (1992)
42. S. Gravel, Hamiltonians separable in Cartesian coordinates and third-order integrals of motion. J. Math. Phys. **45**, 1003–19 (2004)
43. S. Gravel, P. Winternitz, Superintegrability with third order integrals in quantum and classical mechanics. J. Math. Phys. **43**, 5902–5912 (2002)
44. A.N.W. Hone, Painlevé tests, singularity structure and integrability, in Integrability, pp. 245–277 (Springer, Berlin, 2009)
45. M.F. Hoque, Superintegrable systems, polynomial algebra structures and exact derivations of spectra, Ph.D. thesis, School of Mathematics and Physics, The University of Queensland, Australia, January, 175 pages, 2018, arXiv:1802.08410
46. M.F. Hoque, I. Marquette, Y.-Z. Zhang, Quadratic algebra structure in the 5D Kepler system with non-central potentials and Yang-Coulomb monopole interaction. Ann. Phys. **380**, 121–134 (2017)
47. P. Iliev, Symmetry algebra for the generic superintegrable system on the sphere. J. High Energy Phys. **2**, 44, 22 pp. (2018)
48. E.L. Ince, *Ordinary Differential Equations*, 574pp. (Dover, New York, 1956)
49. J.M. Jauch, E.L. Hill, The problem of degeneracy in quantum mechanics. Phys. Rev. **57**, 641–645 (1940)
50. G. Junker, *Supersymmetric Methods in Quantum and Statistical Physics* (Springer, New York, 1995)
51. E.G. Kalnins, *Separation of Variables for Riemannian Spaces of Constant Curvature*, p. 196 (Addison-Wesley, Reading, 1986)
52. E.G. Kalnins, J.M. Kress, W. Miller Jr., A recurrence relation approach to higher order quantum superintegrability. SIGMA **7**, 031 (2011)
53. E.G. Kalnins, J.M. Kress, W. Miller, *Separation of Variables and Superintegrability: The Symmetry of Solvable Systems* (IOP, Bristol, 2018)
54. G.E. Kalnins, W. Miller Jr., S. Post, Contractions of 2D 2nd order quantum superintegrable systems and the Askey scheme for hypergeometric orthogonal polynomials. SIGMA **9**, 057, 28 pp. (2013)
55. M.D. Kruskal, P.A. Clarkson, The Painlevé-Kowalevski and poly-Painlevé tests for integrability. Stud. Appl. Math. **86**, 87–165 (1992)

56. P. Letourneau, L. Vinet, Superintegrable systems, polynomial algebras and quasi-exactly solvable Hamiltonian. Ann. Phys. **243**, 1, 144 (1995)
57. Y. Liao, I. Marquette, Y.-Z. Zhang, Quantum superintegrable system with a novel chain structure of quadratic algebras. J. Phys. A: Math. Theor. **51**, 255201, 13pp. (2018)
58. A. Makarov, J. Smorodinsky, Kh. Valiev, P. Winternitz, A systematic search for non-relativistic systems with dynamical symmetries. Nuovo Cimento A **52**, 1061–1084 (1967)
59. A. Marchesiello, L. Šnobl, Superintegrable 3D systems in a magnetic field corresponding to Cartesian separation of variables. J. Phys. A Math. Theor. **50**, 245202 (2017)
60. A. Marchesiello, L. Šnobl, P. Winternitz, Three-dimensional superintegrable systems in a static electromagnetic field. J. Phys. A **48**, 395206 (2015)
61. A. Marchesiello, L. Šnobl, P. Winternitz, Spherical type integrable classical systems in a magnetic field. J. Phys. A Math. Theor. **51**, 135205 (2018)
62. I. Marquette, Superintegrability with third order integrals of motion, cubic algebras, and supersymmetric quantum mechanics. I. Rational function potentials, J. Math. Phys. **50**, 012101 (2009)
63. I. Marquette, Superintegrability with third order integrals of motion, cubic algebras, and supersymmetric quantum mechanics. II. Painlevé transcendent potentials. J. Math. Phys. **50**, 095202 (2009)
64. I. Marquette, An infinite family of superintegrable systems from higher order ladder operators and supersymmetry. J. Phys. Conf. Ser. **284**, 012047 (2011)
65. I. Marquette, C. Quesne, New families of superintegrable systems from Hermite and Laguerre exceptional orthogonal polynomials. J. Math. Phys. **54**, 042102 (2013)
66. I. Marquette, C. Quesne, Combined state-adding and state-deleting approaches to type III multi-step rationally-extended potentials: applications to ladder operators and superintegrability. J. Math. Phys. **55**, 112103 (2014)
67. I. Marquette, C. Quesne, Connection between quantum systems involving the fourth Painleve transcendent and k-step rational extensions of the harmonic oscillator related to Hermite EOP. J. Math. Phys. **57**, 052101 (2016)
68. I. Marquette, P. Winternitz, Polynomial Poisson algebras for classical superintegrable systems with a third order integral of motion. J. Math. Phys. **48**, 012902, 1–16 (2007). Erratum 49,019907
69. I. Marquette, P. Winternitz, Superintegrable systems with third order integrals of motion. J. Phys. A. Math. Theor. **41**, 303031 (2008)
70. I. Marquette, M. Sajedi, P. Winternitz, Fourth order superintegrable systems separating in Cartesian coordinates I. Exotic quantum potentials. J. Phys. A **50**, 315201 (2017)
71. I. Marquette, M. Sajedi, P. Winternitz, Two-dimensional superintegrable systems from operator algebras in one dimension. J. Phys. A **52**, 115202 (2019)
72. J. Mateo, J. Negro, Third-order differential ladder operators and supersymmetric quantum mechanics. J. Phys. A Math. Theor. **41**, 045204 (2008)
73. W. Miller, *Symmetry and Separation of Variables*, p. 285 (Addison-Wesley, Reading, 1977)
74. W. Miller, S. Post, P. Winternitz. Classical and quantum superintegrability with applications. J. Phys. A **46**, 423001 (2013)
75. M. Moshinsky, Yu.F. Smirnov, *The Harmonic Oscillator in Modern Physics* (Harwood Academic, New York, 1996)
76. N.N. Nekhoroshev, Action-angle variables and their generalizations. Trans. Moscow Math. Soc. **26**, 180 (1972)
77. A.G. Nikitin, Higher-order symmetry operators for Schrödinger equation, in *Superintegrability in Classical and Quantum Systems*. CRM Proceedings and Lecture Notes, vol. 37 (American Mathematical Society, Providence, RI, 2004)
78. A.G. Nikitin, New exactly solvable systems with Fock symmetry. J. Phys. A Math. Theor. **45**, 485204 (2012)
79. A.G. Nikitin, Laplace-Runge-Lenz vector for arbitrary spin. J. Math. Phys. **54**, 123506 (2013)
80. Yu.A. Orlov, E.I. Shulman, Additional symmetries of the nonlinear Schrodinger equation. Theor. Math. Phys. **64**, 862 (1985)

81. Yu.A. Orlov, E.I. Schulman, Additional symmetries for integrable equations and conformal algebra representation. Lett. Math. Phys. **12**, 171 (1986)
82. Yu.A. Orlov, P. Winternitz, Algebra of pseudodifferential operators and symmetries of equations in the Kadomtsev-Petviashvili hierarchy. J. Math. Phys. **38**, 4644 (1997)
83. P. Painlevé, Sur les équations différentielles du second ordre et d'ordre supérieur dont l'intégrale générale est uniforme. Acta Math. **25**, 1–85 (1902)
84. W. Pauli, Uber das wasserstoffspektrum vom Standpunkt der neuen Quantenmechanik. Z. Phys. **36**, 336 (1926)
85. I. Popper, S. Post, P. Winternitz, Third-order superintegrable systems separable in parabolic coordinates. J. Math. Phys. **53**, 062105 (2012)
86. S. Post, P. Winternitz, An infinite family of deformations of the Coulomb potential. J. Phys. A. Math. Gen. **43**, 222001 (2010)
87. S. Post, P. Winternitz, A nonseparable quantum superintegrable system in 2D real Euclidean space. J. Phys. A. Math. Theor. **44**, 162001 (2011)
88. S. Post, P. Winternitz, General Nth order integrals of motion in the Euclidean plane. J. Phys. A **48**, 405201 (2015)
89. S. Post, S. Tsujimoto, L. Vinet, Families of superintegrable Hamiltonians constructed from exceptional polynomials. J. Phys. A. Math. Theor. **45**, 405202 (2012)
90. M.F. Ranada, Higher order superintegrability of separable potentials with a new approach to the Post-Winternitz system. J. Phys. A-Math. Theor. **46**, 125206 (2013)
91. D. Riglioni, O. Gingras, P. Winternitz, Superintegrable systems with spin induced by co-algebra symmetry. J. Phys. A Math. Theor. **47**, 122002 (2014)
92. M.A. Rodriguez, P. Tempesta, P. Winternitz, Reduction of superintegrable systems: the anisotropic harmonic oscillator. Phys. Rev. E **78**, 046608 (2008)
93. M.B. Sheftel, P. Tempesta, P. Winternitz, Recursion operators, higher order symmetries and superintegrability in quantum mechanics. Czech J. Phys. **51**, 392–399 (2001)
94. P. Tempesta, A.V. Turbiner, P. Winternitz, Exact solvability of superintegrable systems. J. Math. Phys. **42**, 4248–4257 (2001)
95. F. Tremblay, P. Winternitz, Third order superintegrable systems separating in polar coordinates. J. Phys. A. Math. Theor. **43**, 175206 (2010)
96. F. Tremblay, A.V. Turbiner, P. Winternitz, An infinite family of solvable and integrable quantum systems on a plane. J. Phys. A. Math. Theor. **42**, 242001 (2009)
97. F. Tremblay, A.V. Turbiner, P. Winternitz, Periodic orbits for a family of classical superintegrable systems. J. Phys. A. Math. Theor. **43**, 015202 (2010)
98. L. Vinet, A. Zhedanov, A "missing" family of classical orthogonal polynomials. J. Phys. A. Math. Theor. **44**, 8, 085201 (2011)
99. P. Winternitz, Superintegrability with second and third order integrals of motion. Phys. Atom. Nuclei **72**, 875–882 (2009)
100. P. Winternitz, J. Smorodinsky, M. Uhliř, I. Friš, Symmetry groups in classical and quantum mechanics. Yad. Fiz **4**, 625–635 (1966). English translation Sov. J. Nucl. Phys. **4**, 444–450 (1967)

Supersymmetries in Schrödinger–Pauli Equations and in Schrödinger Equations with Position Dependent Mass

Anatoly G. Nikitin

Dedicated to Véronique Hussin.

Abstract The contemporary results concerning supersymmetries in generalized Schrödinger equations are presented. Namely, position dependent mass Schrödinger equations are discussed as well as the equations with matrix potentials. An extended number of realistic quantum mechanical problems admitting extended supersymmetries are described.

Keywords Position dependent mass · Schrödinger–Pauli equations · Extended supersymmetries · Matrix potentials · Integrable systems

1 Introduction

In seventieth of the previous century a qualitatively new symmetry in physics had been discovered and called *supersymmetry* (SUSY) see, e.g. [1] but also [2] where the idea of SUSY was formulated in somewhat rudimentary form. Its rather specific property is the existence of symmetry transformations mixing bosonic and fermionic states. In other words transformations which connect fields with different statistics have been introduced.

Among the many attractive features of SUSY is that it provides an effective mechanism for the cancelation of the ultraviolet divergences in quantum field theory. In addition, it opens new ways to unify space-time symmetries (i.e., relativistic invariance) with internal symmetries and to construct unified field theories, including all types of interactions, refer, e.g. [3, 4] and [5].

A. G. Nikitin (✉)
Institute of Mathematics, National Academy of Sciences of Ukraine, Kyiv, Ukraine
e-mail: nikitin@imath.kiev.ua

Ş. Kuru et al. (eds.), *Integrability, Supersymmetry and Coherent States*, CRM Series in Mathematical Physics, https://doi.org/10.1007/978-3-030-20087-9_5

Mathematically, SUSY requests using of the graded Lie algebras instead of the usual ones, and the corresponding group parameters are not numbers but Grassmann variables. The essential progress in the related fields of mathematics was induced exactly by the needs of SUSY.

Unfortunately, till now we do not have convincing experimental arguments for introducing SUSY as a universal symmetry principle realized in Nature. Nevertheless, it is possible to find a number of realistic physical systems which admit this nice symmetry. Moreover, SUSY presents effective tools for understanding the relations between spectra of different Hamiltonians as well as for explaining degeneracy of their spectra, for constructing exactly or quasi-exactly solvable systems, for justifying formulations of initial and boundary problems, etc.; see, e.g., surveys [3, 6] and [7]. In other words, SUSY is realized in Nature at least in a rather extended number of particular physical systems.

The present work is concentrated on quantum mechanical systems since they provide a ground for testing the principal question: whether SUSY is realized in Nature or not, free of the complexities of field theories. Examples of such systems (like interaction of spin 1/2 particle with the Coulomb or constant and homogeneous magnetic field) which admit exact $N = 2$ SUSY are well known [8, 9] (see also Refs. [6, 7] and the references therein). However, we will concentrate on systems admitting more extended SUSY.

Let us remain that the supersymmetric quantum mechanics was created by Witten [10] as a toy model for illustration of global properties of the quantum field theory. But rather quickly it becomes a fundamental field attracting the interest of numerous physicists and mathematicians. In particular the SSQM presents powerful tools for explicit solution of quantum mechanical problems using the shape invariance approach [11]. The number of problems satisfying the shape invariance condition is rather restricted but includes the majority of exactly solvable Schrödinger equations. The well-known exceptions are exactly solvable Schrödinger equations with Natanzon potentials [12] which are formulated in terms of implicit functions.

A very important application of SUSY in quantum mechanics is classification of families of isospectral Hamiltonians. And there is a number of systems isospectral with the basic exactly solvable SEs. In the standard SUSY approach with the first order intertwining operators the problem of description of such families is reduced to constructing general solutions of the Riccati equations. More refined approaches can include intertwining operators of higher order [13], the N-fold supersymmetry [14], and the hidden nonlinear supersymmetry [15]. One more relevant subject of contemporary SUSY are the so-called exceptional orthogonal polynomials [16, 17].

Let us mention that other generalized supersymmetries which include the usual SUSY have been discussed also, among them the so-called parasupersymmetry [18–20], which also has good ruts in real physical problems. However, the standard SUSY is seemed to be more fundamental.

Just in quantum mechanics SUSY presents powerful tools for constructing exact solutions of Schrödinger equation (SE). And we will present a survey of contemporary results belonging to this field. We will not discuss generalizations of the standard SUSY in quantum mechanics like the ones mentioned above, but restrict ourselves to the standard SUSY quantum mechanics with the first order intertwining operators [21]. However, the systems with extended SUSY as well as systems including SEs with Pauli and spin-orbit couplings, with position dependent mass and with abstract matrix potentials will be considered. Notice that just these fields are the subjects of current interest of numerous investigators.

Let us stress that there are two faces of SUSY in quantum mechanics. First, there exist QM systems like the charged particle with spin 1/2 in the constant and homogeneous magnetic field which admit exact SUSY. Such systems admit constants of motion forming superalgebras. Second, it is possible to indicate the QM systems with "hidden" SUSY like the hydrogen atom, and just these systems can be solved exactly using the shape invariance of the related Schrödinger equations. We will discuss both types of SUSY. The realistic physical systems which admit exact SUSY will be considered in the next section, while the shape invariant systems are discussed in Sects. 3–6.

An inspiring example of QM problem with a shape invariant potential was discovered by Pron'ko and Stroganov [22] who studied a motion of a neutral non-relativistic fermion, e.g., neutron, interacting with the magnetic field generated by a current carrying wire. A relativistic version of such problem was discussed in [23].

The specificity of the PS problem is that it includes a *matrix superpotential*, while in the standard SUSY in quantum mechanics the superpotential is a scalar function. Matrix potentials and superpotentials naturally appear in quantum mechanical models including particles with spin (see, e.g. [24], Sections 10 and 11) and in multidimensional models of SSQM [25, 26]. Particular examples of such superpotentials were discussed in [27–31]. In papers [32] such superpotentials were used for modeling the motion of a spin $\frac{1}{2}$ particle in superposed magnetic and scalar fields. In paper [29] a certain class of such superpotentials was described, while more extended classes of them were classified in [33, 34]. In any case just systems matrix superpotentials belong to an interesting research field which makes it possible to find new coupled systems of exactly solvable Schrödinger equations. The contemporary results in this field will be discussed in the following.

In addition to SUSY, some SEs can possess one more nice property called superintegrability (SI). By definition, the quantum system is called superintegrable if it admits more integrals of motion than the degrees of freedom. Like SUSY, the SI can cause the exact solvability of the related SE, especially in the case when it is the maximal SI when the number of integrals of motion is equal to $2n + 1$ where n is the number of degrees of freedom.

There exists a tight connection between the SI and SUSY, and many QM systems are both supersymmetric and superintegrable. In fact the maximal SI induces SUSY and vice versa, in spite of that this fact was never proven for generic QM systems.

The superintegrable systems which are also supersymmetric will be a special subject of our discussion. Moreover, there will be systems with position dependent masses which are discussed in Sect. 6.

2 QM Systems with Exact SUSY

2.1 System with N = 2 SUSY

Let us start with the well-known and important physical system, i.e., the spinning and charged particle interacting with an external magnetic field. The corresponding QM Hamiltonian can be written in the following form:

$$H = \frac{\pi^2}{2m} + \frac{e}{2m}\sigma_i B_i, \tag{2.1}$$

where $\pi^2 = \pi_1^2 + \pi_2^2 + \pi_3^2$, $\pi_i = -\mathrm{i}\frac{\partial}{\partial x_i} - eA_i$, $i = 1, 2, 3$, $B_i = \varepsilon_{ijk}\frac{\partial A_j}{\partial k}$, σ_i are Pauli matrices, B_i and A_i are components of the external magnetic field and the corresponding vector-potential, and summation is imposed over the repeating index i.

In contrast with the standard Schrödinger Hamiltonian, operator (2.1) includes the Pauli term $\frac{e}{2m}\sigma_i B_i$ describing the interaction of the particle spin with the external magnetic field. The related stationary Schrödinger equation has the standard form:

$$H\psi = E\psi \tag{2.2}$$

with E being the Hamiltonian eigenvalues.

In the case of the constant and homogeneous magnetic field directed along the third coordinate axis the vector-potential can be reduced to the form:

$$A_1 = -\frac{1}{2}x_2 B_3, \quad A_2 = \frac{1}{2}x_1 B_3 \quad A_3 = 0, \tag{2.3}$$

and by definition $B_1 = B_2 = 0$, $B_3 = B = const$. Thus Hamiltonian (2.1) can be rewritten in the following form:

$$H = H_1 + H_2, \quad H_1 = \frac{p_3^2}{2m}, \quad H_2 = \frac{(\sigma_1\pi_1 + \sigma_2\pi_2)^2}{2m} \tag{2.4}$$

with $p_3 = -\mathrm{i}\frac{\partial}{\partial x_3}$.

The immediate consequence of representation (2.4) is that our Hamiltonian commutes with the three operators:

$$Q_1 = \sigma_1\pi_1 + \sigma_2\pi_2, \ Q_2 = \mathrm{i}\sigma_3 Q_1, \ Q_3 = p_3, \tag{2.5}$$

which satisfy the following algebraic relations:

$$[Q_3, Q_1] = [Q_3, Q_2] = [Q_3, H] = 0, \tag{2.6}$$

$$\{Q_\mu, Q_\nu\} = 2\delta_{\mu\nu}H_2, \ [Q_\mu, H_2] = 0, \tag{2.7}$$

where μ, ν independently takes the values 1, 2, $\delta_{\mu\nu}$ is the Kronecker delta, and the symbols [.,.] and {.,.} denote the commutator and anticommutator correspondingly.

Thus the considered Hamiltonian admits three constants of motion, one of which, i.e., Q_3, commutes with the two others. On the other hand, Q_1 and Q_2 are not in involution, but satisfy more complicated relations (2.7), which characterize a *Lie superalgebra*.

Just this specific supersymmetry can be treated as the reason of the twofold degeneration of the Landau levels, i.e., the non-ground energy levels of a spin 1/2 particle interacting with the constant and homogeneous magnetic field.

Generally speaking, superalgebra is a graded algebra. In the simplest case of the Z_2 grading the elements of the superalgebra belong to two different classes, say, odd or even. The multiplication laws for even and odd elements are different. In our case Q_1 and Q_2 are odd, while Q_3, H_1, and H_2 are even. The product of two algebra elements is defined as the commutator if at least one of them is even and as the anticommutator if both the elements are odd. In SUSY quantum mechanics the odd elements are called supercharges. Since we have indicated two supercharges then it is possible to say about $N = 2$ SUSY.

2.2 Extended SUSY

The considered system is only a particular (albeit very important) example of realistic physical problem admitting exact supersymmetry. In particular, it is obvious that the presented SUSY is valid for arbitrary Hamiltonian admitting representation (2.1) provided one component of the vector-potential of the external field is identically zero.

We will discuss also another examples, but first let us note that in fact Eq. (2.2) with Hamiltonian (2.4) admits a more extended SUSY.

In analogy with the above we can construct a supercharge valid for Eq. (2.1) in the case of arbitrary external magnetic field:

$$\tilde{Q}_1 = \sigma_i \pi_i \tag{2.8}$$

since $\tilde{Q}_1^2 = H$.

Let us show that it is possible to find three more supercharges provided the external field is given by relations (2.3). To do it we exploit the fact that Eq. (2.4) is invariant w.r.t. the following three discrete transformations:

$$\psi \to R_3\psi, \qquad \psi \to CR_1\psi, \qquad \psi \to CR_2\psi, \tag{2.9}$$

where R_a $(a = 1, 2, 3)$ are the space reflection transformations

$$R_a\psi(\mathbf{x}) = \sigma_a\theta_a\psi(\mathbf{x}), \qquad \theta_a\psi(\mathbf{x}) = \psi(r_a\mathbf{x}). \tag{2.10}$$

Here

$$r_1\mathbf{x} = (-x_1, x_2, x_3), \quad r_2\mathbf{x} = (x_1, -x_2, x_3), \quad r_3\mathbf{x} = (x_1, x_2, -x_3), \tag{2.11}$$

and $C = i\sigma_2 c$, where c is the operator of complex conjugation

$$c\psi(\mathbf{x}) = \psi^*(\mathbf{x}). \tag{2.12}$$

Note that operators (2.9) satisfy the following relations:

$$\begin{aligned} &\{R_a, \sigma_i\pi_i\} = \{R_a, C\} = \{CR_1, \sigma_i\pi_i\} = \{CR_2, \sigma_i\pi_i\} = 0, \\ &R_a^2 = -C^2 = 1, \qquad a = 1, 2, 3. \end{aligned} \tag{2.13}$$

Using (2.8), (2.13) we can see that the operators

$$\tilde{Q}_1 = \sigma_i\pi_i \quad Q_2 = iR_3\tilde{Q}_1, \quad Q_3 = CR_1\tilde{Q}_1, \quad Q_4 = CR_2\tilde{Q}_1 \tag{2.14}$$

fulfill the following relations:

$$\{Q_k, Q_l\} = 2g_{kl}\hat{H}, \qquad \left[Q_k, \hat{H}\right] = 0, \tag{2.15}$$

where $k, l = 1, 2, 3, 4$, $g_{11} = g_{22} = -g_{33} = -g_{44} = 1$; $g_{kl} = 0, k \neq l$. In other words, operators (2.14) are supercharges generating the $N = 4$ extended SUSY.

Let us note that the main trick for constructing the extended SUSY was using the discrete involutive symmetries, i.e., reflections (2.10), (2.11). We will see that in analogous way it is possible to find extended SUSY for rather generic Eqs. (2.2).

2.3 *Extended SUSY with Arbitrary Vector-Potentials*

The results of the previous section can be generalized to extended class of arbitrary potentials with well-defined parities. Starting with reflections (2.10) we find that the corresponding parity properties of vector-function $\mathbf{A}(\mathbf{x})$ (2.3) are of the form:

$$\mathbf{A}(r_1\mathbf{x}) = -r_1\mathbf{A}(\mathbf{x}), \quad \mathbf{A}(r_2\mathbf{x}) = -r_2\mathbf{A}(\mathbf{x}), \quad \mathbf{A}(r_3\mathbf{x}) = r_3\mathbf{A}(\mathbf{x}). \tag{2.16}$$

Relations (2.16) are satisfied by a large class of potentials which includes (2.3) as a particular case. For all such potentials the corresponding Eq. (2.2) is invariant w.r.t. involutions (2.9) and so admits the extended SUSY generated by supercharges (2.14). Moreover, Eq. (2.1) for $g = 2$ and an arbitrary uniform magnetic field, i.e., the field

$$A_1 = A_1(x_1, x_2), \quad A_2 = A_2(x_1, x_2), \quad A_3 = 0, \tag{2.17}$$

admits all internal symmetries described in the previous section provided $\mathbf{A}(\mathbf{x})$ satisfies relations (2.16).

Other systems with extended SUSY can be found by extending reflections (2.11) to the eight-dimensional group of involutions, i.e., by adding the fixed *rotation* transformations

$$\begin{aligned} &r_{12}\mathbf{x} = (-x_1, -x_2, x_3), \quad r_{31}\mathbf{x} = (-x_1, x_2, -x_3), \\ &r_{23}\mathbf{x} = (x_1, -x_2, -x_3), \quad r_{123}\mathbf{x} = (-x_1, -x_2, -x_3), \quad I\mathbf{x} = \mathbf{x}. \end{aligned} \tag{2.18}$$

Let the vector-potential $\mathbf{A}(\mathbf{x})$ has definite parities w.r.t. a subset of transformations (2.11) and (2.18). Then it is possible to construct supercharges which generate extended $N = 4$ and even $N = 5$ SUSY [35].

Thus we present a number of SE admitting extended SUSY. Let us stress then among them there is a lot of systems with a clear exact physical meaning, see [36] for discussion of this aspect.

3 SUSY in One Dimension and Shape Invariance

The models considered in the above were two or three dimensional in spatial variables and include systems of coupled Shcrödinger equations. However, many of them can be reduced to one dimensional systems using the separation of variables. Moreover, these systems can be decoupled.

Returning to Eq. (2.2) for a charged particle interacting with the constant and homogeneous magnetic field we can exploit its rotational invariance and search for solutions in separated radial and angular variables, i.e., to represent the wave function ψ as

$$\psi = \frac{1}{\tilde{r}} R(\tilde{r}) \mathrm{e}^{n\varphi}, \tag{3.1}$$

where $\tilde{r} = \sqrt{x_1^2 + x_2^2}$, $\varphi = \arctan \frac{x_2}{x_1}$. As a result we come to the following equation for the radial functions:

$$\tilde{H} R \equiv \left(-\frac{\partial^2}{\partial \tilde{r}^2} - \frac{m(m+1)}{\tilde{r}^2} + \omega\sigma_3 + \omega^2 \tilde{r}^2 \right) R = \tilde{E} R, \tag{3.2}$$

where $m = n - \frac{1}{2}$, $\omega = 2m\alpha$, and $\tilde{E} = 2mE + q_3^2 + \omega n$.

Alternatively, using the gauge transformation it is possible to pass from vector-potential (2.3) to the following ones: $A_1 = eHx_2$, $A_2 = A_3 = 0$. Then, representing the wave function in the form $\psi = \exp[\mathrm{i}(p_1 x_1 + p_3 x_3)]\phi(x_2)$ and setting $x_2 = \frac{1}{\alpha B}(p_1 + \sqrt{\alpha B} y)$ we obtain the following equation for ϕ:

$$\hat{H}\phi = \hat{E}\phi, \tag{3.3}$$

where

$$\hat{H} = -\frac{\partial^2}{\partial y^2} + \sigma_3\omega + \omega^2 x^2, \quad \hat{E} = 2mE - p_3^2. \tag{3.4}$$

Equation (3.4) defines the supersymmetric oscillator, while (3.2) is rather similar to the "3d supersymmetric oscillator" but includes half-integer parameter m while in the $3d$ oscillator this parameter is integer. Both the mentioned equations are decoupled to direct sums of equations since the related Hamiltonians $\tilde{H}$ and $\hat{H}$ have the following form:

$$\tilde{H} = \begin{pmatrix} \tilde{H}_+ & 0 \\ 0 & \tilde{H}_- \end{pmatrix}, \quad \hat{H} = \begin{pmatrix} \hat{H}_+ & 0 \\ 0 & \hat{H}_- \end{pmatrix}, \tag{3.5}$$

where

$$\tilde{H}_\pm = -\frac{\partial^2}{\partial \tilde{r}^2} + \frac{n(n \mp 1)}{\tilde{r}^2} + \omega^2 \tilde{r}^2 \pm \omega, \quad \hat{H}_\pm = -\frac{\partial^2}{\partial \tilde{y}^2} + \omega^2 \tilde{y}^2 \pm \omega. \tag{3.6}$$

Hamiltonians $\hat{H}_\pm$ have two nice properties. First, they can be factorized:

$$\hat{H}_+ = a^+ a, \quad \hat{H}_- = a a^+, \tag{3.7}$$

where a^+ and a^- are the first order differential operators:

$$a^+ = -\frac{\partial}{\partial y} + W, \quad a^- = \frac{\partial}{\partial y} + W$$

with $W = \omega y$. Second, these Hamiltonians coincide up to a constant term: $\hat{H}_+ = \hat{H}_- + 2\omega$.

Hamiltonians $\tilde{H}_\pm$ are factorizable too:

$$\tilde{H}_- = a_\kappa^+ a_\kappa^- + c_\kappa, \quad \tilde{H}_+ = a_\kappa^- a_\kappa^+ + c_{\kappa+1}, \tag{3.8}$$

where

$$a_\kappa^- = \frac{\partial}{\partial x} + W_\kappa, \quad a_\kappa^+ = -\frac{\partial}{\partial x} + W_\kappa, \tag{3.9}$$

and $c_\kappa = (2\kappa - 1)\omega$. Moreover these Hamiltonians satisfy the following relation:

$$\tilde{H}_+(\kappa) = \tilde{H}_-(\kappa + 1) + C_\kappa \tag{3.10}$$

with $C_\kappa = 2\omega$. In other words, Hamiltonians $\tilde{H}_\pm(\kappa)$ are *shape invariant* [11]. The same is true for Hamiltonians $\hat{H}_\pm(\kappa)$, which, however, do not include variable parameter κ.

Thus our analysis of the realistic quantum mechanical system having a clear physical meaning (charged particle with spin 1/2 interacting with the constant and homogeneous magnetic field) makes it possible to discover its nice hidden symmetry, i.e., the shape invariance. It happens that this symmetry is valid for many other important QM systems like the hydrogen atom, and causes their exact solvability [11].

To be shape invariant, Hamiltonian should be factorizable, i.e., to admit representation (3.8), (3.9) for $\tilde{H}_-(\kappa)$ with some function W called *superpotential.* In addition, it should satisfy condition (3.10) together with the corresponding Hamiltonian $\tilde{H}_+(\kappa)$ which is called *superpartner.* If so, the related eigenvalue problem (2.2) is exactly solvable, and its solutions can be found algorithmically.

The shape invariance condition can be formulated as a condition for the potential. Considering the 1d Hamiltonian $H = -\frac{\partial 2}{\partial x^2} + V(\kappa, x)$ with a given potential V dependent on x and parameter κ and representing $V(\kappa, x)$ as

$$V = W_\kappa^2 + W_\kappa', \tag{3.11}$$

where $W_\kappa' = \frac{\partial W_\kappa}{\partial x}$, and superpotential is a solution of the Riccati equation (3.11). Then we construct a superpartner potential

$$\tilde{V} = W_\kappa^2 - W_\kappa'. \tag{3.12}$$

The corresponding stationary Schrödinger equation is shape invariant provided $\tilde{V}(\kappa, x) = V(\kappa + 1) + C_\kappa$, where C_κ is a constant. In terms of the superpotential this condition looks as follows:

$$W_\kappa^2 - W_\kappa' = W_\kappa^2 + W_\kappa' + C_\kappa. \tag{3.13}$$

A natural question arises whether it is possible to formulate the shape invariance condition with another transformation law for potential parameters. The answer is yes, but the rule $\kappa \to \kappa + 1$ can be treated as general up to redefinition of these parameters. In other words, we always can change these parameters by some functions of them in such a way that their transformations will be reduced to shifts [37].

4 Matrix Superpotentials

4.1 Pron'ko–Stroganov Problem

The supersymmetric systems considered in the above include matrix potentials. However, when speaking about shape invariance, we deal with scalar potentials and superpotentials, refer to Eqs. (3.6). Let us show that the concept of shape invariance can be extended to the case of matrix superpotentials.

Like in Sect. 2 we will start with a well-defined QM system which includes a matrix potential and appears to be shape invariant. Namely, let us consider a neutral QM particle with non-trivial dipole momentum (e.g., neutron), interacting with the magnetic field generated by a straight line current directed along the third coordinate axis (Pron'ko–Stroganov problem [22]) The corresponding Schrodinger–Pauli Hamiltonian looks as follows:

$$\mathcal{H} = \frac{p_1^2 + p_2^2}{2m} + \lambda \frac{\sigma_1 x_2 - \sigma_2 x_1}{\tilde{r}^2}, \tag{4.1}$$

where λ is the integrated coupling constant, and σ_1 and σ_2 are Pauli matrices.

The last term in (4.1) is the Pauli interaction term $\lambda\sigma_i H_i$ where the magnetic field $\mathbf{H}$ has the following components which we write ignoring the constant multiplier included into the parameter λ:

$$H_1 \sim \frac{y}{r^2}, \quad H_2 \sim -\frac{x}{r^2}, \quad H_3 = 0. \tag{4.2}$$

Hamiltonian (4.1) commutes with the third component of the total orbital momentum $J_3 = x_1 p_2 - x_2 p_1 + 1/2\sigma_3$; thus, the corresponding stationary Schrödinger equation (2.2) admits solutions in separated variables. Moreover, the equation for radial functions takes the following form:

$$\hat{H}_\kappa \psi = E_\kappa \psi, \tag{4.3}$$

where $\hat{H}_\kappa$ is a Hamiltonian with a matrix potential, E_κ and ψ are its eigenvalue and eigenfunction correspondingly, moreover, ψ is a two-component spinor. Up to normalization of the radial variable $\tilde{r}$ the Hamiltonian $\hat{H}_\kappa$ can be represented as

$$\hat{H}_\kappa = -\frac{\partial^2}{\partial \tilde{r}^2} + \kappa(\kappa - \sigma_3)\frac{1}{\tilde{r}^2} + \sigma_1 \frac{1}{\tilde{r}}, \tag{4.4}$$

where σ_1 and σ_3 are Pauli matrices and κ is a natural number. In addition, solutions of Eq. (4.3) must be normalizable and vanish at $x = 0$.

Hamiltonian $\hat{H}_\kappa$ can be factorized as in (3.8) where

$$a_\kappa^- = \frac{\partial}{\partial x} + W_\kappa, \quad a_\kappa^+ = -\frac{\partial}{\partial x} + W_\kappa, \quad c_\kappa = -\frac{1}{(2\kappa + 1)^2}$$

and W is a *matrix superpotential*

$$W_\kappa = \frac{1}{2x}\sigma_3 - \frac{1}{2\kappa+1}\sigma_1 - \frac{2\kappa+1}{2x}. \tag{4.5}$$

It is easily verified that the superpartner of Hamiltonian $\hat{H}_\kappa$ satisfies relation (3.10). In other words, Eq. (4.3) admits supersymmetry with shape invariance and can be solved using the standard technique of SSQM exposed, e.g., in survey [24].

4.2 Generic Matrix Superpotentials

Following a natural desire to find other shape invariant matrix potentials we return to conditions (3.13) which should be satisfied by the corresponding matrix superpotentials.

Assume $W_k(x)$ is Hermitian. Then the corresponding potential $V_k(x)$ and its superpartner $V_k^+(x)$ are Hermitian too.

The problem of classification of shape invariant superpotentials, i.e., $n \times n$ matrices whose elements are functions of x, k satisfying conditions (3.13), was formulated and partially solved in papers [33] and [34]. Here we present the completed classification results for a special class of superpotentials being 2×2 matrices.

Consider superpotentials of the following special form:

$$W_k = kQ + \frac{1}{k}R + P, \tag{4.6}$$

where P, R, and Q are Hermitian matrices depending on x.

Substituting (4.6) into (3.13) we obtain the following equations for P, R, and Q:

$$Q' = \alpha(Q^2 + \nu I), \tag{4.7}$$

$$P' - \frac{\alpha}{2}\{Q, P\} + \varkappa I = 0, \tag{4.8}$$

$$\{R, P\} + \lambda I = 0, \tag{4.9}$$

$$R^2 = \omega^2 I, \tag{4.10}$$

where $Q' = \frac{dQ}{dx}$, $\{Q, P\} = QP + PQ$ is an anticommutator of matrices Q and P, I is the unit matrix, and $\varkappa$, λ, ω are constants. Thus the problem of classification of matrix superpotentials is reduced to solution of Eqs. (4.8)–(4.10) for unknown matrices Q and P, R.

4.3 Scalar Superpotentials

First we consider the scalar case when Q, P, and R in (4.6) are 1×1 "matrices." The corresponding Eqs. (4.7)–(4.10) can be integrated rather easily, refer to [34] for detailed calculations. As a result we obtain the well-known list of scalar superpotentials:

$$W = -\frac{\kappa}{x} + \frac{\omega}{\kappa} \; \texttt{(Coulomb)}, \tag{4.11}$$

$$W = \lambda\kappa \tan \lambda x + \frac{\omega}{\kappa} \; \texttt{(Rosen 1)}, \tag{4.12}$$

$$W = \lambda\kappa \tanh \lambda x + \frac{\omega}{\kappa} \; \texttt{(Rosen 2)}, \tag{4.13}$$

$$W = -\lambda\kappa \coth \lambda x + \frac{\omega}{\kappa} \; \texttt{(Eckart)}, \tag{4.14}$$

$$W = \mu x \; \texttt{(Harmonic Oscillator)}, \tag{4.15}$$

$$W = \mu x - \frac{\kappa}{x} \; \texttt{(3D Oscillator)}, \tag{4.16}$$

$$W = \lambda\kappa \tan \lambda x + \mu \sec \lambda x \; \texttt{(Scarf I)}, \tag{4.17}$$

$$W = \lambda\kappa \tanh \lambda x + \mu \mathrm{sech} \lambda x \; \texttt{(Scarf 2)}, \tag{4.18}$$

$$W = \lambda\kappa \coth \lambda x + \mu \mathrm{cosech} \lambda x \; \texttt{(Pöschl-Teller)}, \tag{4.19}$$

$$W = \kappa - \mu \exp(-x) \; \texttt{(Morse)}. \tag{4.20}$$

Thus we recover the known list of superpotentials (4.11)–(4.20) which generate classical additive shape invariant potentials, in a straightforward and very simple way. The corresponding potentials V_κ can be found using definition (3.11).

4.4 Matrix Superpotentials of Dimension 2 × 2

Here we consider the case when superpotentials are x-dependent 2×2 matrices of form (4.6).

Supposing that $Q(x)$ is diagonal (like in (4.5)), it is possible to specify five inequivalent solutions of Eqs. (3.13):

$$W_{\kappa,\mu} = ((2\mu + 1)\sigma_3 - 2\kappa - 1)\frac{1}{2x} + \frac{\omega}{2\kappa + 1}\sigma_1, \quad \mu > -\frac{1}{2}, \tag{4.21}$$

$$W_{\kappa,\mu} = \lambda\left(-\kappa + \mu \exp(-\lambda x)\sigma_1 - \frac{\omega}{\kappa}\sigma_3\right), \tag{4.22}$$

$$W_{\kappa,\mu} = \lambda\left(\kappa \tan \lambda x + \mu \sec \lambda x \sigma_3 + \frac{\omega}{\kappa}\sigma_1\right), \tag{4.23}$$

$$W_{\kappa,\mu} = \lambda\left(-\kappa \coth \lambda x + \mu \operatorname{csch} \lambda x \sigma_3 - \frac{\omega}{\kappa}\sigma_1\right), \quad \mu < 0,\ \omega > 0, \tag{4.24}$$

$$W_{\kappa,\mu} = \lambda\left(-\kappa \tanh \lambda x + \mu \operatorname{sech} \lambda x \sigma_1 - \frac{\omega}{\kappa}\sigma_3\right), \tag{4.25}$$

where we introduce the rescaled parameter $\kappa = \frac{k}{\alpha}$. These superpotentials are defined up to translations $x \to x + c$, $\kappa \to \kappa + \gamma$, and up to unitary transformations $W_{\kappa,\mu} \to U_a W_{\kappa,\mu} U_a^\dagger$, where $U_1 = \sigma_1$, $U_2 = \frac{1}{\sqrt{2}}(1 \pm \mathrm{i}\sigma_2)$, and $U_3 = \sigma_3$. In particular these transformations change signs of parameters μ and ω in (4.22)–(4.25) and of $\mu + \frac{1}{2}$ in (4.21), thus without loss of generality we can set

$$\omega > 0, \quad \mu > 0 \tag{4.26}$$

in superpotentials (4.22)–(4.25).

Notice that the transformations $k \to k' = k + \alpha$ correspond to the following transformations for κ:

$$\kappa \to \kappa' = \kappa + 1. \tag{4.27}$$

If $\mu = 0$ and $\omega = 1$, then operator (4.21) coincides with the superpotential for PS problem given by Eq. (4.5). For $\mu \neq 0$ superpotential (4.21) is not equivalent to (4.5). The other presented matrix superpotentials were found in [33] for the first time.

The corresponding potentials V_κ can be found starting with (4.21)–(4.24) and using definition (3.11):

$$\hat{V}_\kappa = \left(\mu(\mu+1) + \kappa^2 - \kappa(2\mu+1)\sigma_3\right)\frac{1}{x^2} - \frac{\omega}{x}\sigma_1, \tag{4.28}$$

$$\hat{V}_\kappa = \lambda^2\left(\mu^2 \exp(-2\lambda x) - (2\kappa - 1)\mu \exp(-\lambda x)\sigma_1 + 2\omega\sigma_3\right), \tag{4.29}$$

$$\hat{V}_\kappa = \lambda^2\left((\kappa(\kappa-1) + \mu^2)\sec^2 \lambda x + 2\omega \tan \lambda x \sigma_1\right.$$
$$\left. + \mu(2\kappa - 1)\sec \lambda x \tan \lambda x \sigma_3\right), \tag{4.30}$$

$$\hat{V}_\kappa = \lambda^2\left((\kappa(\kappa-1) + \mu^2)\operatorname{csch}^2(\lambda x) + 2\omega \coth \lambda x \sigma_1\right.$$
$$\left. + \mu(1 - 2\kappa)\coth \lambda x \operatorname{csch} \lambda x \sigma_3\right), \tag{4.31}$$

$$\hat{V}_\kappa = \lambda^2\left((\mu^2 - \kappa(\kappa-1))\operatorname{sech}^2 \lambda x + 2\omega \tanh \lambda x \sigma_3\right.$$
$$\left. - \mu(2\kappa - 1)\operatorname{sech} \lambda x \tanh \lambda x \sigma_1\right). \tag{4.32}$$

Potentials (4.28), (4.29), (4.30), (4.31), and (4.32) are generated by superpotentials (4.21), (4.22), (4.23), (4.24), and (4.25), respectively. All the above potentials are shape invariant and give rise to exactly solvable problems for systems of Schrödinger–Pauli type.

It was proven in [33] that $n \times n$ matrix superpotentials of the form (4.6) with a diagonal matrix Q and $n > 2$ can be reduced to direct sums of operators fixed in (4.21) and scalar superpotentials specified in Eqs. (4.11)–(4.20). Thus in fact we present a complete description of superpotentials (4.6) being matrices of arbitrary dimension, provided matrix Q is diagonal.

The case of non-diagonal matrices Q has been examined in paper [34]. The classifying Eqs. (4.7)–(4.10) have been solved for the cases of superpotentials being 2×2 or 3×3 matrices. In the first case the following list of superpotentials was obtained:

$$W_\kappa^{(1)} = \lambda \Big(\kappa \left(\sigma_+ \tan(\lambda x + c) + \sigma_- \tan(\lambda x - c)\right) \tag{4.33}$$

$$+ \mu\sigma_1 \sqrt{\sec(\lambda x - c)\sec(\lambda x + c)} + \frac{1}{\kappa} R \Big), \tag{4.34}$$

$$W_\kappa^{(2)} = \lambda \Big(-\kappa \left(\sigma_+ \coth(\lambda x + c) + \sigma_- \coth(\lambda x - c)\right) \tag{4.35}$$

$$+ \mu\sigma_1 \sqrt{\operatorname{csch}(\lambda x - c)\operatorname{csch}(\lambda x + c)} + \frac{1}{\kappa} R \Big), \tag{4.36}$$

$$W_\kappa^{(3)} = \lambda \Big(-\kappa \left(\sigma_+ \tanh(\lambda x + c) + \sigma_- \tanh(\lambda x - c)\right) \tag{4.37}$$

$$+ \mu\sigma_1 \sqrt{\operatorname{sech}(\lambda x - c)\operatorname{sech}(\lambda x + c)} + \frac{1}{\kappa} R \Big), \tag{4.38}$$

$$W_\kappa^{(4)} = \lambda \Big(-\kappa \left(\sigma_+ \tanh(\lambda x + c) + \sigma_+ \coth(\lambda x - c)\right) \tag{4.39}$$

$$+ \mu\sigma_1 \sqrt{\operatorname{sech}(\lambda x + c)\operatorname{csch}(\lambda x - c)} + \frac{1}{\kappa} R \Big), \tag{4.40}$$

$$W_\kappa^{(5)} = \lambda \left(-\kappa \left(\sigma_+ \tanh(\lambda x) + \sigma_-\right) + \mu\sigma_1 \sqrt{\operatorname{sech}(\lambda x)\exp(-\lambda x)} + \frac{1}{\kappa} R \right),$$

$$W_\kappa^{(6)} = \lambda \left(-\kappa \left(\sigma_+ \coth(\lambda x) + \sigma_-\right) + \mu\sigma_1 \sqrt{\operatorname{csch}(\lambda x)\exp(-\lambda x)} + \frac{1}{\kappa} R \right),$$

$$W_\kappa^{(7)} = -\kappa \left(\frac{\sigma_+}{x + c} + \frac{\sigma_-}{x - c} \right) + \frac{\mu\sigma_1}{\sqrt{x^2 - c^2}} + \frac{1}{\kappa} R, \tag{4.41}$$

$$W_\kappa^{(8)} = -\kappa \frac{\sigma_+}{x} + \mu\sigma_1 \frac{1}{\sqrt{x}} + \frac{1}{\kappa} R, \tag{4.42}$$

$$W_\kappa^{(9)} = \lambda \left(-\kappa I + \mu \exp(-\lambda x)\sigma_1 - \frac{\omega}{\kappa}\sigma_3 \right). \tag{4.43}$$

Here

$$\sigma_{\pm} = \frac{1}{2}(\sigma_0 \pm \sigma_3), \quad R = r_3\sigma_3 + r_2\sigma_2, \tag{4.44}$$

r_a are constants satisfying $r_2^2 + r_3^2 = \omega^2$, and $\kappa,\ \mu,\ \lambda$, and $c \neq 0$ are arbitrary parameters.

4.5 Matrix Superpotentials of Dimension 3 × 3

In analogy with the above we can find superpotentials realized by irreducible 3×3 matrices which are presented in the following formulae:

$$W = (S_1^2 - 1)\frac{\kappa}{x + c_1} + (S_2^2 - 1)\frac{\kappa}{x + c_2} + (S_3^2 - 1)\frac{\kappa}{x}$$
$$+ S_1\frac{\mu_1}{\sqrt{x(x + c_1)}} + S_2\frac{\mu_2}{\sqrt{x(x + c_2)}} + \frac{\omega}{\kappa}(2S_3^2 - 1),$$
$$W = (S_1^2 - 1)\frac{\kappa}{x} + (S_2^2 - 1)\frac{\kappa}{x + c_1} + S_1\frac{\mu_2}{\sqrt{x}} + S_2\frac{\mu_1}{\sqrt{x + c_1}} + \frac{\omega}{\kappa}(2S_3^2 - 1),$$
$$W = (S_1^2 - 1)\frac{\kappa}{x + c_1} + (S_3^2 - 1)\frac{\kappa}{x} + S_1\frac{\mu_2}{\sqrt{x}} + S_3\frac{\mu_1}{\sqrt{x(x + c_1)}} + \frac{\omega}{\kappa}(2S_3^2 - 1),$$
$$W = (S_1^2 - 1)\frac{\kappa}{x} + S_1 c + S_2\frac{\mu_1}{\sqrt{x}} + \frac{\omega}{\kappa}(2S_3^2 - 1),$$
$$W = (S_1^2 - 1)\frac{\kappa}{x + c_1} + (S_2^2 - 1)\frac{\kappa}{x + c_2} + (S_3^2 - 1)\frac{\kappa}{x}$$
$$+ S_1\frac{\mu_1}{\sqrt{x(x + c_1)}} + S_2\frac{\mu_2}{\sqrt{x(x + c_2)}} + S_3\frac{\mu_3}{\sqrt{(x + c_1)(x + c_2)}},$$
$$W = (S_1^2 - 1)\frac{\kappa}{x} + (S_2^2 - 1)\frac{\kappa}{x + c_2} + S_1\frac{\mu_1}{\sqrt{x}} + S_2\frac{\mu_2}{\sqrt{x + c_2}} + S_3\frac{\mu_3}{\sqrt{x(x + c_2)}},$$
$$W = (S_1^2 - 1)\frac{\kappa}{x} + S_1 c + S_3\frac{\mu_1}{\sqrt{x}} + S_2\frac{\mu_2}{\sqrt{x}},$$

where c, c_1, c_2, μ_1, and μ_2 are integration constants, and

$$S_1 = \begin{pmatrix} 0 & 0 & 0 \\ 0 & 0 & -\mathrm{i} \\ 0 & \mathrm{i} & 0 \end{pmatrix}, \quad S_2 = \begin{pmatrix} 0 & 0 & \mathrm{i} \\ 0 & 0 & 0 \\ -\mathrm{i} & 0 & 0 \end{pmatrix}, \quad S_3 = \begin{pmatrix} 0 & -\mathrm{i} & 0 \\ \mathrm{i} & 0 & 0 \\ 0 & 0 & 0 \end{pmatrix} \tag{4.45}$$

are matrices of spin $s = 1$.

The hermiticity condition generates the following restrictions:

$$x > 0, \quad \texttt{if} \quad \mu_1^2 + \mu_2^2 > 0; \qquad c_i < 0 \quad \texttt{if} \quad \mu_i \neq 0. \tag{4.46}$$

Formulae (4.34)–(4.43) give the completed list of the certain class of matrix potentials. Note that they give rise to many realistic QM models described by coupled systems of Schrödinger equations, see the following section.

4.6 *Shape Invariant QM Systems with Matrix Potentials*

The discussed matrix superpotentials naturally appear in realistic QM systems. The entire collection of such system can be found in [38, 39] and [40]. Here we present two examples only.

Consider the following Hamiltonian:

$$H = \frac{p^2}{2m} + \frac{\lambda}{2m}\sigma_i B_i + V, \tag{4.47}$$

where σ_i are Pauli matrices, $B_i = B_i(\mathbf{x})$ are vector components of magnetic field strength, $V = V(\mathbf{x})$ is a potential, and vector $\mathbf{x}$ represents independent variables. In addition, λ denotes the constant of anomalous coupling which is usually represented as $\lambda = g\mu_0$, where μ_0 is the Bohr magneton and g is the Landé factor.

Formula (4.47) presents a generalization of the Pron'ko–Stroganov Hamiltonian for the case of arbitrary external field. And some Schrödinger equations with Hamiltonians (4.47) appear to be shape invariant. The example is given by the following equation:

$$\begin{aligned} H\psi \equiv (-\nabla^2 + \lambda(1-2\kappa)\exp(-x_2)(\sigma_1\cos x_1 - \sigma_2 \sin x_1) \\ + \lambda^2 \exp(-2x_2))\psi = \hat{E}\psi. \end{aligned} \tag{4.48}$$

Here λ is the integrated coupling constant, and independent variables are rescaled to obtain more compact formulae.

Hamiltonian H in (4.48) admits integral of motion $Q = p_1 - \frac{\sigma_3}{2}$. Thus it is possible to expand solutions of (4.48) via eigenvectors of Q which look as follows:

$$\psi_p = \begin{pmatrix} \exp(i(p+\frac{1}{2})x_1)\varphi(x_2) \\ \exp(i(p-\frac{1}{2})x_1)\xi(x_2) \end{pmatrix} \tag{4.49}$$

and satisfy the condition $Q\psi_p = p\psi_p$.

Substituting (4.49) into (4.48) we come to Eq. (5.1) where

$$\hat{V}_\kappa = \lambda^2 \exp(-2y) - \lambda(2\kappa - 1)\exp(-y)\sigma_1 - p\sigma_3,$$
$$y = x_2, \quad E = \tilde{E} - p^2 - \frac{1}{4}, \quad \psi = \begin{pmatrix} \varphi \\ \xi \end{pmatrix}. \tag{4.50}$$

Potential $\hat{V}_\kappa$ (4.50) belongs to the list of shape invariant matrix potentials presented in the above, see Eq. (4.29). Thus Eq. (4.48) can be solved exactly using tools of SUSY quantum mechanics [39]. Notice that this equation is also superintegrable [38].

Let us present an analog of the PS model for particle of spin 1. This model is both superintegrable and shape invariant. It is based on the following Hamiltonian:

$$\mathcal{H}_s = \frac{p_1^2 + p_2^2}{2m} + \frac{1}{r}\mu_s(\mathbf{n}), \tag{4.51}$$

where

$$\mu_s(\mathbf{n}) = \mu_1(\mathbf{n}) = \mu(2(\mathbf{S} \times \mathbf{n})^2 - 1) + \lambda(2(\mathbf{S} \cdot \mathbf{n})^2 - 1). \tag{4.52}$$

Here μ and λ are arbitrary real parameters, $\mathbf{S} \cdot \mathbf{n} = S_1 n_2 + S_2 n_1$, and $\mathbf{S} \times \mathbf{n} = S_1 n_2 - S_2 n_1$, $n_1 = \frac{x_1}{\sqrt{x_1^2 + x_2^2}}$, $n_2 = \frac{x_2}{\sqrt{x_1^2 + x_2^2}}$, S_1 and S_2 are matrices of spin 1 given by formula (4.45).

It is the Hamiltonian defined by Eqs. (4.1) and (4.2) that generalized the Pron'ko–Stroganov model for the case of spin one. This Hamiltonian leads to shape invariant radial equations with matrix potential being the direct sum of a modified Coulomb potential and potential (4.28).

4.7 Dual Shape Invariance

Starting with (4.21)–(4.24) we found the related potentials (4.28)–(4.31) in a unique fashion. But there is an interesting inverse problem: to find possible superpotentials corresponding to given potentials. Formally speaking, this means to find all solutions of the Riccati equation (3.11) for W. However, such solutions depend on two arbitrary parameters (κ and the integration constant), and there is some ambiguity in choosing such of them which should be changing to generate the superpartner potential. Notice that the mentioned inverse problem is very interesting since it opens a way to generate families of isospectral Hamiltonians [24].

In the case of matrix superpotentials this business is even more important since in some cases there exist two superpotentials compatible with the shape invariance condition. And both these superpotential can be requested to generate solutions of the related eigenvalue problem.

To find the mentioned additional superpotentials we use the invariance of potentials (4.28), (4.30), and (4.31) with respect to the simultaneous change of arbitrary parameters:

$$\mu \rightarrow \kappa - \frac{1}{2}, \quad \kappa \rightarrow \mu + \frac{1}{2}. \tag{4.53}$$

This means that in addition to the shape invariance w.r.t. shifts of κ potentials (4.28), (4.30), and (4.31) should be shape invariant w.r.t. shifts of parameter μ too.

Thus, it is possible to represent potentials (4.21), (4.23), and (4.24) in the following alternative form:

$$\widetilde{W}^2_{\mu,\kappa} - \widetilde{W}'_{\mu,\kappa} = \hat{V}_\mu + c_\mu, \tag{4.54}$$

where $\hat{V}_\mu = \hat{V}_\kappa$, and

$$\widetilde{W}_{\mu,\kappa} = \frac{\kappa\sigma_3 - \mu - 1}{x} + \frac{\omega}{2(\mu+1)}\sigma_1, \quad c_\mu = \frac{\omega^2}{4(\mu+1)^2} \tag{4.55}$$

for $\hat{V}_k$ given by Eq. (4.28)

$$\widetilde{W}_{\mu,\kappa} = \frac{\lambda}{2}\left((2\mu+1)\tan\lambda x + (2\kappa - 1)\sec\lambda x\sigma_3 + \frac{4\omega}{2\mu+1}\sigma_1\right) \tag{4.56}$$

for potential (4.30), and

$$\widetilde{W}_{\mu,\kappa} = \frac{\lambda}{2}\left(-(2\mu+1)\coth\lambda x + (2\kappa - 1)\,\mathrm{csch}\,\lambda x\sigma_3 - \frac{4\omega}{2\mu+1}\sigma_1\right) \tag{4.57}$$

for potential (4.31). The related constant c_μ is

$$c_\mu = \lambda^2\left(\pm\frac{1}{4}(2\mu+1)^2 + \frac{4\omega^2}{(2\mu+1)^2}\right), \tag{4.58}$$

where the sign "+" and "−" correspond to the cases (4.56) and (4.57), respectively.

We stress that superpartners of potentials (4.54) constructed using superpotentials $\widetilde{W}_{\mu,\kappa}$, i.e.,

$$\hat{V}^+_\mu = \widetilde{W}^2_{\mu,\kappa} + \widetilde{W}'_{\mu,\kappa}, \tag{4.59}$$

satisfy the shape invariance condition since

$$\hat{V}^+_\mu = \hat{V}_{\mu+1} + C_\mu$$

with $C_\mu = c_{\mu+1} - c_\mu$.

Thus potentials are shape invariant w.r.t. shifts of two parameters, namely, κ and μ. More exactly, superpartners for potentials (4.28), (4.30), and (4.31) can be obtained either by shifts of κ or by shifts of μ, while simultaneous shifts are forbidden. We call this phenomena *dual shape invariance*.

5 Exact Solutions of Shape Invariant Schrödinger Equations

5.1 Generic Approach and Energy Values

An important consequence of the shape invariance is the nice possibility to construct exact solutions of the related stationary Schrödinger equation. The procedure of construction of exact solutions for the case of scalar shape invariant potentials is described in various surveys, see, e.g. [24]. Here we present this procedure for the more general case of matrix potentials.

Consider the stationary Schrödinger equation

$$\hat{H}_\kappa\psi \equiv \left(-\frac{\partial^2}{\partial x^2} + \hat{V}_\kappa\right)\psi = E_\kappa\psi, \tag{5.1}$$

where $\hat{H}_\kappa = a^+_{\kappa,\mu}a^-_{\kappa,\mu} + c_\kappa$ and $\hat{V}_\kappa$ is a shape invariant potential. An algorithm for construction of exact solutions of supersymmetric and shape invariant Schrödinger equations includes the following steps (see, e.g. [24]):

- To find the ground state solutions $\psi_0(\kappa,\mu,x)$ which are proportional to square integrable solutions of the first order equation

$$a^-_{\kappa,\mu}\psi_0(\kappa,\mu,x) \equiv \left(\frac{\partial}{\partial x} + W_{\kappa,\mu}\right)\psi_0(\kappa,\mu,x) = 0. \tag{5.2}$$

  Function $\psi_0(\kappa,\mu,x)$ solves Eq. (5.1) with

$$E_\kappa = E_{\kappa,0} = -c_\kappa. \tag{5.3}$$

- To find a solution $\psi_1(\kappa,\mu,x)$ for the first excited state which is defined by the following relation:

$$\psi_1(\kappa,\mu,x) = a^+_{\kappa,\mu}\psi_0(\kappa+1,\mu,x) \equiv \left(-\frac{\partial}{\partial x} + W_{\kappa,\mu}\right)\psi_0(\kappa+1,\mu,x). \tag{5.4}$$

  Since $a^\pm_\kappa$ and $\hat{H}_\kappa$ satisfy the intertwining relations

$$\hat{H}_\kappa a^+_{\kappa,\mu} = a^+_{\kappa,\mu}\hat{H}_{\kappa+1} \tag{5.5}$$

function (5.4) solves Eq. (5.1) with $E_\kappa = E_{\kappa,1} = -c_{\kappa+1}$.

- Solutions for the second excited state can be found as $\psi_2(\kappa, \mu, x) = a^+_{\kappa,\mu}\psi_1(\kappa + 1, \mu, x)$, etc. Finally, solutions which correspond to nth exited state for any admissible natural number $n > 0$ can be represented as

$$\psi_n(\kappa, \mu, x) = a^+_{\kappa,\mu}a^+_{\kappa+1,\mu} \cdots a^+_{\kappa+n-1,\mu}\psi_0(\kappa + n, \mu, x). \tag{5.6}$$

The corresponding eigenvalue $E_{\kappa,n}$ is equal to $-c_{\kappa+n}$.

- For systems admitting the dual shape invariance it is necessary to repeat the steps enumerated above using alternative (or additional) superpotentials.

All matrix potentials presented in the above generate integrable models with Hamiltonian (5.1). However, it is necessary to examine their consistency, in particular, to verify that there exist square integrable solutions of Eq. (5.2) for the ground states.

In the following sections we find such solutions for all superpotentials given by Eqs. (4.21)–(4.24) and (4.55)–(4.57). However, to obtain normalizable ground state solutions it is necessary to impose certain conditions on parameters of these superpotentials.

Let us present the energy spectra for models (5.1) with potentials (4.28)–(4.31) which can be found by applying the presented algorithm:

$$E = -\frac{\omega^2}{(2N+1)^2} \tag{5.7}$$

for potential (4.28),

$$E = -\lambda^2\left(N^2 + \frac{\omega^2}{N^2}\right) \tag{5.8}$$

for potentials (4.29), (4.31), (4.32), and

$$E = \lambda^2\left(N^2 - \frac{\omega^2}{N^2}\right) \tag{5.9}$$

for potentials (4.30).

Here N is the spectral parameter which can take the following values:

$$N = n + \kappa, \tag{5.10}$$

and (or)

$$N = n + \mu + \frac{1}{2}, \tag{5.11}$$

where $n = 0, 1, 2, \ldots$ are natural numbers which can take any values for potentials (4.28)–(4.30). For potentials (4.29), (4.32), and (4.31) *with a fixed* $k < 0$ the admissible values of n are bound by the condition $(k + n)^2 > |\omega|$.

5.2 Ground State Solutions

To find the ground state solutions for Eqs. (5.1) with potentials (4.28)–(4.31) it is sufficient to solve Eqs. (5.2), where $W_{\kappa,\mu}$ are superpotentials (4.21)–(4.24), and analogous equation with superpotentials (4.55)–(4.57). This can be done for all the mentioned cases, but we present here only two of them.

The corresponding solutions should be square integrable two-component functions which we denote as

$$\psi_0(\kappa, \mu, x) = \begin{pmatrix} \varphi \\ \xi \end{pmatrix}. \tag{5.12}$$

Consider the superpotential defined by Eq. (4.21). Substituting (4.21) and (5.12) into (5.2) we obtain

$$\frac{\partial \varphi}{\partial x} + (\mu - \kappa)\frac{\varphi}{x} + \frac{\omega}{2\kappa + 1}\xi = 0, \tag{5.13}$$

$$\frac{\partial \xi}{\partial x} - (\mu + \kappa + 1)\frac{\xi}{x} + \frac{\omega}{2\kappa + 1}\varphi = 0. \tag{5.14}$$

Solving (5.14) for φ, substituting the solution into (5.13) and making the change

$$\xi = y^{\kappa+1}\hat{\xi}(y), \quad y = \frac{\omega x}{2k + 1}, \tag{5.15}$$

we obtain the equation

$$y^2\frac{\partial^2 \hat{\xi}}{\partial y^2} + y\frac{\partial \hat{\xi}}{\partial y} - \left(y^2 + \mu^2\right)\hat{\xi} = 0, \tag{5.16}$$

whose square integrable solution is proportional to the modified Bessel function:

$$\hat{\xi} = cK_\mu(y). \tag{5.17}$$

Substituting (5.17) into (5.15) and using (5.14) we obtain

$$\varphi = y^{\kappa+1}K_{\mu+1}(y), \quad \xi = y^{\kappa+1}K_{|\mu|}(y), \tag{5.18}$$

where y is the variable defined in (5.15), $\omega x/(2\kappa + 1) \geq 0$.

Functions (5.18) are square integrable provided parameter κ is positive and satisfies the following relation:

$$\kappa - \mu > 0. \tag{5.19}$$

If this condition is violated, i.e., $\kappa - \mu \leq 0$ solutions (5.18) are not square integrable. But since potential (4.28) admits the dual shape invariance, it is possible to make an alternative factorization of Eq. (5.1) using superpotential (4.55) and search for normalizable solutions of the following equation:

$$\tilde{a}^-_{\mu,\kappa}\tilde{\psi}_0(\mu,\kappa,x)\tilde{\psi}_0(\mu,\kappa,x) = 0. \tag{5.20}$$

where $\tilde{a}^-_{\mu,\kappa} = \frac{\partial}{\partial x} + \widetilde{W}_{\mu,\kappa}$. Indeed, solving (5.20) we obtain a perfect ground state vector:

$$\tilde{\psi}_0(\mu,\kappa,x) = \begin{pmatrix} \tilde{\varphi} \\ \tilde{\xi} \end{pmatrix}, \quad \tilde{\varphi} = y^{\mu+\frac{3}{2}} K_{|\nu|}(y), \quad \tilde{\xi} = y^{\mu+\frac{3}{2}} K_{|\nu-1|}(y), \tag{5.21}$$

where $y = \frac{\omega x}{2(\mu+1)}$ and $\nu = \kappa + 1/2$. The normalizability conditions for solution (5.21) are

$$\kappa - \mu < 1, \quad \text{if} \quad \kappa \geq 0, \text{ and } \quad \kappa + \mu > 1, \quad \text{if} \quad \kappa < 0. \tag{5.22}$$

Analogously, considering Eq. (5.2) with superpotential (4.22) and representing its solution in the form (5.12) with

$$\xi = y^{\frac{1}{2}-\kappa}\hat{\xi}(y), \quad \varphi = y^{\frac{1}{2}-\kappa}\hat{\varphi}(y), \quad y = \mu\exp(-\lambda x),$$

we find the following solutions:

$$\varphi = y^{\frac{1}{2}-\kappa}K_{|\nu|}(y), \quad \xi = -y^{\frac{1}{2}-\kappa}K_{|\nu-1|}(y) \tag{5.23}$$

where $\nu = \omega/\kappa + 1/2$ and parameters ω and κ should satisfy the conditions

$$\kappa < 0, \quad \kappa^2 > \omega. \tag{5.24}$$

Since potential (4.29) does not admit the dual shape invariance, there are no other ground state solutions.

In analogous manner we find solutions of Eqs. (5.2) and (5.20) for the remaining superpotentials (4.22)–(4.24), refer to [33] for details. Solutions which correspond to nth energy level can be obtained by applying Eq. (5.6). Under certain conditions on spectral parameters all such solutions are square integrable and reduce to zero at $x = 0$ [33].

5.3 *Isospectrality*

Let us note that for some values of parameters μ and κ potentials (4.28)–(4.32) are isospectral with direct sums of known scalar potentials.

Considering potential (4.28) and using its dual shape invariance it is possible to show that for half-integer μ V_κ can be transformed to a direct sum of scalar Coulomb potentials. In analogous way we can show that potentials (4.30) with half-integer κ or integer μ is isospectral with the potential

$$\hat{V}_\kappa = \lambda^2 \left(r(r-1)\sec^2 \lambda x + 2\omega \tan \lambda x \sigma_1 \right), \quad r = \frac{1}{2} \pm \mu \quad \text{or} \quad r = \kappa, \tag{5.25}$$

which is equivalent to the direct sum of two trigonometric Rosen–Morse potentials. Under the same conditions for parameters μ and κ potential (4.32) is isospectral with the direct sum of two Eckart potentials. Finally, potential (4.32) is isospectral with direct sum of two hyperbolic Rosen–Morse potentials.

In other words, for some special values of parameters μ and κ there exist the isospectrality relations of matrix potentials (4.28)–(4.32) with well-known scalar potentials. However, for another values of these parameters such relations do not exist.

6 Shape Invariant Systems with Position Dependent Mass

SE with position dependent mass are requested for description of various condensed-matter systems such as semiconductors, quantum liquids and metal clusters, quantum dots, etc. However, in contrast with standard QM systems, their symmetries, supersymmetries, and integrals of motion were never investigated systematically.

The systematic study of symmetries of the position dependent mass SEs was started recently. In particular, the completed group classification of such equations in two and three dimensions has been carried out in [41, 42] and [43]. Here we present the classification of all rotationally invariant systems admitting second order integrals of motion [44] which appear to be shape invariant and exactly solvable.

6.1 *Rotationally Invariant Systems*

We will study stationary Schrödinger equations with position dependent mass, which formally coincide with (4.3), but include Hamiltonians with variable mass parameters:

$$\hat{H} = p_a f(\mathbf{x}) p_a + \tilde{V}(\mathbf{x}). \tag{6.1}$$

Here $V(\mathbf{x})$ and $f(\mathbf{x}) = \frac{1}{2m(\mathbf{x})}$ are arbitrary functions associated with the effective potential and inverse effective PDM, and summation from 1 to 3 is imposed over the repeating index a. In addition, $\mathbf{x} = (x^1, x^2, x^3)$ denotes a 3d space vector.

In paper [41] all Hamiltonians (6.1) admitting first order integrals of motion are classified. In particular, the rotationally invariant systems include the following functions f and V:

$$f = f(x), \quad \tilde{V} = \tilde{V}(x), \quad x = \sqrt{x_1^2 + x_2^2 + x_3^2}. \tag{6.2}$$

In accordance with [41] there are four Hamiltonians with a more extended symmetry. They are specified by the following inverse masses and potentials:

$$f = x^2, \quad \tilde{V} = 0, \tag{6.3}$$

$$f = (1 + x^2)^2, \quad \tilde{V} = -6x^2, \tag{6.4}$$

$$f = (1 - x^2)^2, \quad \tilde{V} = -6x^2, \tag{6.5}$$

$$f = x^4, \quad \tilde{V} = -6x^2. \tag{6.6}$$

PDM systems admitting second order integrals of motion are classified in [44]. There are two subclasses of such systems. One class includes the systems admitting vector integrals of motion while in the second one we have the tensor integrals. All these systems are shape invariant, and are presented in the classification Tables 1 and 2.

In the third columns of the tables the effective radial potentials are indicated which appear after the separation of variables. All radial potentials are scalar and shape invariant, i.e., can be expressed in the form (3.11) where the related superpotentials W_κ are enumerated in formulae (4.11)–(4.20). The kinds of the superpotentials is fixed in the fifth columns. The content of the terms presented in the fourth columns is explained in the next section.

We see that there exist exactly 20 superintegrable systems invariant with respect to 3d rotations. Moreover, the majority of them is defined up to one arbitrary parameter, while there exist four systems dependent on two parameters, see Items 9 and 10.

6.2 Two Strategies in Construction of Exact Solutions

Let us consider Eqs. (4.3) where H are Hamiltonians (6.1) whose mass and potential terms are specified in the presented tables. We will search for square integrable solutions of these systems vanishing at $x = 0$.

First let us transform (4.3) to the following equivalent form:

$$\tilde{H}\Psi = E\Psi, \tag{6.7}$$

Table 1 Functions f and V specifying non-equivalent Hamiltonians (6.1)

No	f	V	Solution approach	Effective potentials
1.	x	αx	Direct or two-step	3d oscillator or Coulomb
2.	x^4	αx	Direct or two-step	Coulomb or 3d oscillator
3.	$x(x-1)^2$	$\frac{\alpha x}{(x+1)^2}$	Direct or two-step	Eckart or hyperbolic Pöschl–Teller
4.	$x(x+1)^2$	$\frac{\alpha x}{(x-1)^2}$	Direct or two-step	Eckart or trigonometric Pöschl–Teller
5.	$(1+x^2)^2$	$\frac{\alpha(1-x^2)}{x}$	Direct	Trigonometric Rosen–Morse
6.	$(1-x^2)^2$	$\frac{\alpha(1+x^2)}{x}$	Direct	Eckart
7.	$\frac{x}{x+1}$	$\frac{\alpha x}{x+1}$	Two-step	Coulomb
8.	$\frac{x}{x-1}$	$\frac{\alpha x}{x-1}$	Two-step	Coulomb
9.	$\frac{(x^2-1)^2 x}{x^2-2\kappa x+1}$	$\frac{\alpha x}{x^2-2\kappa x+1}$	Two-step	Eckart
10.	$\frac{(x^2+1)^2 x}{x^2-2\kappa x-1}$	$\frac{\alpha x}{x^2-2\kappa x-1}$	Two-step	Trigonometric Rosen–Morse

where

$$\tilde{H} = \sqrt{f} H \frac{1}{\sqrt{f}} = f p^2 + V, \quad \Psi = \sqrt{f}\psi. \tag{6.8}$$

Then, introducing spherical variables and expanding solutions via spherical functions Y_m^l

$$\Psi = \frac{1}{x} \sum_{l,m} \phi_{lm}(x) Y_m^l, \tag{6.9}$$

we obtain the following equation for radial functions:

$$-f \frac{\partial^2 \phi_{lm}}{\partial x^2} + \left(\frac{f l(l+1)}{x^2} + V \right) \phi_{lm} = E \phi_{lm}. \tag{6.10}$$

Let us present two possible ways to solve Eq. (6.10). They can be treated as particular cases of Liouville transformation (refer to [45] for definitions) and include

Table 2 Functions f and V specifying non-equivalent Hamiltonians (6.1) which admit tensor integrals of motion

No	f	V	Solution approach	Effective radial potential
1.	$\frac{1}{x^2}$	$\frac{\alpha}{x^2}$	Direct or two-step	Coulomb or 3d oscillator
2.	x^4	$-\frac{\alpha}{x^2}$	Direct or two-step	3d oscillator or Coulomb
3.	$(x^2-1)^2$	$\frac{\alpha x^2}{(x^2+1)^2}$	Direct or two-step	Eckart or hyperbolic Pöschl–Teller
4.	$(x^2+1)^2$	$\frac{\alpha x^2}{(x^2-1)^2}$	Direct or two-step	Eckart or trigonometric Pöschl–Teller
5.	$\frac{(x^4-1)^2}{x^2}$	$\frac{\alpha(x^4+1)}{x^2}$	Direct	Eckart
6.	$\frac{(x^4+1)^2}{x^2}$	$\frac{\alpha(x^4-1)}{x^2}$	Direct	Trigonometric Rosen–Morse
7.	$\frac{1}{x^2+1}$	$\frac{\alpha}{x^2+1}$	Two-step	3d oscillator
8.	$\frac{1}{x^2-1}$	$\frac{\alpha}{x^2-1}$	Two-step	3d oscillator
9.	$\frac{(x^4-1)^2}{x^4-2\kappa x^2+1}$	$\frac{\alpha x^2}{x^4-2\kappa x^2+1}$	Two-step	Eckart
10.	$\frac{(x^4+1)^2}{x^4-2\kappa x^2-1}$	$\frac{\alpha x^2}{x^4-2\kappa x^2-1}$	Two-step	Trigonometric Rosen–Morse

commonly known steps. But it is necessary to fix them as concrete algorithms to obtain shape invariant potentials presented in the tables.

The first way (which we call direct) includes consequent changes of independent and dependent variables:

$$\phi_{lm} \to \Phi_{lm} = f^{\frac{1}{4}}\phi_{lm}, \quad \frac{\partial}{\partial x} \to f^{\frac{1}{4}}\frac{\partial}{\partial x}f^{-\frac{1}{4}} = \frac{\partial}{\partial x} + \frac{f'}{4f} \tag{6.11}$$

and then

$$x \to y(x), \tag{6.12}$$

where y solves the equation $\frac{\partial y}{\partial x} = \frac{1}{\sqrt{f}}$. As a result Eq. (6.9) will be reduced to a more customary form:

$$-\frac{\partial^2 \Phi_{lm}}{\partial y^2} + \tilde{V}\Phi_{lm} = E\Phi_{lm}, \tag{6.13}$$

where $\tilde{V}$ is an effective potential

$$\tilde{V} = V + f\left(\frac{l(l+1)}{x^2} - \left(\frac{f'}{4f}\right)^2 - \left(\frac{f'}{4f}\right)'\right), \quad x = x(y). \tag{6.14}$$

Equations (6.7), (6.8) with functions f and V specified in Items 1–6 of both Tables 1 and 2 can be effectively solved using the presented reduction to radial Eq. (6.13). All the corresponding potentials (6.14) appear to be shape invariant, and just these potentials are indicated in the fifth columns of the tables. The related Eqs. (6.13) are shape invariant too and can be solved using the SUSY routine.

However, if we apply the direct approach to the remaining systems (indicated in Items 7–10 of both tables), we come to Eqs. (6.13) which are not shape invariant and are hardly solvable, if at all. To solve these systems we need a more sophisticated procedure which we call two-step approach. To apply it we multiply (6.10) by αV^{-1} and obtain the following equation:

$$-\tilde{f}\frac{\partial^2 \phi_{lm}}{\partial x^2} + \left(\frac{\tilde{f}l(l+1)}{x^2} + \tilde{V}\right)\phi_{lm} = \mathcal{E}\phi_{lm}, \tag{6.15}$$

where $\tilde{f} = \frac{\alpha f}{V}$, $\tilde{V} = -\frac{\alpha E}{V}$, and $\mathcal{E} = -\alpha$. Then treating $\mathcal{E}$ as an eigenvalue and solving Eq. (6.15) we can find α as a function of E, which defines admissible energy values at least implicitly. To do it, it is convenient to make changes (6.11) and (6.12), where $f \rightarrow \tilde{f}$.

The presented trick with a formal changing the roles of constants α and E is well known. Our point is that *any of the presented superintegrable systems can be effectively solved using either the direct approach presented in Eqs.* (6.8)–(6.14) *or the two-step approach*. Moreover, some of the presented systems can be solved using both the direct and two-step approaches, as indicated in the fourth columns of Tables 1 and 2. In all cases we obtain shape invariant effective potentials and can use tools of SUSY quantum mechanics.

6.3 System Dependent on Two Parameters

Let us consider the systems specified in Item 10 of Table 2. The corresponding Hamiltonian (6.8) and radial Eq. (6.10) have the following form:

$$H = \frac{(x^4+1)^2}{x^4 - 2\kappa x^2 - 1}p^2 + \frac{\alpha x^2}{x^4 - 2\kappa x^2 - 1}$$

and

$$\left(-\frac{(x^4+1)^2}{x^4-2\kappa x^2-1}\left(\frac{\partial^2}{\partial x^2}-\frac{l(l+1)}{x^2}\right)+\frac{\alpha x^2}{x^4-2\kappa x^2-1}\right)\phi_{lm}=E\phi_{lm}. \tag{6.16}$$

Multiplying (6.16) from the left by $\frac{x^4-2\kappa x^2-1}{x^2}$ we come to the following equation:

$$\left(-\frac{(x^4+1)^2}{x^2}\left(\frac{\partial^2}{\partial x^2}-\frac{l(l+1)}{x^2}\right)+\frac{\tilde{\alpha}(x^4-1)}{x^2}\right)\phi_{lm}=\mathcal{E}\phi_{lm}, \tag{6.17}$$

where

$$\tilde{\alpha}=-E \quad \text{and} \quad \mathcal{E}=-\alpha-2\kappa E. \tag{6.18}$$

Notice that Eq. (6.17) with $\tilde{\alpha}\to\alpha$ and $\mathcal{E}\to E$ is needed also to find eigenvectors of the Hamiltonian whose mass and potential terms are specified in Item 6 of Table 2.

Making transformations (6.11) and (6.12) with $f=\frac{(x^4+1)^2}{x^2}$ and $y=\frac{1}{2}\arctan(x^2)$ we reduce Eq. (6.17) to the following form:

$$-\frac{\partial^2\Phi_{lm}}{\partial y^2}+\left(\mu(\mu-4)\csc^2(4y)+2\tilde{\alpha}\cot(4y)\right)\Phi_{lm}=\tilde{\mathcal{E}}\Phi_{lm}, \tag{6.19}$$

where

$$\tilde{\mathcal{E}}=\mathcal{E}+4, \quad \mu=2l+3. \tag{6.20}$$

Thus we come to equation with a shape invariant (Rosen–Morse I) potential. It is consistent provided parameters $\tilde{\alpha}$ and μ are positive. Solving this equation using the standard tools of SUSY QM we can easily find its eigenfunctions and eigenvalues; the corresponding eigenvalues for Eq. (6.16) are given by the following formula [44]:

$$E_n=(2l+3+4n)^2\left(\kappa-\sqrt{\kappa^2+1+\frac{\alpha-4}{(2l+3+4n)^2}}\right), \tag{6.21}$$

where both n and l are integers.

7 Discussion

To construct QM systems with extended SUSY we essentially use discrete symmetries, i.e., reflections and rotations to the fixed angles.

The idea itself to apply reflections to construct $N = 2$ SUSY was proposed in paper [46]. Then it was applied to generate extended supersymmetries [35, 36, 47, 48], moreover, in the latter paper the discrete rotations were applied also. In addition, using these discrete symmetries, it is possible to make a reduction of SUSY algebras as it was shown in paper [49] and some others.

We start our discussion with presenting of these old results in order to stress that SUSY has strong roots in quantum mechanics since a lot of important QM models do be supersymmetric. Moreover, even the simplest SUSY model, i.e., the charged particle interacting with the uniform magnetic field, in fact admits the extended supersymmetry with four supercharges [35].

But the main content of the present survey are some modern trends in SUSY quantum mechanics. They are the matrix formulation of the shape invariance which is requested for description of QM particles with spin interacting with external fields, and supersymmetries of Schrödinger equations with position dependent masses. And we believe that the presented results can be treated as a challenge to generalize various branches of SUSY to the case of matrix superpotentials. And it is nice that some elements of such generalizations can be already recognized in literature, see, e.g. [50–55].

References

1. Y.A. Gol'fand, E.P. Lichtman, Sov. Phys. JETP Lett. **13**, 452 (1971); D.V. Volkov, V.P. Akulov, Phys. Lett. B **46**, 109 (1973); J. Wess, B. Zumino, Nucl. Phys. B **70**, 39 (1974)
2. J. Lipkin, Phys. Lett. **9**, 203 (1964); J. Schwinger, Phys. Rev. **152**, 1219 (1966); G.L. Stavraki, in *High Energy Physics and the Theory of Elementary Particles*, Naukova Dumka, Kiev (1966), p. 296 (in Russian); Preprint ITP 67-21, Kiev, 1967; H. Migazawa, Progr. Theor. Phys. **36**, 1266 (1968), Phys. Rev. **170**, 1586 (1968); M. Flato; P. Hillon, Phys. Rev. D **1**, 1667 (1970); A. Neveu, J.M. Schwartz, Nucl. Phys B **31**, 86 (1971); J.L. Gervais, B. Sakita, Nucl. Phys. B **34**, 633 (1971); A. Joseph, Nuovo Cimento A **8**, 217 (1972); Y. Aharonov, A. Casher, L. Susskind, Phys. Lett. B **35**, 512 (1974)
3. V.A. Kostelecky, D.K. Campbell, Phys. D **15**, 3 (1985)
4. M.B. Green, J.H. Schwartz, E. Witten, Superstring theory, in 2 vols (Cambridge University Press, Cambridge, 1987); M. Kaku, *Strings, Conformal Field Theory and Topology* (Springer, New York, 1989)
5. N. Seiberg, E. Witten, Nucl. Phys. B **426**, 19 (1994); N. Seiberg, Phys. Rev. D **49**, 6857 (1994)
6. L.E.Gendenshtein, I.V. Krive, Usp. Fiz. Nauk **146**, 553 (1985)
7. F. Cooper, A. Khare, U. Sukhatme, Phys. Rep. **211**, 268 (1995)
8. S.V. Sukhumar, J. Phys. A **18**, L697 (1985)
9. F. Ravndal, Phys. Rev. D **21**, 2461 (1980); A. Khare, J. Maharana, Nucl. Phys. B **244**, 409 (1984)
10. E. Witten, Nucl. Phys. B **185**, 513 (1981); **202**, 253 (1982)
11. L. Gendenshtein, JETP Lett. **38**, 356 (1983)

12. G.A. Natanzon, Vestnik Leningrad Univ. **10**, 22 (1971); Teor. Mat. Fiz. **38**, 146 (1979)
13. A.A. Andrianov, M.V. Ioffe, J. Phys. A **45**, 503001 (2012)
14. T. Tanaka, Nucl. Phys. B **662**, 413 (2003)
15. F. Correa, V. Jakubsky, M.S. Plyushchay, J. Phys. A: Math. Theor. **41**, 485303 (2008)
16. C. Quesne, SIGMA **3**, 067 (2007)
17. V.Y. Novokshenov, SIGMA **14**, 106 (2018)
18. V.A. Rubakov, V.P. Spiridonov, Mod. Phys. Lett. A **3**, 1337 (1988)
19. J. Beckers, N. Debergh, Nucl. Phys. B **340**, 767 (1990)
20. J, Beckers, N. Debergh, A.G. Nikitin, Mod. Phys. Let A **8**, 435 (1993)
21. J. Beckers, N. Debergh, A.G. Nikitin, J. Math. Phys. **33**, 152 (1992)
22. G.P. Pron'ko, Y.G. Stroganov, Sov. Phys. JETP **45**, 1075 (1977)
23. E. Ferraro, N. Messina, A.G. Nikitin, Phys. Rev. A **81**, 042108 (2010)
24. F. Cooper, A. Khare, U. Sukhatme, Phys. Rep. **251**, 267 (1995)
25. A.A. Andrianov, N.V. Borisov, M. V. Ioffe, Theor. Math. Phys. **61**, 1078 (1984)
26. M.V. Ioffe, SIGMA **6**, 075 (2010)
27. A.A. Andrianov, M.V. Ioffe, Phys. Lett. B **255**, 543 (1991); A.A. Andrianov, M.V. Ioffe, V.P. Spiridonov, L. Vinet, Phys. Lett. B **272**, 297 (1991)
28. A.A. Andrianov, F. Cannata, M.V. Ioffe, D.N. Nishnianidze, J. Phys. A: Math. Gen. **30**, 5037 (1997)
29. T. Fukui, Phys. Lett. A **178**, 1 (1993)
30. M.V. Ioffe, S. Kuru, J. Negro, L.M. Nieto, J. Phys. A **39**, 6987 (2006)
31. R. de Lima Rodrigues, V.B. Bezerra, A.N. Vaidyac, Phys. Lett. A **287**, 45 (2001)
32. V.M. Tkachuk, P. Roy, Phys. Lett. A **263**, 245 (1999); V.M. Tkachuk, P. Roy, J. Phys. A **33**, 4159 (2000)
33. A.G. Nikitin, Y. Karadzhov, Matrix superpotentials. J. Phys. A Math. Theor. **44**, 305204 (2011)
34. A.G. Nikitin, Y. Karadzhov, Enhanced classification of matrix superpotentials. J. Phys. A Math. Theor. **44**, 445202 (2011)
35. J. Niederle, A.G. Nikitin, J. Math. Phys. **40**, 1280 (1999)
36. A.G. Nikitin, J. Mod. Phys. A **14**, 885 (1999)
37. Y. Karadzhov, Matrix superpotentials, Thesis, Kiev, Institute of Mathematics, 2015
38. A.G. Nikitin, J. Math. Phys. **53**, 122103 (2012)
39. A.G. Nikitin, Superintegrable and supersymmetric systems of Schrödinger equations, in *Proceedings of the Sixth International Workshop on Group Analysis of Differential Equations and Integrable Systems*, June 17–21, 2012, Protaras, Cyprus. University of Cyprus, Nikosia, 2013, pp.154–169
40. A.G. Nikitin, J. Phys. A: Math. Theor. **45**, 225205 (2012)
41. A.G. Nikitin, T.M. Zasadko, J. Math. Phys. 56, 042101 (2015)
42. A.G. Nikitin, T.M. Zasadko, J. Phys. A **49**, 365204 (2016)
43. A.G. Nikitin, J. Math. Phys. **58**, 083508 (2017)
44. A.G. Nikitin, J. Phys. A **48**, 335201 (2015)
45. F.W. Olver , *Asymptotics and Special Functions* (Academic Press, New York, 1974)
46. L.E. Gendenshtein, JETP Lett. **39**, 234 (1984)
47. V.M. Tkachuk, S.I. Vakarchuk, Phys. Lett. A **228**, 141 (1997)
48. A.G. Nikitin, in Problems of quantum field theory, *JINR E2-96-369* (Dubna, 1996), p. 509
49. J. Beckers, N. Debergh, A.G. Nikitin, Int. J. Theor. Phys. **36**, 1991 (1997)
50. T. Tanaka, Mod. Phys. Lett. A **27**, 1250051 (2012)
51. A.V. Sokolov, J. Phys. A **48**, 085202 (2015)
52. A.V. Sokolov, Phys. Lett. A **377**, 655 (2013)
53. A.A. Andrianov, A.V. Sokolov, Phys. Lett. A **379**, 279 (2015)
54. A.A. Andrianov, A.V. Sokolov, Theor. Math. Phys. **186**, 2 (2016)
55. M.V. Ioffe, E.V. Kolevatova, D.N. Nishnianidze, Phys. Lett. A **380**, 3349 (2016)

Nonlinear Supersymmetry as a Hidden Symmetry

Mikhail S. Plyushchay

Abstract Nonlinear supersymmetry is characterized by supercharges to be higher order in bosonic momenta of a system, and thus has a nature of a hidden symmetry. We review some aspects of nonlinear supersymmetry and related to it exotic supersymmetry and nonlinear superconformal symmetry. Examples of reflectionless, finite-gap and perfectly invisible $\mathcal{PT}$-symmetric zero-gap systems, as well as rational deformations of the quantum harmonic oscillator and conformal mechanics, are considered, in which such symmetries are realized.

Keywords Hidden symmetry · Exotic supersymmetry · Nonlinear superconformal symmetry · Reflectionless and finite-gap systems · Perfect invisibility

1 Introduction

Hidden symmetries are associated with integrals of motion of higher-order in momenta. They mix the coordinate and momenta variables in the phase space of a system, and generate a nonlinear, W-type algebras [1]. The best known examples of hidden symmetries are provided by the Laplace–Runge–Lenz vector integral in the Kepler–Coulomb problem, and the Fradkin–Jauch–Hill tensor in isotropic harmonic oscillator systems. Hidden symmetries also appear in anisotropic oscillator with commensurable frequencies, where they underlie the closed nature of classical trajectories and specific degeneration of the quantum energy levels. Hidden symmetry is responsible for complete integrability of geodesic motion of a test particle in the background of the vacuum solution to the Einstein's equation represented by the Kerr metric of the rotating black hole and its generalizations in the form of the Kerr-NUT-(A)dS solutions of the Einstein–Maxwell equations

M. S. Plyushchay (✉)
Departamento de Física, Universidad de Santiago de Chile, Casilla, Santiago, Chile
e-mail: mikhail.plyushchay@usach.cl

Ş. Kuru et al. (eds.), *Integrability, Supersymmetry and Coherent States*, CRM Series in Mathematical Physics, https://doi.org/10.1007/978-3-030-20087-9_6

[2]. Another class of hidden symmetries underlies a complete integrability of the field systems described by nonlinear wave equations such as the Korteweg–de Vries (KdV) equation. Those symmetries are responsible for peculiar properties of the soliton and finite-gap solutions of the KdV system, whose equation of motion can be regarded as a geodesic flow on the Virasoro-Bott group [3, 4].

Nonlinear supersymmetry [5–45] is characterized by supercharges to be higher order in even (bosonic) momenta of a system, and thus has a nature of hidden symmetry. Here, we review some aspects of nonlinear supersymmetry, and related to it exotic supersymmetry and nonlinear superconformal symmetry.

Nonlinear supersymmetry appears, particularly, in purely parabosonic harmonic oscillator systems generated by the deformed Heisenberg algebra with reflection [12] as well as in a generalized Landau problem [15]. The peculiarity of supersymmetric parabosonic systems shows up in the nonlocal nature of supercharges to be of infinite order in the momentum operator as well as in the ladder operators but anti-commuting for a polynomial in Hamiltonian being quadratic in creation-annihilation operators. Similar peculiarities characterize hidden supersymmetry and hidden superconformal symmetry appearing in some usual quantum bosonic systems with a local Hamiltonian operator [20, 21, 24–26, 30–32, 35, 46–51]. Exotic supersymmetry emerges in superextensions of the quantum systems described by soliton and finite-gap potentials, in which the key role is played by the Lax–Novikov integrals of motion [30–33, 42]. A structure similar to that of the exotic supersymmetry of reflectionless and finite-gap quantum systems can also be identified in the "SUSY in the sky" type supersymmetry [52–55] based on the presence of the Killing–Yano tensors in the abovementioned class of the black hole solutions to the Einstein–Maxwell equations. Nonlinear superconformal symmetry appears in rational deformations of the quantum harmonic oscillator and conformal mechanics systems [49, 51]. Both exotic supersymmetry and nonlinear superconformal symmetry characterize the interesting class of the perfectly invisible zero-gap $\mathcal{PT}$-symmetric systems, which includes the $\mathcal{PT}$-regularized two-particle Calogero systems and their rational extensions with potentials satisfying the equations of the KdV hierarchy and exhibiting a behavior of extreme (rogue) waves [56, 57].

2 Nonlinear Supersymmetry and Quantum Anomaly

Classical analog of the Witten's supersymmetric quantum mechanics [58–61] is described by the Hamiltonian

$$\mathcal{H} = p^2 + W^2 + W'N\,, \tag{2.1}$$

where $N = \theta^+\theta^- - \theta^-\theta^+$, $W = W(x)$ is a superpotential, x and p are even canonical variables, $\{x, p\} = 1$, and $\theta^+, \theta^- = (\theta^+)^*$ are Grassmann variables with the only nonzero Poisson bracket $\{\theta^+, \theta^-\} = -i$. System (2.1) is characterized by the even, N, and odd, $Q_+ = (W + ip)\theta^+$ and $Q_- = (Q_+)^*$, integrals of motion satisfying the algebra of $\mathcal{N}=2$ Poincaré supersymmetry

$$\{Q_+, Q_-\} = -i\mathcal{H}\,, \quad \{\mathcal{H}, Q_\pm\} = 0\,, \quad \{N, \mathcal{H}\} = 0\,, \quad \{N, Q_\pm\} = \pm 2i\, Q_\pm\,. \tag{2.2}$$

For any choice of the superpotential, canonical quantization of this classical system gives rise to the supersymmetric quantum system in which quantum supercharges and Hamiltonian satisfy the $\mathcal{N}=2$ superalgebra given by a direct quantum analog of the corresponding Poisson bracket relations, with the quantum analog of the integral N playing simultaneously the role of the $\mathbb{Z}_2$-grading operator $\Gamma = \sigma_3$ of the Lie superalgebra.

A simple change of the last term in (2.1) for $n\mathcal{W}'N$ with n taking any integer value yields a system characterized by a nonlinear supersymmetry of order n generated by the supercharges $S_+ = (\mathcal{W}+ip)^n\theta^+$ and $S_- = (S_+)^*$ being the integrals of order n in the momentum p. Their Poisson bracket $\{S_+, S_-\} = -i(\mathcal{H})^n$ has order n in the Hamiltonian [12, 14, 62]

$$\mathcal{H} = p^2 + \mathcal{W}^2 + n\mathcal{W}'N\,. \tag{2.3}$$

System (2.3) can be regarded as a kind of the classical supersymmetric analog of the planar anisotropic oscillator with commensurable frequencies [63, 64]. Unlike a linear case (2.1) with $n=1$, canonical quantization of the system (2.3) with $n = 2, 3, \ldots$ faces, however, the problem of quantum anomaly: for arbitrary form of the superpotential, quantum analogs of the classical odd integrals $S_\pm$ cease to commute with the quantum analog of the Hamiltonian (2.3). In [14], it was found a certain class of superpotentials $\mathcal{W}(x)$ for which the supercharge S_+ has a polynomial structure in $z = \mathcal{W}+ip$ instead of monomial one so that the corresponding systems admit an anomaly-free quantization giving rise to quasi-exactly solvable systems [65–67].

If instead of the "holomorphic" dependence of the supercharge S_+ on the complex variable z we consider the supercharges with polynomial dependence on the momentum variable p, the case of quadratic supersymmetry turns out to be a special one. The Hamiltonian and supercharges then can be presented in the most general form

$$\mathcal{H} = zz^* - \frac{C}{\mathcal{W}^2} + 4\mathcal{W}'N + a\,, \tag{2.4}$$

$$S_+ = \left(z^2 + \frac{C}{\mathcal{W}^2}\right)\theta^+\,, \qquad S_- = (S_+)^*\,. \tag{2.5}$$

Here a and C are real constants, and we have

$$\{S_+, S_-\} = -i\left((\mathcal{H}-a)^2 + C\right)\,. \tag{2.6}$$

Supersymmetry of the system (2.4), (2.5), (2.6) with an arbitrary superpotential can be preserved at the quantum level if to correct the direct quantum analog of the Hamiltonian and supercharges by adding to them the term quadratic in Plank constant [14, 62]:

$$\hat{\mathcal{H}} - a = -\hbar^2 \frac{d^2}{dx^2} + \mathcal{W}^2 - 2\hbar\sigma_3\mathcal{W}' - \frac{C}{\mathcal{W}^2} + \Delta(\mathcal{W}), \tag{2.7}$$

$$\hat{S}_+ = \hat{s}_+\sigma_+, \qquad \hat{s}_+ = \left(\hbar\frac{d}{dx} + \mathcal{W}\right)^2 + \frac{C}{\mathcal{W}^2} - \Delta(\mathcal{W}), \tag{2.8}$$

$$\Delta(\mathcal{W}) = \frac{1}{2}\hbar^2\left(\frac{\mathcal{W}''}{\mathcal{W}} - \frac{1}{2}\left(\frac{\mathcal{W}'}{\mathcal{W}}\right)^2\right) = \hbar^2\frac{1}{\sqrt{\mathcal{W}}}\left(\sqrt{\mathcal{W}}\right)'', \tag{2.9}$$

where $\sigma_+ = \frac{1}{2}(\sigma_1 + i\sigma_2)$. The quantum term $\Delta(\mathcal{W})$ can be presented as a Schwarzian, $\Delta = -\frac{1}{2}\hbar^2 S(\omega(x))$, $S(\omega(x)) = (\omega''/\omega')' - \frac{1}{2}(\omega''/\omega')^2$, of the function $\omega(x) = \int^x dy/\mathcal{W}(y)$. The quadratic in $\hbar$ terms in the quantum Hamiltonian (2.7) can be unified and presented in a form similar to that of the kinetic term of the quantum particle in a curved space: $-\hbar^2\frac{d^2}{dx^2} + \Delta(\mathcal{W}) = \hat{\mathcal{P}}^\dagger\hat{\mathcal{P}}$, where $\hat{\mathcal{P}} = \hbar\zeta^{-1}\frac{d}{dx}\zeta$, $\zeta = 1/\sqrt{\mathcal{W}}$. Analogously, the first and third terms in $\hat{s}_+$ in (2.8) can be collected and presented in the form $\hat{z}^2 - \Delta(W) = (\zeta\hat{z}\zeta^{-1})(\zeta^{-1}\hat{z}\zeta)$, where $\hat{z} = \hbar\frac{d}{dx} + \mathcal{W}$ [62].

3 Exotic Nonlinear $\mathcal{N} = 4$ Supersymmetry

The anomaly-free prescription for quantization of the classical systems (2.3) with supersymmetry of order higher than two in general case is unknown, but there exist infinite families of the quantum systems described by supersymmetries of arbitrary order. They can be generated easily by applying the higher order Darboux–Crum (DC) transformations [68–70] to a given, for instance, exactly solvable quantum system instead of starting from a classical supersymmetric system of the form (2.3) followed by a search for the anomaly-free quantization scheme.

In general case the DC transformation of a given system described by the Hamiltonian operator $\hat{H}_- = -\frac{d^2}{dx^2} + V_-(x)$ is generated by selection of the set of physical or non-physical eigenstates $(\psi_1, \psi_2, \ldots, \psi_n)$ of $\hat{H}_-$ as the seed states. Here and below we put $\hbar = 1$. If they are chosen in such a way that their Wronskian $\mathbb{W}(\psi_1, \ldots, \psi_n)$ takes nonzero values in the region where $V_-(x)$ is defined, then the new potential

$$V_+ = V_- - 2(\ln \mathbb{W}(\psi_1, \ldots, \psi_n))'' \tag{3.1}$$

will be regular in the same region as V_-. Physical and non-physical eigenstates of the new Hamiltonian operator $\hat{H}_+ = -\frac{d^2}{dx^2} + V_+$ are obtained from those of the original system $\hat{H}_-$ by the transformation

$$\psi_{+,\lambda} = \frac{\mathbb{W}(\psi_1, \ldots, \psi_n, \psi_\lambda)}{\mathbb{W}(\psi_1, \ldots, \psi_n)} = \mathbb{A}_n \psi_\lambda , \tag{3.2}$$

where ψ_λ is an eigenstate of $\hat{H}_-$ different from eigenstates in the set of the seed states with eigenvalue $E_\lambda \neq E_j$, $j = 1, \ldots, n$. The state $\psi_{+,\lambda}$ is of the same eigenvalue of $\hat{H}_+$ as ψ_λ of $\hat{H}_-$, $\hat{H}_-\psi_\lambda = \lambda\psi_\lambda \Rightarrow \hat{H}_+\psi_{+,\lambda} = \lambda\psi_{+,\lambda}$, and vice versa, from $\hat{H}_+\psi_{+,\lambda} = \lambda\psi_{+,\lambda}$ it follows that $\hat{H}_-\psi_\lambda = \lambda\psi_\lambda$. Operator $\mathbb{A}_n$ in (3.2) is a differential operator of order n,

$$\mathbb{A}_n = A_n \ldots A_1 , \quad A_j = (\mathbb{A}_{j-1}\psi_j)\frac{d}{dx}(\mathbb{A}_{j-1}\psi_j)^{-1} , \quad j = 1, \ldots, n , \quad \mathbb{A}_0 = 1 , \tag{3.3}$$

which is constructed recursively from the selected seed states. Operators $\mathbb{A}_n$ and $\mathbb{A}_n^\dagger$ intertwine Hamiltonian operators $\hat{H}_-$ and $\hat{H}_+$,

$$\mathbb{A}_n \hat{H}_- = \hat{H}_+ \mathbb{A}_n , \qquad \mathbb{A}_n^\dagger \hat{H}_+ = \hat{H}_- \mathbb{A}_n^\dagger , \tag{3.4}$$

and satisfy relations

$$\mathbb{A}_n^\dagger \mathbb{A}_n = \prod_{j=1}^{n}(\hat{H}_- - E_j) , \qquad \mathbb{A}_n \mathbb{A}_n^\dagger = \prod_{j=1}^{n}(\hat{H}_+ - E_j) , \tag{3.5}$$

where E_j is eigenvalue of the seed eigenstate ψ_j. Relations (3.4) and (3.5) underlie nonlinear supersymmetry of the extended system $\hat{\mathcal{H}} = \text{diag}\,(\hat{H}_+, \hat{H}_-)$, the supercharges of which are constructed from the operators $\mathbb{A}_n$ and $\mathbb{A}_n^\dagger$.

Using Eq. (3.2), one can prove the relation [71]

$$\mathbb{W}(\psi_*, \widetilde{\psi_*}, \psi_1, \ldots, \psi_n) = \mathbb{W}(\psi_1, \ldots, \psi_n) . \tag{3.6}$$

Here and in what follows equality between Wronskians is implied up to inessential multiplicative constant; ψ_* is some eigenstate of $\hat{H}_-$ with eigenvalue E_* different from E_j, $j = 1, \ldots, n$, and $\widetilde{\psi_*} = \psi_* \int^x dy/(\psi_*(y))^2$ is a linear independent eigenstate with the same eigenvalue E_* so that $\mathbb{W}(\psi_*, \widetilde{\psi_*}) = 1$.

Among supersymmetric quantum systems generated by DC transformations, there exists special class of infinite subfamilies in which the corresponding superextended systems are characterized simultaneously by supersymmetries of two different orders, one of which is of even order $n = 2l$, while another has some odd order $n = 2k + 1$ [30–32, 36, 37, 42, 56]. This corresponds to supersymmetrically extended finite-gap or reflectionless systems, which can be regarded as "instant photos" of solutions to the KdV equation [72] and are characterized by the presence of a nontrivial Lax–Novikov integrals to be operators of the odd differential order $n = 2\ell + 1 \geq 3$ with $\ell = l + k$. Factorization of Lax–Novikov integrals into two differential operators of orders $2l$ and $2k+1$ is reflected in the presence of the exotic nonlinear $\mathcal{N} = 4$ Poincaré supersymmetry generated by supercharges of orders $2l$ and $2k + 1$ instead of linear or nonlinear $\mathcal{N} = 2$ Poincaré supersymmetry obtained usually via the Darboux or Darboux–Crum transformation construction.

A simple example of a system with exotic nonlinear $\mathcal{N} = 4$ supersymmetry is generated via the construction of Witten's supersymmetric quantum mechanics with superpotential $W(x) = \kappa \tanh \kappa x$, where κ is a parameter of dimension of inverse length. The corresponding superextended system is described by the Hamiltonian $\hat{\mathcal{H}} = \text{diag}\,(\hat{H}_+, \hat{H}_-)$ with $\hat{H}_- = -\frac{d^2}{dx^2} + \kappa^2$, $\hat{H}_+ = \hat{H}_- - 2\kappa^2/\cosh^2 \kappa x$, and first order supercharges $\hat{Q}_+ = (\frac{d}{dx} - W(x))\sigma_+$, $\hat{Q}_- = (\hat{Q}_+)^\dagger$. They generate the $\mathcal{N} = 2$ Poincaré superalgebra via the (anti)commutation relations

$$\{\hat{Q}_+, \hat{Q}_-\} = \hat{\mathcal{H}}\,, \qquad [\hat{\mathcal{H}}, \hat{Q}_\pm] = 0\,. \tag{3.7}$$

This system can also be obtained via the construction of the $n = 2$ supersymmetry by choosing $\mathcal{W}(x) = -\frac{1}{2}\kappa \tanh \kappa x$ and $C = -\frac{1}{16}\kappa^4$ [62]. In this case $\Delta = -\frac{\kappa^2}{\cosh^2 \kappa x}(1 + \frac{1}{4\sinh^2 \kappa x})$, and the operator in the second order supercharge (2.8) is factorized in the form

$$\hat{s}_+ = \left(\frac{d}{dx} - \kappa \tanh \kappa x\right)\frac{d}{dx}\,. \tag{3.8}$$

We have here

$$\{\hat{S}_+, \hat{S}_-\} = \left(\hat{\mathcal{H}} - \frac{1}{2}\kappa^2\right)^2 - \frac{1}{16}\kappa^4\,, \qquad [\hat{\mathcal{H}}, \hat{S}_\pm] = 0\,. \tag{3.9}$$

The anti-commutators of the first and second order supercharges generate a nontrivial even integral of motion,

$$\{\hat{S}_+, \hat{Q}_-\} = -\{\hat{S}_-, \hat{Q}_+\} = i\hat{\mathcal{L}}\,, \tag{3.10}$$

$$\hat{\mathcal{L}} = \begin{pmatrix} \hat{q}_+ \hat{p}\, \hat{q}_+^\dagger & 0 \\ 0 & \hat{H}_- \hat{p} \end{pmatrix}, \tag{3.11}$$

where $\hat{q}_+ = \frac{d}{dx} - \kappa \tanh \kappa x$. Operator (3.11) satisfies the commutation relations

$$[\hat{\mathcal{L}}, \hat{Q}_\pm] = [\hat{\mathcal{L}}, \hat{S}_\pm] = [\hat{\mathcal{L}}, \hat{\mathcal{H}}] = 0\,, \tag{3.12}$$

which mean that the integral $\hat{\mathcal{L}}$ is the central element of the nonlinear superalgebra generated by $\hat{\mathcal{H}}$, $\hat{Q}_\pm$, $\hat{S}_\pm$, and $\hat{\mathcal{L}}$. The lower term in the diagonal operator $\hat{\mathcal{L}}$ is the momentum operator of a free quantum particle multiplied by $\hat{H}_-$, while the third order differential operator $\hat{q}_+\hat{p}\,\hat{q}_+^\dagger$ is the Lax–Novikov integral of reflectionless system described by the Hamiltonian operator $\hat{H}_+$.

Operator $\hat{\mathcal{L}}$ plays essential role in the description of the system $\hat{\mathcal{H}}$: it detects and annihilates a unique bound state in the spectrum of reflectionless subsystem $\hat{H}_+$, which is described by the wave function $\Psi_0 = \left(\sqrt{2\kappa^{-1}}\cosh\kappa x, 0\right)^t$ of zero energy. It also annihilates the doublet of states $\Psi_+ = (\tanh\kappa x,\ 0)^t$ and $\Psi_- = (0,\ 1)^t$ of the system $\hat{\mathcal{H}}$ of energy $E = \kappa^2$. Besides, operator $\hat{\mathcal{L}}$ distinguishes (with the aid of the integral σ_3) the states $\Psi_+^{\pm k} = (\pm ikx - \kappa\tanh\kappa x)e^{\pm ikx},\ 0)^t$ and $\Psi_-^{\pm k} = (0,\ e^{\pm ikx})^t$ in the four-fold degenerate scattering part of the spectrum of $\hat{\mathcal{H}}$: $\hat{\mathcal{L}}\Psi_+^{\pm k} = \pm k(\kappa^2 + k^2)\Psi_+^{\pm k}$, $\hat{\mathcal{L}}\Psi_-^{\pm k} = \pm k(\kappa^2 + k^2)\Psi_-^{\pm k}$. Zero energy state Ψ_0 is annihilated here by all the supercharges and by the Lax–Novikov integral $\hat{\mathcal{L}}$, and thus the system realizes exotic supersymmetry in the unbroken phase [36, 42].

Within the framework of the Darboux–Crum construction, the described reflectionless system $\hat{H}_+$ is obtained from the free particle system $\hat{H}_0 = -\frac{d^2}{dx^2}$ by taking its non-physical eigenstate $\psi_1(x) = \cosh\kappa x$ of eigenvalue $-\kappa^2$ as the seed state by constructing the operator

$$\hat{H}_+ = \hat{H}_- - 2(\ln \mathbb{W})'' , \tag{3.13}$$

where $\hat{H}_- = \hat{H}_0 + \kappa^2$ and $\mathbb{W} = \psi_1(x)$. The supercharge $\hat{Q}_+$ is constructed then from the operator $\hat{q}_+ = \psi_1\frac{d}{dx}\frac{1}{\psi_1(x)} = \frac{d}{dx} - \kappa\tanh\kappa x$. The same superpartner system $\hat{H}_+$ can be generated via relation (3.13) by changing $\mathbb{W} = \psi_1(x)$ in it for Wronskian of the set of eigenstates $\psi_0 = 1$ and $\psi_1 = \sinh\kappa x$, which is equal, up to inessential multiplicative constant, to the same function $\mathbb{W} = \psi_1(x)$: $\mathbb{W}(1, \sinh\kappa x) = \kappa\cosh\kappa x$. This second DC scheme generates the intertwining operator (3.8) corresponding to the second order supercharge $\hat{S}_+$ via the chain of relations $\hat{s}_+ = A_2A_1$, where $A_1 = \psi_0\frac{d}{dx}\frac{1}{\psi_0} = \frac{d}{dx}$, $A_2 = (A_1\psi_1)\frac{d}{dx}\frac{1}{(A_1\psi_1)} = \hat{q}_+$. In this construction the third order Lax–Novikov integral $\hat{q}_+\hat{p}\,\hat{q}_+^\dagger$ of the subsystem $\hat{H}_+$ is the Darboux-dressed momentum operator of the free particle.

The described DC construction of superextended systems described by exotic $\mathcal{N} = 4$ supersymmetry is generalized for arbitrary case of the system of the form $\hat{\mathcal{H}} = \text{diag}\,(\hat{H}_+, \hat{H}_-)$, with reflectionless subsystems $\hat{H}_+$ and $\hat{H}_-$ having an arbitrary number and energies of bound states, but with identical continuous parts of their spectra [42]. The key point underlying the appearance of the two supersymmetries of different orders by means of which the partner systems $\hat{H}_+$ and $\hat{H}_-$ are related is that the same reflectionless system can be generated by two different Darboux–Crum transformations. One transformation is generated by the choice of the set of non-physical eigenstates

$$\psi_1 = \cosh\kappa_1(x + \tau_1),\ \psi_2 = \sinh\kappa_2(x + \tau_2),\ \ldots,\ \psi_n \tag{3.14}$$

of the free particle system taken as the seed states. Here $\psi_{2l+1} = \cosh\kappa_{2l+1}(x + \tau_{2l+1})$, $\psi_{2l} = \sinh\kappa_{2l}(x + \tau_{2l})$, $1 \leq 2l < 2l + 1 \leq n$, and κ_j and τ_j, $j = 1, \ldots, n$, are arbitrary real parameters with restriction $0 < \kappa_j < \kappa_{j+1}$. The indicated choice of eigenstates guarantees that the Wronskian of these states takes nonzero values, and the potential produced via the Wronskian construction, $V(x) = -2(\ln \mathbb{W}(\psi_1, \ldots, \psi_n))''$, will be nonsingular reflectionless potential maintaining n bound states. The choice of the translation parameters τ_j in the form $\tau_j = x_{0j} - 4\kappa_j^2 t$ promotes the potential into the n-soliton solution to the KdV equation [43, 73]

$$u_t = 6uu_x - u_{xxx} \,. \tag{3.15}$$

Exactly the same reflectionless potential $V(x)$ is generated by taking the following set of eigenstates of the free particle Hamiltonian operator:

$$\phi_0 = 1,\ \phi_1 = \sinh\kappa_1(x + \tau_1),\ \phi_2 = \cosh\kappa_2(x + \tau_2),\ \ldots,\ \phi_n \,, \tag{3.16}$$

as the seed states for the Darboux–Crum transformation. Here

$$\phi_{2l+1} = \sinh\kappa_{2l+1}(x + \tau_{2l+1}), \qquad \phi_{2l} = \cosh\kappa_{2l}(x + \tau_{2l}),$$

and modulo the unimportant multiplicative constant, we have

$$\mathbb{W}(\psi_1, \ldots, \psi_n) = \mathbb{W}(1, \psi_1' \ldots, \psi_n') \,. \tag{3.17}$$

When the number of bound states n in each partner reflectionless system $\hat{H}_+$ and $\hat{H}_-$ is the same but all the discrete energies of one subsystem are different from those of another subsystem, one pair of supercharges will have differential order $2n$ while another pair will have differential order $2n + 1$ independently on the values of translation parameters τ_j of subsystems. This corresponds to the nature of the described Darboux–Crum transformations. In this case one pair of the supercharges is constructed from intertwining operators which relate the partner system $\hat{H}_+$ via the "virtual" free particle system $\hat{H}_0$, and then $\hat{H}_0$ to $\hat{H}_-$. The corresponding intertwining operators are composed from intertwining operators obtained from the sets of the seed states of the form (3.14) used for the construction of each partner system. Another pair of supercharges of differential order $2n + 1$ is constructed from the intertwining operators of a similar form but with inserted in the middle free particle integral $\frac{d}{dx}$. This corresponds to the use of the set of the seed states of the form (3.16) for one of the partner subsystems. The Lax–Novikov integral being even generator of the exotic supersymmetry and having differential order $2n + 1$ is produced via anti-commutation of the supercharges of different differential orders. It, however, is not a central charge of the nonlinear superalgebra: commuting with one pair of supercharges it transforms them into another pair of supercharges multiplied by certain polynomials in Hamiltonian $\hat{\mathcal{H}}$ of corresponding orders [42]. The structure of exotic supersymmetry undergoes a reduction each time when some r discrete energies of one subsystem coincide with any r discrete energies of another subsystem. In this case the sum of differential orders of two pairs of supercharges reduces from $4n+1$ to $4n-2r+1$, and nonlinear superlagebraic structure acquires a dependence on r relative translation parameters $\tau_j^+ - \tau_{j'}^-$ whose indexes j and j' correspond to coinciding discrete energy levels. When all the discrete energy levels of one subsystem coincide with those of the partner system, the Lax integral transforms into the bosonic central charge of the corresponding nonlinear superalgebra [42].

Different supersymmetric systems of the described nature can also be related by sending some of the translation parameters τ_j to infinity. In such a procedure exotic supersymmetry undergoes certain transmutations, particularly, between the unbroken and broken phases, and admits an interpretation in terms of the picture of soliton scattering [74].

In the interesting case of a superextended system unifying two finite-gap periodic partners described by the associated Lamé potentials shifted mutually for the half of the period of their potentials, the two corresponding Darboux–Crum transformations are constructed on the two sets of the seed states which correspond to the edges of the valence and conduction bands, one of which is composed from periodic states while another consists from antiperiodic states. One of such sets corresponding to antiperiodic wave functions has even dimension, while another that includes wave functions with the same period as the potentials has odd dimension. These sets generate the pairs of supercharges of the corresponding even and odd differential orders. On these sets of the states, certain finite-dimensional non-unitary representations of the $sl(2,\mathbb{R})$ algebra are realized of the same even and odd dimensions [30]. Lax–Novikov integral in such finite-gap systems with exotic nonlinear $\mathcal{N}=4$ supersymmetry has a nature of the bosonic central charge and differential order equal to $2g+1$, where g is the number of gaps in the spectrum of completely isospectral partners. The indicated class of the supersymmetric finite-gap systems admits an interpretation as a planar model of a non-relativistic electron in periodic magnetic and electric fields that produce a one-dimensional crystal for two spin components separated by a half-period spacing [30]. Exotic supersymmetry in such systems is in the unbroken phase with two ground states having the same zero energy, particularly, in the case when one pair of the supercharges has differential order one and corresponds to the construction of the Witten's supersymmetric quantum mechanics. The simplest case of such a system is given by the pair of the mutually shifted for the half-period one-gap Lamé systems,

$$\hat{H}_\pm = -\frac{d^2}{dx^2} + V_\pm(x), \quad V_-(x) = 2\mathrm{sn}^2(x|k) - k^2, \quad V_+(x) = V_-(x+\mathbf{K}), \tag{3.18}$$

where k is the modular parameter and $4\mathbf{K}$ is the period of the Jacobi elliptic function $\mathrm{sn}\,(x|k)$. The extended matrix system $\hat{\mathcal{H}}$ is described by the first order supercharges constructed on the base of the superpotential $W(x) = -(\ln \mathrm{dn}\, x)'$ generated by the ground state $\mathrm{dn}\, x$ of the subsystem $\hat{H}_-$ which has the same period $2\mathbf{K}$ as the potential $V_-(x)$. The second order supercharges are generated via the Darboux–Crum construction on the base of the seed states $\mathrm{cn}\, x$ and $\mathrm{sn}\, x$ which change sign under the shift for 2**K**, and describe the states of energies $1 - k^2$ and 1 at the edges of valence and conduction bands of $\hat{H}_-$, respectively.

The superextended system composed from the same one-gap systems but shifted mutually for the distance less than half-period of their potentials is described by exotic nonlinear $\mathcal{N} = 4$ supersymmetry with supercharges to be differential operators of the same first and second orders, and Lax–Novikov integral having differential order three. But in this case supersymmetry is broken, the positive energy of the doublet of the ground states depends on the value of the mutual shift, and though the Lax–Novikov integral is the bosonic central charge, the structure coefficients of the nonlinear superalgebra depend on the value of the shift parameter [37].

As was shown in [45], reflectionless and finite-gap periodic systems described by exotic nonlinear supersymmetry can also be generated in quantum systems with a position-dependent mass [75–78].

Very interesting physical properties are exhibited in the systems with the exotic nonlinear $\mathcal{N} = 4$ supersymmetry realized on finite-gap systems with soliton defects [73, 79]. By applying Darboux–Crum transformations to a Lax pair formulation of the KdV equation, one can construct multi-soliton solutions to this equation as well as to the modified Korteweg–de Vries equation which represent different types of defects in crystalline background of the pulse and compression modulation types. These periodicity defects reveal a chiral asymmetry in their propagation. Exotic nonlinear supersymmetric structure in such systems unifies solutions to the KdV and modified KdV equations, it detects the presence of soliton defects in them, distinguishes their types, and identifies the types of crystalline backgrounds [73].

4 Perfectly Invisible $\mathcal{PT}$-Symmetric Zero-Gap Systems

Darboux–Crum transformations can be realized not only on the base of the physical or non-physical eigenstates of a system, but also by including into the set of the seed states of Jordan and generalized Jordan states [56, 57, 80–82], which, in turn, can be obtained by certain limit procedures from eigenstates of a system. For instance, one can start from the free quantum particle, and choose the set of the states $(x, x^2, x^3, \ldots x^n)$, $x^n = \lim_{k\to 0}(\sin kx/k)^n$. The first state x is a non-physical eigenstate of $\hat{H}_0 = -\frac{d^2}{dx^2}$ of zero eigenvalue. The states x^{2l}, x^{2l+1}, $l \geq 1$, are the Jordan states of order l of $\hat{H}_0$: $(\hat{H}_0)^l$ acting on both states transforms them into zero energy eigenstates $\psi_0 = 1$ and $\psi_1 = x = \widetilde{\psi_0}$, respectively. The Wronskian of these states is $\mathbb{W}(x, x^2, x^3, \ldots, x^n) = const \cdot x^n$, and the system generated via the corresponding Darboux–Crum transformation is $\hat{H}_n = -\frac{d^2}{dx^2} + \frac{n(n+1)}{x^2}$. Operator $\hat{H}_n$, however, is singular on the whole real line, and can be identified with the Hamiltonian of the two-particle Calogero [83, 84] model with the omitted center of mass coordinate, which requires for definition of its domain with $x \in (0, +\infty)$ the introduction of the Dirichlet boundary condition $\psi(0^+) = 0$. Systems $\hat{H}_0$ and $\hat{H}_n$ are intertwined by differential operators $\mathbb{A}_n = A_n \ldots A_1$ and $\mathbb{A}_n^\dagger$, $\mathbb{A}_n \hat{H}_0 = \hat{H}_n \mathbb{A}_n$,

$\mathbb{A}_n^\dagger \hat{H}_n = \hat{H}_0 \mathbb{A}_n^\dagger$ where $A_l = \frac{d}{dx} - \frac{l}{x}$, and construction of $\mathbb{A}_n$ corresponds to Eq. (3.3). The systems $\hat{H}_0$ and $\hat{H}_n$ can also be intertwined by the operators $\mathbb{B}_n = A_n \dots A_1 A_0$ and $\mathbb{B}_n^\dagger$, where $A_0 = \frac{d}{dx}$, which are obtained by realizing the Darboux–Crum transformation constructed on the base of the set of the states $(x^2, \dots, x^{n+1})$ extended with the state $\psi_0 = 1$. One could take then the extended system composed from $\hat{H}_+ = \hat{H}_n$ and $\hat{H}_- = \hat{H}_0$ with $\hat{H}_0$ restricted to the same domain as $\hat{H}_n$, and construct the supercharge operators of differential orders n and $n+1$ from the introduced intertwining operators. However, we find that the supercharge constructed on the base of the intertwining operators $\mathbb{B}_n$ and $\mathbb{B}_n^\dagger$ will be non-physical as the intertwining operator $\mathbb{B}_n$ acting on physical eigenstates $\sin kx$ of $\hat{H}_-$ of energy k^2 will transform them into non-physical eigenstates $\mathbb{B}_n \sin kx$ of the system $\hat{H}_+$ of the same energy but not satisfying the boundary condition $\psi(0^+) = 0$. In correspondence with this, differential operator of order $2n+1$, $\hat{\mathcal{L}} = \text{diag}\,(\hat{\mathcal{L}}_+, \hat{\mathcal{L}}_-)$, with $\hat{\mathcal{L}}_+ = \mathbb{B}_n \mathbb{A}_n^\dagger = \mathbb{A}_n \frac{d}{dx} \mathbb{A}_n^\dagger$ and $\hat{\mathcal{L}}_- = \mathbb{A}_n^\dagger \mathbb{B}_n = (\hat{H}_-)^n \frac{d}{dx}$ formally commutes with $\hat{\mathcal{H}}$, but it is not a physical operator for the system $\hat{\mathcal{H}}$ as acting on its physical eigenstates satisfying boundary condition at $x = 0^+$, it transforms them into non-physical eigenstates not satisfying the boundary condition. The situation can be "$\mathcal{PT}$-regularized" by shifting the variable x: $x \to \xi = x + i\alpha$, where α is a nonzero real parameter [56]. The obtained in such a way superextended system can be considered on the whole real line $x \in \mathbb{R}$, and boundary condition at $x = 0$ can be omitted. The system $\hat{H}_+(\xi)$ is $\mathcal{PT}$-symmetric [85–91]: $[PT, \hat{H}_+(\xi)] = 0$, where P is a space reflection operator, $Px = -Px$, and T is the operator defined by $T(x + i\alpha) = (x - i\alpha)T$. Subsystem $\hat{H}_+(\xi)$ has one bound eigenstate of zero eigenvalue described by quadratically integrable on the whole real line function $\psi_0^+ = \xi^{-n}$, which lies at the very edge of the continuous spectrum with $E > 0$. System $\hat{H}_+(\xi)$ therefore can be identified as $\mathcal{PT}$-symmetric zero-gap system. Moreover, it turns out that the transmission amplitude for this system is equal to one as for the free particle system, and $\hat{H}_+(\xi)$ can be regarded as a perfectly invisible $\mathcal{PT}$-symmetric zero-gap system. Exotic nonlinear supersymmetry of the system $\hat{\mathcal{H}}(\xi)$ will be described by two supercharges of differential order n constructed from the intertwining operators $\mathbb{A}_n(\xi)$ and $\mathbb{A}_n^\#(\xi) = A_1^\# \dots A_n^\#$, $A_j^\# = -\frac{d}{dx} - \frac{j}{\xi}$, by supercharges of the order $n+1$ constructed from the intertwining operators $\mathbb{B}_n(\xi)$ and $\mathbb{B}_n^\#(\xi)$, and by the Lax–Novikov integral $\hat{\mathcal{L}}(\xi)$ to be differential operator of order $2n+1$. Operator $\hat{\mathcal{L}}(\xi)$ annihilates the unique bound state of the system $\hat{\mathcal{H}}(\xi)$ and the state $\psi_0 = 1$ of zero energy in the spectrum of the free particle subsystem, and distinguishes plane waves e^{ikx} in the spectrum of the free particle subsystem and deformed plane waves $\mathbb{A}_n(\xi)e^{ik\xi}$ in the spectrum of the superpartner system $\hat{H}_+(\xi)$.

In the simplest case $n = 1$, the supercharges have the form

$$\hat{Q}_1 = \begin{pmatrix} 0 & A_1(\xi) \\ A_1^\#(\xi) & 0 \end{pmatrix}, \qquad \hat{Q}_2 = i\sigma_3 \hat{Q}_1\,, \tag{4.1}$$

$$\hat{S}_1 = \begin{pmatrix} 0 & -A_1(\xi)\frac{d}{dx} \\ \frac{d}{dx}A_1^\#(\xi) & 0 \end{pmatrix}, \qquad \hat{S}_2 = i\sigma_3 \hat{S}_1\,, \tag{4.2}$$

where $\hat{Q}_1 = \hat{Q}_+ + \hat{Q}_-$, $\hat{S}_1 = \hat{S}_+ + \hat{S}_-$. The Lax–Novikov integral is

$$\hat{\mathcal{L}} = \begin{pmatrix} -iA_1(\xi)\frac{d}{dx}A_1^{\#}(\xi) & 0 \\ 0 & -i\frac{d}{dx}\hat{H}_0 \end{pmatrix}. \tag{4.3}$$

Together with Hamiltonian $\hat{\mathcal{H}} = \text{diag}\,(\hat{H}_1(\xi), \hat{H}_0)$ they satisfy the following nonlinear superalgebra [56]:

$$[\hat{\mathcal{H}}, \hat{Q}_a] = 0\,, \qquad [\hat{\mathcal{H}}, \hat{S}_a] = 0\,, \tag{4.4}$$

$$\{\hat{Q}_a, \hat{Q}_b\} = 2\delta_{ab}\hat{\mathcal{H}}\,, \qquad \{\hat{S}_a, \hat{S}_b\} = 2\delta_{ab}\hat{\mathcal{H}}^2\,, \tag{4.5}$$

$$\{\hat{Q}_a, \hat{S}_b\} = 2\epsilon_{ab}\hat{\mathcal{L}}\,. \tag{4.6}$$

$$[\hat{\mathcal{L}}, \hat{\mathcal{H}}] = 0\,, \qquad [\hat{\mathcal{L}}, \hat{Q}_a] = 0\,, \qquad [\hat{\mathcal{L}}, S_a] = 0\,. \tag{4.7}$$

In the case of the superextended system $\hat{\mathcal{H}} = \text{diag}\,(\hat{H}_1(\xi_2), \hat{H}_1(\xi_1))$, where $\xi_j = x + i\alpha_j$, $j = 1, 2$, and $\alpha_1 \neq \alpha_2$, exotic nonlinear supersymmetry is partially broken: the doublet of zero energy bound states is annihilated by the second order supercharges $\hat{S}_a$ and by the Lax–Novikov integral $\hat{\mathcal{L}}$, but they are not annihilated by the first order supercharges $\hat{Q}_a$ [56]. The first order supercharges $\hat{Q}_a$ are constructed in this case from the intertwining operators $A = \frac{d}{dx} + \mathcal{W}$, $\mathcal{W} = \xi_1^{-1} - \xi_2^{-1} - (\xi_1 - \xi_2)^{-1}$, and $A^{\#} = -\frac{d}{dx} + \mathcal{W}$. The second order supercharges $\hat{S}_a$ are composed from the intertwining operators $A_1(\xi_2)A_1^{\#}(\xi_1)$ and $A_1(\xi_1)A_1^{\#}(\xi_2)$. In the limit $\alpha_1 \to \infty$, the system $\hat{\mathcal{H}} = \text{diag}\,(\hat{H}_1(\xi_2), \hat{H}_1(\xi_1)$ transforms into the system given by the $\mathcal{PT}$-symmetric Hamiltonian $\hat{\mathcal{H}} = \text{diag}\,(\hat{H}_1(\xi_2), \hat{H}_0)$, and exotic nonlinear supersymmetry in the partially broken phase transmutes into the supersymmetric structure corresponding to the unbroken phase [56].

It is interesting to note that if to use the appropriate linear combinations of the Jordan states of the quantum free particle as the seed states for the Darboux–Crum transformations, one can construct $\mathcal{PT}$-symmetric time-dependent potentials which will satisfy equations of the KdV hierarchy and will exhibit a behavior typical for extreme (rogue) waves [56].

5 Nonlinear Superconformal Symmetry of the $\mathcal{PT}$-Symmetric Zero-Gap Calogero Systems

Free particle system is characterized by the Schrödinger symmetry generated by the first order integrals $\hat{P}_0 = \hat{p} = -i\frac{d}{dx}$ and $\hat{G}_0 = x + 2it\frac{d}{dx}$, and the second order integrals $\hat{H}_0 = -\frac{d^2}{dx^2}$, $\hat{D}_0 = \frac{1}{4}\{\hat{P}_0, \hat{G}_0\}$ and $\hat{K}_0 = \hat{G}_0^2$. Operators $\hat{G}_0$ as well as $\hat{D}_0$ and $\hat{K}_0$ are dynamical integrals of motion satisfying the equation of motion of the form $\frac{d}{dt}\hat{I} = \frac{\partial}{\partial t}\hat{I} - [\hat{H}_0, \hat{I}] = 0$. These time-independent and dynamical integrals generate the Schrödinger algebra

$$[\hat{D}_0, H_0] = i\hat{H}_0\,, \qquad [\hat{D}_0, \hat{K}_0] = -i\hat{K}_0\,, \qquad [\hat{K}_0, \hat{H}_0] = 8i\hat{D}_0\,, \tag{5.1}$$

$$[\hat{D}_0, \hat{P}_0] = \tfrac{i}{2}\hat{P}_0\,, \qquad \hat{[}D_0, \hat{G}_0] = -\tfrac{i}{2}\hat{G}_0\,, \tag{5.2}$$

$$[\hat{H}_0, \hat{G}_0] = -2i\hat{P}_0\,, \qquad [\hat{H}_0, \hat{P}_0] = 0\,, \tag{5.3}$$

$$[\hat{K}_0, \hat{P}_0] = 2i\hat{G}_0\,, \qquad [\hat{K}_0, \hat{G}_0] = 0\,, \tag{5.4}$$

$$[\hat{G}_0, \hat{P}_0] = i\,\mathbb{I}\,. \tag{5.5}$$

Equations (5.1) and (5.5) correspond to the $sl(2,\mathbb{R})$ and Heisenberg subalgebras, respectively. If we make a shift $x \rightarrow \xi = x + i\alpha$, and make Darboux-dressing of operators $\hat{P}_0$, $\hat{G}_0$, $\hat{D}_0$, and $\hat{K}_0$, we find the integrals of motion for the perfectly invisible zero-gap $\mathcal{PT}$-symmetric system $\hat{H}_1(\xi)$. These are $\hat{P}_1(\xi) = A_1(\xi)\hat{P}_0 A_1^{\#}(\xi)$, $\hat{G}_1(\xi) = A_1(\xi)\hat{G}_0 A_1^{\#}(\xi)$, and

$$\hat{D}_1(\xi) = -\frac{i}{2}\left(\xi\frac{d}{dx} + \frac{1}{2}\right) - t\hat{H}_1(\xi)\,, \tag{5.6}$$

$$\hat{K}_1(\xi) = \xi^2 - 8t\hat{D}_1(\xi) - 4t^2\hat{H}_1(\xi)\,, \tag{5.7}$$

where the dynamical integrals $\hat{D}_1(\xi)$ and $\hat{K}_1(\xi)$ have been extracted from the corresponding Darboux-dressed operators by omitting in them the operator factor $\hat{H}_1(\xi)$ [57]. Operators $\hat{H}_1(\xi)$, $\hat{D}_1(\xi)$, and $\hat{K}_1(\xi)$ generate the same $sl(2,\mathbb{R})$ algebra as in the case of the free particle. But now we have relations

$$[D_1, P_1] = \frac{3}{2}iP_1\,, \qquad [D_1, G_1] = \frac{i}{2}G_1\,, \qquad [K_1, P_1] = 6iG_1\,, \tag{5.8}$$

$$[G_1, P_1] = 3i(H_1)^2 \tag{5.9}$$

instead of the corresponding relations of the free particle system. In addition, two new dynamical integrals of motion,

$$V_1(\xi) = i\xi^2 A_1^{\#}(\xi) - 4tG_1(\xi) - 4t^2P_1(\xi) \tag{5.10}$$

and

$$R_1(\xi) = \xi^3 - 6tV_1(\xi) - 12t^2G_1(\xi) - 8t^3\xi_1\,, \tag{5.11}$$

are generated via the commutation relations

$$[\hat{K}_1, \hat{G}_1] = -4i\hat{V}_1\,, \qquad [\hat{K}_1, \hat{V}_1] = -2i\hat{R}_1\,, \tag{5.12}$$

and we obtain additionally the commutation relations

$$[\hat{V}_1, \hat{H}_1] = 4i\hat{G}_1\,, \qquad [\hat{V}_1, \hat{D}_1] = \tfrac{i}{2}\hat{V}_1\,,$$

$$[\hat{V}_1, \hat{P}_1] = 12i\hat{H}_1\hat{D}_1 - 6\hat{H}_1\,, \qquad [\hat{V}_1, \hat{G}_1] = 12i(\hat{D}_1)^2 + \tfrac{3}{4}i\,\mathbb{I}\,,$$

$$[\hat{R}_1, \hat{H}_1] = 6i\hat{V}_1\,, \qquad [\hat{R}_1, \hat{D}_1] = \tfrac{3}{2}i\hat{R}_1\,, \quad [\hat{R}_1, \hat{K}_1] = 0\,,$$

$$[\hat{R}_1, \hat{P}_1] = 36i\,\hat{D}_1^2 + \tfrac{21}{4}i\,\mathbb{I}, \quad [\hat{R}_1, \hat{G}_1] = 12i\,\hat{D}_1\hat{K}_1 - 6\hat{K}_1\,, \quad [\hat{R}_1, \hat{V}_1] = 3i\,\hat{K}_1^2\,.$$

The Schrödinger algebra of the free particle is extended for its nonlinear generalization in the case of the $\mathcal{PT}$-symmetric system $\hat{H}_1(\xi)$, which is generated by the operators $\hat{H}_1(\xi)$, $\hat{P}_1(\xi)$, $\hat{G}_1(\xi)$, $\hat{D}_1(\xi)$, $\hat{K}_1(\xi)$, $\hat{V}_1(\xi)$, $\hat{R}_1(\xi)$, and central charge $\mathbb{I}$ (equals to 1 in the chosen system of units). All these integrals are eigenstates of the dilatation operator $\hat{D}_1(\xi)$ with respect to its adjoint action.

Now we can consider the generalized and extended superconformal symmetry of the system described by the matrix Hamiltonian operator $\hat{\mathcal{H}} = \text{diag}\,(\hat{H}_1(\xi), \hat{H}_0)$. Supplying the Hamiltonian $\hat{\mathcal{H}}$ and Lax–Novikov integral (4.3) with the bosonic integrals $\hat{\mathcal{D}} = \text{diag}\,(\hat{D}_1(\xi), \hat{D}_0(\xi))$, $\hat{\mathcal{K}} = \text{diag}\,(\hat{K}_1(\xi), \hat{K}_0(\xi))$, and commuting them with supercharges (4.1) and (4.2), we obtain a nonlinear superalgebra that describes the symmetry of the system $\hat{\mathcal{H}}$, which corresponds to some nonlinear extension of the super-Schrödinger algebra. It is generated by the set of the even (bosonic) integrals $\hat{\mathcal{H}}$, $\hat{\mathcal{D}}$, $\hat{\mathcal{K}}$, $\hat{\mathcal{L}}$, $\hat{\mathcal{G}}$, $\hat{\mathcal{V}}$, $\hat{\mathcal{R}}$, $\hat{\mathcal{P}}_-$, $\hat{\mathcal{G}}_-$, $\Sigma = \sigma_3$, $\hat{\mathcal{I}} = \text{diag}\,(1, 1)$, and by the odd (fermionic) integrals $\hat{\mathcal{Q}}_a$, $\hat{\mathcal{S}}_a$, and $\hat{\lambda}_a$, $\hat{\mu}_a$ and $\hat{\kappa}_a$, $a = 1, 2$, where

$$\hat{\mathcal{G}} = \text{diag}\,\left(\hat{G}_1(\xi),\, \tfrac{1}{2}\{\hat{G}_0(\xi), \hat{H}_0\}\right)\,, \qquad \mathcal{V} = i\xi^2 A_1^{\alpha\#}\mathcal{I} - 4t\mathcal{G} - 4t^2\mathcal{L}\,, \tag{5.13}$$

$$\hat{\mathcal{R}} = \xi^3\mathcal{I} - 6t\hat{\mathcal{V}} - 12t^2\hat{\mathcal{G}} - 8t^3\hat{\mathcal{L}}\,, \tag{5.14}$$

$$\hat{\mathcal{P}}_- = \tfrac{1}{2}(1 - \sigma_3)\hat{P}_0\,, \qquad \hat{\mathcal{G}}_- = \tfrac{1}{2}(1 - \sigma_3)\hat{G}_0(\xi), \tag{5.15}$$

$$\hat{\lambda}_1 = \begin{pmatrix} 0 & i\xi \\ -i\xi & 0 \end{pmatrix} - 2t\hat{\mathcal{Q}}_1\,, \qquad \hat{\lambda}_2 = i\sigma_3\hat{\lambda}_1\,, \tag{5.16}$$

$$\hat{\mu}_1 = \begin{pmatrix} 0 & \xi\hat{P}_0 \\ \hat{P}_0\xi & 0 \end{pmatrix} - 2t\hat{\mathcal{S}}_1\,, \qquad \hat{\mu}_2 = i\sigma_3\hat{\mu}_1\,, \tag{5.17}$$

$$\hat{\kappa}_1 = \begin{pmatrix} 0 & \xi^2 \\ \xi^2 & 0 \end{pmatrix} - 4t\hat{\mu}_1 - 4t^2\hat{\mathcal{S}}_1\,, \qquad \hat{\kappa}_2 = i\sigma_3\hat{\kappa}_1\,, \tag{5.18}$$

and we use the notation $\hat{G}_0(\xi) = \hat{G}_0(x + i\alpha)$. Explicit form of the nonlinear superalgebra generated by these integrals of motion of the system $\hat{\mathcal{H}}$ is presented in [57]. All the even and odd integrals here are eigenstates of the matrix dilatation operator $\hat{\mathcal{D}}$.

Essentially different generalized nonlinear superconformal structure appears in the system described by the matrix Hamiltonian

$$\hat{\mathcal{H}} = \mathrm{diag}\,(\hat{H}_1(\xi_2), \hat{H}_1(\xi_1))$$

and characterized by the partially broken exotic nonlinear $\mathcal{N} = 4$ supersymmetry. In that case the number of the even and odd integrals of motion is the same as in the system $\hat{\mathcal{H}} = \mathrm{diag}\,(\hat{H}_1(\xi), \hat{H}_0)$ in the phase with unbroken supersymmetry. However, no odd (fermionic) integral of motion is eigenstate of the matrix dilatation operator $\hat{\mathcal{D}} = \mathrm{diag}\,(\hat{D}_1(\xi_2), \hat{D}_1(\xi_1))$, and, as a result, the structure of the nonlinear superalgebra has more complicated form. When one of the shift parameters, α_1, is sent to infinity, the system $\hat{\mathcal{H}} = \mathrm{diag}\,(\hat{H}_1(\xi_2), \hat{H}_1(\xi_1))$ transforms into the system $\hat{\mathcal{H}} = \mathrm{diag}\,(\hat{H}_1(\xi), \hat{H}_0)$ in the unbroken phase of the exotic nonlinear $\mathcal{N} = 4$ super-Poincaré symmetry, and all the integrals of the latter system can be reproduced from the integrals of the former system. The relation between the integrals turns out to be rather nontrivial and requires some sort of a "renormalization" [57].

6 Rationally Extended Harmonic Oscillator and Conformal Mechanics Systems

Quantum harmonic oscillator (QHO) and conformal mechanics systems [92–122] described by de Alfaro-Fubini-Furlan (AFF) model [92] are characterized by conformal symmetry. In the case of harmonic oscillator, like in the free particle case, it extends to the Schrödinger symmetry [93–95, 123]. Heisenberg subalgebra in the free particle system is generated by the momentum operator being time-independent integral of motion, and by generator of the Galilean boosts $\hat{G}_0$, which is a dynamical integral of motion. In the case of the QHO, Heisenberg subalgebra is generated by two dynamical integrals of motion to be linear in the ladder operators. In correspondence with this, ladder operators are the spectrum-generating operators of the QHO having discrete equidistant spectrum instead of the continuous spectrum of the free particle. As a consequence of these similarities and differences between the free particle and QHO, exotic supersymmetry can also be generated by Darboux–Crum transformations applied to the latter system. Instead of the two pairs of time-independent supercharge generators in superextended reflectionless systems, in superextended systems constructed from the pairs of the rational extensions of the QHO, only two supercharges are time-independent integrals, while other two odd generators are dynamical integrals of motion. As a result, instead of the exotic nonlinear $\mathcal{N} = 4$ supersymmetry of the paired reflectionless (and finite-gap) systems, in the case of the deformed oscillator systems there appear some nonlinearly deformed and generalized super-Schrödinger symmetry. The superextended systems composed from the AFF model (with special values $g = n(n + 1)$ of the coupling constant in its additional potential term g/x^2) and its rational extensions are described by the nonlinearly deformed and generalized superconformal symmetry [51].

Let us consider first in more detail the case of rational deformations of the QHO system [5, 6, 49, 51, 71, 124–128]. To generate a rational deformation of the QHO, it is necessary to choose the set of its physical or non-physical eigenstates as seed states for the Darboux–Crum transformation so that their Wronskian will take nonzero values. In this way we generate an almost isospectral quantum system with difference only in finite number of added or eliminated energy levels. The QHO Hamiltonian $\hat{H}_{osc} = -\frac{d^2}{dx^2} + x^2$ possesses the same symmetry under the Wick rotation as the quantum free particle system: if $\psi(x)$ is a solution of the time-independent Schrödinger equation $\hat{H}_{osc}(x)\psi(x) = E\psi(x)$, then $\psi(ix)$ is a solution of equation $\hat{H}_{osc}(x)\psi(ix) = -E\psi(ix)$. To construct a rational deformation of the QHO described by a nonsingular on the whole real line potential, one can take the following set of the non-physical eigenstates of $\hat{H}_{osc}$ as the seed states for the Darboux–Crum transformation:

$$(\psi^-_{j_1}, \ldots, \psi^-_{j_1+l_1}),\ (\psi^-_{j_2}, \ldots, \psi^-_{j_2+l_2}),\ \ldots,\ (\psi^-_{j_r}, \ldots, \psi^-_{j_r+l_r}), \tag{6.1}$$

where $j_1 = 2g_1$, $j_{k+1} = j_k + l_k + 2g_{k+1}$, $g_k = 1, \ldots$, $l_k = 0, 1, \ldots$, $k = 1, \ldots, r-1$. Here $\psi^-_n(x) = \psi_n(ix)$, $n = 0, \ldots$, is a non-physical eigenstate of $\hat{H}_{osc}$ of eigenvalue $E^-_n = -(2n+1)$, obtained by Wick rotation from a (non-normalized) physical eigenstate $\psi_n(x) = H_n(x)e^{-x^2/2}$ of energy $E_n = 2n+1$, where $H_n(x)$ is Hermite polynomial of order n. The indicated set of non-physical eigenstates of $\hat{H}_{osc}$ guarantees that the Wronskian of the chosen seed states, $\mathbb{W} = \mathbb{W}(-n_m, \ldots, -n_1)$, takes nonzero values for all $x \in \mathbb{R}$ [129]. Here we assume that $n_m > \ldots > n_1 > 0$, and in what follows we use the notation for physical and non-physical eigenstates $n = \psi_n$ and $-n = \psi^-_n$, respectively. The DC scheme based on the set of the non-physical states having negative eigenvalues was called "negative" in [71]. Wronskian $\mathbb{W} = \mathbb{W}(-n_m, \ldots, -n_1)$ is equal to some polynomial multiplied by $\exp(n_- x^2/2)$, where $n_- = (l_1+1) + \cdots + (l_r+1)$ is the number of the chosen seed states, and according to Eq. (3.1), the DC transformation generates the system described by the harmonic term x^2 extended by some rational in x term. Transformation based on the negative scheme $(-n_m, \ldots, -n_1)$ introduces effectively into the spectrum of the QHO the n_- bound states of energy levels $-2n_m-1, \ldots, -2n_1-1$. These additional energy levels are grouped into r "valence" bands with $l_k + 1$ levels in the band with index k, which are separated by gaps of the size $4g_k$, with the first valence band separated from the infinite equidistant part of the spectrum by the gap of the size $4g_1$. The same structure of the spectrum can be achieved alternatively by eliminating $n_+ = 2(g_1 + \cdots + g_r)$ energy levels from the spectrum of the QHO by taking n_+ physical states

$$(\psi_{l_r+1}, \ldots, \psi_{l_r+2g_r}),\ \ldots,\ (\psi_{n_m-2g_1+1}, \ldots, \psi_{n_m}), \tag{6.2}$$

organized into n_- groups.

The duality of the negative and positive schemes based on the sets of the seed states (6.1) and (6.2) can be established as follows. Applying Eq. (3.6) with $\psi_* = \widetilde{-0}$, and equalities $\psi_0^- \frac{d}{dx} \frac{1}{\psi_0^-} = -a^+$, $a^+ \widetilde{\psi_0^-} = \psi_0$, $a^+(-n) = -(n-1)$, where $a^+ = -\frac{d}{dx} + x$ is the raising ladder operator of the QHO, we obtain the relation [71]

$$\begin{aligned} \mathbb{W}(-n_m, \ldots, -n_1) &= \mathbb{W}(-0, \widetilde{-0}, -n_m, \ldots, -n_1) \\ &= e^{x^2/2} \mathbb{W}(0, -(n_m - 1), -(n-1)). \end{aligned} \tag{6.3}$$

It means that the negative scheme generated by the set of the n_- non-physical seed states $(-n_m, \ldots, -n_1)$ and the "mixed" scheme based on the set of the seed states $(0, -(n_m - 1), -(n-1))$ involving the ground eigenstate generate, according to Eq. (3.1), the same quantum system but given by the Hamiltonian operator shifted for the additive constant term: the potential obtained on the base of the indicated mixed scheme will be shifted for the constant $+4$ in comparison with the potential generated via the DC transformation based on the negative scheme. Eq. (6.3) is analogous to the Wronskian relation (3.17) for the free particle states, with the state $\psi_0 = 1$ and operator $\psi_0 \frac{d}{dx} \frac{1}{\psi_0} = \frac{d}{dx}$ there to be analogous to the ground state and raising ladder operator of the QHO here. In (3.17), however, the Wronskian equality does not contain any nontrivial functional factor in comparison with the exponential multiplier appearing in (6.3). As a result, as we saw before, in the case of the free particle any reflectionless system can be generated from it by means of the two DC transformations, which produce exactly the same Hamiltonian operator. Consequently, we construct there two pairs of the supercharges for the corresponding superextended system which are the integrals of motion not depending explicitly on time. On the other hand, in the case of a superextended system produced from the QHO we shall have two fermionic integrals to be true, time-independent integrals of motion, but two other odd generators of the superalgebra will be time-dependent, dynamical integrals of motion.

Applying repeatedly the procedure of Eq. (6.3), we obtain finally the relation [71]

$$\mathbb{W}(-n_m, \ldots, -n_1) = e^{(n_m+1)x^2/2} \mathbb{W}(n'_1, \ldots, n'_m = n_m), \tag{6.4}$$

where $0 < n'_1 < \cdots < n'_m = n_m$. This relation means that the negative scheme $(-n_m, \ldots, -n_1)$ with n_- seed states is dual to the positive scheme $(n'_1, \ldots, n'_m = n_m)$ with $n_+ = n_m + 1 - n_- = 2(g_1 + \cdots + g_k)$ seed states representing physical eigenstates of the QHO. The two dual schemes can be unified in one "mirror" diagram, in which any of the two schemes can be obtained from another by a kind of a "charge conjugation," see ref. [71]. In this way we obtain, as an example, the pairs of dual schemes $(-2) \sim (1, 2)$ and $(-2, -3) \sim (2, 3)$. Eq. (6.4) means that the dual schemes generate the same rationally extended QHO system but the Hamiltonian corresponding to the positive scheme will be shifted in comparison to the Hamiltonian produced on the base of the negative scheme for additive constant equal to $2(n_+ + n_-) = 2(n_m + 1)$. One can also note that in comparison with the free particle case, the total number of the seed states in both dual schemes can be odd or even.

We denote by $\mathbb{A}^-_{(-)}$ the intertwining operator $\mathbb{A}_{n_-}$ constructed on the base of the negative scheme, and $\mathbb{A}^+_{(-)} \equiv (\mathbb{A}^-_{(-)})^\dagger$, see Eq. (3.3). These are differential operators of order n_-. Analogously, the intertwining operators constructed by employing the dual positive scheme we denote as $\mathbb{A}^-_{(+)}$, and $\mathbb{A}^+_{(+)} \equiv (\mathbb{A}^-_{(+)})^\dagger$; they are differential operators of order n_+. We denote by $\hat{L}_{(-)}$ and $\hat{L}_{(+)}$ the Hamiltonian operators generated from the QHO Hamiltonian $\hat{H}_- = \hat{H}_{osc}$ by means of the DC transformation realized on the base of the negative and positive dual schemes, respectively. Then $\hat{L}_{(+)} = \hat{L}_{(-)} + 2(n_+ + n_-)$, $\mathbb{A}^-_{(-)}\hat{H}_- = \hat{L}_{(-)}$, $\mathbb{A}^-_{(+)}\hat{H}_- = \hat{L}_{(+)}$. For the rationally deformed QHO system $\hat{L}_{(-)}$ one can construct three pairs of the ladder operators, two of which are obtained by Darboux-dressing of the ladder operators of the QHO system $\mathcal{A}^\pm = \mathbb{A}^-_{(-)}a^\pm\mathbb{A}^+_{(-)}$, and $\mathcal{B}^\pm = \mathbb{A}^-_{(+)}a^\pm\mathbb{A}^+_{(+)}$, while the third pair is obtained by gluing different intertwining operators, $\mathcal{C}^- = \mathbb{A}^-_{(+)}\mathbb{A}^+_{(-)}$, $\mathcal{C}^+ = \mathbb{A}^-_{(-)}\mathbb{A}^+_{(+)}$. These ladder operators detect all the separated states in the rationally deformed QHO system $\hat{L}_{(-)}$ (or $\hat{L}_{(+)}$) organized into the valence bands; they also distinguish the valence bands themselves, and any of the two sets $(\mathcal{C}^\pm, \mathcal{A}^\pm)$ or $(\mathcal{C}^\pm, \mathcal{B}^\pm)$ represents the complete spectrum-generating set of the ladder operators of the system $\hat{L}_{(-)}$. The operators $\mathcal{A}^\pm e^{\mp 2it}$, $\mathcal{B}^\pm e^{\mp 2it}$, $\mathcal{C}^\pm e^{\pm 2(n_+ + n_-)it}$ are the dynamical integrals of motion of the system $L_{(-)}$. Being higher derivative differential operators, they have a nature of generators of a hidden symmetry. If we construct now the extended system $\hat{\mathcal{H}} = \text{diag}\,(\hat{L}_{(-)}, \hat{H}_-)$, the pair of the supercharges constructed from the intertwining operators $\mathbb{A}^\pm_{(-)}$ will be its time-independent odd integrals of motion, while from the intertwining operators $\mathbb{A}^\pm_{(+)}$ we obtain a pair of the fermionic dynamical integrals of motion. Proceeding from these odd integrals of motion and the Hamiltonian $\hat{\mathcal{H}}$, one can generate a nonlinearly deformed generalized super-Schrödinger symmetry of the superextended system $\hat{\mathcal{H}}$. In the superextended system $\hat{\mathcal{H}} = \text{diag}\,(\hat{L}_{(+)}, \hat{H}_-)$, the pair of the time-independent supercharges is constructed from the pair of intertwining operators $\mathbb{A}^\pm_{(+)}$, while the dynamical fermionic integrals of motion are obtained from the intertwining operators $\mathbb{A}^\pm_{(-)}$. This picture with the nonlinearly deformed generalized super-Schrödinger symmetry can also be extended for the case of a superextended system $\hat{\mathcal{H}}$ composed from any pair of the rationally deformed quantum harmonic oscillator systems.

In [71], it was shown that the AFF model $\hat{H}_g = -\frac{d^2}{dx^2} + x^2 + \frac{g}{x^2}$ with special values $g = n(n+1)$ of the coupling constant can be obtained by applying the appropriate CD transformation to the half-harmonic oscillator obtained from the QHO by introducing the infinite potential barrier at $x = 0$. As a consequence, rational deformations of the AFF conformal mechanics model can be obtained by employing some modification of the described DC transformations based on the dual schemes applied to the QHO system. The corresponding superextended systems composed from rationally deformed versions of the conformal mechanics are described by the nonlinearly deformed generalized superconformal symmetry instead of the nonlinearly deformed generalized super-Schrödinger symmetry appearing in the case of the superextended rationally deformed QHO systems, see [51]. The construction of rational deformations for the AFF model can be generalized for the case of arbitrary values of the coupling constant $g = \nu(\nu + 1)$ [130].

7 Conclusion

We considered nonlinear supersymmetry of one-dimensional mechanical systems which has the nature of the hidden symmetry generated by supercharges of higher order in momentum. In the case of reflectionless, finite-gap, rationally deformed oscillator and conformal mechanics systems, as well as in a special class of the $\mathcal{PT}$-regularized Calogero systems, the nonlinear $\mathcal{N} = 2$ Poincaré supersymmetry expands up to exotic nonlinear $\mathcal{N} = 4$ supersymmetric and nonlinearly deformed generalized super-Schrödinger or superconformal structures.

Classical symmetries described by the linear Lie algebraic structures are promoted by geometric quantization to the quantum level [131, 132]. Though nonlinear symmetries described by W-type algebras can be produced from linear symmetries via some reduction procedure [64], the problem of generation of nonlinear quantum mechanical supersymmetries from the linear ones was not studied in a systematic way. It would be interesting to investigate this problem bearing particularly in mind the problem of the quantum anomaly associated with nonlinear supersymmetry [14]. Some first steps were realized in this direction in [62] in the light of the so-called coupling constant metamorphosis mechanism [133]. Note also that, as was shown in [12], nonlinear supersymmetry of purely parabosonic systems can be obtained by reduction of parasupersymmetric systems.

Hidden symmetries can be associated with the presence of the peculiar geometric structures in the corresponding systems [1, 2, 134]. It would be interesting to investigate nonlinear supersymmetry and related exotic nonlinear supersymmetric and superconformal structures from a similar perspective.

Acknowledgements Financial support from research projects Convenio Marco Universidades del Estado (Project USA1555) and FONDECYT Project 1190842, Chile, and MINECO (Project MTM2014-57129-C2-1-P), Spain, is acknowledged.

References

1. M. Cariglia, Hidden symmetries of dynamics in classical and quantum physics. Rev. Mod. Phys. **86**, 1283 (2014)
2. V. Frolov, P. Krtous, D. Kubiznak, Black holes, hidden symmetries, and complete integrability. Living Rev. Relativ. **20**(1), 6 (2017)

3. B. Khesin, G. Misolek, Euler equations on homogeneous spaces and Virasoro orbits. Adv. Math. **176**, 116 (2003)
4. B.A. Khesin, R. Wendt, *The Geometry of Infinite-Dimensional Groups* (Springer, Berlin, 2009)
5. S. Yu. Dubov, V.M. Eleonskii, N. E. Kulagin, Equidistant spectra of anharmonic oscillators. Zh. Eksp. Teor. Fiz. **102**, 814 (1992)
6. S.Y. Dubov, V.M. Eleonskii, N.E. Kulagin, Equidistant spectra of anharmonic oscillators. Chaos **4**, 47 (1994)
7. A.P.Veselov, A.B. Shabat, Dressing chains and the spectral theory of the Schrödinger operator. Funct. Anal. Appl. **27**, 81 (1993)
8. A.A. Andrianov, M.V. Ioffe, V.P. Spiridonov, Higher derivative supersymmetry and the Witten index. Phys. Lett. A **174**, 273 (1993)
9. D.J. Fernandez C, SUSUSY quantum mechanics. Int. J. Mod. Phys.A **12**, 171 (1997)
10. D.J. Fernandez C, V. Hussin, Higher-order SUSY, linearized nonlinear Heisenberg algebras and coherent states. J. Phys. A: Math. Gen. **32**, 3603 (1999)
11. B. Bagchi, A. Ganguly, D. Bhaumik, A. Mitra, Higher derivative supersymmetry, a modified Crum–Darboux transformation and coherent state. Mod. Phys. Lett. A **14**, 27 (1999)
12. M. Plyushchay, Hidden nonlinear supersymmetries in pure parabosonic systems. Int. J. Mod. Phys. A **15**, 3679 (2000)
13. D.J. Fernandez, J. Negro, L.M. Nieto, Second-order supersymmetric periodic potentials. Phys. Lett. A **275**, 338 (2000)
14. S.M. Klishevich, M.S. Plyushchay, Nonlinear supersymmetry, quantum anomaly and quasi-exactly solvable systems. Nucl. Phys. B **606**, 583 (2001)
15. S.M. Klishevich, M.S. Plyushchay, Nonlinear supersymmetry on the plane in magnetic field and quasi-exactly solvable systems. Nucl. Phys. B **616**, 403 (2001)
16. S.M. Klishevich, M.S. Plyushchay, Nonlinear holomorphic supersymmetry, Dolan–Grady relations and Onsager algebra. Nucl. Phys. B **628**, 217 (2002)
17. S.M. Klishevich, M.S. Plyushchay, Nonlinear holomorphic supersymmetry on Riemann surfaces. Nucl. Phys. B **640**, 481 (2002)
18. D.J. Fernandez C, B. Mielnik, O. Rosas-Ortiz, B. F. Samsonov, Nonlocal supersymmetric deformations of periodic potentials. J. Phys. A **35**, 4279 (2002)
19. R. de Lima Rodrigues, The Quantum mechanics SUSY algebra: An Introductory review. arXiv: hep-th/0205017 (2002)
20. C. Leiva, M.S. Plyushchay, Superconformal mechanics and nonlinear supersymmetry. JHEP **0310**, 069 (2003)
21. A. Anabalon, M.S. Plyushchay, Interaction via reduction and nonlinear superconformal symmetry. Phys. Lett. B **572**, 202 (2003)
22. B. Mielnik, O. Rosas-Ortiz, Factorization: little or great algorithm? J. Phys. A **37**, 10007 (2004)
23. M.V. Ioffe, D.N. Nishnianidze, SUSY intertwining relations of third order in derivatives. Phys. Lett. A **327**, 425 (2004)
24. F. Correa, M.A. del Olmo, M.S. Plyushchay, On hidden broken nonlinear superconformal symmetry of conformal mechanics and nature of double nonlinear superconformal symmetry. Phys. Lett. B **628**, 157 (2005)
25. F. Correa, M.S. Plyushchay, Hidden supersymmetry in quantum bosonic systems. Ann. Phys. **322**, 2493 (2007)
26. F. Correa, L.M. Nieto, M.S. Plyushchay, Hidden nonlinear supersymmetry of finite-gap Lamé equation. Phys. Lett. B **644**, 94 (2007)
27. F. Correa, M.S. Plyushchay, Peculiarities of the hidden nonlinear supersymmetry of Pöschl–Teller system in the light of Lamé equation. J. Phys. A **40**, 14403 (2007)
28. A. Ganguly, L.M. Nieto, Shape-invariant quantum Hamiltonian with position-dependent effective mass through second order supersymmetry. J. Phys. A **40**, 7265 (2007)
29. F. Correa, L.M. Nieto, M.S. Plyushchay, Hidden nonlinear $su(2|2)$ superunitary symmetry of $N = 2$ superextended 1D Dirac delta potential problem. Phys. Lett. B **659**, 746 (2008)

30. F. Correa, V. Jakubsky, L.M. Nieto, M.S. Plyushchay, Self-isospectrality, special supersymmetry, and their effect on the band structure. Phys. Rev. Lett. **101**, 030403 (2008)
31. F. Correa, V. Jakubsky, M.S. Plyushchay, Finite-gap systems, tri-supersymmetry and self-isospectrality. J. Phys. A **41**, 485303 (2008)
32. F. Correa, V. Jakubsky, M.S. Plyushchay, Aharonov-Bohm effect on AdS(2) and nonlinear supersymmetry of reflectionless Pöschl–Teller system. Ann. Phys. **324**, 1078 (2009)
33. F. Correa, G.V. Dunne, M. S. Plyushchay, The Bogoliubov/de Gennes system, the AKNS hierarchy, and nonlinear quantum mechanical supersymmetry. Ann. Phys. **324**, 2522 (2009)
34. F. Correa, H. Falomir, V. Jakubsky, M.S. Plyushchay, Supersymmetries of the spin-1/2 particle in the field of magnetic vortex, and anyons. Ann. Phys. **325**, 2653 (2010)
35. V. Jakubsky, L.M. Nieto, M.S. Plyushchay, The origin of the hidden supersymmetry. Phys. Lett. B **692**, 51 (2010)
36. M.S. Plyushchay, L.M. Nieto, Self-isospectrality, mirror symmetry, and exotic nonlinear supersymmetry. Phys. Rev. D **82**, 065022 (2010)
37. M.S. Plyushchay, A. Arancibia, L.M. Nieto, Exotic supersymmetry of the kink-antikink crystal, and the infinite period limit. Phys. Rev. D **83**, 065025 (2011)
38. V. Jakubsky, M.S. Plyushchay, Supersymmetric twisting of carbon nanotubes. Phys. Rev. D **85**, 045035 (2012)
39. F. Correa, M. S. Plyushchay, Self-isospectral tri-supersymmetry in PT-symmetric quantum systems with pure imaginary periodicity. Ann. Phys. **327**, 1761 (2012)
40. F. Correa, M.S. Plyushchay, Spectral singularities in PT-symmetric periodic finite-gap systems. Phys. Rev. D **86**, 085028 (2012)
41. A.A. Andrianov, M.V. Ioffe, Nonlinear supersymmetric quantum mechanics: concepts and realizations. J. Phys. A **45**, 503001 (2012)
42. A. Arancibia, J. Mateos Guilarte, M.S. Plyushchay, Effect of scalings and translations on the supersymmetric quantum mechanical structure of soliton systems. Phys. Rev. D **87**(4), 045009 (2013)
43. A. Arancibia, J. Mateos Guilarte, M.S. Plyushchay, Fermion in a multi-kink-antikink soliton background, and exotic supersymmetry. Phys. Rev. D **88**, 085034 (2013)
44. F. Correa, O. Lechtenfeld, M. Plyushchay, Nonlinear supersymmetry in the quantum Calogero model. JHEP **1404**, 151 (2014)
45. R. Bravo, M.S. Plyushchay, Position-dependent mass, finite-gap systems, and supersymmetry. Phys. Rev. D **93**(10), 105023 (2016)
46. M.S. Plyushchay, Supersymmetry without fermions. arXiv:hep-th/9404081 (1994)
47. M.S. Plyushchay, Deformed Heisenberg algebra, fractional spin fields and supersymmetry without fermions. Ann. Phys. **245**, 339 (1996)
48. J. Gamboa, M. Plyushchay, J. Zanelli, Three aspects of bosonized supersymmetry and linear differential field equation with reflection. Nucl. Phys. B **543**, 447 (1999)
49. J.F. Cariñena, M.S. Plyushchay, Ground-state isolation and discrete flows in a rationally extended quantum harmonic oscillator. Phys. Rev. D **94**(10), 105022 (2016)
50. L. Inzunza, M. S. Plyushchay, Hidden superconformal symmetry: where does it come from? Phys. Rev. D **97**(4), 045002 (2018)
51. L. Inzunza, M.S. Plyushchay, Hidden symmetries of rationally deformed superconformal mechanics. Phys. Rev. D **99**(2), 025001 (2019)
52. G.W. Gibbons, R.H. Rietdijk, J.W. van Holten, SUSY in the sky. Nucl. Phys. B **404**, 42 (1993)
53. M. Tanimoto, The Role of Killing–Yano tensors in supersymmetric mechanics on a curved manifold. Nucl. Phys. B **442**, 549 (1995)
54. F. De Jonghe, A.J. Macfarlane, K. Peeters, J.W. van Holten, New supersymmetry of the monopole. Phys. Lett. B **359**, 114 (1995)
55. M.S. Plyushchay, On the nature of fermion-monopole supersymmetry. Phys. Lett. B **485**, 187 (2000)
56. J. Mateos Guilarte, M.S. Plyushchay, Perfectly invisible $\mathcal{PT}$-symmetric zero-gap systems, conformal field theoretical kinks, and exotic nonlinear supersymmetry. JHEP **1712**, 061 (2017)

57. J. Mateos Guilarte, M.S. Plyushchay, Nonlinear symmetries of perfectly invisible PT-regularized conformal and superconformal mechanics systems. J. High Energy Phys. **2019**(1), 194 (2019)
58. E. Witten, Dynamical breaking of supersymmetry. Nucl. Phys. B **188**, 513 (1981)
59. E. Witten, Constraints on supersymmetry breaking. Nucl. Phys. B **202**, 253 (1982)
60. F. Cooper, A. Khare, U. Sukhatme, Supersymmetry and quantum mechanics. Phys. Rept. **251**, 267 (1995)
61. G. Junker, *Supersymmetric Methods in Quantum, Statistical and Solid State Physics, Revised and Enlarged Edition* (IOP Publishing, Bristol, 2019)
62. M.S. Plyushchay, Schwarzian derivative treatment of the quantum second-order supersymmetry anomaly, and coupling-constant metamorphosis. Ann. Phys. **377**, 164 (2017)
63. D. Bonatsos, C. Daskaloyannis, P. Kolokotronis, D. Lenis, The symmetry algebra of the N-dimensional anisotropic quantum harmonic oscillator with rational ratios of frequencies and the Nilsson model. arXiv preprint hep-th/9411218 (1994)
64. J. de Boer, F. Harmsze, T. Tjin, Nonlinear finite W symmetries and applications in elementary systems. Phys. Rept. **272**, 139 (1996)
65. A.V. Turbiner, Quasiexactly solvable problems and SL(2) group. Commun. Math. Phys. **118**, 467 (1988)
66. F. Finkel, A. Gonzalez-Lopez, N. Kamran, P.J. Olver, M.A. Rodriguez, Lie algebras of differential operators and partial integrability. arXiv preprint hep-th/9603139 (1996)
67. M.A. Shifman, New findings in quantum mechanics (partial algebraization of the spectral problem). Int. J. Mod. Phys.A **4**, 2897 (1989)
68. V.B. Matveev, M.A. Salle, *Darboux Transformations and Solitons* (Springer, Berlin, 1991)
69. M.G. Krein, On a continuous analogue of a Christoffel formula from the theory of orthogonal polynomials. Dokl. Akad. Nauk SSSR **113**, 970 (1957)
70. V.E. Adler, A modification of Crum's method. Theor. Math. Phys. **101**, 1381 (1994)
71. J.F. Cariñena, L. Inzunza, M.S. Plyushchay, Rational deformations of conformal mechanics. Phys. Rev. D **98**, 026017 (2018)
72. S.P. Novikov, S.V. Manakov, L.P. Pitaevskii, V.E. Zakharov, *Theory of Solitons* (Plenum, New York, 1984)
73. A. Arancibia, M.S. Plyushchay, Chiral asymmetry in propagation of soliton defects in crystalline backgrounds. Phys. Rev. D **92**(10), 105009 (2015)
74. A. Arancibia, M.S. Plyushchay, Transmutations of supersymmetry through soliton scattering, and self-consistent condensates. Phys. Rev. D **90**(2), 025008 (2014)
75. C. Quesne, V.M. Tkachuk, Deformed algebras, position dependent effective masses and curved spaces: an exactly solvable Coulomb problem. J. Phys. A **37**, 4267 (2004)
76. A. Ganguly, S. Kuru, J. Negro, L. M. Nieto, A Study of the bound states for square potential wells with position-dependent mass. Phys. Lett. A **360**, 228 (2006)
77. S.C. y Cruz, J. Negro, L.M. Nieto, Classical and quantum position-dependent mass harmonic oscillators. Phys. Lett. A **369**, 400 (2007)
78. S.C. y Cruz, O. Rosas-Ortiz, Position dependent mass oscillators and coherent states. J. Phys. A **42**, 185205 (2009)
79. A. Arancibia, F. Correa, V. Jakubský, J. Mateos Guilarte, M.S. Plyushchay, Soliton defects in one-gap periodic system and exotic supersymmetry. Phys. Rev. D **90**(12), 125041 (2014)
80. A. Schulze-Halberg, Wronskian representation for confluent supersymmetric transformation chains of arbitrary order. Eur. Phys. J. Plus **128**, 68 (2013)
81. F. Correa, V. Jakubsky, M.S. Plyushchay, PT-symmetric invisible defects and confluent Darboux–Crum transformations. Phys. Rev. A **92**(2), 023839 (2015)
82. A. Contreras-Astorga, A. Schulze-Halberg, Recursive representation of Wronskians in confluent supersymmetric quantum mechanics. J. Phys. A **50**(10), 105301 (2017)
83. F. Calogero, Solution of the one-dimensional N body problems with quadratic and/or inversely quadratic pair potentials. J. Math. Phys. **12**, 419 (1971)
84. A.P. Polychronakos, Physics and mathematics of Calogero particles. J. Phys. A **39**, 12793 (2006)

85. C.M. Bender, Making sense of non-Hermitian Hamiltonians. Rept. Prog. Phys. **70**, 947 (2007)
86. A. Mostafazadeh, Pseudo-Hermitian representation of quantum mechanics. Int. J. Geom. Meth. Mod. Phys. **7**, 1191 (2010)
87. P. Dorey, C. Dunning, R. Tateo, Spectral equivalences, Bethe ansatz equations, and reality properties in PT-symmetric quantum mechanics. J. Phys. A **34**, 5679 (2001)
88. P. Dorey, C. Dunning, R. Tateo, Supersymmetry and the spontaneous breakdown of PT symmetry. J. Phys. A **34**, L391 (2001)
89. A. Fring, M. Znojil, PT-symmetric deformations of Calogero models. J. Phys. A **41**, 194010 (2008)
90. A. Fring, PT-symmetric deformations of integrable models. Philos. Trans. R. Soc. Lond. A **371**, 20120046 (2013)
91. R. El-Ganainy, K.G. Makris, M. Khajavikhan, Z.H. Musslimani, S. Rotter, D.N. Christodoulides, Non-Hermitian physics and PT symmetry. Nat. Phys. **14**, 11 (2018)
92. V. de Alfaro, S. Fubini, G. Furlan, Conformal invariance in quantum mechanics. Nuovo Cimento **34A**, 569 (1976)
93. J. Beckers, V. Hussin, Dynamical supersymmetries of the harmonic oscillator. Phys. Lett. A **118**, 319 (1986)
94. J. Beckers, D. Dehin, V, Hussin, Symmetries and supersymmetries of the quantum harmonic oscillator. J. Phys. A **20**, 1137 (1987)
95. J. Beckers, D. Dehin, V, Hussin, On the Heisenberg and orthosymplectic superalgebras of the harmonic oscillator. J. Math. Phys. **29**, 1705 (1988)
96. E.A. Ivanov, S.O. Krivonos, V.M. Leviant, Geometry of conformal mechanics. J. Phys. A **22**, 345 (1989)
97. C. Duval, P.A. Horvathy, On Schrödinger superalgebras. J. Math. Phys. **35**, 2516 (1994)
98. P. Claus, M. Derix, R. Kallosh, J. Kumar, P.K. Townsend, A. Van Proeyen, Black holes and superconformal mechanics. Phys. Rev. Lett. **81**, 4553 (1998)
99. J.A. de Azcarraga, J.M. Izquierdo, J.C. Perez Bueno, P.K. Townsend, Superconformal mechanics and nonlinear realizations. Phys. Rev. D **59**, 084015 (1999)
100. G.W. Gibbons, P.K. Townsend, Black holes and Calogero models. Phys. Lett. B **454**, 187 (1999)
101. J. Beckers, Y. Brihaye, N. Debergh, On realizations of 'nonlinear' Lie algebras by differential operators. J. Phys. A **32**, 2791 (1999)
102. J. Michelson, A. Strominger, The geometry of (super)conformal quantum mechanics. Commun. Math. Phys. **213**, 1 (2000)
103. S. Cacciatori, D. Klemm, D. Zanon, W(infinity) algebras, conformal mechanics, and black holes. Classical Quantum Gravity **17**, 1731 (2000)
104. G. Papadopoulos, Conformal and superconformal mechanics. Classical Quantum Gravity **17**, 3715 (2000)
105. E.E. Donets, A. Pashnev, V.O. Rivelles, D.P. Sorokin, M. Tsulaia, $N = 4$ superconformal mechanics and the potential structure of AdS spaces. Phys. Lett. B **484**, 337 (2000)
106. B. Pioline and A. Waldron, Quantum cosmology and conformal invariance. Phys. Rev. Lett. **90**, 031302 (2003)
107. H.E. Camblong, C.R. Ordonez, Anomaly in conformal quantum mechanics: From molecular physics to black holes. Phys. Rev. D **68**, 125013 (2003)
108. C. Leiva, M.S. Plyushchay, Conformal symmetry of relativistic and nonrelativistic systems and AdS/CFT correspondence. Ann. Phys. **307**, 372 (2003)
109. C. Duval, G.W. Gibbons, P. Horvathy, Celestial mechanics, conformal structures and gravitational waves. Phys. Rev.D **43**, 3907 (1991)
110. P.D. Alvarez, J.L. Cortes, P.A. Horvathy, M.S. Plyushchay, Super-extended noncommutative Landau problem and conformal symmetry. JHEP **0903**, 034 (2009)
111. F. Correa, H. Falomir, V. Jakubsky, M.S. Plyushchay, Hidden superconformal symmetry of spinless Aharonov-Bohm system. J. Phys. A **43**, 075202 (2010)
112. T. Hakobyan, S. Krivonos, O. Lechtenfeld, A. Nersessian, Hidden symmetries of integrable conformal mechanical systems. Phys. Lett. A **374**, 801 (2010)

113. C. Chamon, R. Jackiw, S. Y. Pi, L. Santos, Conformal quantum mechanics as the CFT_1 dual to AdS_2. Phys. Lett. B **701**, 503 (2011)
114. Z. Kuznetsova, F. Toppan, D-module representations of $N = 2, 4, 8$ superconformal algebras and their superconformal mechanics. J. Math. Phys. **53**, 043513 (2012)
115. K. Andrzejewski, J. Gonera, P. Kosinski, P. Maslanka, On dynamical realizations of l-conformal Galilei groups. Nucl. Phys. B **876**, 309 (2013)
116. M.S. Plyushchay, A. Wipf, Particle in a self-dual dyon background: hidden free nature, and exotic superconformal symmetry. Phys. Rev. D **89**(4), 045017 (2014)
117. S.J. Brodsky, G.F. de Teramond, H.G. Dosch, J. Erlich, Light-front holographic QCD and emerging confinement. Phys. Rept. **584**, 1 (2015)
118. M. Masuku, J.P. Rodrigues, De Alfaro, Fubini and Furlan from multi matrix systems. JHEP **1512**, 175 (2015)
119. O. Evnin, R. Nivesvivat, Hidden symmetries of the Higgs oscillator and the conformal algebra. J. Phys. A **50**(1), 015202 (2017)
120. I. Masterov, Remark on higher-derivative mechanics with l-conformal Galilei symmetry. J. Math. Phys. **57**(9), 092901 (2016)
121. K. Ohashi, T. Fujimori, M. Nitta, Conformal symmetry of trapped Bose-Einstein condensates and massive Nambu-Goldstone modes. Phys. Rev. A **96**(5), 051601 (2017)
122. R. Bonezzi, O. Corradini, E. Latini, A. Waldron, Quantum mechanics and hidden superconformal symmetry. Phys. Rev. D **96**(12), 126005 (2017)
123. U. Niederer, The maximal kinematical invariance group of the harmonic oscillator. Helv. Phys. Acta **46**, 191 (1973)
124. J.F. Cariñena, A.M. Perelomov, M.F. Rañada, M. Santander, A quantum exactly solvable nonlinear oscillator related to the isotonic oscillator. J. Phys. A Math. Theor. **41**, 085301 (2008)
125. J.M. Fellows, R.A. Smith, Factorization solution of a family of quantum nonlinear oscillators. J. Phys. A **42**, 335303 (2009)
126. D. Gómez-Ullate, N. Kamran, R. Milson, An extension of Bochner's problem: exceptional invariant subspaces. J. Approx. Theory **162**, 897 (2010)
127. I. Marquette, C. Quesne, New ladder operators for a rational extension of the harmonic oscillator and superintegrability of some two-dimensional systems. J. Math. Phys. **54**, 102102 (2013)
128. I. Marquette, New families of superintegrable systems from k-step rational extensions, polynomial algebras and degeneracies. J. Phys. Conf. Ser. **597**, 012057 (2015)
129. J.F. Cariñena, M.S. Plyushchay, ABC of ladder operators for rationally extended quantum harmonic oscillator systems. J. Phys. A **50**(27), 275202 (2017)
130. L. Inzunza, M.S. Plyushchay, Klein four-group and Darboux duality in conformal mechanics. arXiv preprint arXiv:1902.00538 (2019)
131. G.P. Dzhordzhadze, I.T. Sarishvili, Symmetry groups in the extended quantization scheme. Theor. Math. Phys. **93**, 1239 (1992)
132. G. Jorjadze, Constrained quantization on symplectic manifolds and quantum distribution functions. J. Math. Phys. **38**, 2851 (1997)
133. J. Hietarinta, B. Grammaticos, B. Dorizzi, A. Ramani, Coupling constant metamorphosis and duality between integrable Hamiltonian systems. Phys. Rev. Lett. **53**, 1707 (1984)
134. M. Cariglia, A. Galajinsky, G.W. Gibbons, P.A. Horvathy, Cosmological aspects of the Eisenhart-Duval lift. Eur. Phys. J. C **78**(4), 314 (2018)

Coherent and Squeezed States: Introductory Review of Basic Notions, Properties, and Generalizations

Oscar Rosas-Ortiz

To Prof. Veronique Hussin on his 60th birthday with friendship and scientific admiration.

Abstract A short review of the main properties of coherent and squeezed states is given in the introductory form. The efforts are addressed to clarify concepts and notions, including some passages of the history of science, with the aim of facilitating the subject for nonspecialists. In this sense, the present work is intended to be complementary to other papers of the same nature and subject in current circulation.

Keywords Coherent states · Squeezed states · Nonclassical states · Optical detection · Optical coherence · Wave packets · Minimum uncertainty · Harmonic oscillator · Riccati equation · Ermakov equation

1 Introduction

Optical coherence refers to the correlation between the fluctuations at different space-time points in a given electromagnetic field. The related phenomena are described in statistical form by necessity, and include interference as the simplest case in which correlations between light beams are revealed [1]. Until the first half of the last century the classification of coherence was somehow based on the averaged intensity of field superpositions. Indeed, with the usual conditions of stationarity and ergodicity, the familiar concept of coherence is associated with the sinusoidal modulation of the averaged intensity that arises when two fields are superposed. Such a modulation produces the extremal values $\langle I_{max} \rangle_{av}$ and $\langle I_{min} \rangle_{av}$, which are used to define the visibility of interference fringes

O. Rosas-Ortiz (✉)
Physics Department, Cinvestav, México City, Mexico
e-mail: orosas@fis.cinvestav.mx

Ş. Kuru et al. (eds.), *Integrability, Supersymmetry and Coherent States*, CRM Series in Mathematical Physics, https://doi.org/10.1007/978-3-030-20087-9_7

$$\mathcal{V} = \frac{\langle I_{max} \rangle_{av} - \langle I_{min} \rangle_{av}}{\langle I_{max} \rangle_{av} + \langle I_{min} \rangle_{av}}.$$

The visibility is higher as larger is the difference $\langle I_{max} \rangle_{av} - \langle I_{min} \rangle_{av} \geq 0$. At the limit $\langle I_{min} \rangle_{av} \to 0$ we find $\mathcal{V} \to 1$. If no modulation is produced, then $\langle I_{max} \rangle_{av} = \langle I_{min} \rangle_{av}$, and no fringes are observed ($\mathcal{V} = 0$). The fields producing no interference fringes are called *incoherent*. In turn, the highest order of coherence is traditionally assigned to the fields that produce fringes with maximum visibility.

The Young's experiment is archetypal to introduce the above concepts. Let us write $2|\Lambda_{(1,2)}| \cos \theta_{(1,2)}$ for the sinusoidal modulation of the averaged intensity at the detection screen. Then

$$\langle I_{min} \rangle_{av} = \langle I_1 \rangle_{av} + \langle I_2 \rangle_{av} - 2|\Lambda_{(1,2)}|, \qquad \langle I_{max} \rangle_{av} = \langle I_{min} \rangle_{av} + 4|\Lambda_{(1,2)}|,$$

with $\langle I_1 \rangle_{av}$ and $\langle I_2 \rangle_{av}$ the average intensities that would be contributed by either pinhole in the absence of the other. In general $\langle I_1 \rangle_{av}$, $\langle I_2 \rangle_{av}$, and $|\Lambda_{(1,2)}|$ depend on the geometry of the experimental setup. Therefore we can write

$$\mathcal{V}_Y = \left(\frac{2\sqrt{\langle I_1 \rangle_{av} \langle I_2 \rangle_{av}}}{\langle I_1 \rangle_{av} + \langle I_2 \rangle_{av}} \right) \lambda_{(1,2)}, \quad \text{with} \quad \lambda_{(1,2)} = \frac{|\Lambda_{(1,2)}|}{\sqrt{\langle I_1 \rangle_{av} \langle I_2 \rangle_{av}}}. \tag{1}$$

The expression $\lambda_{(1,2)}$ is the *correlation function* of the variable fields associated with the averages $\langle I_1 \rangle_{av}$ and $\langle I_2 \rangle_{av}$. If $|\Lambda_{(1,2)}| = 0$, then $\lambda_{(1,2)} = 0$, and no fringes are observed ($\mathcal{V}_Y = 0$). The fields incident on the pinhole-screen are in this case *incoherent*. On the other hand, if the fields emerging from the pinholes have equal intensity, the visibility $\mathcal{V}_Y$ will be equal to 1 only when $\lambda_{(1,2)} = 1$. The simplest form to obtain such a result is by considering the factorized form $|\Lambda_{(1,2)}| = \sqrt{\langle I_1 \rangle_{av} \langle I_2 \rangle_{av}}$, together with $\langle I_1 \rangle_{av} = \langle I_2 \rangle_{av}$.

The change of paradigm emerged in 1955, after the Brown and Twiss experiments oriented to measure correlations between quadratic forms of the field variables [2–4]. Unlike ordinary interferometer outcomes, the results of Brown and Twiss demanded the average of square intensities for their explanation. In other words, to embrace the new phenomenology, the concept of coherence as well as the *first-order* correlation function $\lambda_{(1,2)}$ needed a generalization. It was also clear that not all the fields described as "coherent" in the traditional approach would end up satisfying the new definitions of coherence. Thus, the Brown–Twiss results opened up the way to the quantitative investigation of higher-order forms of coherence [1, 5–7], though most of the light studied at the time was mainly produced by thermal sources. The development of new sources of light (minimizing noise generation) and new detectors (strongly sensitive to individual quanta of light) represented the experimental sustenance for the study of such concepts. However, the latter implied that the formal structure of optical coherence should be constructed on the basis of two mutually opposed features of light. While interference has long been regarded as a signature of the wavelike nature of electromagnetic radiation (Maxwell theory), photodetection implies the annihilation of individual photons to release photoelectrons from a given material (Einstein description of the photoelectric effect). These contradictory aspects were reconciled by Glauber in 1963, with his quantum approach to optical coherence, after considering expectation values of normal-ordered products of the (boson) creation and annihilation operators as quantum analogs of the classical correlation functions [8–10]. The expectation value of a single normal-ordered product corresponds to the first-order correlation function, that of two products corresponds to the second-order correlation function, and so on. Basically, Glauber formulated the theory of quantum optical detection for which the Young and Brown–Twiss experiments correspond to the measurement of first- and second-order correlation functions, respectively [11].

According to Glauber, most of the fields generated from ordinary sources lack second- and higher-order coherence, though they may be considered "coherent" in the traditional sense (i.e., they are only first-order coherent in Glauber's approach) [12]. Partial coherence means that there exist correlations up to a finite order only, and that the correlation functions of such order and all the lower orders are normalized. Full coherence implies "partial coherence" at all orders (the complete compilation of the Glauber contributions to quantum theory of optical coherence can be found in [13]). In concordance with $\lambda_{(1,2)}$ in the Young's interferometer, the Glauber's approach adopts the factorization of the n-th-order correlation function as a condition for coherence. Each of the factors corresponds to the probability of detecting a single photon at a given position at a given time. Factorization represents independence in the single-photon detection process. As recording a photon means its annihilation, the factors are indeed the squared norm of the vector state after the action of an ideal detector. At this stage, the brilliant contribution of Glauber was to notice that the simplest form of satisfying full coherence is by asking the quantum state of the field to be an eigenvector of the boson-annihilation operator with complex eigenvalue. Quite interestingly, this eigenvalue is a solution (written in complex form) of the corresponding Maxwell equations. In this form, the Glauber's fully coherent states of the quantized electromagnetic radiation are directly associated with the conventional electromagnetic theory. A classical description is then feasible for such a special class of quantum states of light.

It is apparent that the Fock (or number) states $|n\rangle$ arising from the quantization of (single-mode) electromagnetic fields are not fully coherent for $n \neq 0$. Namely, with exception of $|n = 0\rangle$, the states $|n \geq 1\rangle$ are not eigenvectors of the boson-annihilation operator. It may be proved that $|n = 1\rangle$ is first-order coherent, but it lacks second and higher-order coherence. The states $|n \geq 2\rangle$ are also first-order coherent but they do not factorize the second-order correlation function. Then, the number states $|n \geq 1\rangle$ are *nonclassical* in the sense that they are not fully coherent, so that no Maxwell field can be attached to them, and no classical description is possible. However, the "classical" property of states $|n \geq 1\rangle$ to be first-order coherent justifies their recurrent use as the incoming signal in contemporary versions of the Young's experiment [14–16]. On the other hand, the vacuum state

$|n = 0\rangle$ belongs to the eigenvalue 0 of the annihilation operator. As such eigenvalue is the trivial solution of the Maxwell equations, the zero-photon state $|n = 0\rangle$ corresponds to the classical notion of the absence of any field.

Despite the above remarks, the marvellous properties of quantum systems offer the possibility of using linear combinations of number states (rather than a given number state alone) to represent nontrivial eigenvectors of the boson-annihilation operator. Denoting by $|\alpha\rangle$ one of such vectors, the square modulus $|\alpha|^2$ of the complex eigenvalue α provides the expectation value of the number of photons in the superposition. In turn, the real and imaginary parts of α supply the expectation values of the field-quadratures $\langle \hat{x}_1 \rangle = \sqrt{2}\text{Re}(\alpha)$, $\langle \hat{x}_2 \rangle = \sqrt{2}\text{Im}(\alpha)$. Therefore, the variances are equal $(\Delta\hat{x}_1)^2 = (\Delta\hat{x}_2)^2 = 1/2$ and the related uncertainty is minimized $\Delta\hat{x}_1\Delta\hat{x}_2 = 1/2$. In other words, the vectors $|\alpha\rangle$ represent the closest quantum states to the Maxwell description of single-mode electromagnetic radiation (similar conclusions hold for the multi-mode case). A very important feature of the set $\{|\alpha\rangle\}$ is that, although it is not orthogonal, this satisfies the resolution of the identity [17]. Thus, $\{|\alpha\rangle\}$ is an overcomplete basis of states for the quantized single-mode electromagnetic fields. This property was used by Glauber [10] and Sudarshan [18] to introduce a criterion to classify the fields as those that admit a classical description (like the fully coherent states) and the ones which are nonclassical (like the number states $|n \geq 1\rangle$). The former can be written as a mixture of pure states $|\alpha\rangle\langle\alpha|$ for which the weights $P(\alpha)$ are admissible as conventional probabilities, the Dirac delta distribution $P(\alpha) = \delta(\alpha)$ included. The nonclassical fields are such that $P(\alpha)$ is not a conventional probability.

Over the time, states other than $|\alpha\rangle$ were found to minimize the quadrature-uncertainty [19–28]. In contraposition with $|\alpha\rangle$, such states lead to $(\Delta\hat{x}_1)^2 \neq (\Delta\hat{x}_2)^2$ and, depending on a complex parameter ξ, one of the quadrature-variances can be squeezed by preserving the product $\Delta\hat{x}_1\Delta\hat{x}_2 = 1/2$. Accordingly, the complementary variance is stretched. These properties found immediate applications in optical communication [29–31] and interferometry [32, 33], including the detection of gravitational waves [34–40]. The ξ-parameterized minimal uncertainty states $|\alpha, \xi\rangle$ are called *squeezed* [34] (see also [41–43]) and, like the number states $|n \geq 1\rangle$, they admit no description in terms of the Maxwell theory. That is, the squeezed states $|\alpha, \xi\rangle$ are nonclassical.

As it can be seen, we have three different basis sets to represent the quantum states of single-mode (and multi-mode) electromagnetic radiation, namely the number states $|n\rangle$, the fully coherent states (hereafter *coherent states* for short) $|\alpha\rangle$, and the squeezed states $|\alpha, \xi\rangle$. The former and last states (with exception of $n = 0$) are nonclassical while the coherent states may be described within the Maxwell theory. Depending on the optical field under study, we can use either of the above basis to make predictions and to explain experimental outcomes. Pretty interestingly, the classicalness of a given state is not invariant under linear superpositions. As immediate example recall that the "classical" state $|\alpha\rangle$ is a superposition of the nonclassical number states $|n\rangle$. In turn, it may be shown that nonclassical states $|\alpha, \xi\rangle$ can be expressed as a superposition of coherent states $|\alpha\rangle$ [44–47]. The "mystery" is hidden in the relative phases occurring as a consequence of any superposition of quantum states. According to Dirac, the reason because people had not thought of quantum mechanics much earlier is that the phase quantity was very well hidden in nature [48, p. 218]. Indeed, it is the probability amplitude of the entire superposition which expresses the difference between quantum and classical behavior. Thus, in quantum mechanics, not probabilities but probability amplitudes are summed up to make predictions! In the Young's experiment discussed above, for example, the sinusoidal modulation is the result of calculating the complete probability amplitude $\psi = \psi_1 + \psi_2$, with ψ_1 and ψ_2 the amplitudes relative to either pinhole. The modulation term $2\text{Re}(\psi_1 \psi_2^*)$ of the entire probability $|\psi|^2$ is different from zero only when we have no information about the pinhole that actually emitted the detected photon. It is then relevant to find a form to measure classicalness in quantum states [45, 49–55]. Besides the Glauber–Sudarshan P-representation [10, 18], the main criteria include the negativity of the Wigner function [49–51], some asymmetries in the Wigner and other pseudo-probability distributions [52], the identification of sub-Poissonian statistics (Mandel parameter) [53], and the presence of entanglement in the outcomes of a beam splitter (Knight conjecture) [54, 55].

The nonclassical properties of light have received a great deal of attention in recent years, mainly in connection with quantum optics [56], quantum information [57], and the principles of quantum mechanics [33]. Pure states representing fields occupied by a finite number of photons $n \neq 0$ exhibit nonclassical properties. The same holds for squeezed states and any other field state having sub-Poissonian statistics [53]. Using some deformations of the algebra generated by the boson operators, other states have been constructed to represent photons with "unusual properties" [58, 59], which may be applied in photon counting statistics, squeezing, and signal-to-quantum noise ratio [60]. Immediate generalizations [61, 62] motivated the development of the subject as an important branch of quantum optics [56]. Other deformations of the boson algebra include supersymmetric structures [63–75] for which the so-called polynomial Heisenberg algebras are quite natural [70, 76–82]. Recently, some non-Hermitian models have been shown to obey the distortions of the boson algebra that arise in the conventional supersymmetric approaches [83–87]. The deformed oscillator algebras have been used to construct the corresponding generalized (also called *nonlinear*) coherent states [61, 62, 75–82, 86–108]. Most of these states exhibit nonclassical properties that distinguish them from the coherent states of the conventional boson algebra.

The aim of the present work is to provide materials addressed to introduce the subject of coherent and squeezed states. The contents have been prepared for nonspecialists, so particular attention is given to the basic concepts as well as to their historical development. I preliminary apologize many authors because I have surely missed some fundamental references. In Sect. 2 the fundamentals of optical detection and coherence are revisited. The meaning of the affirmation that the photon "interferes only with itself" and that two different photons cannot interfere is analyzed in detail (see Sect. 2.1). The conditions for fully coherence are then given for fields of any number of modes and polarization. In Sect. 3 the coherent states for single-mode field are analyzed at the time that their fundamental properties are revisited. Spatial attention is addressed to the wave packets of minimum uncertainty that can be constructed for the conventional oscillator (Sects. 3.5 and 3.6), where the historical development of related ideas and concepts is overview. Some generalizations for wave packets with widths that depend on time are given in Sect. 3.7. The discussion about the wave packets for the hydrogen atom, as well as its historical development and controversies, is given in Sects. 3.8 and 3.9. In Sect. 4 the connection with representations of Lie groups in terms of generalized coherent states is reviewed in introductory form. The notions of generalized coherent states are discussed in Sect. 5, some analogies with classical systems are also indicated. Final comments are given in Sect. 6.

2 Basics of Quantum Optical Detection and Coherence

The quantized electric field is represented by the Hermitian operator

$$\mathbf{E}(\mathbf{r},t) = \mathbf{E}^{(+)}(\mathbf{r},t) + \mathbf{E}^{(-)}(\mathbf{r},t), \tag{2}$$

where its positive and negative frequency parts, $\mathbf{E}^{(+)}(\mathbf{r},t)$ and $\mathbf{E}^{(-)}(\mathbf{r},t)$, are mutually adjoint

$$\mathbf{E}^{(+)}(\mathbf{r},t) = \mathbf{E}^{(-)\dagger}(\mathbf{r},t). \tag{3}$$

Details concerning field quantization can be consulted in, e.g., [6]. The positive frequency part $\mathbf{E}^{(+)}(\mathbf{r},t)$ is a photon annihilation operator [9], so it is bounded from below $\mathbf{E}^{(+)}(\mathbf{r},t)|\text{vac}\rangle = 0$, with $|\text{vac}\rangle$ the state in which the field is empty of all photons. In turn, $\mathbf{E}^{(-)}(\mathbf{r},t)$ is a photon creation operator, with no upper bound. In particular, $\mathbf{E}^{(-)}(\mathbf{r},t)|\text{vac}\rangle$ represents a one-photon state of the field.

Following Glauber [13], let us associate the action of an ideal photodetector with the operator $\mathbf{E}^{(+)}(\mathbf{r},t)$. Assuming that the field is in state $|i\rangle$, and that one photon (polarized in the μ-direction) has been absorbed, after the photo-detection we have $\mathbf{E}_{\mu}^{(+)}(\mathbf{r},t)|i\rangle$. The probability that such a result coincides with the state $|f\rangle$ is regulated by the probability amplitude

$$\mathcal{A}_{i\to f}^{(1)} = \langle f|\mathbf{E}_{\mu}^{(+)}(\mathbf{r},t)|i\rangle, \tag{4}$$

which is a complex number in general. Notice that we do not require to know which of the possible states of the field is $|f\rangle$. The only requirement is that $|f\rangle$ be a physically admissible state. Then

$$\mathcal{P}_{i\to f}^{(1)} = |\mathcal{A}_{i\to f}^{(1)}|^2 = \langle i|\mathbf{E}_{\mu}^{(-)}(\mathbf{r},t)|f\rangle\langle f|\mathbf{E}_{\mu}^{(+)}(\mathbf{r},t)|i\rangle \tag{5}$$

is the probability we are looking for. To obtain the probability per unit time $\mathcal{P}_{det}^{(1)}$ that an individual photon be absorbed by the ideal detector at point $\mathbf{r}$ at time t, we have to sum over all possible (admissible) states

$$\mathcal{P}_{det}^{(1)}(\mathbf{r},t) = \sum_f \mathcal{P}_{i\to f}^{(1)} = \langle i|\mathbf{E}_\mu^{(-)}(\mathbf{r},t)\left[\sum_f |f\rangle\langle f|\right]\mathbf{E}_\mu^{(+)}(\mathbf{r},t)|i\rangle. \tag{6}$$

Now, it is quite natural to assume that the (admissible) final states form a complete orthonormal set. Therefore, the sum of projector operators $|f\rangle\langle f|$ between brackets in (6) can be substituted by the identity operator $\mathbb{I}$ to get

$$\mathcal{P}_{det}^{(1)}(\mathbf{r},t) = \langle i|\mathbf{E}_\mu^{(-)}(\mathbf{r},t)\mathbf{E}_\mu^{(+)}(\mathbf{r},t)|i\rangle = ||\mathbf{E}_\mu^{(+)}(\mathbf{r},t)|i\rangle||^2. \tag{7}$$

That is, probability $\mathcal{P}_{det}^{(1)}$ coincides with the expectation value of the Hermitian product $\mathbf{E}_\mu^{(-)}(\mathbf{r},t)\mathbf{E}_\mu^{(+)}(\mathbf{r},t)$, evaluated at the initial state $|i\rangle$ of the field. Equivalently, this is equal to the square norm of the vector $\mathbf{E}_\mu^{(+)}(\mathbf{r},t)|i\rangle$, which represents the state of the field just after the action of the ideal detector. These results show that the detector we have in mind measures the average value of the product $\mathbf{E}_\mu^{(-)}(\mathbf{r},t)\mathbf{E}_\mu^{(+)}(\mathbf{r},t)$, and not the average of the square of the Hermitian operator (2) representing the field [9]. Thus, the field intensity I, as a quantum observable, is represented by the operator $\mathbf{E}^{(-)}\mathbf{E}^{(+)}$, and not by the operator $\mathbf{E}^2$. It is illustrative to rewrite (7) as follows:

$$\mathcal{P}_{det}^{(1)}(\mathbf{r},t) = \langle i|\hat{I}_\mu(\mathbf{r},t)|i\rangle, \quad \hat{I}(\mathbf{r},t) = \mathbf{E}^{(-)}(\mathbf{r},t)\mathbf{E}^{(+)}(\mathbf{r},t), \tag{8}$$

which makes evident that $\mathcal{P}_{det}^{(1)}$ is the expectation value of the intensity $\hat{I}_\mu(\mathbf{r},t)$. Notice that $|i\rangle = |\text{vac}\rangle$ produces $\mathcal{P}_{det}^{(1)} = 0$, as this would be expected. The above results can be easily extended to arbitrary initial states (either pure or mixed) represented by ρ as follows:

$$\mathcal{P}_{det}^{(1)}(y) = \text{Tr}\left\{\rho\mathbf{E}_\mu^{(-)}(y)\mathbf{E}_\mu^{(+)}(y)\right\}, \quad y \equiv (\mathbf{r},t). \tag{9}$$

A lucky researcher has at his disposal more than one photodetector in his laboratory. He can use two detectors situated at different space-time points y_1 and y_2 to detect photon (delayed) coincidences. The probability amplitude associated with his predictions is of the form

$$\mathcal{A}_{i\to f}^{(2)} = \langle f|\mathbf{E}_\mu^{(+)}(y_2)\mathbf{E}_\mu^{(+)}(y_1)|i\rangle, \tag{10}$$

so the probability per unit (time)^2 that one photon is recorded at y_1 and another at y_2 is given by the expression

$$\mathcal{P}_{det}^{(2)}(y_1, y_2) = \text{Tr}\left\{\rho \mathbf{E}_{\mu}^{(-)}(y_1)\mathbf{E}_{\mu}^{(-)}(y_2)\mathbf{E}_{\mu}^{(+)}(y_2)\mathbf{E}_{\mu}^{(+)}(y_1)\right\}. \tag{11}$$

To rewrite (11) in terms of the intensity operator $\hat{I}(y)$, it is customary to use the normally ordered notation

$$:\mathbf{E}^{(-)}\mathbf{E}^{(+)}\mathbf{E}^{(-)}\mathbf{E}^{(+)}: = \mathbf{E}^{(-)}\mathbf{E}^{(-)}\mathbf{E}^{(+)}\mathbf{E}^{(+)}. \tag{12}$$

Therefore

$$\mathcal{P}_{det}^{(2)}(y_1, y_2) = \langle i| : \hat{I}_{\mu}(y_2)\hat{I}_{\mu}(y_1) : |i\rangle \tag{13}$$

corresponds to the expectation value of the square-intensity observable, which formalizes the experimental outcomes obtained by Brown and Twiss [11].

The above results can be generalized at will to include an arbitrary number of photodetectors (I have in mind a researcher even more fortunate than the previous one!). Another generalization may be addressed to investigate the correlations of the fields at separated positions and times. In this context, Glauber introduced the first-order correlation function

$$G^{(1)}(y_1, y_2) = \text{Tr}\left\{\rho \mathbf{E}_{\mu}^{(-)}(y_1)\mathbf{E}_{\mu}^{(+)}(y_2)\right\}, \tag{14}$$

which is complex-valued in general and satisfies $G^{(1)}(y, y) = \mathcal{P}_{det}^{(1)}(y)$. The expression for the n-th-order correlation function

$$G^{(n)}(y_1, \ldots, y_n, y_{n+1}, \ldots, y_{2n}) = \text{Tr}\left\{\rho \prod_{k=1}^{n} \mathbf{E}_{\mu}^{(-)}(y_k) \prod_{\ell=n+1}^{2n} \mathbf{E}_{\mu}^{(+)}(y_{\ell})\right\} \tag{15}$$

is now clear. The normalized form of the above formula is defined as

$$g^{(n)}(y_1, \ldots, y_{2n}) = \frac{G^{(n)}(y_1, \ldots, y_{2n})}{\prod_{k=1}^{2n}\left\{G^{(1)}(y_k, y_k)\right\}^{1/2}} \equiv \frac{G^{(n)}(y_1, \ldots, y_{2n})}{\prod_{k=1}^{2n}\left\{\mathcal{P}_{det}^{(1)}(y_k)\right\}^{1/2}}. \tag{16}$$

Thus, $g^{(n)}$ is the n-th-order correlation function $G^{(n)}$, weighted by the root-squared product of the probabilities that one photon is detected at y_1, another at y_2, and so on until all the $2n$ space-time points y_k have been exhausted. Notice that the product of probabilities $\mathcal{P}_{det}^{(1)}(y_k)$ means independence in detecting the individual photons.

Glauber found that $|g^{(n)}| = 1$, $n = 1, 2, \ldots$, is a necessary condition for coherence. The simplest way to satisfy such a requirement is by demanding the factorization of $G^{(n)}$ as follows:

$$\left|G^{(n)}(y_1,\ldots,y_{2n})\right| = \prod_{k=1}^{2n}\left\{G^{(1)}(y_k,y_k)\right\}^{1/2}. \tag{17}$$

In other words, if the correlations of a given field at $2n$ space-time points y_k can be expressed, up to a phase, as the root-squared product of the one-photon detection probabilities $\mathcal{P}_{det}^{(1)}(y_k)$, then the field is n-th-order coherent. If the latter condition is fulfilled for all orders, the field is fully coherent. General properties of the functions $G^{(n)}$ and $g^{(n)}$ can be consulted in [13].

2.1 *Self-interference of Single Photons*

To provide an immediate example let us compare the Glauber's first-order correlation function $g^{(1)}(y_1, y_2)$ with its counterpart $\lambda_{(1,2)}$ in the Young's experiment. As a first conclusion we have $G^{(1)}(y_k, y_k) = \mathcal{P}_{det}^{(1)}(y_k) = \langle \hat{I}(y_k)\rangle$, $k = 1, 2$, which verifies that the Hermitian operator $\mathbf{E}^{(-)}\mathbf{E}^{(+)}$ represents the "quantum observable" of field intensity $\hat{I}$. Second, the normalization condition $|g^{(1)}(y_1, y_2)| = 1$ means $g^{(1)}(y_1, y_2) = \exp[i\varphi(y_1, y_2)]$, so that $\varphi(y_1, y_2) = \theta_{(1,2)}$. Using these results the visibility (1) can be rewritten in the form

$$\mathcal{V}_Y = \frac{2\sqrt{G^{(1)}(y_1,y_1)G^{(1)}(y_2,y_2)}}{G^{(1)}(y_1,y_1)+G^{(1)}(y_2,y_2)}\left|g^{(1)}(y_1,y_2)\right|, \tag{18}$$

which is simplified to

$$\mathcal{V}_Y = \frac{2\sqrt{G^{(1)}(y_1,y_1)G^{(1)}(y_2,y_2)}}{G^{(1)}(y_1,y_1)+G^{(1)}(y_2,y_2)} \tag{19}$$

for first-order coherent fields (i.e., if $|g^{(1)}(y_1, y_2)| = 1$). In such case, the Young's visibility (19) is equal to 1 whenever $G^{(1)}(y_1, y_1) = G^{(1)}(y_2, y_2)$, which is equivalent to $\mathcal{P}_{det}^{(1)}(y_1) = \mathcal{P}_{det}^{(1)}(y_2)$. On the other hand, given $\mathcal{P}_{det}^{(1)}(y_k) = \langle \hat{I}(y_k)\rangle$, we may interpret $\mathcal{P}_{det}^{(1)}(y_k)$ as the probability that one photon emitted from the k-th pinhole has been recorded by the detector. In this sense the identity $\mathcal{P}_{det}^{(1)}(y_1) = \mathcal{P}_{det}^{(1)}(y_2)$ means that we cannot determine which of the two pinholes is the one that emitted such a photon. Thus, in the Young's experiment for a first-order coherent field, if the detection of an individual photon implies lack of knowledge about the source, interference fringes will be produced with maximum visibility. Our affirmation is particularly relevant for a single-photon wave packet that impinges on the Young's interferometer. As "any pure state in which the field is occupied by a single photon possesses first order coherence" [13, p. 62], the single-photon wave packet is able to produce interference fringes with maximum visibility, so this may be classified as highly coherent in the ordinary sense.

We would like to emphasize that, although the above results might be put in correspondence with the very famous sentence of Dirac that "each photon then interferes only with itself. Interference between two different photons can never occur," we must take it with a grain of salt. On the one hand, the origin of the sentence can be traced back to the first edition of the Dirac's book, published in 1930 [109], long before sources of coherent light like the maser (1953) or the laser (1960) were built. Then, by necessity, Dirac used the ordinary notion of coherence to formulate his sentence. The correlations of photons discussed in [109, Ch. I], are thus of the first order in the Glauber sense. In other words, the phrase "each photon then interferes only with itself" applies to conventional interferometry only (in the Young's experiment, for example). On the other hand, although the Brown–Twiss results and the Glauber theory were published much later than the first edition of the Dirac's book, it is also true that the sentence we are dealing with survived, with minimal modifications, until the fourth edition (revised) of the book, published in 1967 [110]. Therefore, it seems that even in 1967 Dirac was not aware that single-photon fields lack second- and higher-order coherence. Other option is that he was not interested in making the appropriate adjustments to his manuscript. In my opinion, the latter option is in opposition to the Dirac's perfectionism, so it can be discarded. The former option is viable but unlikely for somebody as learned as Dirac. A third option is that Dirac was aware of the Brown–Twiss and Glauber works but considered them as not definitive. To me, this last is the most reasonable since many people were reluctant to accept the Brown–Twiss results [111]. Besides, the Glauber theory, albeit corroborated on the blackboard, was far from being experimentally confirmed at the time. In any case, phenomena associated with the second-order correlation function (including the Brown–Twiss effect) are experimentally observed over and over in quantum-optics laboratories around the world. So the second part of Dirac's sentence "interference between two different photons can never occur" is also currently defeated. Other remarks in the same direction can be found in [112].

Nevertheless, it is remarkable that efforts to produce interference with "feeble light" were reported as early as 1905 by Taylor [113]. Fundamental advances on the single-photon interference arrived up to 1986, with the experimental results of Grangier, Roger, and Aspect about the anticorrelation effect produced on individual photons by a beam splitter [14].

2.2 *Fully Coherent States of Quantized Radiation Fields*

The touchstone used by Glauber to determine the quantum states that satisfy the factorization property (17) is reduced to the eigenvalue equation

$$\mathbf{E}_{\mu}^{(+)}(y)|?\rangle = \mathcal{E}_{\mu}(y)|?\rangle. \tag{20}$$

That is, the states $|?\rangle$ which Glauber was looking for should be eigenvectors of the positive frequency operator $\mathbf{E}_{\mu}^{(+)}(y)$. As the latter is not self-adjoint, two features of the solutions to (20) are easily recognized. First, the eigenvalues $\mathcal{E}_{\mu}(y)$ are complex numbers in general. Second, the (possible) orthogonality of the set of solutions $|?\rangle$ is not (automatically) granted. Nevertheless, assuming (20) is fulfilled, the introduction of the state $\rho = |?\rangle\langle ?|$ into (15) gives

$$G^{(n)}(y_1, \ldots, y_{2n}) = \prod_{k=1}^{n} \mathcal{E}_{\mu}^{*}(y_k) \prod_{\ell=n+1}^{2n} \mathcal{E}_{\mu}(y_\ell), \tag{21}$$

where z^* stands for the complex-conjugate of $z \in \mathbb{C}$. Clearly, the above expression satisfies (17) and produces $|g^{(n)}| = 1$. Thus, the eigenvectors of $\mathbf{E}_{\mu}^{(+)}(y)$ belonging to complex eigenvalues are the most suitable to represent fully coherent states.

3 Single-Mode Coherent States

The electric-field operator (2) for a single-mode of frequency ω, linearly polarized in the x-direction, with z-spatial dependence, can be written as

$$E(z,t) = \mathcal{E}_F \left[\frac{a(t) + a^{\dagger}(t)}{\sqrt{2}} \right], \qquad \mathcal{E}_F = \mathcal{E}_{vac} \sin(kz), \tag{22}$$

where $k = \omega/c$ is the wave vector and $\mathcal{E}_{vac}$ (expressed in electric field units) is a measure of the minimum size of the quantum optical noise that is inherent to the field [42]. The latter is associated with the vacuum fluctuations of the field since it is the same for any strength of excitation (even in the absence of any excitation) of the mode. The mutually adjoint time-dependent operators, $a(t)$ and $a^{\dagger}(t)$, are defined in terms of the boson ladder operators $[a, a^{\dagger}] = 1$ as usual $a(t) = a \exp(-i\omega t)$. It is useful to introduce the field quadratures

$$\hat{x}_1 = \frac{1}{\sqrt{2}}(a^{\dagger} + a), \quad \hat{x}_2 = \frac{i}{\sqrt{2}}(a^{\dagger} - a), \quad [\hat{x}_1, \hat{x}_2] = i, \tag{23}$$

to write

$$E(z,t) = \mathcal{E}_F (\hat{x}_1 \cos \omega t + \hat{x}_2 \sin \omega t). \tag{24}$$

At $t = 0$ we have $E(z, 0) = \mathcal{E}_F \hat{x}_1$. That is, the quadrature $\hat{x}_1$ represents the (initial) electric field. It is not difficult to show that the conjugate quadrature $\hat{x}_2$ corresponds to the (initial) magnetic field [42].

Equipped with physical units of position and momentum, the quadratures $\hat{x}_1$ and $\hat{x}_2$ can be put in correspondence with a pair of phase-space operators

$$\hat{x}_1 = \sqrt{\frac{m\omega}{\hbar}}\hat{q}, \quad \hat{x}_2 = \frac{1}{\sqrt{m\hbar\omega}}\hat{p}, \quad [\hat{q}, \hat{p}] = i\hbar, \tag{25}$$

so they define the oscillator-like Hamiltonian $\hat{H} = \hbar\omega H$, with H the dimensionless quadratic operator

$$H = \frac{1}{2}\left(\hat{x}_1^2 + \hat{x}_2^2\right) = \frac{1}{2}\left(\frac{m\omega}{\hbar}\hat{q}^2 + \frac{1}{m\hbar\omega}\hat{p}^2\right). \tag{26}$$

3.1 Field Correlations

Assuming we are interested in photon delayed coincidences, we may compute field correlations at t and $t + \tau$, measured at the same space point. Introducing (22) into (16) we obtain

$$g^{(1)}(t, t+\tau) = e^{i\omega\tau}, \tag{27}$$

and

$$g^{(2)}(t, t+\tau) = \frac{\langle i|a^\dagger a^\dagger a a|i\rangle}{\langle i|a^\dagger a|i\rangle^2} = 1 - \frac{1}{\langle i|\hat{n}|i\rangle}, \quad \langle i|\hat{n}|i\rangle \neq 0, \tag{28}$$

where we have used the photon-number operator $a^\dagger a = \hat{n}$. Notice that $g^{(1)}$ depends on the delay τ between detections and not on the initial time t, so we write $g^{(1)}(t, t+\tau) = g^{(1)}(\tau)$. Besides, $g^{(2)}$ does not depend on any time variable, so we can write $g^{(2)}(t, t+\tau) = g^{(2)}(0)$.

The first-order correlation function (27) shows that any field represented by the operator (22) is first-order coherent. That is, any quantized single-mode field $E(z, t)$ is coherent in the ordinary sense! In particular, the single-photon fields thought by Dirac in his book belong to the class $g^{(1)} = \exp(i\omega\tau)$, see our discussion on the matter in Sect. 2.1. In turn, the second-order correlation function (28) shows that the single-photon fields $|i\rangle = |n = 1\rangle$ produce the trivial result $g^{(2)} = 0$, so they lack second- and higher-order coherence, as we have already indicated. On the other hand, the field occupied by two or more photons $|i\rangle = |n \geq 2\rangle$ leads to $g^{(2)} = 1 - 1/n$. As the state $|n \geq 2\rangle$ produces $g^{(2)} < 1$ for a finite number of photons, we know that the factorization (17) is not admissible if $n \geq 2$. Of course, $g^{(2)} \to 1$ as $n \to \infty$. Note also that the probabilities of detecting a photon at time t, and another one at time $t + \tau$, produce the same result $\mathcal{P}_{det}^{(1)}(t) = \mathcal{P}_{det}^{(1)}(t+\tau) = |\mathcal{E}_F|^2 n/2$, $n = 2, 3, \ldots$, which does not depend on any time variable.

3.2 Mandel Parameter

To get additional information about the initial state $|i\rangle$ of the field, let us rewrite (28) as follows:

$$g^{(2)}(0) = 1 + \frac{(\Delta\hat{n})^2 - \langle i|\hat{n}|i\rangle}{\langle i|\hat{n}|i\rangle^2}, \tag{29}$$

where we have added a zero to complete the photon-number variance $(\Delta\hat{n})^2 = \langle i|\hat{n}^2|i\rangle - \langle i|\hat{n}|i\rangle^2$. Introducing the Mandel parameter [53]:

$$Q_M = \frac{(\Delta\hat{n})^2 - \langle i|\hat{n}|i\rangle}{\langle i|\hat{n}|i\rangle} = \frac{(\Delta\hat{n})^2}{\langle i|\hat{n}|i\rangle} - 1, \quad \langle i|\hat{n}|i\rangle \neq 0, \tag{30}$$

we arrive at the relationships

$$g^{(2)}(0) = 1 + \frac{Q_M}{\langle i|\hat{n}|i\rangle}, \quad Q_M = \left[g^{(2)}(0) - 1\right]\langle i|\hat{n}|i\rangle, \qquad \langle i|\hat{n}|i\rangle \neq 0. \tag{31}$$

For any number state $|n\rangle$ we clearly have $(\Delta\hat{n})^2 = 0$, so that $g^{(2)} = 1 - 1/n$ and $Q_M = -1$ for $n \geq 1$.

3.3 Klauder–Glauber–Sudarshan States

Using $\mathcal{E}_F$ as the natural unit of the electric field strength [42], and dropping the phase-time dependence of $a(t)$, the introduction of (22) into the eigenvalue equation (20) gives $a|\alpha\rangle = \alpha|\alpha\rangle$, the solution of which can be written as the normalized superposition

$$|\alpha\rangle = e^{-|\alpha|^2/2}\sum_{n=0}^{\infty}\frac{\alpha^n}{\sqrt{n!}}|n\rangle, \quad \alpha \in \mathbb{C}. \tag{32}$$

The above result was reported by Glauber in his quantum theory of optical coherence [13]. Remarkably, the superposition (32) was previously used (in implicit form) by Schwinger in his studies on quantum electrodynamics [114], and introduced by Klauder as the generator of an *overcomplete family of states* which is very appropriate to study the Feynman quantization of the harmonic oscillator [17]. Indeed, Klauder realized that $\langle\alpha|a^\dagger a|\alpha\rangle = \alpha^*\alpha$, and proved that the set of states $|\alpha\rangle$ forms a "basis" for the oscillator's Hilbert space

$$\frac{1}{\pi}\int d^2\alpha|\alpha\rangle\langle\alpha| = \sum_{n=0}^{\infty}|n\rangle\langle n| = \mathbb{I}. \tag{33}$$

Notwithstanding, the basis elements are not mutually orthogonal

$$\langle\beta|\alpha\rangle = \exp\left[-\frac{|\alpha-\beta|^2}{2} + i\mathrm{Im}(\beta^*\alpha)\right], \quad |\langle\beta|\alpha\rangle|^2 = e^{-|\alpha-\beta|^2}. \tag{34}$$

Further improvements of the mathematical structure associated with the superpositions (32) were provided by Klauder himself in his continuous representation theory [115, 116]. In addition to the Glauber contributions [13], fundamental properties of these states addressed to the characterization of light beams were also discussed by Sudarshan [18]. Hereafter the superpositions (32) will be called "Klauder–Glauber–Sudarshan states" or "*canonical* coherent states" (KGS-states or coherent states for short).

The probability that the field represented by $|\alpha\rangle$ is occupied by n-photons is given by the Poisson distribution:

$$\mathcal{P}_{\alpha\to n} = |\langle n|\alpha\rangle|^2 = \frac{(\overline{n})^n e^{-\overline{n}}}{n!}, \tag{35}$$

where $\overline{n} \equiv \langle\alpha|\hat{n}|\alpha\rangle = |\alpha|^2$. As it is well known, Poisson distributions like (35) are useful for describing random events that occur at some known average rate. In the present case, the rate is ruled by the mean value of the photon-number $\overline{n}$. Mandel and Wolf provided a very pretty physical interpretation: "when light from a single-mode laser falls on a photodetector, photoelectric pulses are produced at random at an average rate proportional to the mean light intensity, and the number of pulses emitted within a given time interval therefore obeys a Poisson distribution" [6, pp. 23–24]. In addition, the distribution we are dealing with is characterized by the fact that the variance $(\Delta\hat{n})^2$ is equal to the mean value $\langle\alpha|\hat{n}|\alpha\rangle$, which can be easily verified. Therefore, if the initial state of the field in the relationships (31) is a KGS-state $|\alpha\rangle$, then $(\Delta\hat{n})^2 = \langle\alpha|\hat{n}|\alpha\rangle = |\alpha|^2$, and $Q_M = 0$. This result is quite natural by noticing that the normalization condition $|g^{(2)}| = 1$ is automatically fulfilled by $|\alpha\rangle$.

On the other hand, the straightforward calculation shows that:

(I) States $|\alpha\rangle$ evolve in time by preserving their form (i.e., they have temporal stability):

$$|\alpha(t)\rangle = e^{-iHt}|\alpha\rangle = e^{-i\omega t/2}|\alpha e^{-i\omega t}\rangle. \tag{36}$$

(II) They are displaced versions of the vacuum state, $|\alpha\rangle = D(\alpha)|n=0\rangle$, with

$$D(\alpha) = \exp\left(\alpha a^\dagger - \alpha^* a\right) = \exp\left(-|\alpha|^2\right)\exp\left(\alpha a^\dagger\right)\exp\left(-\alpha^* a\right) \tag{37}$$

the unitary displacement operator fulfilling

$$D(\alpha) a D^\dagger(\alpha) = a - \alpha, \quad D^\dagger(\alpha) a^\dagger D(\alpha) = a^\dagger - \alpha^*. \tag{38}$$

(III) They are such that $\langle \hat{x}_1 \rangle = \sqrt{2}\mathrm{Re}(\alpha)$, $\langle \hat{x}_2 \rangle = \sqrt{2}\mathrm{Im}(\alpha)$, and $\langle \hat{x}_k^2 \rangle = \langle \hat{x}_k \rangle^2 + \frac{1}{2}$, $k = 1, 2$. That is, the states $|\alpha\rangle$ minimize the uncertainty of quadratures

$$(\Delta \hat{x}_1)^2 = (\Delta \hat{x}_2)^2 = \frac{1}{2}, \quad \Delta \hat{x}_1 \Delta \hat{x}_2 = \frac{1}{2}. \tag{39}$$

Any of the properties (I)–(III), including the eigenvalue equation $a|\alpha\rangle = \alpha|\alpha\rangle$ and the identity resolution (33), referred to as properties (A) and (B) in the sequel, can be assumed as the definition of the canonical coherent states. Then, the remaining properties may be derived as a consequence. For systems other than the harmonic oscillator it is very well known that not all the properties (A), (B), and (I)–(III) are simultaneously satisfied. It is then customary to construct states by using either of the above properties and to call them *generalized coherent states* (see Sect. 5). If by chance the states so constructed exhibit any other property of the KGS-states, the "coherence criterion" can be refined in each case. Not all generalized coherent states are *classical* in the sense established by the Glauber theory, so they usually deserve a study of their classicalness to be classified. After all: coherent states are superpositions of basis elements to which some specific properties are requested on demand (R.J. Glauber, private communication, Cocoyoc, Mexico, 1994).

3.4 Glauber–Sudarshan *P*- and Fock–Bargmann Representations

One of the most remarkable benefits offered by the KGS-states is the possibility of expressing any state of the radiation field as follows [10, 18]:

$$\rho = \int P(\alpha)|\alpha\rangle\langle\alpha| d^2\alpha. \tag{40}$$

The above "diagonal" form of the density operator ρ expresses the idea of having a mixed state, even though the basis defined by $|\alpha\rangle$ is not orthogonal. As the superposition of pure states $|\alpha\rangle\langle\alpha|$ defined above must be convex, the following condition is imposed

$$\int P(\alpha) d^2\alpha = 1. \tag{41}$$

Glauber introduced (40) to study thermal fields [10] and coined the term *P*-representation for it. In turn, Sudarshan argued that such representation is valid provided that $P(\alpha)$ is a conventional probability distribution [18]. It is easy to see that the state $\rho = |\beta\rangle\langle\beta|$, with $|\beta\rangle$ a GKS-state, implies $P(\alpha) = \delta(\alpha - \beta)$ by necessity. Then, the *P*-function is permitted to be as singular as the Dirac's delta distribution. It may be shown that the number states $|n \geq 1\rangle$ are P-represented by the n-th derivatives of the delta distribution, so the latter is stronger singular than $\delta(z)$ and are not allowed as probabilities. Therefore, the states $|n \geq 1\rangle$ admit no classical description in terms of the convex superposition (40). For states other than the oscillator ones, the criterion applies in similar form. Details and general properties of the P-function can be consulted in the book by Klauder and Sudarshan [117].

Using the identity (33), one can write any element $|\psi\rangle$ of the Hilbert space $\mathcal{H}$ as the superposition

$$|\psi\rangle = \frac{1}{\pi} \int d^2\alpha \psi(\alpha)|\alpha\rangle. \tag{42}$$

The Fourier coefficients $\psi(\alpha)$ are analytic over the whole complex α-plane. Indeed, as these functions are holomorphic and are in one-to-one correspondence with the number eigenstates, they are elements of a Hilbert space of entire functions $\mathcal{F}$ named after Fock [118] and Bargmann [119]. The representation of the ladder operators a and $a^\dagger$ in the $\mathcal{F}$-space corresponds to the derivative with respect to α and the multiplication by α, respectively. The properties of the Fock–Bargmann functions $\psi(\alpha)$ and the related representations are studied by Saxon, under the name "creation operator representation," in his Book on quantum mechanics (1968) [120, Ch. VI.e] (where no reference is given to neither Fock nor Bargmann works!).

3.5 Oscillator Wave Packets

Let us calculate the superposition (32) in the x_1-quadrature representation

$$\psi_\alpha(x) := \langle x|\alpha\rangle = e^{-|\alpha|^2/2} \sum_{n=0}^{\infty} \frac{\alpha^n}{\sqrt{n!}} \varphi_n(x), \tag{43}$$

where $\hat{x}_1|x\rangle = x|x\rangle$, $x \in \mathbb{R}$, $\int_{\mathbb{R}} dx|x\rangle\langle x| = \mathbb{I}$, and

$$\varphi_n(x) := \langle x|n\rangle = \frac{e^{-x^2/2}}{\pi^{1/4}(2^n n!)^{1/2}} H_n(x), \quad n = 0, 1, \ldots \tag{44}$$

The expression $H_n(x)$ stands for the Hermite-polynomials [121], and $\varphi_n(x)$ represents the wave functions of the harmonic oscillator [120, 122]. After introducing (44) into (43), and using the generating function of $H_n(x)$, Eq. 22.9.17 of Ref. [121], we arrive at the Gaussian-like expression

$$\psi_\alpha(x) = \pi^{-1/4} \exp\left[-\frac{(x - \langle \hat{x}_1 \rangle)^2}{2} + i\langle \hat{x}_2 \rangle (x - \langle \hat{x}_1 \rangle) \right]. \tag{45}$$

Using (39) the above result acquires the familiar form of a wave packet

$$\psi_\alpha(x) = \frac{1}{[2\pi(\Delta\hat{x}_1)^2]^{1/4}} \exp\left[-\frac{(x - \langle\hat{x}_1\rangle)^2}{4(\Delta\hat{x}_1)^2} + i\langle\hat{x}_2\rangle(x - \langle\hat{x}_1\rangle) \right]. \tag{46}$$

The Gaussian wave packet $\psi_\alpha(x)$ is localized about the point $x = \langle\hat{x}_1\rangle = \sqrt{2}\text{Re}(\alpha)$, within a neighborhood defined by $\Delta\hat{x}_1 = 1/\sqrt{2}$. Finally, equipped with physical units of position and momentum, the function (46) is given by

$$\psi_\alpha(x) = \left(\frac{\hbar}{m\omega}\right)^{1/4} \frac{1}{[2\pi(\Delta\hat{q})^2]^{1/4}} \exp\left[-\frac{(q - \langle\hat{q}\rangle)^2}{4(\Delta\hat{q})^2} + i\frac{\langle\hat{p}\rangle(q - \langle\hat{q}\rangle)}{\hbar} \right]. \tag{47}$$

For arbitrary values of $\Delta\hat{q}$, expression (47) coincides with the normalized *minimum wave packets* reported by Schiff in 1949 [122]. The smaller $\Delta\hat{q}$ is, the more localized the wave packet. Although Schiff derived (47) by thinking on the free particle motion, he was certain that "the structure of this minimum packet is the same whether or not the particle is free, since this form can be regarded simply as the initial condition on the solution of the Schrödinger equation for any V" [122, p. 54], where V stands for the potential defining the Schrödinger equation. Then, considering the harmonic oscillator, he realized that arbitrary superpositions of the related solutions are periodic functions of t, with the period of the classical oscillator $\tau_{osc} = 2\pi/\omega$. His conclusion on the matter is very interesting to our review purposes: "this suggest that it might be possible to find a solution in the form of a wave packet whose center of gravity oscillates with the period of the classical motion" [122, p. 67]. After some calculations, Schiff finally proved that such time-dependent wave packet is viable actually. However, the oscillator wave packet of Schiff was reported by Schrödinger 23 years earlier [123], as a minor result of his wave-formulation of quantum mechanics [124], in a try to give a physical meaning to the function that bears his name (see Sect. 3.8). It seems that Schiff was not aware of the Schrödinger results at the time of the first edition of his book [122] since the paper [123] is not mentioned until the third edition [125], published in 1968. It is also to be noted that the minimum wave packet (47) is as well discussed in the Saxon's book [120], where the celebrated Schrödinger's solution is not mentioned either. According to Saxon, supposing $x = f(t)$ is an integral of the classical equations of motion, it is "tempting to guess that a suitable form of the corresponding quantum mechanical probability function in the classical limit is $\psi_\alpha^*\psi_\alpha$—with $\langle\hat{q}\rangle$ substituted by $f(t)$ and $\langle\hat{p}\rangle = 0$ in our notation—for sufficiently small $\Delta\hat{q}$". He added, "this expression represents a wave packet of width $\Delta\hat{x}_1$ moving along the classical trajectory in accordance with the classical equations of motion" [120, pp. 26–27]. Saxon also proved that (47) is *of minimum uncertainty* [120, pp. 109–110], and then studied the motion of such wave packet in the harmonic oscillator potential [120, Ch. VI.6]. The analysis by Saxon is very close to that of Schwinger [114]. The above discussion is addressed to emphasize that in the late 1960s the currently famous paper of Schrödinger [123] was not central in quantum theory yet. Even more, neither the Brown–Twiss experimental results nor the Glauber theory had impacted with enough strength in the literature.

3.6 Schrödinger's Wave Packets of Minimum Uncertainty

To recover the Schrödinger's wave packet [123] let us reproduce the steps that led us to Eq. (45), but this time using the time-dependent KGS-state $\vert\alpha(t)\rangle$ as point of departure. With the help of (36), it is easy to verify the following result:

$$\psi_\alpha(x,t) = \frac{e^{-i\omega t}}{\pi^{1/4}} \exp\left\{ -\frac{[x-\lambda_1(t)]^2}{2} + i\lambda_2(t)[x-\lambda_1(t)]t \right\}, \tag{48}$$

where $\lambda_1(t)$ and $\lambda_2(t)$ are the real and imaginary parts of $\alpha \exp(-i\omega t)$, written in short notation as

$$\boldsymbol{\lambda}(t) = \frac{1}{\sqrt{2}} R(t) \langle \hat{\boldsymbol{x}} \rangle, \quad R(t) = \begin{pmatrix} \cos\omega t & \sin\omega t \\ -\sin\omega t & \cos\omega t \end{pmatrix}, \quad \boldsymbol{A} = \begin{pmatrix} A_1 \\ A_2 \end{pmatrix}. \tag{49}$$

The rotation matrix $R(t)$ has the classical oscillator period $\tau_{osc} = 2\pi/\omega$, so the point $x = \lambda_1(t)$ describes a circumference of radius $\Delta\hat{x}_1 = \Delta\hat{x}_2 = 1/\sqrt{2}$, centered on the origin of the quadrature phase-space.

Both, Schrödinger and Schiff considered a wave packet (48) with $\langle\hat{x}_2\rangle = 0$ (the expected value of the initial magnetic-field quadrature is equal to zero!). Notedly, Schrödinger was only interested on the real part of his wave packet $\psi_\alpha(x,t)$. After dropping the imaginary part of (48) he wrote (in our notation):

$$\psi_\alpha(x,t) = \frac{e^{-\frac{1}{2}\left(x - \frac{\langle\hat{x}_1\rangle}{\sqrt{2}}\cos\omega t\right)^2}}{\pi^{1/4}} \cos\left[\omega t + \left(\frac{\langle\hat{x}_1\rangle}{\sqrt{2}}\sin\omega t\right)\left(x - \frac{\langle\hat{x}_1\rangle}{\sqrt{2}}\cos\omega t\right)\right]. \tag{50}$$

Schrödinger realized that the first factor of (50) represents "a relatively tall and narrow *hump*, of the form of a *Gaussian error-curve*, which at a given moment lies in the neighborhood of the position" $x = \frac{\langle\hat{x}_1\rangle}{\sqrt{2}}\cos\omega t$ [124, p. 43]. Accordingly, he insisted, "the hump oscillates under exactly the same law as would operate in the usual mechanics for a particle having (26) as its energy function."

The reason for which Schrödinger discarded the imaginary part of $\psi_\alpha(x,t)$ is that, initially, he considered the solutions of his equation to be real. Indeed, the factor $i = \sqrt{-1}$ was missing in the first three papers of his celebrated series on *quantization as a problem of proper values* [124]. The purely imaginary number was introduced in his fourth paper on quantization. The latter influenced Born to formulate the appropriate probability interpretation of $\psi^*\psi$ [126]. In his first intuitive intent, Born postulated the probabilities proportional to the probability-amplitudes, just because the papers published by Schrödinger at the time formulated ψ to be real. An improved, though still imprecise version (probabilities = amplitude squares), was included at the last moment, in his note added in proof [127]. The argument for such a correction was that (real) amplitudes may be either positive or negative, and the latter cannot be associated with probabilities. Once Born was aware of the complex-valued wave functions, introduced in the Schrödinger's fourth paper on quantization, the correct formula was finally provided in [128], with some additional precisions by Pauli [129].

Coming back to the wave packet $\psi_\alpha(x,t)$, Schiff based his analysis on the probability density:

$$|\psi_\alpha(x,t)|^2 = \pi^{-1/2} \exp\left\{ -\frac{\left[\sqrt{2}x - \langle \hat{x}_1 \rangle \cos\omega t - \langle \hat{x}_2 \rangle \sin\omega t\right]^2}{2} \right\}, \tag{51}$$

which in our case corresponds to the probability of finding the field state at a given eigenvalue x of the (electric-field) quadrature $\hat{x}_1$. Clearly, the wave packet $\psi_\alpha(x,t)$ oscillates without change of shape about $x = 0$, with (classical) frequency $\nu_{osc} = \omega/(2\pi)$.

3.7 *Time-Dependent Oscillator Wave Packets*

The study of "minimum wave packets" includes a big number of relevant works throughout different stages of modern quantum theory. A nonexhaustive list comprehends the pioneering paper of Schrödinger [123], followed almost immediately by Kennard [130], where a wave packet is constructed to follow the classical motion with a Gaussian profile but the width of which oscillates with time (the Kennard's state, obtained in 1927, is in this form the first antecedent of what is nowadays called squeezed state!). A very useful ansatz to construct Gaussian wave packets characterized by the exponentiation of quadratic forms was given in 1953 by Husimi [131, 132]. It was also found that displaced versions of the number states are able to follow the classical motion by keeping their shape [133–138] but, unlike the KGS-states, they produce uncertainty products that "can be arbitrarily large, showing that the classical motion is not necessarily linked with minimum uncertainty" [139] (see also [140]). A very interesting class of minimum wave packets was exhaustively studied by Nieto [141–146]. On the other hand, the dynamics of many physical systems can be described by using the quantum time-dependent harmonic oscillator [147–157], where the construction of minimum wave packets is relevant [158–165]. In a more general situation, wave packets with time-dependent width may occur for systems with different initial conditions, time-dependent frequency, or in contact with a dissipative environment [166–169]. In all these cases, the corresponding coherent states, position and momentum uncertainties, as well as the quantum mechanical energy contributions, can be obtained in the same form if the creation and annihilation operators are expressed in terms of a complex variable that fulfills a nonlinear Riccati equation, which determines the time-evolution of the wave packet width. Explicitly, the wave packet (48) may be generalized to the form

$$\Psi(x,t) = N(t) \exp\left\{ i \left[y(t)\widetilde{x}^2 + \langle \hat{x}_2 \rangle \widetilde{x} + K(t) \right] \right\}, \tag{52}$$

where $y(t) = y_R(t) + i y_I(t)$ is a time-dependent complex-valued function, $\widetilde{x} = x - \langle \hat{x}_1 \rangle(t) = x - \eta(t)$, with $\eta(t)$ the (dimensionless) position of the wave packet maximum describing a classical trajectory determined by the Newtonian equation

$$\ddot{\eta}(t) + \omega^2(t)\eta(t) = 0, \tag{53}$$

and $\langle \hat{x}_2 \rangle = \dot{\eta}(t)$ the (dimensionless) classical momentum [166–169]. The time-dependent coefficient of the quadratic term obeys the Riccati equation

$$\dot{y}(t) + y^2(t) + \omega^2(t) = 0. \tag{54}$$

The concrete form of the normalization factor $N(t)$ and the purely time-dependent phase $K(t)$ are determined in each case. The straightforward calculation shows that (54) is solved by the function $y = \frac{\dot{\alpha}}{\alpha} + \frac{i}{\alpha}$, where α is a solution of the Ermakov equation

$$\ddot{\alpha}(t) + \omega^2(t)\alpha(t) = \frac{1}{\alpha^3(t)}. \tag{55}$$

The solutions of the above equations depend on the physical system under consideration and on the (complex) initial conditions. Besides, they have close formal similarities with general superpotentials leading to isospectral potentials in supersymmetric quantum mechanics [63–69]. Recent applications include propagation of optical beams in parabolic media [170–172] and Kerr media [173–176] as well, studies of the geometry of the Riccati equation [177] and the fourth-order Schrödinger equation with the energy spectrum of the Pöschl–Teller system [178]. Further discussion on the subject can be found in the Schuch's book on a nonlinear perspective to quantum theory [179].

3.8 A Quantum-Family Portrait

The development of modern quantum mechanics started in 1925, when Heisenberg conceived the idea that rather constructing a theory from quantities which could not be observed (like the electron orbits inside the atom), one should try to use quantities that are provided by experiment (like the frequencies and amplitudes associated with emission radiation) [180]. Shortly afterwards, during 1925–1927, great progress was made in developing matrix-mechanics (Heisenberg, Born, Jordan), wave-mechanics (Schrödinger), and quantum-algebra (Dirac). The Born's probability interpretation [126, 128, 129] and the uncertainty principle of Heisenberg [181] completed the foundations of what would become the most successful branch of physics throughout the past century. Notwithstanding, interpretative problems arose as soon as the first of the above formulations came to light. Schrödinger, for example, was "discouraged" with respect to the "very difficult methods of transcendental algebra" of the Heisenberg's theory [123]. Like Einstein and others, he did not feel comfortable with the quantum jumps involved in the matrix picture. Inspired by the de Broglie formulation of matter waves (1924), Schrödinger introduced a formulation based on finite single-valued functions obeying an "enigmatic" time-dependent wave equation that is characterized by linear (instead of quadratic) variations in the time-variable [124]. Remarkably, Schrödinger was firmly convinced that the solutions of his equation should represent something physically real. In the best of his attempts to provide the wave function with a physical meaning, for the charge density of an electron as a function of the space and time variables, Schrödinger suggested the squared modulus of the wave function multiplied by the total charge e (see paper IV on quantization in [124]). From the Schrödinger's perspective, the substitution of discrete energies by wave eigenfrequencies would eradicate "quantum jumps" from physics forever. Certainly, he was not immediately aware that not only the Heisenberg's theory and his own formalism are just two faces of the same coin, but that the discreteness of quantum energies as well as quantum jumps arrived to stay. The former was discovered by Schrödinger himself (see third paper in [124]), and the latter is a natural consequence of the equivalence between both approaches. Ironically, very far from having solved the "problems" of the Heisenberg's picture, the Schrödinger's formulation added a number of elements to the quantum theory that are unclear even today, such as the wave function. A main example of the latter can be found in a paper by Dirac [182], published in 1965. After a detailed comparison between the advantages and disadvantages of the Heisenberg and Schrödinger pictures to set up quantum electrodynamics, Dirac found that the two pictures are not equivalent. His opinion is rather clear "we now see that, if we want a logical quantum electrodynamics, we must work entirely with q numbers in the Heisenberg picture. All references to Schrödinger wave functions must be cut out as dead wood. The Schrödinger wave functions involve infinities, associated with v-v Feynman diagrams, which destroy all hope of logic." Nevertheless, the above expression requires some caution, as Dirac indicates "of course the development of quantum theory proposed here should not be considered as detracting from the value of Schrödinger's work... Only when one goes to an infinite number of degrees of freedom does one find that the Schrödinger picture is inadequate and that the Heisenberg picture has more fundamental validity."

3.9 The Quantum-Pictures Controversy: Hydrogen Atom Wave Packets

The controversy on the Schrödinger's quantization papers was immediate. After receiving copies of the first three papers, Lorentz wrote to Schrödinger on May 27, 1926, expressing "if I had to choose now between your wave mechanics and the matrix mechanics, I would give the preference to the former, because of its greater intuitive clarity, so long as one only has to deal with the three coordinates x, y, z. If, however, there are more degrees of freedom, then I cannot interpret the waves and vibrations physically, and I must therefore decide in favor of matrix mechanics" [183]. Clearly, Lorentz was foreseeing the Dirac's arguments on the infinities derived from the Schrödinger picture. In addition, among other points, Lorentz remarked that the equation proposed by Schrödinger in his first three papers did not contain time-derivatives (namely, it was stationary only), and expressed some doubts about the Schrödinger's functions "if I have understood you correctly, then a *particle*, an electron, for example, would be comparable to a wave packet which moves with the group velocity. But a wave packet can never stay together and remain confined to a small volume in the long run. The slightest dispersion in the medium will pull it apart in the direction of propagation, and even without that dispersion it will always spread more and more in the transverse direction. Because of this unavoidable blurring a wave packet does not seem to me to be very suitable for representing things to which we want to ascribe a rather permanent individual existence" [183]. Lorentz also emphasized some difficulties arising in the case of the hydrogen atom and included some calculations on the matter.

On June 6, Schrödinger replied that he had found a form to include time-derivatives in his equation. Concerning the difficulties about his functions, he added "allow me to send you, in an enclosure, a copy of a short note in which something is carried through for the simple case of the oscillator which is also an urgent requirement for all more complicated cases, where however it encounters great computational difficulties. (It would be nicest if it could be carried through in general, but for the present that is hopeless.) It is a question of really establishing the wave groups (or wave packets) which mediate the transition to macroscopic mechanics when one goes to large quantum numbers. You see from the text of the note, which was written before I received your letter, how much I too was concerned about the "staying together" of these wave packets. I am very fortunate that now I can at least point to a simple example where, contrary to all reasonable conjectures, it still proves right" [183].

By "the short note" Schrödinger meant his wave packet paper [123] which, at the very end, included a controversial statement "we can definitely foresee that, in a similar way, wave groups can be constructed which move around highly quantized Kepler ellipses and are the representation by wave mechanics of the hydrogen electron. But the technical difficulties in the calculation are greater than in the especially simple case which we have treated here" [124, p. 44]. The latter attracted Lorentz' attention who, on 19 June, replied "you gave me a great deal of pleasure by sending me your note, *the continuous transition from micro- to macro-mechanics*, and as soon as I had read it my first thought was: one must be on the right track with a theory that can refute an objection in such a surprising and beautiful way. Unfortunately my joy immediately dimmed again; namely, I cannot comprehend how, e.g., in the case of the hydrogen atom, you can construct wave packets that move like the electron (I am now thinking of the very high Bohr orbits). The short waves required for doing this are not at your disposal." Then Lorentz recalled that some lines were wrote by him in his previous communication and proceeded to extend his arguments on the subject. The *short note* of Schrödinger [123] was published on July, 1926.

The first published criticism appeared in a paper by Heisenberg [181], received on March 23, 1927. Yes, it is the work in which Heisenberg introduced the uncertainty principle of quantum mechanics! In his own words "the transition from micro to macro mechanics has already been dealt with by Schrödinger [123], but I do not believe that Schrödinger's considerations address the essence of the problem" [181, p. 184]. Heisenberg based his argument on the fact that, unlike the harmonic oscillator, "the frequencies of the spectral lines emitted by the atom are never integer multiples of a fundamental frequency, according with quantum mechanics with the exception of the special case of the harmonic oscillator. Thus, Schrödinger's consideration is applicable only to the harmonic oscillator considered by him, while other cases in the course of time the wave packet spreads over all space surrounding the atom" [181, p. 185]. That is, Heisenberg criticism was in complete agreement with the doubts expressed by Lorentz. According to Moore [48, p. 216], Schrödinger soon de-emphasized the wave packet picture while Lorentz thought that the demise of wave packets also meant the end of the analogy between wave mechanics and wave optics.

Taking into account the historical development, we have to say that the work of Schrödinger on the oscillator wave packets withstood the test of time just because the Glauber theory came to light. Although Schrödinger foresaw some possibilities for his wave packets to be applied in optics, his main efforts were addressed not to solve a practical problem originated in optics, but to provide the solutions of his equation with a physical meaning. The fantastic coincidence of the Schrödinger's wave packet and the x_1-quadrature representation of the fully coherent states of Glauber is due to the fact that, in both cases, connection with the classical world was looked for the quantum states of the harmonic oscillator. However, the states of Schrödinger and those of Glauber were originated by different reasons and obeying different approaches. In this form, the voices declaring that Schrödinger "discovered" the coherent states mislead the physical meaning of quantum optical coherence. Simply, it was not possible for Schrödinger to guess in any way that his wave packets would be connected with interference phenomena, not even with the Young's interferometer, since no experimental evidence of higher-order coherence (like the Brown–Twiss effect) was available at the time. The connection between the Schrödinger's wave packet and the "position" representation of the KGS-states is merely mathematical in origin. Nevertheless, the above does not discard the brilliant intuition of Schrödinger as far as quantum physics is concerned.

The works quoted in Sect. 3.5 testify the correctness of the Lorentz–Heisenberg argument. Although almost all of them are based on the oscillator number states, only the Schrödinger wave packets (i.e., the KSG-states) satisfy the properties of preserving their initial Gaussian profile by following the classical oscillator's trajectory. The most "exotic" of the aforementioned oscillator-like systems is the forced oscillator discussed by Husimi [131, 132], which was recovered by Carruthers and Nieto as the first application of the KGS-states to study quantum systems other than the harmonic oscillator [184]. Nevertheless, attracted by the Schrödinger's affirmation, some authors have presented different proposals to solve the problem of finding non-deformable wave packets for the hydrogen atom (see, e.g., [185–192]). Current interest on the subject is addressed to the potential applications of the Rydberg atoms in micro-wave cavities [193]. Notably, most of such proposals are based on the dynamics (in the sense of Lie theory) associated with the Coulomb problem. However, as Heisenberg indicated, the states so constructed do not go into a state of the same class under time evolution. The problem was partially solved by Klauder in 1996 [194], after relaxing some of the properties associated with the KGS-states to construct the appropriate generalized coherent states (see also [195]).

4 Group Approach and Squeezed States

The foundations of property (II) rest on the Lie group-representation of the Heisenberg–Weyl algebra. Namely, the operator $D(\alpha)$ that generates the displaced state $|\alpha\rangle = D(\alpha)|0\rangle$ is obtained by the product of exponentiated forms of the algebra generators $\mathbb{I}$, $a^{\dagger}$, and a. For if we take the exponentiation of $\alpha a^{\dagger}$ and $-\alpha^* a$, using the Baker–Campbell–Hausdorff formula [196], the related product can be simplified as follows:

$$e^{\alpha a^{\dagger}} e^{-\alpha^* a} = e^{\alpha a^{\dagger} - \alpha^* a} e^{\frac{1}{2}[\alpha^* a, \alpha a^{\dagger}]} = e^{\alpha a^{\dagger} - \alpha^* a} e^{|\alpha|^2 \mathbb{I}}. \tag{56}$$

Therefore

$$e^{-|\alpha|^2 \mathbb{I}} e^{\alpha a^{\dagger}} e^{-\alpha^* a} = e^{\alpha a^{\dagger} - \alpha^* a}. \tag{57}$$

The factors at the left-hand side of (57) are the result of a parametric exponentiation of the generators of the Heisenberg–Weyl algebra. In turn, the element at the right-hand side is the exponentiated form of the operator $\alpha a^{\dagger} - \alpha^* a$, which is also a member of the algebra. Notably, all the exponentiated forms included in (57) are elements of the Heisenberg–Weyl group, with $\exp(\alpha a^{\dagger} - \alpha^* a)$ the result of multiplying the group basis elements properly parameterized. Comparing with (37) we find the origin of the displacement operator $D(\alpha)$. On the other hand, expression (57) is a *disentangling formula* [197–199] that permits the factorization of $D(\alpha)$ into a normally ordered form, see Eq. (12). The factorization in antinormal-order is also feasible [197–199]. As the zero-photon state $|0\rangle$ is annihilated by a, it is immediate to see that the action of the group-element $\exp(-\alpha^* a)$ leaves $|0\rangle$ invariant. Hence $|0\rangle$ is called *fiducial state* with respect to $\exp(-\alpha^* a)$ [200–204]. The above results can be abridged by saying that the Heisenberg–Weyl algebra rules the dynamics of the quantum harmonic oscillator, and that the basis elements of the related Lie group define a mechanism to get the coherent state $|\alpha\rangle$, with the vacuum $|0\rangle$ as fiducial state. Then we have at hand a recipe to generalize the notion of coherent states for which the symmetric properties of the system under study are relevant. It is just required the identification of the dynamical group that defines the properties of the system we are interested in, as well as the related fiducial state [200–204].

4.1 The Versatile Representations of $SU(1,1)$ and $SU(2)$ Lie Groups

To give an example let us consider the operators $K_\pm$ and K_0 that satisfy the commutation relations

$$[K_-, K_+] = 2\sigma_\pm K_0, \quad [K_0, K_\pm] = \pm K_\pm, \quad \sigma_\pm = \pm 1, \tag{58}$$

as well as the operator

$$K^2 = K_0^2 - \sigma_\pm^{1/2}(K_+K_- + K_-K_+), \tag{59}$$

which satisfies $[K^2, K_\pm] = [K^2, K_0] = 0$. The above expressions correspond to the $su(1,1)$ Lie algebra for σ_+, and to the $su(2)$ Lie algebra for σ_-, with K^2 the Casimir operator. The normal order disentangling formula is in this case as follows [198]:

$$e^{A_+K_+} e^{(\ln A_0)K_0} e^{A_-K_-} = e^{a_+K_+ + a_0K_0 + a_-K_-}, \tag{60}$$

with

$$A_\pm = \frac{a_\pm \sinh\phi}{\phi\sqrt{A_0}}, \quad A_0 = \frac{1}{[\cosh\phi - \frac{a_0 \sinh\phi}{2\phi}]^2}, \quad \phi = \left[\frac{a_0^2}{4} - \sigma_\pm a_+ a_-\right]^{1/2}.$$

For the $su(1,1)$ Lie algebra we have two immediate bosonic representations. Namely, the single-mode one

$$K_+ = \frac{1}{2}(a^\dagger)^2, \quad K_- = \frac{1}{2}a^2, \quad K_0 = \frac{1}{2}\left(a^\dagger a + \frac{1}{2}\right), \tag{61}$$

and the two-mode representation

$$K_+ = a^\dagger b^\dagger, \quad K_- = ab, \quad K_0 = \frac{1}{2}\left(a^\dagger a + b^\dagger b + 1\right), \tag{62}$$

where $[a, a^\dagger] = [b, b^\dagger] = 1$, $[a, b^\dagger] = 0$, etc.

4.1.1 Vacuum Squeezed States

From (60) and (61), with $a_0 = 0$, $a_+ = \xi = -a_-^*$, and $\xi = re^{i\varphi} \in \mathbb{C}$, we obtain a new operator $S_s(\xi) = \exp[\xi(a^\dagger)^2 - \xi^* a^2]/2$ [21, 22] such that

$$|\xi_s\rangle := S_s(\xi)|0\rangle = \frac{1}{\sqrt{\cosh r}} \sum_{n=0}^{\infty} \frac{e^{in\varphi} \tanh^n r}{2^n n!} \sqrt{(2n)!}|2n\rangle. \tag{63}$$

The superposition $|\xi_s\rangle$ is known as the vacuum squeezed state [34] (see also [41–43]). The first antecedents of these states can be found in the paper of Kennard [130], and those of Infeld and Plebánski [134–137]. Noticeably, Cachill and Glauber used a mixed state prepared with even number states $|2n\rangle\langle 2n|$ to explore a class of displaced thermal states [13, Ch. 10]. The pure state $|\xi_s\rangle\langle\xi_s|$ includes populations $|2n\rangle\langle 2n|$ as well as coherences $|2n\rangle\langle 2m|$, $n \neq m$, and the former may share some properties with the Cachill–Glauber states since it corresponds to the "classical" part of the vacuum squeezed state (63).

On the other hand, it may be proved that the Mandel parameter (30) is in this case $Q_M = 2\langle\hat{n}\rangle + 1 = 2\sinh^2 r + 1$, and that the variances (evaluated with arbitrary φ) are parameterized by the modulus of ξ as follows $(\Delta\hat{x}_1)^2 = e^{2r}/2$, $(\Delta\hat{x}_2)^2 = e^{-2r}/2$. Clearly, although $\hat{x}_1$ is stretched at the time that $\hat{x}_2$ is squeezed, the related uncertainty is minimized. The roles are interchanged for other values of φ. From (31) we also see that $g^{(2)} = 3 + 1/\sinh^2 r$. Therefore, as $g^{(2)} > 3$, the vacuum squeezed state $|\xi_s\rangle$ is nonclassical. The latter is enforced by noticing that $Q_M > 1$. Of course, $r = 0$ is forbidden to evaluate either $g^{(2)}$ or Q_M since this produces the vacuum state $|\xi_s = 0\rangle$. Additionally, the state $|\xi_s\rangle$ is temporally stable $|\xi_s(t)\rangle = e^{-i\omega t/2}|\xi e^{-2i\omega t}\rangle$. On the other hand, the expectation value $\langle a^2\rangle = \sinh(2r)e^{i\varphi}/2$ leads to the probability $|\langle a^2\rangle|^2 = \sinh^2(2r)/4$ that two photons have been detected at the same space point by preserving the field state. This probability is always different from zero and increases with the mean number of photons.

4.1.2 Even and Odd Coherent States

Remarkably, Dodonov et al. introduced in 1974 a pair of states $|\alpha_\pm\rangle$, called *even and odd coherent states* [44], which include (63) as the even case $|\alpha_+\rangle$. These states result from the superpositions

$$|\alpha_{\pm}\rangle = \left[2\left(1 \pm e^{-2|\alpha|^2}\right)\right]^{-1/2} (|\alpha\rangle \pm |-\alpha\rangle). \tag{64}$$

The properties of $|\alpha_{\pm}\rangle$ can be consulted in the review by Man'ko [56, Ch. 4]. For historical details about general squeezed states see the pretty review prepared by Dodonov and Man'ko [56, Ch. 1].

4.1.3 Squeezed Coherent States

Applying the displacement operator $D(\alpha)$ on $|\xi_s\rangle$ we obtain the squeezed coherent state $|\alpha, \xi_s\rangle = D(\alpha)S_s(\xi)|0\rangle$ [25, 42], which produces the same variances as $|\xi_s\rangle$, with a modification in the expected number of photons $\langle\hat{n}\rangle = |\alpha|^2 + \sinh^2 r$. Thus, the displacement produced by $D(\alpha)$ does not modify the squeezing properties. However, depending on the combination of α and r, the Mandel parameter Q_M can be negative and $g^{(2)} \leq 1$ as well. In other words, for such values of (r, α) the states $|\alpha, \xi_s\rangle$ are very close in properties to the number states $|n \geq 1\rangle$.

4.1.4 Two-Mode Squeezed States

If we now use (60) and (62) with the same parameters a_0, $a_{\pm}$ as before, the action of the resulting operator $S_{two}(\xi) = \exp(\xi a^{\dagger} b^{\dagger} - \xi^* ab)$ on the two-mode vacuum state $|0, 0\rangle$ produces [205, 206]:

$$|\xi_{two}\rangle = \frac{1}{\cosh r} \sum_{n=0}^{\infty} e^{in\varphi} \tanh^n r |n, n\rangle. \tag{65}$$

The mean number of photons in any mode is the same $\langle\hat{n}_a\rangle = \langle\hat{n}_b\rangle = \sinh^2 r$, with correlation between modes defined by $\langle ab\rangle = \sinh(2r)e^{i\varphi}/2$ and $\langle a^{\dagger}b\rangle = 0$. Besides $\langle a^2\rangle = \langle b^2\rangle = 0$, and $\langle a\rangle = \langle b\rangle = 0$. The latter means that detection of only one photon, no matter the mode, is forbidden. Indeed, as $|\langle a\rangle|^2 = |\langle b\rangle|^2 = 0$, we see that the state of the field is inevitably changed to an orthogonal configuration after detecting a single photon! The same holds for the probabilities $|\langle a^2\rangle|^2 = |\langle b^2\rangle|^2 = 0$, so the field is changed to an orthogonal configuration after detecting two photons of the same mode. The situation is different for the square modulus of the correlation $|\langle ab\rangle|^2 = \sinh^2(2r)/4$, which means that only detections of one photon in a given mode AND one photon in the complementary mode are allowed. These properties are such that the modes a and b themselves are not squeezed, and suggest the symmetrization $(a \pm b)/\sqrt{2}$ to obtain the properly defined quadratures (a 50–50 beam splitter is useful in this subject [52, Ch. 3.2]). Then, it may be shown that the squeezing operator $S_{two}(\xi)$ can be factorized as the product of two-single mode squeezing operators. It is also possible to define two-mode squeezed coherent states $|\alpha, \beta, \xi\rangle = D(\alpha)D(\beta)S_{two}(\xi)|0, 0\rangle$. Additional properties of these states can be consulted in the book by Agarwal [52].

5 Generalized Coherent States

The bare essentials of coherent states can be expressed as a linear superposition

$$|\alpha_{CS}\rangle = \sum_{k\in\mathcal{I}} f_n(\alpha)|\psi_n\rangle, \quad \alpha \in \mathbb{C}, \tag{66}$$

where the vectors $|\psi_n\rangle$ generate a (separable) Hilbert space $\mathcal{H}$ (R.J. Glauber, private communication, Cocoyoc, Mexico, 1994), $\mathcal{I} \subseteq \mathbb{R}$ is an appropriate set of indices, and $f_n(\alpha)$ is a set of analytical functions permitting normalization. The superposition (66) must exhibit some specific properties that are determined by the "user" with basis on either the phenomenology under study or theoretical arguments.

Although the Klauder–Glauber–Sudarshan states (32) are the only superpositions that posses all the properties (A), (B), (I)–(III), and many other, the term *coherent states* (CS) has been used for a wide class of mathematical objects over the years. Nowadays, the overcomplete bases of states constructed to include at least one of the properties discussed in Sect. 3.3 are called *generalized coherent states*. The most valuable profile of the latter is that they can be studied for many systems in terms of the definition leading to the desirable result. For instance, the generalized CS studied by Barut and Girardello [207] and by Perelomov [200–202] are respectively based on properties (A) and (II) of the KGS-states. Property (III) is incidentally found as a secondary result for some special systems. Indeed, besides the Schrödinger [123], Kennard [130], Schiff [122], Husimi [131, 132], and Saxon [120] contributions, the construction of wave packets addressed to minimize the uncertainty relation of a pair of observables has been rarely reported in the literature. An exception is represented by the Nieto's results [141–146]. Of course, we have in mind that squeezed states can be considered as deformations of generalized CS for which the quadratures are not equal but minimize the related uncertainty. Klauder, for instance, constructed generalized CS for the hydrogen atom [194] such that they have temporal estability (i.e., they satisfy property I), are normalized and parametrized continuously, like it is defined in Eq. (66), and admit a resolution of unity with a positive measure (a fundamental property of the KGS-states proved in advance by Klauder himself [115, 116]). Thus, as a basis to construct his states for the hydrogen atom, Klauder used the concept of "coherent state" introduced in his compilation book, signed together with Skagerstam [208]. Further improvements were given in [209]. Additional generalizations have been discussed in, e.g., [210–212].

As a general rule, it is expected that the set $\{|\alpha_{CS}\rangle\}$ forms an overcomplete basis of the corresponding Hilbert space $\mathcal{H}$. In turn, the superpositions $|\alpha_{CS}\rangle$ are wished to be temporally stable.

5.1 Generalized Oscillator Algebras

The deformed oscillator (or boson) algebras can be encoded in a global symbolic expression that facilitates their study. One of the advantages of working in symbolic form is that the related nonlinear coherent states can be written in the same mathematical context [107]. The main idea is to consider (66) with

$$f_n(\alpha) = \frac{\alpha^n}{\sqrt{E(n)!}}, \quad E(n)! = E(1)E(2)\cdots E(n), \quad E(0)! := 1, \tag{67}$$

where E is a nonnegative function, $\mathcal{I} = \{0, 1, 2, \ldots\}$, and $|\psi_n\rangle \equiv |n\rangle \, \forall n \in \mathcal{I}$. The normalization requires

$$\mathcal{N}_E(|\alpha|) = \left[\sum_{n=0}^{\infty} \frac{|\alpha|^{2n}}{E(n)!} \right]^{-1/2} \tag{68}$$

to be finite, so that not any E and $|\alpha|$ are allowed. Assuming E and α properly chosen, the normalized vectors $|\alpha_{CS}\rangle_N = \mathcal{N}_E(\alpha)|\alpha_{CS}\rangle$ satisfy the closure relation

$$\mathbb{I} = \int |\alpha_{CS}\rangle_N \langle \alpha_{CS}| d\sigma_E(\alpha); \qquad d\sigma_E(\alpha) = \frac{d^2\alpha}{\pi} \Lambda_E\left(|\alpha|^2\right), \tag{69}$$

with $\Lambda_E(\alpha)$ an additional function to be determined such that

$$\int_0^{\infty} \Lambda_E(x) x^n dx = E(n)!, \quad \alpha = re^{i\theta}, \quad x = r^2. \tag{70}$$

After the change $b \to m - 1$, integral equation (70) coincides with the Mellin transform [213] of $\Lambda_E(x)$. The simplest form to obtain (67) is by a modification of the oscillator algebra that preserves the number operator $\hat{n}$ but changes the ladder operators, now written a_E and $a_E^{\dagger}$, as the set of generators. That is, one has

$$[\hat{n}, a_E] = -a_E, \quad [\hat{n}, a_E^{\dagger}] = a_E^{\dagger}, \tag{71}$$

so that the product $a_E^{\dagger} a_E$ preserves the number of quanta provided it is equal to the function $E(\hat{n})$. Equivalently, $a_E a_E^{\dagger} = E(\hat{n} + 1)$, so that $a_E|n\rangle = \sqrt{E(n)}|n - 1\rangle$, $a_E^{\dagger}|n\rangle = \sqrt{E(n+1)}|n + 1\rangle$, and

$$[a_E, a_E^{\dagger}] - E(\hat{n} + 1) = E(\hat{n}).$$

As the vacuum state $|0\rangle$ does not contain quanta we shall assume $E(0) = 0$ in order to have a bounded from below annihilation operator $a_E|0\rangle = 0$. It is say that any system obeying the new algebra is a generalized oscillator. It may be shown that when the function $E(n)$ is a real polynomial of degree $\ell \geq 1$, the determination of the measure (70) is reduced to a moment problem that is solved by the Meijer G-function [107]. This property automatically defines the delta distribution as the P-representation of the generalized CS $|\alpha_{CS}\rangle_N$ so defined. Then, in principle, there must be a classical analogy for them. However, in [107] it has been shown that they exhibit properties like antibunching and that they lack second-order coherence. That is, although they are P-represented by a delta function, they are not fully coherent. Therefore, the systems associated with the generalized oscillator algebras cannot be considered "classical" in the context of the quantum theory of optical coherence. Examples include the f-oscillators of Man'ko and co-workers [62], q-deformed oscillators [58, 59, 214], deformed photon phenomenology [60], the $su(1,1)$ oscillators applied to the study of the Jaynes-Cummings model [215] for intensity dependent interactions [216, 217] as well as the supersymmetric and nonlinear models [61, 62, 75–82, 86–108].

5.2 Position-Dependent Mass Systems and Quantum-Classical Analogies

The problem of calculating the energies of quantum systems endowed with position-dependent mass $m(x)$ has been a subject of increasing interest in recent years [218–246]. This model represents an interface between theoretical and applied physics, with analogies in geometric optics where the position dependent refractive index can be interpreted as a variable mass [247]. Its main characteristic is that the conventional expression for the kinetic term $\frac{\hat{p}^2}{2m}$ is not self-adjoint [218, 219], so that the Hermiticity of the Hamiltonian is a part of the problem if the mass is not a constant. Nevertheless, different generalized CS have been constructed [248–256]. The above is remarkable since well-known quantum-classical analogies [38, 40, 247, 257–259] can be exploited to test quantum-theoretical predictions in the laboratory. That is, although we nowadays have at hand precise forms to produce single photons on demand (see, e.g., [260–262] and the review paper [263]), relevant information is accessible from the optical analogies of quantum behavior [257, 259]. A first example is offered by the propagation of signals in optical waveguides [264], which can be used to test important predictions dealing with quantum resonances and leaky electromagnetic modes [265, 266], solitons [267, 268], and supersymmetry [269, 270]. Another example arises by recalling that classic optics includes an uncertainty relation between position and momentum, with Planck's constant $\hbar$ replaced by the light wave length λ [257]. The latter analogies connect, as we have seen, Gaussian light beams with fully coherent states. However, it may be also useful in the study of multilevel quantum systems [271]. Of course, the studies on the propagation of optical beams in parabolic [170–172] and Kerr [173–176] media include a refractive index with special properties that can be expressed as a concrete parametrization of the quantum states of light. On the other hand, quantum field theory in curved space-time also permits classical analogies [259], so that Hawking radiation can be studied in nonlinear Kerr media (including the analysis of the vacuum state for a star collapsing to a black hole which leads to the controversial effect named after Unruh [272–275]). At the moment, the analogies of Hawking radiation seem to be feasible [276–278].

5.3 Two Faces of the Same Coin

Considering the above remarks, it must be clear that squeezed and coherent states can be treated as different faces of a superposition $|\alpha_{CS}\rangle$ that minimizes the uncertainty of the appropriate quadratures. Indeed, they appear at different times when repeated measurements are made on a system [279], in the context of quantum nondemolition measurements [35–37] required to detect gravitational waves [38–40], where the expression *squeezed state* was coined [34] (see also [280–282]). Applications include quantum gravity [283], gauge field theory [284–287], and Yang–Mills theory [288]. Addressed to the interest on Rydberg atoms [193, 289], the generalized CS [290–293] are also relevant in manipulating atom–photon interactions [193, 294–297]. Special properties are revealed in the Jaynes–Cummings model [298–301], and for dynamical systems ruled by either the $su(1,1)$ or the $su(2)$ Lie algebras [256, 302–309]. The coherent states can be entangled [310–313], superposed [44–47, 314], and constructed for non-Hermitian operators [86, 87, 315] in terms of either a bi-orthogonal basis [84, 86, 87, 316] or noncommutative spaces [312, 317, 318], for which nonclassical properties can be found [87, 107, 319, 320]. They have been also associated with super algebraic structures [313, 321–325], nonlinear oscillators [326–328], and solvable models [329–341].

6 Conclusion

I have revisited the profiles of coherent and squeezed states that, in my opinion, have been overpassed in other works of the same nature and subject. The efforts have been addressed to clarify the main concepts and notions, including some passages of the history of science, with the aim of facilitating the subject for nonspecialists. In this sense, the present work must be treated as complementary to the reviews already reported by other authors. Clearly, it is not possible to scan all the literature on the matter so that, by necessity, any review is an incomplete work. The papers included in the references cannot cover all the relevant contributions on the matter, so the bibliography is, after all, the imperfect selection of the author. I apologize for the missed fundamental references as well as for the imprecise quotations (if any). I would conclude the work by addressing the readers attention to the books by Mandel and Wolf [6], Klauder and Sudarshan [117], Glauber [13], Perelomov [202], Ali et al. [212], Dodonov and Man'ko [56], Combescure and Robert [160], Agarwal

[52], Schuch [179], and Gazeau [342]. The reviews of Gilmore [203], Walls [41], Loudon and Knight [42], Fonda et al. [210], Zhang et al. [204], and Ali et al. [211] are also terrific to initiate the study of the subject.

Acknowledgements Financial support from Ministerio de Economía y Competitividad (Spain) grant number MTM2014-57129-C2-1-P, Consejería de Educación, Junta de Castilla y León (Spain) grant number VA057U16, and Consejo Nacional de Ciencia y Tecnología (México) project A1-S-24569 is acknowledged.

References

1. L. Mandel, E. Wolf, Coherence properties of optical fields. Rev. Mod. Phys. **37**, 231 (1965)
2. R.H. Brown, R.Q. Twiss, Correlation between photons in two coherent beams of light. Nature **177**, 27 (1956)
3. R.H. Brown, R.Q. Twiss, Interferometry of the intensity fluctuations in light I. Basic theory: the correlation between photons in coherent beams of radiation. Proc. R. Soc. Lond. **A242**, 300 (1957)
4. R.H. Brown, R.Q. Twiss, Interferometry of the intensity fluctuations in light II. An experimental test of the theory for partially coherent light. Proc. R. Soc. Lond. **A243**, 291 (1958)
5. G.J. Troup, R.G. Turner, Optical coherence theory. Rep. Prog. Phys. **37**, 771 (1974)
6. L. Mandel, E. Wolf, *Optical Coherence and Quantum Optics* (Cambridge University Press, New York, 1995)
7. M. Born, E. Wolf, *Principles of Optics* (Cambridge University Press, Cambridge, 1999)
8. R.J. Glauber, Photon correlations. Phys. Rev. Lett. **10**, 84 (1963)
9. R.J. Glauber, The quantum theory of optical coherence. Phys. Rev. **130**, 2529 (1963)
10. R.J. Glauber, Coherent and incoherent states of the radiation field. Phys. Rev. **131**, 2766 (1963)
11. D.F. Walls, Evidence for the quantum nature of light. Nature **280**, 451 (1979)
12. R.J. Glauber, Optical coherence and photon statistics, in *Quantum Optics and Electronics*, ed. by C. DeWitt, A. Blandin, C. Cohen-Tannoudji (Gordon and Breach, New York, 1964), pp. 65–185
13. R.J. Glauber, *Quantum Theory of Optical Coherence*. Selected Papers and Lectures (Wiley-VCH, Weinheim, 2007)
14. P. Grangier, G. Roger, A. Aspect, Experimental evidence for a photon anticorrelation effect on a beam splitter: a new light on single-photon interferences. Europhys. Lett. **1**, 173 (1986)
15. W. Rueckner, J. Peidle, Young's double-slit experiment with single photons and quantum eraser. Am. J. Phys. **81**, 951 (2013)
16. R.S. Aspden, M.J. Padgett, G.C. Spalding, Video recording true single-photon double-slit interference. Am. J. Phys. **84**, 671 (2016)
17. J.R. Klauder, The action option and a Feynman quantization of spinor fields in terms of ordinary *c*-numbers. Ann. Phys. **11**, 123 (1960)
18. E.C.G. Sudarshan, Equivalence of semiclassical and quantum mechanical descriptions of statistical light beams. Phys. Rev. Lett. **10**, 277 (1963)
19. H. Takahasi, Information theory of quantum-mechanical channels. Adv. Commun. Syst. **1**, 227 (1965)
20. D.R. Robinson, The ground state of the Bose gas. Commun. Math. Phys. **1**, 159 (1965)
21. D. Stoler, Equivalence classes of minimum uncertainty packets. Phys. Rev. D **1**, 3217 (1970)
22. D. Stoler, Equivalence classes of minimum uncertainty packets II. Phys. Rev. D **4**, 1925 (1971)

23. E.Y.C. Lu, New coherent states of the electromagnetic field. Lett. Nuovo Cimento **2**, 1241 (1971)
24. E.Y.C. Lu, Quantum correlations in two-photon amplification. Lett. Nuovo Cimento **2**, 585 (1972)
25. H.P. Yuen, Two-photon coherent states of the radiation field. Phys. Rev. A **13**, 2226 (1976)
26. V.V. Dodonov, E.V. Kurmyshev, V.I. Man'ko, Generalized uncertainty relation and correlated coherent states. Phys. Lett. A. **79**, 150 (1980)
27. A.K. Rajagopal, J.T. Marshall, New coherent states with applications to time-dependent systems. Phys. Rev. A **26**, 2977 (1982)
28. H.P. Yuen, Contractive states and the standard quantum limit for monitoring free-mass positions. Phys. Rev. Lett. **51**, 719 (1983)
29. H.P. Yuen, J.H. Shapiro, Optical communication with two-photon coherent states–Part I: quantum-state propagation and quantum-noise. IEEE Trans. Inf. Theory **24**, 657 (1978)
30. J.H. Shapiro, H.P. Yuen, J.A. Machado Mata, Optical communication with two-photon coherent states–Part II: photoemissive detection and structured receiver performance. IEEE Trans. Inf. Theory **25**, 179 (1979)
31. H.P. Yuen, J.H. Shapiro, Optical communication with two-photon coherent states–Part III: quantum measurements realizable with photoemissive detectors. IEEE Trans. Inf. Theory **26**, 78 (1980)
32. P. Hariharan, *Optical Interferometry* (Academic Press, San Diego, 2003)
33. M.P. Silverman, *Quantum Superposition. Counterintuitive Consequences of Coherence, Entanglement, and Interference* (Springer, Berlin, 2008)
34. J.N. Hollenhorst, Quantum limits on resonant-mass gravitational-radiation detectors. Phys. Rev. D **19**, 1669 (1979)
35. V.B. Braginsky, Y.I. Vorontsov, K.S. Thorne, Quantum nondemolition measurements. Science **209**, 547 (1980)
36. C.M. Caves, Quantum-mechanical radiation-pressure fluctuations in an interferometer. Phys. Rev. Lett. **45**, 75 (1980)
37. C.M. Caves, K.S. Thorne, R.W.P. Drever, V.D. Sandberg, M. Zimmerman, On the measurement of a weak classical force coupled to a quantum-mechanical oscillator. I. Issues of principle. Rev. Mod. Phys. **52**, 341 (1980)
38. K.S. Thorne, *Black Holes and Time Warps. Einstein's Outrageous Legacy* (W.W. Norton & Company, New York, 1994)
39. R. Schnabel, N. Mavalvala, D.E. McClelland, P.K. Lam, Quantum metrology for gravitational wave astronomy. Nat. Commun. **1**, 121 (2010)
40. P.R. Saulson, *Fundamentals of Interferometric Gravitational Wave Detectors*, 2nd edn. (World Scientific, Singapore, 2017)
41. D.F. Walls, Squeezed states of light. Nature **306**, 141 (1983)
42. R. Loudon, P.L. Knight, Squeezed light. J. Mod. Opt. **34**, 709 (1987)
43. M.C. Teich, B.E.A. Saleh, Squeezed states of light. Quantum Opt. **1**, 153 (1989)
44. V.V. Dodonov, I.A. Malkin, V.I. Man'ko, Even and odd coherent states and excitations of a singular oscillator. Physica **72**, 597 (1974)
45. C.C. Gerry, Non-classical properties of even and odd coherent states. J. Mod. Opt. **40**, 1053 (1993)
46. R.L. de Matos Filho, W. Vogel, Even and odd coherent states of the motion of a trapped ion. Phys. Rev. Lett. **76**, 608 (1996)
47. B. Roy, P. Roy, Coherent states, even and odd coherent states in a finite-dimensional Hilbert space and their properties. J. Phys. A Math. Gen. **31**, 1307 (1998)
48. W. Moore, *Schrödinger. Life and Tough* (Cambridge University Press, Cambridge, 1987)
49. E.P. Wigner, On the quantum correction for thermodynamic equilibrium. Phys. Rev. **40**, 749 (1932)
50. T.L. Curtright, D.B. Fairlie, C.K. Zachos, *A Concise Treatise on Quantum Mechanics in Phase Space* (World Scientific, Singapore, 2014)

51. A. Kenfack, K. Zyczkowski, Negativity of the Wigner function as an indicator of non-classicality. J. Opt. B Quantum Semiclass. Opt. **6**, 396 (2004)
52. G.H. Agarwal, *Quantum Optics* (Cambridge University Press, Cambridge, 2013)
53. L. Mandel, Sub-Poissonian photon statistics in resonance fluorescence. Opt. Express **4**, 205 (1979)
54. M.S. Kim, W. Son, V. Buzek, P.L. Knight, Entanglement by a beam splitter: nonclassicality as a prerequisite for entanglement. Phys. Rev. A **65**, 032323 (2002)
55. X.-b. Wang, Theorem for the beam-splitter entangler. Phys. Rev. A **66**, 024303 (2002)
56. V.V. Dodonov, V.I. Man'ko (eds.), *Theory of Nonclassical States of Light* (Taylor and Francis, London, 2003)
57. D. Bouwmeester, A. Ekert, A. Zeilinger, *The Physics of Quantum Information: Quantum Cryptography, Quantum Teleportation, Quantum Computation* (Springer, New York, 2000)
58. L.C. Biedenharn, The quantum group $SU_q(2)$ and a q-analogue of the boson operators. J. Phys. A Math. Gen. **22**, L873 (1989)
59. A.J. Macfarlane, On q-analogues of the quantum harmonic oscillator and the quantum group $SU(2)_q$. J. Phys. A Math. Gen. **22**, 4581 (1989)
60. A.I. Solomon, A characteristic functional for deformed photon phenomenology. Phys. Lett. A **196**, 29 (1994)
61. R.L. de Matos Filho, W. Vogel, Nonlinear coherent states. Phys. Rev. A **54**, 4560 (1996)
62. V.I. Ma'ko, G. Marmo, E.C.G. Sudarshan, F. Zaccaria, f-oscillators and nonlinear coherent states. Phys. Scrip. **55**, 528 (1997)
63. F. Cooper, A. Khare, U. Sukhatme, Supersymmetry and quantum mechanics. Phys. Rep. **251**, 267 (1995)
64. B.K. Bagchi, *Supersymmetry in Quantum and Classical Mechanics* (Chapman & Hall, Boca Raton, 2001)
65. H. Aoyama, M. Sato, T. Tanaka, General forms of a N-fold supersymmetric family. Phys. Lett. B **503**, 423 (2001)
66. H. Aoyama, M. Sato, T. Tanaka, N-fold supersymmetry in quantum mechanics: general formalism. Nucl. Phys. B **619**, 105 (2001)
67. B. Mielnik, O. Rosas-Ortiz, Factorization: little or great algorithm? J. Phys. A Math. Gen. **37**, 10007 (2004)
68. A.A. Andrianov, M.V. Ioffe, Nonlinear supersymmetric quantum mechanics: concepts and realizations. J. Phys. A Math. Gen. **45**, 503001 (2012)
69. A. Gangopadhyaya, J. Mallow, C. Rasinari, *Supersymmetric Quantum Mechanics: An Introduction*, 2nd edn. (World Scientific, Singapore, 2018)
70. B. Mielnik, Factorization method and new potentials with the oscillator spectrum. J. Math. Phys. **25**, 3387 (1984)
71. J. Beckers, D. Dehin, V. Hussin, Dynamical and kinematical supersymmetries of the quantum harmonic oscillator and the motion in a constant magnetic field. J. Phys. A Math. Gen. **21**, 651 (1988)
72. V.M. Eleonsky, V.G. Korolev, Isospectral deformation of quantum potentials and the Liouville equation. Phys. Rev. A **55**, 2580 (1997)
73. S. Seshadri, V. Balakrishnan, S. Lakshmibala, Ladder operators for isospectral oscillators. J. Math. Phys. **39**, 838 (1998)
74. C.L. Williams, N.N. Pandya, B.G. Bodmann, Coupled supersymmetry and ladder structures beyond the harmonic oscillator. Mol. Phys. **116**, 2599 (2018)
75. Y. Berube-Lauziere, V. Hussin, Comments of the definitions of coherent states for the SUSY harmonic oscillator. J. Phys. A Math. Gen. **26**, 6271 (1993)
76. D.J. Fernandez, V. Hussin, L.M. Nieto, Coherent states for isospectral oscillator Hamiltonians. J. Phys. A Math. Gen. **27**, 3547 (1994)
77. D.J. Fernandez, L. M. Nieto, O. Rosas-Ortiz, Distorted Heisenberg algebra and coherent states for isospectral oscillator Hamiltonians. J. Phys. A Math. Gen. **28**, 2693 (1995)
78. J.O. Rosas-Ortiz, Fock-Bargmann representation of the distorted Heisenberg algebra. J. Phys. A Math. Gen. **29**, 3281 (1996)

79. M.S. Kumar, A. Khare, Coherent states for isospectral Hamiltonians. Phys. Lett. A **217**, 73 (1996)
80. D.J. Fernandez, V. Hussin, Higher-order SUSY, linearized nonlinear Heisenberg algebras and coherent states. J. Phys. A Math. Gen. **32**, 3603 (1999)
81. D.J. Fernandez, V. Hussin, O. Rosas-Ortiz, Coherent states for Hamiltonians generated by supersymmetry. J. Phys. A Math. Teor. **40**, 6491 (2007)
82. D.J. Fernandez, O. Rosas-Ortiz, V. Hussin, Coherent states for SUSY partner Hamiltonians. J. Phys. Conf. Ser. **128**, 012023 (2008)
83. O. Rosas-Ortiz, O. Castaños, D. Schuch, New supersymmetry-generated complex potentials with real spectra. J. Phys. A Math. Theor. **48**, 445302 (2015)
84. A. Jaimes-Najera, O. Rosas-Ortiz, Interlace properties for the real and imaginary parts of the wave functions of complex-valued potentials with real spectrum. Ann. Phys. **376**, 126 (2017)
85. Z. Blanco-Garcia, O. Rosas-Ortiz, K. Zelaya, Interplay between Riccati, Ermakov and Schrödinger equations to produce complex-valued potentials with real energy spectrum. Math. Methods Appl. Sci. 1–14 (2018). https://doi.org/10.1002/mma.5069; arXiv:1804.05799
86. O. Rosas-Ortiz, K. Zelaya, Bi-orthogonal approach to non-Hermitian Hamiltonians with the oscillator spectrum: generalized coherent states for nonlinear algebras. Ann. Phys. **388**, 26 (2018)
87. K. Zelaya, S. Dey, V. Hussin, O. Rosas-Ortiz, Nonclassical states for non-Hermitian Hamiltonians with the oscillator spectrum. arXiv:1707.05367
88. M. Orzag, S. Salamo, Squeezing and minimum uncertainty states in the supersymmetric harmonic oscillator. J. Phys. A Math. Gen. **21**, L1059 (1988)
89. G. Junker, P. Roy, Non-linear coherent states associated with conditionally exactly solvable problems. Phys. Lett. A **257**, 113 (1999)
90. B. Roy, P. Roy, New nonlinear coherent states and some of their nonclassical properties. J. Opt. B Quantum Semiclass. Opt. **2**, 65 (2000)
91. R. Roknizadeh, M.K. Tavassoly, The construction of some important classes of generalized coherent states: the nonlinear coherent states method. J. Phys. A Math. Gen. **37**, 8111 (2004)
92. R. Roknizadeh, M.K. Tavassoly, Representations of coherent and squeezed states in a f-deformed Fock space. J. Phys. A Math. Gen. **37**, 5649 (2004)
93. M.K. Tavassoly, New nonlinear coherent states associated with inverse bosonic and f-deformed ladder operators. J. Phys. A Math. Theor. **41**, 285305 (2008)
94. F. Bagarello, Extended SUSY quantum mechanics, intertwining operators and coherent states. Phys. Lett. A **372**, 6226 (2008)
95. S. Twareque Ali, F. Bagarello, Supersymmetric associated vector coherent states and generalized Landau levels arising from two-dimensional supersymmetry. J. Math. Phys. **49**, 032110 (2008)
96. F. Bagarello, Vector coherent states and intertwining operators. J. Phys. A Math. Theor. **42**, 075302 (2009)
97. F. Bagarello, Quons, coherent states and intertwining operators. Phys. Lett. A **373**, 2637 (2009)
98. G.R. Honarasa, M.K. Tavassoly, M. Hatami, Quantum phase properties associated to solvable quantum systems using the nonlinear coherent states approach. Opt. Commun. **282**, 2192 (2009)
99. O. Abbasi, M.K. Tavassoly, Superposition of two nonlinear coherent states $\frac{\pi}{2}$ out of phase and their nonclassical properties. Opt. Commun. **282**, 3737 (2009)
100. M.K. Tavassoly, On the non-classicality features of new classes of nonlinear coherent states. Opt. Commun. **283**, 5081 (2010)
101. O. Safaeian, M.K. Tavassoly, Deformed photon-added nonlinear coherent states and their non-classical properties. J. Phys. A Math. Theor. **44**, 225301 (2011)
102. M. Kornbluth, F. Zypman, Uncertainties of coherent states for a generalized supersymmetric annihilation operator. J. Math. Phys. **54**, 012101 (2013)

103. A. NoormandiPour, M.K. Tavassoly, f-deformed squeezed vacuum and first excited states, their superposition and corresponding nonclassical properties. Commun. Theor. Phys. **61**, 521 (2014)
104. V. Hussin, I. Marquette, Generalized Heisenberg algebras, SUSYQM and degeneracies: infinite well and Morse potential. SIGMA **7**, 024 (2011)
105. M. Angelova, A. Hertz, V. Hussin, Squeezed coherent states and the one-dimensional Morse quantum system. J. Phys. A Math. Theor. **45**, 244007 (2012)
106. M. Angelova, A. Hertz, V. Hussin, Corrigendum: squeezed coherent states and the one-dimensional Morse quantum system. J. Phys. A Math. Theor. **46**, 129501 (2013)
107. K. Zelaya, O. Rosas-Ortiz, Z. Blanco-Garcia, S. Cruz y Cruz, Completeness and nonclassicality of coherent states for generalized oscillator algebras. Adv. Math. Phys. **2017**, 7168592 (2017)
108. B. Mojarevi, A. Dehghani, R.J. Bahrbeig, Excitation on the para-Bose states: nonclassical properties. Eur. Phys. J. Plus **133**, 34 (2018)
109. P.A.M. Dirac, *The Principles of Quantum Mechanics* (Oxford University Press, Oxford, 1930)
110. P.A.M. Dirac, *The Principles of Quantum Mechanics*, 4th edn revised (Oxford University Press, Oxford, 1967)
111. R.H. Brown, *The Intensity Interferometer. Its Application to Astronomy* (Taylor and Francis Ltd, New York, 1974)
112. E.T. Jaynes, Quantum beats, in *Foundations of Radiation Theory and Quantum Electrodynamics*, ed. by A.O. Barut (Plenum Press, New York, 1980), pp. 37–43
113. G.I. Taylor, Interference fringes with feeble light. Proc. Camb. Philos. Soc. Math. Phys. Sci. **15**, 114 (1909)
114. J. Schwinger, The theory of quantized fields III. Phys. Rev. **91**, 728 (1953)
115. J.R. Klauder, Continuous-representation theory I. Postulates of continuous-representation theory. J. Math. Phys. **4**, 1055 (1963)
116. J.R. Klauder, Continuous-representation theory II. Generalized relation between quantum and classical dynamics. J. Math. Phys. **4**, 1058 (1963)
117. J.R. Klauder, E.C.G. Sudarshan, *Fundamentals of Quantum Optics* (Benjamin, New York, 1968)
118. V. Fock, Verallgemeinerung und Loösung der Diracschen statistischen Gleichung. Z. Phys. **49**, 39 (1928)
119. V. Bargmann, On a Hilbert space of analytic functions and an associated integral transform. Commun. Pure Appl. Math. **14**, 187 (1961)
120. D.S. Saxon, *Elementary Quantum Mechanics* (Holden-Day, Inc., San Francisco, 1968)
121. M. Abramowitz, I.A. Stegun, *Handbook of Mathematical Functions* (Dover Publications, New York, 1964)
122. L.I. Schiff, *Quantum Mechanics* (McGraw-Hill Book Company, New York, 1949)
123. E. Schrödinger, Der stetige Übergang von der Mikro-zur Makromechanik. Naturwissenschaften **14**, 664 (1926); English translation in E. Schrödinger, *Collected Papers on Wave Mechanics*. AMS/Chelsea Series, vol. 302, 3rd (Augmented) edn. (American Mathematical Society, Providence, 2003)
124. E. Schrödinger, *Collected Papers on Wave Mechanics*. AMS/Chelsea Series, vol. 302, 3rd (Augmented) edn. (American Mathematical Society, Providence, 2003)
125. L.I. Schiff, *Quantum Mechanics*, 3rd edn. (McGraw-Hill, Singapore, 1968)
126. M. Born, Zur quantenmechanik der stossvorgänge. Z. Phys. **37**, 863 (1926); English translation in *Quantum Theory and Measurement*, ed. by J. Wheeler, W. Zurek (Princeton University Press, Princeton, 1983)
127. B. Mielnik, O. Rosas-Ortiz, Quantum mechanical laws, in *Fundamental of Physics*, vol 1. Encyclopedia of Life Support Systems (UNESCO, Paris, 2009), pp. 255–326
128. M. Born, Quantenmechanik der stossvorgänge. Z. Phys. **38**, 803 (1926)
129. W. Pauli, Über gasentartung und paramagnetismus. Z. Phys. **43**, 81 (1927)
130. E.H. Kennard, Zur Quantenmechanik einfacher Bewegungstypen. Z. Phys. **44**, 326 (1927)
131. K. Husimi, Miscellanea in elementary quantum mechanics I. Prog. Theor. Phys. **9**, 238 (1953)

132. K. Husimi, Miscellanea in elementary quantum mechanics II. Prog. Theor. Phys. **9**, 381 (1953)
133. I.R. Senitzky, Harmonic oscillator wave functions. Phys. Rev. **95**, 1115 (1954)
134. J. Plebański, Classical properties of oscillator wave packets. Bull. Acad. Polon. Sci. Cl. III **11**, 213 (1954)
135. J. Plebański, On certain wave-packets. Acta Phys. Polon. **XIV**, 275 (1955)
136. L. Infeld, J. Plebański, On a certain class of unitary transformations. Acta Phys. Pol. **XIV**, 41 (1955)
137. J. Plebański, Wave functions of a harmonic oscillator. Phys. Rev. **101**, 1825 (1956)
138. S.T. Epstein, Harmonic oscillator wave packets. Am. J. Phys. **27**, 291 (1959)
139. S.M. Roy, V. Singh, Generalized coherent states and the uncertainty principle. Phys. Rev. D **25**, 3413 (1982)
140. M.M. Nieto, Displaced and squeezed number states. Phys. Lett. A **229**, 135 (1997)
141. M.M. Nieto, L.M. Simmons, Coherent states for general potentials I. Formalism. Phys. Rev. D **20**, 1321 (1979)
142. M.M. Nieto, L.M. Simmons, Coherent states for general potentials II. Confining one-dimensional examples. Phys. Rev. D **20**, 1332 (1979)
143. M.M. Nieto, L.M. Simmons, Coherent states for general potentials III. Nonconfining one-dimensional examples. Phys. Rev. D **20**, 1342 (1979)
144. M.M. Nieto, Coherent states for general potentials IV. Three-dimensional systems. Phys. Rev. D **22**, 391 (1980)
145. V.P. Gutschick, M.M. Nieto, Coherent states for general potentials V. Time evolution. Phys. Rev. D **22**, 403 (1980)
146. M.M. Nieto, L.M. Simmons, V.P. Gutschick, Coherent states for general potentials VI. Conclusions about the classical motion and the WKB approximation. Phys. Rev. D **23**, 927 (1981)
147. M. Combescure, A quantum particle in a quadrupole radio-frequency trap. Ann. Inst. Henri Poincare A **44**, 293 (1986)
148. M. Combescure, The quantum stability problem for some class of time-dependent Hamiltonians. Ann. Phys. **185**, 86 (1988)
149. V.N. Gheorghe, F. Vedel, Quantum dynamics of trapped ions. Phys. Rev. A **45**, 4828 (1992)
150. B.M. Mihalcea, A quantum parametric oscillator in a radiofrequency trap. Phys. Scr. **2009**, 014006 (2009)
151. F.G. Major, V.N. Gheorghe, G. Werth, *Charged Particle Traps: Physics and Techniques of Charged Particle Field Confinement* (Springer, Berlin, 2005)
152. R.d.J. León-Montiel, H.M. Moya-Cessa, Exact solution to laser rate equations: three-level laser as a Morse-like oscillator. J. Mod. Opt. **63**, 1521 (2016)
153. L. Zhang, W. Zhang, Lie transformation method on quantum state evolution of a general time-dependent driven and damped parametric oscillator. Ann. Phys. **373**, 424 (2016)
154. A. Contreras-Astorga, J. Negro, S. Tristao, Confinement of an electron in a non-homogeneous magnetic field: integrable vs superintegrable quantum systems. Phys. Lett. A **380**, 48 (2016)
155. K. Zelaya, O. Rosas-Ortiz, Exactly solvable time-dependent oscillator-like potentials generated by Darboux transformations. J. Phys. Conf. Ser. **839**, 012018 (2017)
156. A. Contreras-Astorga, A time-dependent anharmonic oscillator. J. Phys. Conf. Ser. **839**, 012019 (2017)
157. H. Cruz, M. Bermúdez-Montaña, R. Lemus, Time-dependent local-to-normal mode transition in triatomic molecules. Mol. Phys. **116**, 77 (2018)
158. J.G. Hartley, J.R. Ray, Coherent states for the time-dependent harmonic oscillator. Phys. Rev. D **25**, 382 (1982)
159. A. Contreras-Astorga, D. Fernandez, Coherent states for the asymmetric penning trap. Int. J. Theor. Phys. **50**, 2085 (2011)
160. M. Combescure, D. Robert, *Coherent States and Applications in Mathematical Physics* (Springer, Netherlands, 2012)

161. D. Afshar, S. Mehrabankar, F. Abbasnezhad, Entanglement evolution in the open quantum systems consisting of asymmetric oscillators. Eur. Phys. J. D **70**, 64 (2016)
162. B. Mihalcea, Squeezed coherent states of motion for ions confined in quadrupole and octupole ion traps. Ann. Phys. **388**, 100 (2018)
163. N. Ünal, Quasi-coherent states for the Hermite Oscillator. J. Math. Phys. **59**, 062104 (2018)
164. K. Zelaya, O. Rosas-Ortiz, Comment on "Quasi-coherent states for the Hermite oscillator" [J. Math. Phys. 59, 062104 (2018)]. J. Math. Phys. **60**, 054101 (2019). arXiv:1810.03662
165. R. Razo, K. Zelaya, S. Cruz y Cruz, O. Rosas-Ortiz, in preparation
166. O. Castaños, D. Schuch, O. Rosas-Ortiz, Generalized coherent states for time-dependent and nonlinear Hamiltonians via complex Riccati equations. J. Phys. A Math. Theor. **46**, 075304 (2013)
167. D. Schuch, O. Castaños, O. Rosas-Ortiz, Generalized creation and annihilation operators via complex nonlinear Riccati equations. J. Phys. Conf. Ser. **442**, 012058 (2013)
168. H. Cruz, D. Schuch, O Castaños, O. Rosas-Ortiz, Time-evolution of quantum systems via a complex nonlinear Riccati equation I. Conservative systems with time-independent Hamiltonian. Ann. Phys. **360**, 44 (2015)
169. H. Cruz, D. Schuch, O Castaños, O. Rosas-Ortiz, Time-evolution of quantum systems via a complex nonlinear Riccati equation II. Dissipative systems. Ann. Phys. **373**, 609 (2016)
170. S. Cruz y Cruz, Z. Gress, Group approach to the paraxial propagation of Hermite-Gaussian modes in a parabolic medium. Ann. Phys. **383**, 257 (2017)
171. Z. Gress, S. Cruz y Cruz, A note on the off-axis Gaussian beams propagation in parabolic media. J. Phys. Conf. Ser. **839**, 012024 (2017)
172. R. Razo, S. Cruz y Cruz, New confining optical media generated by Darboux transformations. J. Phys. Conf. Ser. 1194, 012091 (2019)
173. R. Román-Ancheyta, M. Berrondo, J. Récamier, Parametric oscillator in a Kerr medium: evolution of coherent states. J. Opt. Soc. Am. B **32**, 1651 (2015)
174. R. Román-Ancheyta, M. Berrondo, J. Récamier, Approximate yet confident solution for a parametric oscillator in a Kerr medium. J. Phys. Conf. Ser. **698**, 012008 (2016)
175. R. Román-Ancheyta, C. González-Gutiérrez, J. Récamier, Influence of the Kerr nonlinearity in a single nonstationary cavity mode. J. Opt. Soc. Am. B **34**, 1170 (2017)
176. R.d.J. León-Montiel, H.M. Moya-Cessa, Generation of squeezed Schrödinger cats in a tunable cavity filled with a Kerr medium. J. Opt. **17**, 065202 (2015)
177. J. de Lucas, M. Tobolski, S. Vilariño, Geometry of Riccati equations over normed division algebras. J. Math. Anal. Appl. **440**, 394 (2016)
178. M.A. Rego-Monteiro, E.M.F. Curado, L.M.C.S. Rodrigues, Time evolution of linear and generalized Heisenberg algebra nonlinear Pöschl-Teller coherent states. Phys. Rev. A **96**, 052122 (2017)
179. D. Schuch, *Quantum Theory from a Nonlinear Perspective*. Riccati Equations in Fundamental Physics (Springer, Switzerland, 2018)
180. W. Heisenberg, *Physics and Beyond, Encounters and Conversations* (Harper and Row, Publishers, Inc., New York, 1971)
181. W. Heisenberg, Über den anschaulichen Inhalt der quantentheoretischen Kinematik und Mechanik. Z. Phys. **43**, 172 (1927); English translation in NASA Technical Report Server, Document ID: 19840008978, web page https://ntrs.nasa.gov/search.jsp?R=19840008978, consulted December 2018
182. P.A.M. Dirac, Quantum electrodynamics without dead wood. Phys. Rev. **139**, B684 (1965)
183. K. Przibram (ed.), *Albert Einstein: Letters on Wave Mechanics. Correspondence with H.A. Lorentz, Max Planck, and Erwin Schrödinger* (Philosophical Library, New York, 1967)
184. P. Carruthers, M.M. Nieto, Coherent states and the forced quantum oscillator. Am. J. Phys. **33**, 537 (1965)
185. L.S. Brown, Classical limit of the hydrogen atom. Am. J. Phys. **41**, 525 (1973)
186. J. Mostowski, On the classical limit of the Kepler problem. Lett. Math. Phys. **2**, 1 (1977)
187. D. Bhaumik, B. Dutta-Roy, G. Ghosh, Classical limit of the hydrogen atom. J. Phys. A Math. Gen. **19**, 1355 (1986)

188. J.-C. Gay, D. Delande, A. Bommier, Atomic quantum states with maximum localization on classical elliptical orbits. Phys. Rev. A **39**, 6587 (1989)
189. M. Nauenberg, Quantum wave packets on Kepler elliptic orbits. Phys. Rev. A **40**, 1133 (1989)
190. Z.D. Gaeta, C.R. Stroud, Jr., Classical and quantum-mechanical dynamics of a quasiclassical state of the hydrogen atom. Phys. Rev. A **42**, 6308 (1990)
191. J.A. Yeazell, C.R. Stroud, Jr., Observation of fractional revivals in the evolution of a Rydberg atomic wave packet. Phys. Rev. A **43**, 5153 (1991)
192. I. Zaltev, W.M. Zhang, D.H. Feng, Possibility that Schrödinger's conjecture for the hydrogen-atom coherent states is not attainable. Phys. Rev. A **50**, R1973 (1994)
193. S. Haroche, J.-M. Raimond, *Exploring the Quantum: Atoms, Cavities, and Photons* (Oxford University Press, Oxford, 2006)
194. J.R. Klauder, Coherent states for the hydrogen atom. J. Phys. A Math. Gen. **29**, L293 (1996)
195. P. Majumdar, H.S. Sharatchandra, Coherent states for the hydrogen atom. Phys. Rev. A **56**, R3322 (1997)
196. B. Mielnik, J. Plebánski, Combinatorial approach to Baker-Campbell-Hausdorff exponents. Annales de l'I.H.P. Physique théorique **12**, 215 (1970)
197. R. Gilmore, Baker-Campbell-Hausdorff formulas. J. Math. Phys. **15**, 2090 (1974)
198. M. Ban, Decomposition formulas for $su(1,1)$ and $su(2)$ Lie algebras and their applications in quantum optics. J. Opt. Soc. Am. B **10**, 1347 (1993)
199. A. DasGupta, Disentanglement formulas: an alternative derivation and some applications to squeezed coherent states. Am. J. Phys. **64**, 1422 (1996)
200. A.M. Perelomov, Coherent states for arbitrary Lie group. Commun. Math. Phys. **26**, 222 (1972)
201. A.M. Perelomov, Generalized coherent states and some of their applications (in Russian). Moscow Usp. Fiz. Nauk. **123**, 23 (1977); English translation in Sov. Phys. Usp. **20**, 703 (1977)
202. A.M. Perelomov, *Generalized Coherent States and Their Applications* (Springer, Heidelberg, 1986)
203. R. Gilmore, On the properties of coherent states. Rev. Mex. Fis. **23**, 143 (1974)
204. W.M. Zhang, D.H. Feng, R. Gilmore, Coherent states-theory and some applications. Rev. Mod. Phys. **62**, 867 (1990)
205. C.M. Caves, B.L. Schumaker, New formalism for two-photon quantum optics I. Quadrature phases and squeezed states. Phys. Rev. A **31**, 3068 (1985)
206. B.L. Schumaker, C.M. Caves, New formalism for two-photon quantum optics II. Mathematical foundation and compact notation. Phys. Rev. A **31**, 3093 (1985)
207. A.O. Barut, L. Girardello, New 'coherent' states associated with noncompact groups. Commun. Math. Phys. **21**, 41 (1971)
208. J.R. Klauder, B.-S. Skagerstam, *Coherent States: Applications in Physics and Mathematical Physics* (World Scientific, Singapore, 1985)
209. J.-P. Gazeau, J.R. Klauder, Coherent states for systems with discrete and continuous spectrum. J. Phys. A Math. Gen. **32**, 123 (1999)
210. L. Fonda, N. Mankoc-Brostnik, M. Rosina, Coherent rotational states: their formation and detection. Phys. Rep. **158**, 159 (1988)
211. S.T. Ali, J.-P. Antoine, J.-P. Gazeau, U.A. Mueller, Coherent states and their generalizations: a mathematical overview. Rev. Math. Phys. **7**, 1013 (1995)
212. S.T. Ali, J.-P. Antoine, J.-P. Gazeau, *Coherent States, Wavelets, and Their Generalizations*, 2nd edn. (Springer, New York, 1999)
213. J. Bertrand, P. Bertrand, J. Ovarlez, The Mellin transform, in *The Transforms and Applications Handbook*, ed. by A.D. Poularikas (CRC Press, Boca Raton, 2000)
214. H. Rosu, C. Castro, q-deformation by intertwining with application to the singular oscillator. Phys. Lett. A **264**, 350 (2000)
215. E.T. Jaynes, F.W. Cummings, Comparison of quantum and semiclassical radiation theories with application to the beam maser. Proc. IEEE **51**, 89 (1963)
216. B. Buck, C.V. Sukumar, Exactly soluble model of atom-phonon coupling showing periodic decay and revival. Phys. Lett. A **81**, 132 (1981)

217. C.V. Sukumar, B. Buck, Multi-phonon generalisation of the Jaynes-Cummings model. Phys. Lett. A **83**, 211 (1981)
218. G. Bastard, Superlattice band structure in the envelope-function approximation. Phys. Rev. B **24**, 5693 (1981)
219. G. Bastard, Theoretical investigations of superlattice band structure in the envelope-function approximation. Phys. Rev. B **24**, 5693 (1982)
220. V. Milanović, Z. Ikonić, Generation of isospectral combinations of the potential and the effective-mass variations by supersymmetric quantum mechanics. J. Phys. A Math. Gen. **32**, 7001 (1999)
221. A. de Souza Dutra, C.A.S. Almeida, Exact solvability of potentials with spatially dependent effective masses. Phys. Lett. A **275**, 25 (2000)
222. B. Roy, Lie algebraic approach to singular oscillator with a position-dependent mass. Eur. Phys. Lett. **72**, 1 (2005)
223. A. Ganguly, M.V. Ioffe, L.M. Nieto, A new effective mass Hamiltonian and associated Lamé equation: bound states. J. Phys. A Math. Gen. **39**, 14659 (2006)
224. O. Mustafa, S.H. Mazharimousavi, Quantum particles trapped in a position-dependent mass barrier; a d-dimensional recipe. Phys. Lett. A **358**, 259 (2006)
225. S. Cruz y Cruz, J. Negro, L.M. Nieto, Classical and quantum position-dependent mass harmonic oscillators. Phys. Lett. A **369**, 400 (2007)
226. S. Cruz y Cruz, S. Kuru, J. Negro, Classical motion and coherent states for Pöschl-Teller potentials. Phys. Lett. A **372**, 1391 (2008)
227. S. Cruz y Cruz, J. Negro, L.M. Nieto, On position-dependent mass harmonic oscillators. J. Phys. Conf. Ser. **128**, 012053 (2008)
228. O. Mustafa, S.H. Mazharimousavi, A singular position-dependent mass particle in an infinite potential well. Phys. Lett. A **373**, 325 (2009)
229. S. Cruz y Cruz, O. Rosas-Ortiz, Dynamical equations, invariants and spectrum generating algebras of mechanical systems with position-dependent mass. SIGMA **9**, 004 (2013)
230. B. Bagchi, S. Das, S. Ghosh, S. Poria, Nonlinear dynamics of a position-dependent mass-driven Duffing-type oscillator. J. Phys. A Math. Theor. **46**, 032001 (2013)
231. M. Lakshmanan, K. Chandrasekar, Generating finite dimensional integrable nonlinear dynamical systems. Eur. Phys. J. Spec. Top. **222**, 665 (2013)
232. S. Cruz y Cruz, Factorization method and the position-dependent mass problem, in *Geometric Methods in Physics*, ed. by P. Kielanowski, S. Ali, A. Odzijewicz, M. Schlichenmaier, T. Voronov. Trends in Mathematics (Birkhäuser, Basel, 2013), pp. 229–237
233. B.G. da Costa, E. Borges, Generalized space and linear momentum operators in quantum mechanics. J. Math. Phys. **55**, 062105 (2014)
234. S. Cruz y Cruz, C. Santiago-Cruz, Bounded motion for classical systems with position-dependent mass. J. Phys. Conf. Ser. **538**, 012006 (2014)
235. A.G. Nikitin, T.M. Zasadko, Superintegrable systems with position dependent mass. J. Math. Phys. **56**, 042101 (2015)
236. O. Mustafa, Position-dependent mass Lagrangians: nonlocal transformations, Euler-Lagrange invariance and exact solvability. J. Phys. A Math. Theor. **48**, 225206 (2015)
237. D. Ghosch, B. Roy, Nonlinear dynamics of classical counterpart of the generalized quantum nonlinear oscillator driven by position dependent mass. Ann. Phys. **353**, 222 (2015)
238. B. Bagchi, A.G. Choudhury, P. Guha, On quantized Liénard oscillator and momentum dependent mass. J. Math. Phys. **56**, 012105 (2015)
239. N. Amir, S. Iqbal, Algebraic solutions of shape-invariant position-dependent effective mass systems. J. Math. Phys. **57**, 062105 (2016)
240. C. Santiago-Cruz, Isospectral trigonometric Pöschl-Teller potentials with position dependent mass generated by supersymmetry. J. Phys. Conf. Ser. **698**, 012028 (2016)
241. R. Bravo, M.S. Plyushchay, Position-dependent mass, finite-gap systems, and supersymmetry. Phys. Rev. D **93**, 105023 (2016)
242. A.G. Nikitin, Kinematical invariance groups of the 3d Schrödinger equations with position dependent masses. J. Math. Phys. **58**, 083508 (2017)

243. B.G. da Costa, E.P. Borges, A position-dependent mass harmonic oscillator and deformed space. J. Math. Phys. **59**, 042101 (2018)
244. A.G.M. Schmidt, A.L. de Jesus, Mapping between charge-monopole and position-dependent mass system. J. Math. Phys. **59**, 102101 (2018)
245. O.Cherroud, A.-A. Yahiaoui, M. Bentaiba, Generalized Laguerre polynomials with position-dependent effective mass visualized via Wigner's distribution functions. J. Math. Phys. **58**, 063503 (2017)
246. S. Cruz y Cruz, C. Santiago-Cruz, Position dependent mass Scarf Hamiltonians generated via the Riccati equation. Math. Methods Appl. Sci. 1–16 (2018). https://doi.org/10.1002/mma.5068
247. K.B. Wolf, *Geometric Optics on Phase Space*. Texts and Monographs in Physics (Springer, Berlin, 2004)
248. A. de Souza Dutra, J.A. de Oliviera, Wigner distribution for a class of isospectral position-dependent mass systems. Phys. Scr. **78**, 035009 (2008)
249. S. Cruz y Cruz, O. Rosas-Ortiz, Position dependent mass oscillators and coherent states. J. Phys. A Math. Theor. **42**, 185205 (2009)
250. C. Chithiika Ruby, M. Senthilvelan, On the construction of coherent states of position dependent mass Schrödinger equation endowed with effective potential. J. Math. Phys. **51**, 052106 (2010)
251. S.-A. Yahiaoui, M. Bentaiba, Pseudo-Hermitian coherent states under the generalized quantum condition with position-dependent mass. J. Phys. A Math. Gen. **45**, 444034 (2012)
252. S.-A. Yahiaoui, M. Bentaiba, New $SU(1,1)$ position-dependent effective mass coherent states for a generalized shifted harmonic oscillator. J. Phys. A Math. Gen. **47**, 025301 (2014)
253. N. Amir, S. Iqbal, Barut-Girardello coherent states for nonlinear oscillator with position-dependent mass. Commun. Theor. Phys. **66**, 41 (2016)
254. S.-A. Yahiaoui, M. Bentaiba, Isospectral Hamiltonian for position-dependent mass for an arbitrary quantum system and coherent states. J. Math. Phys. **58**, 063507 (2017)
255. N. Amir, S. Iqbal, Coherent states of nonlinear oscillators with position-dependent mass: temporal stability and fractional revivals. Commun. Theor. Phys. **68**, 181 (2017)
256. S. Cruz y Cruz, O. Rosas-Ortiz, $SU(1,1)$ coherent states for position-dependent mass singular oscillators. Int. J. Theor. Phys. **50**, 2201 (2011)
257. D. Dragoman, M. Dragoman, *Quantum–Classical Analogies* (Springer, New York, 2004)
258. H. Rauch, S.A. Werner, *Neutron Interferometry: Lessons in Experimental Quantum Mechanics, Wave-Particle Duality, and Entanglement*, 2nd edn. (Oxford University Press, Oxford, 2015)
259. F.D. Belgiorno, S.L. Cacciatori, D. Faccio, *Hawking Radiation. From Astrophysical Black Holes to Analogous Systems in Lab* (World Scientific, Singapore, 2019)
260. L.M. Procopio, O. Rosas-Ortiz, V. Velázquez, On the geometry of spatial biphoton correlation in spontaneous parametric down conversion. Math. Methods Appl. Sci. **38**, 2053 (2015)
261. O. Calderón-Losada, J. Flóres, J.P. Villabona-Monsalve, A. Valencia, Measuring different types of transverse momentum correlations in the biphoton's Fourier plane. Opt. Lett. **41**, 1165 (2016)
262. J. López-Durán, O. Rosas-Ortiz, Quantum and Classical Correlations in the Production of Photon-Pairs with Nonlinear Crystals. J. Phys. Conf. Ser. **839**, 012022 (2017)
263. C. Couteau, Spontaneous parametric down-conversion. Contemp. Phys. **59**, 291 (2018)
264. K. Okamoto, *Fundamentals of Optical Waveguides*, 2nd edn. (Elsevier, California, 2011)
265. S. Cruz y Cruz, R. Razo, Wave propagation in the presence of a dielectric slab: the paraxial approximation. J. Phys. Conf. Ser. **624**, 012018 (2015)
266. S. Cruz y Cruz, O. Rosas-Ortiz, Leaky modes of waveguides as a classical optics analogy of quantum resonances. Adv. Math. Phys. **2015**, 281472 (2015)
267. G.R. Honarasa, M. Hatami, M.K. Tavassoly, Quantum squeezing of dark solitons in optical fibers. Commun. Theor. Phys. **56**, 322 (2011)
268. O. Rosas-Ortiz, S. Cruz y Cruz, Superpositions of bright and dark solitons supporting the creation of balanced gain and loss optical potentials, arXiv:1805.00058

269. W. Walasik, B. Midya, L. Feng, N.M. Litchinitser, Supersymmetry-guided method for mode selection and optimization in coupled systems. Opt. Lett. **43**, 3758 (2018)
270. A. Contreras-Astorga, V. Jakubský, Photonic systems with two-dimensional landscapes of complex refractive index via time-dependent supersymmetry. Phys. Rev. A **99**, 053812 (2019). arXiv:1808.08225
271. R. Gilmore, C.M. Bowden, L.M. Narducci, Classical-quantum correspondence for multilevel systems. Phys. Rev. A **12**, 1019 (1975)
272. S. Cruz y Cruz, B. Mielnik, Non-inertial quantization: truth or illusion? J. Phys. Conf. Ser. **698**, 012002 (2016)
273. G. Cozzella, A.G.S. Landulfo, G.E.A. Matsas, D.A.T. Vanzella, Proposal for observing the Unruh effect using classical electrodynamics. Phys. Rev. Lett. **118**, 161102 (2017)
274. A.M. Cetto, L. de la Peña, Real vacuum fluctuations and virtual Unruh radiation. Fortschr. Phys. **65**, 1600039 (2017)
275. J.A. Rosabal, New perspective on the Unruh effect. Phys. Rev. D **98**, 056015 (2018)
276. D. Bermudez, U. Leonhardt, Hawking spectrum for a fiber-optical analog of the event horizon. Phys. Rev. A **93**, 053820 (2016)
277. D. Bermudez, U. Leonhardt, Resonant Hawking radiation as an instability, arXiv:1808.02210
278. J. Drori, Y. Rosenberg, D. Bermudez, Y. Silbergerg, U. Leonhardt, Observation of stimulated hawking radiation in optics, arXiv:1808.09244
279. M.F. Bocko, R. Onofrio, On the measurement of a weak classical force coupled to a harmonic oscillator: experimental progress. Rev. Mod. Phys. **68**, 755 (1996)
280. A.A. Clerk, M.H. Devoret, S.M. Girvin, F. Marquardt, R.J. Schoelkopf, Introduction to quantum noise, measurement, and amplification. Rev. Mod. Phys. **82**, 1155 (2010)
281. C. Guerlin, J. Bernu, S. Deléglise, C. Sayrin, S. Gleyzes, S. Kuhr, M. Brune, J.-M. Raimond, S. Haroche, Progressive field-state collapse and quantum non-demolition photon counting. Nature **448**, 889 (2007)
282. C. Sayrin, I. Dotsenko, X. Zhou, B. Peaudecerf, T. Rybarczyk, S. Gleyzes, P. Rouchon, M. Mirrahimi, H. Amini, M. Brune, J.-M. Raimond, S. Haroche, Real-time quantum feedback prepares and stabilizes photon number states. Nature **477**, 73 (2011)
283. D. Oriti, R. Pereira, L. Sindoni, Coherent states in quantum gravity: a construction based on the flux representation of loop quantum gravity. J. Phys. A Math. Theor. **45**, 244004 (2012)
284. T. Thiemann, Gauge field theory coherent states (GCS): I. General properties. Class. Quantum Grav. **18**, 2025 (2001)
285. T. Thiemann, O. Winkler, Gauge field theory coherent states (GCS): II. Peakedness properties. Class. Quantum Grav. **18**, 2561 (2001)
286. T. Thiemann, O. Winkler, Gauge field theory coherent states (GCS): III. Ehrenfest theorems. Class. Quantum Grav. **18**, 4629 (2001)
287. T. Thiemann, O. Winkler, Gauge field theory coherent states (GCS): IV. Infinite tensor product and thermodynamical limit. Class. Quantum Grav. **18**, 4997 (2001)
288. B.C. Hall, Coherent states and the quantization of (1+1)-dimensional Yang-Mills theory. Rev. Math. Phys. **13**, 1281 (2001)
289. J.G. Leopold, I.C. Percival, Microwave ionization and excitation of Rydberg atoms. Phys. Rev. Lett. **41**, 944 (1978)
290. E. Lee, A.F. Brunello, D. Farrelly, Coherent states in a Rydberg atom: classical mechanics. Phys. Rev. A **55**, 2203 (1997)
291. D. Meschede, H. Walther, G. Müller, One-atom maser. Phys. Rev. Lett. **54**, 551 (1985)
292. M. Brune, J.M. Raimond, P. Goy, L. Davidovich, S. Haroche, Realization of a two-photon maser oscillator. Phys. Rev. Lett. **59**, 1899 (1987)
293. E.J. Galvez, B.E. Sauer, L. Moorman, P.M. Koch, D. Richards, Microwave ionization of H atoms: breakdown of classical dynamics for high frequencies. Phys. Rev. Lett. **61**, 2011 (1988)
294. D.J. Wineland, C. Monroe, W.M. Itano, D. Leibfried, B.E. King, D.M. Meekhof, Experimental issues in coherent quantum-state manipulation of trapped atomic ions. J. Res. Natl. Inst. Stand. Technol. **103**, 259 (1998)

295. L. Davidovich, Quantum optics in cavities and the classical limit of quantum mechanics. AIP Conf. Proc. **464**, 3 (1999)
296. J.M. Raimond, M. Brune, S. Haroche, Manipulating quantum entanglement with atoms and photons in a cavity. Rev. Mod. Phys. **73**, 565 (2001)
297. H. Moya-Cessa, Decoherence in atom-field interactions: a treatment using superoperator techniques. Phys. Rep. **432**, 1 (2006)
298. Y. Berubelauzire, V. Hussin, L.M. Nieto, Annihilation operators and coherent states for the Jaynes-Cummings model. Phys. Rev. A **50**, 1725 (1994)
299. M. Daoud, V. Hussin, General sets of coherent states and the Jaynes-Cummings model. J. Phys. A Math. Gen. **35**, 7381 (2002)
300. V. Hussin, L.M. Nieto, Ladder operators and coherent states for the Jaynes-Cummings model in the rotating-wave approximation. J. Math. Phys. **46**, 122102 (2005)
301. J.M. Fink, M. Göpl, M. Baur, R. Bianchetti, P.J. Leek, A. Blais, A. Wallraff, Climbing the Jaynes-Cummings ladder and observing its $\sqrt{n}$ nonlinearity in a cavity QED system. Nature **454**, 315 (2008)
302. F.T. Arecchi, E. Courtens, R. Gilmore, H. Thomas, Atomic coherent state in quantum optics. Phys. Rev. A **6**, 2211 (1974)
303. J.-P. Gazeau, V. Hussin, Poincaré contraction of $SU(1,l)$ Fock-Bargmann structure. J. Phys. A Math. Gen **25**, 1549 (1992)
304. N. Alvarez, V. Hussin, Generalized coherent and squeezed states based on the $h(1) \oplus su(2)$ algebra. J. Math. Phys. **43**, 2063 (2002)
305. A. Wünsche, Duality of two types of $SU(1,1)$ coherent states and an intermediate type. J. Opt. B Quantum Semiclass. Opt. **5**, S429 (2003)
306. S.R. Miry, M.K. Tavassoly, Generation of a class of $SU(1,1)$ coherent states of the Gilmore-Perelomov type and a class of $SU(2)$ coherent states and their superposition. Phys. Scr. **85**, 035404 (2012)
307. A. Karimi, M.K. Tavassoly, Quantum engineering and nonclassical properties of $SU(1,1)$ and $SU(2)$ entangled nonlinear coherent states. J. Opt. Soc. Am. B **31**, 2345 (2014)
308. D.N. Daneshmand, M.K. Tavassoly, Generation of $SU(1,1)$ and $SU(2)$ entangled states in a quantized cavity field by strong-driving-assisted classical field approach. Laser Phys. **25**, 055203 (2015)
309. O. Rosas-Ortiz, S. Cruz y Cruz, M. Enríquez, $SU(1,1)$ and $SU(2)$ approaches to the radial oscillator: generalized coherent states and squeezing of variances. Ann. Phys. **346**, 373 (2016)
310. B.C. Sanders, Review of entangled coherent states. J. Phys. A Math. Theor. **45**, 244002 (2012)
311. L. Li, Y.O. Dudin, A. Kuzmich, Entanglement between light and an optical atomic excitation. Nature **498**, 466 (2013)
312. S. Dey, V. Hussin, Entangled squeezed states in noncommutative spaces with minimal length uncertainty relations. Phys. Rev. D **91**, 124017 (2015)
313. A. Motamedinasab, D. Afshar, M. Jafarpour, Entanglement and non-classical properties of generalized supercoherent states. Optik **157**, 1166 (2018)
314. V. Spiridonov, Universal superpositions of coherent states and self-similar potentials. Phys. Rev. A **52**, 1909 (1995)
315. S. Dey, A. Fring, V. Hussin, A squeezed review on coherent states and nonclassicality for non-Hermitian systems with minimal length, arXiv:1801.01139
316. S.T. Ali, R. Roknizadeh, M.K. Tavassoly, Representations of coherent states in non-orthogonal bases. J. Phys. A Math. Gen. **37**, 4407 (2004)
317. S. Dey, V. Hussin, Noncommutative q-photon-added coherent states. Phys. Rev. A **93**, 053824 (2016)
318. S. Dey, A. Fring, V. Hussin, Nonclassicality versus entanglement in a noncommutative space. Int. J. Mod. Phys. B **31**, 1650248 (2017)
319. A. Hertz, S. Dey, V. Hussin, H. Eleuch, Higher order nonclassicality from nonlinear coherent states for models with quadratic spectrum. Symmetry-Basel **8**, 36 (2016)
320. K. Zelaya, S. Dey, V. Hussin, Generalized squeezed states. Phys. Lett. A **382**, 3369 (2018)
321. C. Aragone, F. Zypman, Supercoherent states. J. Phys. A Math. Gen. **19**, 2267 (1986)

322. A.M. El Gradechi, L.M. Nieto, Supercoherent states, super-Kähler geometry and geometric quantization. Commun. Math. Phys. **175**, 521 (1996)
323. J. Jayaraman, R. de Lima Rodrigues, A SUSY formulation ála Witten for the SUSY isotonic oscillator canonical supercoherent states. J. Phys. A Math. Gen. **32**, 6643 (1999)
324. N. Alvarez-Moraga, V. Hussin, $sh(2/2)$ superalgebra eigenstates and generalized supercoherent and supersqueezed states. Int. J. Theor. Phys. **43**, 179 (2004)
325. L.M. Nieto, Coherent and supercoherent states with some recent applications. AIP Conf. Proc. **809**, 3 (2006)
326. A.H. El Kinani, M. Daoud, Generalized intelligent states for nonlinear oscillators. Int. J. Mod. Phys. B **15**, 2465 (2001)
327. B. Midya, B. Roy, A. Biswas, Coherent state of a nonlinear oscillator and its revival dynamics. Phys. Scr. **79**, 065003 (2009)
328. V.C. Ruby, S. Karthiga, M. Senthilevelan, Ladder operators and squeezed coherent states of a three-dimensional generalized isotonic nonlinear oscillator. J. Phys. A Math. Theor. **46**, 025305 (2013)
329. B. Roy, P. Roy, Gazeau-Klauder coherent state for the Morse potential and some of its properties. Phys. Lett. A **296**, 187 (2002)
330. A. Wünsche, Higher-order uncertainty relations. J. Mod. Opt. **53**, 931 (2006)
331. M.K. Tavassoly, Gazeau-Klauder squeezed states associated with solvable quantum systems. J. Phys. A Math. Gen. **39**, 11583 (2006)
332. L. Dello Sbarba, V. Hussin, Degenerate discrete energy spectra and associated coherent states. J. Math. Phys. **48**, 012110 (2007)
333. M. Angelova, V. Hussin, Generalized and Gaussian coherent states for the Morse potential. J. Phys. A Mat. Theor. **41**, 304016 (2008)
334. G.R. Honarasa, M.K. Tavassoly, M. Hatami, R. Roknizadeh, Generalized coherent states for solvable quantum systems with degenerate discrete spectra and their nonclassical properties. Phys. A **390**, 1381 (2011)
335. M.K. Tavassoly, H.R. Jalai, Barut-Girardello and Gilmore-Perelomov coherent states for pseudoharmonic oscillators and their nonclassical properties: factorization method. Chin. Phys. B **22**, 084202 (2013)
336. M.N. Hounkonnoua, S. Arjikab, E. Baloïtcha, Pšchl-Teller Hamiltonian: Gazeau-Klauder type coherent states, related statistics, and geometry. J. Math. Phys. **55**, 123502 (2014)
337. S.E. Hoffmann, V. Hussin, I. Marquette, Y.-Z. Zhang, Non-classical behaviour of coherent states for systems constructed using exceptional orthogonal polynomials. J. Phys. A Math. Theor. **51**, 085202 (2018)
338. S.E. Hoffmann, V. Hussin, I. Marquette, Y.-Z. Zhang, Coherent states for ladder operators of general order related to exceptional orthogonal polynomials. J. Phys. A Math. Theor. **51**, 315203 (2018). arXiv:1803.01318
339. J.-P. Antoine, J.-P. Gazeau, P. Monceau, J.R. Klauder, K.A. Penson, Temporally stable coherent states for infinite well and Pöschl–Teller potentials. J. Math. Phys. **42**, 2349 (2001)
340. H. Bergeron, J.-P. Gazeau, P. Siegl, A. Youssef, Semi-classical behavior of Pöschl-Teller coherent states. Europhys. Lett. **92**, 60003 (2011)
341. J.-P. Gazeau, M. del Olmo, Pissot q-coherent states quantization of the harmonic oscillator. Ann. Phys. **330**, 220 (2012)
342. J.-P. Gazeau, *Coherent States in Quantum Physics* (Wiley, Weinheim, 2009)

Trace Formulas Applied to the Riemann ζ-Function

Mark S. Ashbaugh, Fritz Gesztesy, Lotfi Hermi, Klaus Kirsten, Lance Littlejohn, and Hagop Tossounian

Abstract We use a spectral theory perspective to reconsider properties of the Riemann zeta function. In particular, new integral representations are derived and used to present its value at odd positive integers.

Keywords Dirichlet Laplacian · Trace class operators · Trace formulas · Riemann zeta function · Spectral theory

---

M. S. Ashbaugh
Department of Mathematics, University of Missouri, Columbia, MO, USA
e-mail: ashbaughm@missouri.edu
https://www.math.missouri.edu/people/ashbaugh

F. Gesztesy
Department of Mathematics, Baylor University, Waco, TX, USA
e-mail: fritz_gesztesy@baylor.edu
http://www.baylor.edu/math/index.php?id=935340

H. Tossounian
Department of Mathematics, Baylor University, Waco, TX, USA

Center for Mathematical Modeling, Universidad de Chile, Santiago, Chile
e-mail: htossounian@dim.uchile.cl; Hagop.Tossounian@gmail.com

L. Hermi
Department of Mathematics and Statistics, Florida International University, Miami, FL, USA
e-mail: lhermi@fiu.edu

K. Kirsten (✉)
GCAP-CASPER, Department of Mathematics, Baylor University, Waco, TX, USA
e-mail: Klaus_Kirsten@baylor.edu
http://www.baylor.edu/math/index.php?id=54012

L. Littlejohn
Department of Mathematics, Baylor University, Waco, TX, USA
e-mail: Lance_Littlejohn@baylor.edu
http://www.baylor.edu/math/index.php?id=53980

Ş. Kuru et al. (eds.), *Integrability, Supersymmetry and Coherent States*, CRM Series in Mathematical Physics, https://doi.org/10.1007/978-3-030-20087-9_8

1 Introduction

Spectral zeta functions associated with eigenvalue problems of (partial) differential operators are of relevance in a wide array of topics [3–8, 10, 11, 16, 18, 25, 26, 28]. As an example consider the Dirichlet boundary value problem

$$-\Delta_D f = -f'', \quad f(0) = f(1) = 0, \tag{1.1}$$

where $-\Delta_D$ denotes the Dirichlet Laplacian in the Hilbert space $L^2((0,1); dx)$ (cf. (2.12)), with purely discrete and simple spectrum,

$$\sigma(-\Delta_D) = \{\lambda_k = (k\pi)^2\}_{k\in\mathbb{N}}. \tag{1.2}$$

In particular, the spectral zeta function associated with $-\Delta_D$,

$$\zeta(z; -\Delta_D) = \sum_{k\in\mathbb{N}} \lambda_k^{-z} = \pi^{-2z}\zeta(2z), \quad \mathrm{Re}(z) > 1/2, \tag{1.3}$$

is basically given by the Riemann zeta function

$$\zeta(z) = \sum_{k\in\mathbb{N}} k^{-z}, \quad \mathrm{Re}(z) > 1. \tag{1.4}$$

This elementary and well-known observation identifies the Riemann zeta function as a spectral zeta function and hence spectral theoretic techniques for their analysis can be applied to it. This is the perspective taken in this article. In Sect. 2 we briefly review representations for spectral zeta functions as derived in [13] and we apply them to the zeta function of Riemann. New integral representations for the Riemann zeta function are found and the well-known properties, namely values at even negative and positive integers are easily reproduced. In addition, we derive new representations for the value of the Riemann zeta function at positive odd integers. Typical examples we derive are

$$\begin{aligned} \zeta(z) = \sin(\pi z/2)\pi^{z-1} \int_0^\infty ds\, s^{-z}[\coth(s) - \coth_n(s)], \\ \mathrm{Re}(z) \in (\max(1, 2n), 2n+2),\ n \in \mathbb{N}_0, \end{aligned} \tag{1.5}$$

where

$$\begin{aligned} \coth_0(z) &= \frac{1}{z}, \quad z \in \mathbb{C}\backslash\{0\}, \\ \coth_n(z) &= \frac{1}{z} + \sum_{k=1}^{n} \frac{2^{2k} B_{2k}}{(2k)!} z^{2k-1}, \quad z \in \mathbb{C}\backslash\{0\},\ n \in \mathbb{N}, \end{aligned} \tag{1.6}$$

with B_m the Bernoulli numbers (cf. (A.30)–(A.33)), implying

$$\zeta(3) = -\pi^2 \int_0^\infty ds\, s^{-3}[\coth(s) - (1/s) - (s/3)], \tag{1.7}$$

$$\zeta(5) = \pi^4 \int_0^\infty ds\, s^{-5}\big[\coth(s) - (1/s) - (s/3) + \big(s^3/45\big)\big], \tag{1.8}$$

$$\zeta(7) = -\pi^6 \int_0^\infty ds\, s^{-7}\big[\coth(s) - (1/s) - (s/3) + \big(s^3/45\big) - \big(2s^5/945\big)\big],$$
$$\text{etc.} \tag{1.9}$$

Finally, the Appendix summarizes known results about the Riemann zeta function putting the results we found in some perspective.

2 Computing Traces and the Riemann ζ-Function

After a brief discussion of spectral zeta functions associated with self-adjoint operators with purely discrete spectra, we turn to applications of spectral trace formulas to the Riemann zeta function.

We start by following the recent paper [13] and briefly discuss spectral ζ-functions of self-adjoint operators S with a trace class resolvent (and hence a purely discrete spectrum).

Below we will employ the following notational conventions: A separable, complex Hilbert space is denoted by $\mathcal{H}$, $I_{\mathcal{H}}$ represents the identity operator in $\mathcal{H}$; the resolvent set and spectrum of a closed operator T in $\mathcal{H}$ are abbreviated by $\rho(T)$ and $\sigma(T)$, respectively; the Banach space of trace class operators on $\mathcal{H}$ is denoted by $\mathcal{B}_1(\mathcal{H})$, and the trace of a trace class operator $A \in \mathcal{B}_1(\mathcal{H})$ is abbreviated by $\operatorname{tr}_{\mathcal{H}}(A)$.

Hypothesis 2.1 *Suppose S is a self-adjoint operator in $\mathcal{H}$, bounded from below, satisfying*

$$(S - zI_{\mathcal{H}})^{-1} \in \mathcal{B}_1(\mathcal{H}) \tag{2.2}$$

for some (and hence for all) $z \in \rho(S)$. We denote the spectrum of S by $\sigma(S) = \{\lambda_j\}_{j\in J}$ (with $J \subset \mathbb{Z}$ an appropriate index set), with every eigenvalue repeated according to its multiplicity.

Given Hypothesis 2.1, the spectral zeta function of S is then defined by

$$\zeta(z; S) = \sum_{\substack{j\in J\\ \lambda_j\neq 0}} \lambda_j^{-z} \tag{2.3}$$

for $\mathrm{Re}(z) > 0$ sufficiently large such that (2.3) converges absolutely.

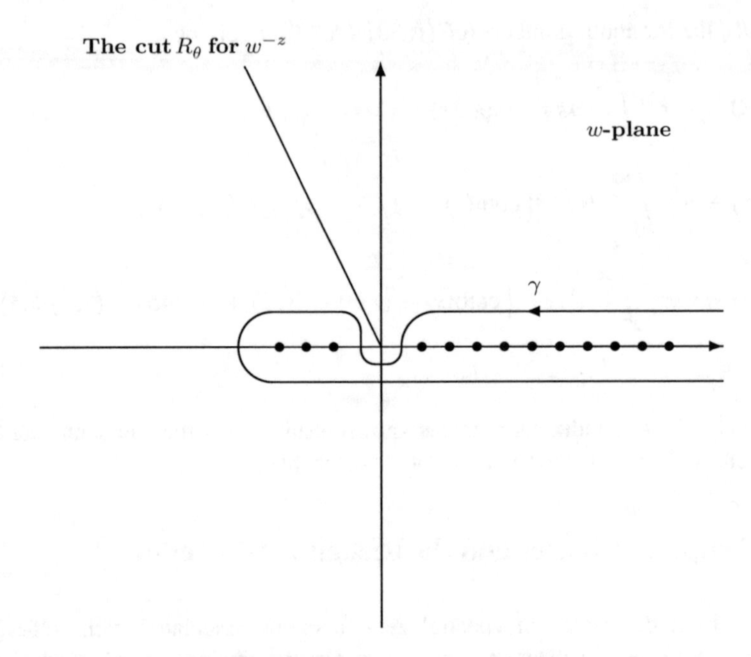

Fig. 1 Contour γ in the complex w-plane

Next, let $P(0; S)$ be the spectral projection of S corresponding to the eigenvalue 0 and denote by $m(\lambda_0; S)$ the multiplicity of the eigenvalue λ_0 of S, in particular,

$$m(0; S) = \dim(\ker(S)). \tag{2.4}$$

In addition, we introduce the simple contour γ encircling $\sigma(S)\backslash\{0\}$ in a counterclockwise manner so as to dip under (and hence avoid) the point 0 (cf. Fig. 1). In fact, following [20] (see also [19]), we will henceforth choose as the branch cut of w^{-z} the ray

$$R_\theta = \{w = te^{i\theta} \,|\, t \in [0, \infty)\} \quad \theta \in (\pi/2, \pi), \tag{2.5}$$

and note that the contour γ avoids any contact with R_θ (cf. Fig. 1).

Lemma 2.6 *In addition to Hypothesis 2.1 and the counterclockwise oriented contour γ just described (cf. Fig. 1), suppose that* $\big|\mathrm{tr}_{\mathcal{H}}\big((S-zI_{\mathcal{H}})^{-1}[I_{\mathcal{H}}-P(0; S)]\big)\big|$ *is polynomially bounded with respect to z on γ. Then*

$$\zeta(z; S) = -(2\pi i)^{-1} \oint_\gamma dw\, w^{-z}\big[\mathrm{tr}_{\mathcal{H}}\big((S - wI_{\mathcal{H}})^{-1}\big) + w^{-1} m(0; S)\big] \tag{2.7}$$

for $\mathrm{Re}(z) > 0$ *sufficiently large.*

We note in passing that one could also use a semigroup approach via

$$\begin{aligned}\zeta(z;S) &= \Gamma(z)^{-1}\int_0^\infty dt\, t^{z-1}\,\mathrm{tr}_{\mathcal{H}}\big(e^{-tS}[I_{\mathcal{H}}-P(0;S)]\big)\\ &= \Gamma(z)^{-1}\int_0^\infty dt\, t^{z-1}\big[\mathrm{tr}_{\mathcal{H}}\big(e^{-tS}\big)-m(0;S)\big],\end{aligned} \tag{2.8}$$

for $\mathrm{Re}(z) > 0$ sufficiently large.

It is natural to continue the computation leading to (2.7) and now deform the contour γ so as to "hug" the branch cut R_θ, but this requires the right asymptotic behavior of $\mathrm{tr}_{\mathcal{H}}\big((S-wI_{\mathcal{H}})^{-1}[I_{\mathcal{H}}-P(0;S)]\big)$ as $|w|\to\infty$ as well as $|w|\to 0$. This applies, in particular, to cases where S is strictly positive and one thus chooses the branch cut along the negative axis, that is, it employs the cut R_π, where

$$R_\pi = (-\infty, 0], \tag{2.9}$$

and choosing the contour γ to encircle R_π clockwise. This renders (2.7) into the following expression:

$$\begin{aligned}\zeta(z;S) &= \frac{\sin(\pi z)}{\pi}\int_0^\infty dt\, t^{-z}\,\mathrm{tr}_{\mathcal{H}}\big((S+tI_{\mathcal{H}})^{-1}\big)\\ &= \mathrm{tr}_{\mathcal{H}}\bigg(\frac{\sin(\pi z)}{\pi}\int_0^\infty dt\, t^{-z}(S+tI_{\mathcal{H}})^{-1}\bigg) = \mathrm{tr}_{\mathcal{H}}\big(S^{-z}\big),\end{aligned} \tag{2.10}$$

employing the fact,

$$S^{-z} = \frac{\sin(\pi z)}{\pi}\int_0^\infty dt\, t^{-z}(S+tI_{\mathcal{H}})^{-1}, \quad \mathrm{Re}(z)\in(0,1), \tag{2.11}$$

whenever $S \geq 0$ in $\mathcal{H}$, with $\ker(S) = \{0\}$ (see, e.g., [15, Proposition 3.2.1 d)]).

Note While (2.11) is rigorous, the manipulations in (2.10) are formal and subject to appropriate convergence and trace class hypotheses which will affect the possible range of $\mathrm{Re}(z)$. ◇

These hypotheses are easily shown to be satisfied when discussing the one-dimensional Dirichlet Laplacian $-\Delta_D$ in $L^2((0,1);dx)$,

$$\begin{aligned}&-\Delta_D = -\frac{d^2}{dx^2},\\ &\mathrm{dom}(-\Delta_D) = \big\{u\in L^2((0,1);dx)\,\big|\, u, u' \in AC_{loc}([0,1]);\ u(0)=0=u(1);\\ &\qquad\qquad u''\in L^2((0,1);dx)\big\}\end{aligned} \tag{2.12}$$

(here $AC([0,1])$ denotes the set of absolutely continuous functions on $[0,1]$).

Recalling Riemann's celebrated zeta function (see the Appendix for more details),

$$\zeta(z) = \sum_{k\in\mathbb{N}} k^{-z}, \quad z \in \mathbb{C},\ \mathrm{Re}(z) > 1. \tag{2.13}$$

we start with the following result.

Lemma 2.14 *Let* $\mathrm{Re}(z) \in (1, 2)$*, then*

$$\zeta(z) = \sin(\pi z/2)\pi^{z-1} \int_0^\infty ds\, s^{-z-1}[s\coth(s) - 1]. \tag{2.15}$$

Proof Since the eigenvalue problem for $-\Delta_D$ reads

$$-\Delta_D u_k = \lambda_k u_k, \quad k \in \mathbb{N}, \tag{2.16}$$

with

$$u_k(x) = 2^{1/2}\sin(k\pi x),\ \|u_k\|_{L^2((0,1);dx)} = 1, \quad \lambda_k = (\pi k)^2, \quad k \in \mathbb{N}, \tag{2.17}$$

and all eigenvalues λ_k, $k \in \mathbb{N}$, are simple, (2.10) works as long as $\mathrm{Re}(z) \in ((1/2), 1)$ and one obtains (see also [7, p. 94])

$$\begin{aligned}
\zeta(z; -\Delta_D) &= \mathrm{tr}_{L^2((0,1);dx)}\big((-\Delta_D)^{-z}\big) = \sum_{k\in\mathbb{N}} (\pi k)^{-2z} = \pi^{-2z}\zeta(2z) \\
&= \frac{\sin(\pi z)}{\pi} \int_0^\infty dt\, t^{-z}\, \mathrm{tr}_{L^2((0,1);dx)}\big((-\Delta_D + tI_{L^2((0,1);dx)})^{-1}\big) \\
&= \frac{\sin(\pi z)}{\pi} \int_0^\infty dt\, t^{-z} \int_0^1 dx\, t^{-1/2}\big[\sinh(t^{1/2})\big]^{-1} \sinh\big(t^{1/2}x\big) \\
&\qquad \times \sinh\big(t^{1/2}(1-x)\big) \\
&= \frac{\sin(\pi z)}{2\pi} \int_0^\infty dt\, t^{-z-1}\big[t^{1/2}\coth\big(t^{1/2}\big) - 1\big], \quad \mathrm{Re}(z) \in ((1/2), 1).
\end{aligned} \tag{2.18}$$

Here we used

$$\begin{aligned}
&(-\Delta_D - zI_{L^2((0,1);dx)})^{-1}(z, x, x') \\
&= \frac{1}{z^{1/2}\sin(z^{1/2})} \begin{cases} \sin(z^{1/2}x)\sin(z^{1/2}(1-x')), & 0 \le x \le x' \le 1, \\ \sin(z^{1/2}x')\sin(z^{1/2}(1-x)), & 0 \le x' \le x \le 1, \end{cases} \\
&\qquad z \in \mathbb{C}\backslash\{\pi^2k^2\}_{k\in\mathbb{N}},
\end{aligned} \tag{2.19}$$

and [14, 2.4254]

$$\int^x dt\, \sinh(at+b)\sinh(at+c)$$
$$= -(x/2)\cosh(b-c) + (4a)^{-1}\sinh(2ax+b+c) + C, \quad a \neq 0. \tag{2.20}$$

Since

$$\left[t^{1/2}\coth\left(t^{1/2}\right) - 1\right] = \begin{cases} O(t), & t \downarrow 0, \\ O\left(t^{1/2}\right), & t \uparrow \infty, \end{cases} \tag{2.21}$$

(2.18) is well-defined for $\mathrm{Re}(z) \in ((1/2), 1)$. Thus, the elementary change of variables $t = s^2$ yields (2.15).

Remark 2.22 Representation (2.15) is suitable to observe the well-known properties

$$\zeta(0) = -1/2, \quad \zeta(-2n) = 0, \ n \in \mathbb{N}. \tag{2.23}$$

To this end, one notes that the restriction $\mathrm{Re}(z) > 1$ results from the $s \to \infty$ behavior of the integrand in (2.15). Explicitly, one has

$$s\coth(s) - 1 \underset{s\to\infty}{=} s - 1 + O\left(e^{-2s}\right), \tag{2.24}$$

from which one infers that

$$\zeta(z) = \sin(\pi z/2)\pi^{z-1}\int_1^\infty ds\, s^{-z-1}(s-1) + E(z)$$
$$= \sin(\pi z/2)\pi^{z-1}\left[\frac{1}{z-1} - \frac{1}{z}\right] + E(z), \tag{2.25}$$

where $E(\cdot)$ is entire and

$$E(-2n) = 0, \quad n \in \mathbb{N}_0. \tag{2.26}$$

This immediately implies (2.23).

The values at positive even integers, $\zeta(2m)$ for $m \in \mathbb{N}$, are best obtained using representation (2.7); see Remark 2.41. ◇

One can generalize (2.11) as follows:

$$S^{-z} = \frac{\Gamma(m)}{\Gamma(n-z)\Gamma(m-n+z)} S^{m-n} \int_0^\infty dt\, t^{n-1-z}(S + tI_{\mathcal{H}})^{-m},$$
$$\mathrm{Re}(z) \in (n-m, n), \ m, n \in \mathbb{N}. \tag{2.27}$$

Formally, this now yields

$$\begin{aligned}\zeta(z;S) &= \frac{\Gamma(m)}{\Gamma(n-z)\Gamma(m-n+z)}\int_0^\infty dt\, t^{n-1-z}\,\mathrm{tr}_{\mathcal{H}}\big(S^{m-n}(S+tI_{\mathcal{H}})^{-m}\big)\\ &= \mathrm{tr}_{\mathcal{H}}\bigg(\frac{\Gamma(m)}{\Gamma(n-z)\Gamma(m-n+z)}S^{m-n}\int_0^\infty dt\, t^{n-1-z}(S+tI_{\mathcal{H}})^{-m}\bigg)\\ &= \mathrm{tr}_{\mathcal{H}}\big(S^{-z}\big). \end{aligned} \tag{2.28}$$

The case $m = n$ appears to be the simplest and yields

$$\begin{aligned}\zeta(z;S) &= \frac{\Gamma(n)}{\Gamma(n-z)\Gamma(z)}\int_0^\infty dt\, t^{n-1-z}\,\mathrm{tr}_{\mathcal{H}}\big((S+tI_{\mathcal{H}})^{-n}\big) \\ &= \mathrm{tr}_{\mathcal{H}}\bigg(\frac{\Gamma(n)}{\Gamma(n-z)\Gamma(z)}\int_0^\infty dt\, t^{n-1-z}(S+tI_{\mathcal{H}})^{-n}\bigg) = \mathrm{tr}_{\mathcal{H}}\big(S^{-z}\big).\end{aligned} \tag{2.29}$$

Note Again, (2.27) is rigorous, but (2.28), (2.29) are subject to "appropriate" convergence and trace class hypotheses. ◇

For $-\Delta_D$, (2.29) indeed works for $n \in \mathbb{N}$ as long as $\mathrm{Re}(z) \in ((1/2), n)$ and one obtains

$$\begin{aligned}\zeta(z;-\Delta_D) &= \mathrm{tr}_{L^2((0,1);dx)}\big((-\Delta_D)^{-z}\big) = \sum_{k\in\mathbb{N}}(\pi k)^{-2z} = \pi^{-2z}\zeta(2z)\\ &= \frac{\Gamma(n)}{\Gamma(n-z)\Gamma(z)}\int_0^\infty dt\, t^{n-1-z}\,\mathrm{tr}_{L^2((0,1);dx)}\big((-\Delta_D+tI_{L^2((0,1);dx)})^{-n}\big).\end{aligned} \tag{2.30}$$

For $n = 2$ this includes $z = 3/2$ and hence leads to a formula for $\zeta(3)$. However, we prefer an alternative approach based on [13, Theorem 3.4 (i)] that applies to $-\Delta_D$ and yields the following results.

Lemma 2.31 *Let* $\mathrm{Re}(z) \in (1, 4)$*, then*

$$\zeta(z) = \frac{\sin(\pi z/2)}{(2-z)}\pi^{z-1}\int_0^\infty ds\, s^{-z-1}\Big[s\coth(s) + s^2[\sinh(s)]^{-2} - 2\Big]. \tag{2.32}$$

In particular,

$$\zeta(2) = \frac{\pi^2}{2}\int_0^\infty ds\, s^{-3}\Big[s\coth(s) + s^2[\sinh(s)]^{-2} - 2\Big] = \ldots. = \pi^2/6, \tag{2.33}$$

$$\zeta(3) = \pi^2\int_0^\infty ds\, s^{-4}\Big[s\coth(s) + s^2[\sinh(s)]^{-2} - 2\Big]. \tag{2.34}$$

Proof Employing [13, Theorem 3.4 (i)],

$$\operatorname{tr}_{L^2((0,1);dx)}\big((-\Delta_D - zI_{L^2((0,1);dx)})^{-1}\big) = -(d/dz)\ln\big(z^{-1/2}\sin\big(z^{1/2}\big)\big),$$
$$z \in \mathbb{C}\backslash\big\{\pi^2k^2\big\}_{k\in\mathbb{N}} \tag{2.35}$$

(see also (A.35)), one confirms that

$$\zeta(2)/\pi^2 = \lim_{z\to 0}\sum_{k\in\mathbb{N}}\big(\pi^2k^2 - z\big)^{-1} = 1/6. \tag{2.36}$$

Actually, setting $z = -t$ in (2.35) yields

$$\begin{aligned}\operatorname{tr}_{L^2((0,1);dx)}\big((-\Delta_D + tI_{L^2((0,1);dx)})^{-1}\big) &= (d/dt)\ln\big(t^{-1/2}\sinh\big(t^{1/2}\big)\big)\\ &= (2t)^{-1}\big[t^{1/2}\coth\big(t^{1/2}\big) - 1\big], \quad t > 0,\end{aligned} \tag{2.37}$$

and hence confirms (2.18). Continuing that process, one notes that

$$\begin{aligned}\operatorname{tr}_{L^2((0,1);dx)}\big((-\Delta_D + tI_{L^2((0,1);dx)})^{-2}\big) &= -\big(d^2/dt^2\big)\ln\big(t^{-1/2}\sinh\big(t^{1/2}\big)\big)\\ &= \big(4t^2\big)^{-1}\Big[t^{1/2}\coth\big(t^{1/2}\big) + t\big[\sinh\big(t^{1/2}\big)\big]^{-2} - 2\Big], \quad t > 0.\end{aligned} \tag{2.38}$$

Insertion of (2.38) into (2.30) taking $n = 2$ then yields

$$\begin{aligned}\zeta(z; -\Delta_D) &= \operatorname{tr}_{L^2((0,1);dx)}\big((-\Delta_D)^{-z}\big) = \sum_{k\in\mathbb{N}}(\pi k)^{-2z} = \pi^{-2z}\zeta(2z)\\ &= \frac{1}{\Gamma(2-z)\Gamma(z)}\int_0^\infty dt\, t^{1-z}\operatorname{tr}_{L^2((0,1);dx)}\big((-\Delta_D + tI_{L^2((0,1);dx)})^{-2}\big)\\ &= \frac{\sin(\pi z)}{4\pi(1-z)}\int_0^\infty dt\, t^{-1-z}\Big[t^{1/2}\coth\big(t^{1/2}\big) + t\big[\sinh\big(t^{1/2}\big)\big]^{-2} - 2\Big],\end{aligned} \tag{2.39}$$
$$\operatorname{Re}(z) \in ((1/2), 2).$$

Since

$$\Big[t^{1/2}\coth\big(t^{1/2}\big) + t\big[\sinh\big(t^{1/2}\big)\big]^{-2} - 2\Big] = \begin{cases} O\big(t^2\big), & t \downarrow 0,\\ O\big(t^{1/2}\big), & t \uparrow \infty,\end{cases} \tag{2.40}$$

(2.39) is well-defined for $\operatorname{Re}(z) \in ((1/2), 2)$. Thus, the elementary change of variables $t = s^2$ yields (2.32)–(2.34).

Remark 2.41 Employing (2.35) in (2.7), one finds the representation

$$\zeta(z; -\Delta_D) = (2\pi i)^{-1} \oint_{\gamma} dw\, w^{-z} \frac{1 - w^{1/2} \cot\left(w^{1/2}\right)}{2w}, \tag{2.42}$$

where the counterclockwise contour γ can be chosen to consist of a circle γ_ϵ of radius $\epsilon < \pi$ and straight lines γ_1, respectively γ_2, just above, respectively just below, the negative x-axis. For $z = m$, $m \in \mathbb{N}$, contributions from γ_1 and γ_2 cancel each other and thus

$$\zeta(m; -\Delta_D) = (2\pi i)^{-1} \oint_{\gamma_\epsilon} dw\, w^{-m} \frac{1 - w^{1/2} \cot\left(w^{1/2}\right)}{2w}. \tag{2.43}$$

This integral is easily computed using the residue theorem. From the Taylor series [14]

$$\frac{1 - w^{1/2} \cot\left(w^{1/2}\right)}{2w} = \sum_{k=1}^{\infty} \frac{2^{2k-1} |B_{2k}|}{(2k)!} w^{k-1}, \quad w \in \mathbb{C},\ 0 < |w| < \pi^2 \tag{2.44}$$

(with B_m the Bernoulli numbers, cf. (A.30)–(A.33)), the relevant term is $k = m$ and

$$\zeta(m; -\Delta_D) = \frac{2^{2m-1} |B_{2m}|}{(2m)!}, \tag{2.45}$$

implying Euler's celebrated result,

$$\zeta(2m) = \frac{2^{2m-1} \pi^{2m} |B_{2m}|}{(2m)!}, \quad m \in \mathbb{N}. \tag{2.46}$$

This procedure works in a much more general context and allows for the computation of traces of powers of Sturm–Liouville operators in a straightforward fashion; this will be revisited elsewhere. ◇

Remark 2.47 An elementary integration by parts of the term $s^{-2} \coth(s)$ in (2.33) indeed verifies once more that $\zeta(2) = \pi^2/6$. The same integration by parts in (2.34) fails to render the integral trivial (as it obviously should not be trivial). Indeed,

$$\begin{aligned} &\int_{\varepsilon}^{R} ds \left\{s^{-2} \coth(s) + s^{-1} [\sinh(s)]^{-2} - 2s^{-3}\right\} \\ &= \int_{\varepsilon}^{R} ds \left\{ - \left[(d/ds)s^{-1}\right] \coth(s) + s^{-1} [\sinh(s)]^{-2} - 2s^{-3}\right\} \\ &= -s^{-1} \coth(s)\Big|_{\varepsilon}^{R} + \int_{\varepsilon}^{R} ds\, (-2)s^{-3} \underset{\varepsilon\downarrow 0,\, R\uparrow\infty}{\longrightarrow} = \frac{1}{3}. \end{aligned} \tag{2.48}$$

Applying the same strategy to (2.34) yields

$$\begin{aligned}
&\int_{\varepsilon}^{R} ds \Big[s^{-3}\coth(s) + s^{-2}[\sinh(s)]^{-2} - 2s^{-4}\Big] \\
&= \int_{\varepsilon}^{R} ds \Big[-(1/2)\big[(d/ds)s^{-2}\big]\coth(s) + s^{-2}[\sinh(s)]^{-2} - 2s^{-4}\Big] \\
&= -(1/2)s^{-2}\coth(s)\Big|_{\varepsilon}^{R} + \int_{\varepsilon}^{R} ds \Big[(1/2)s^{-2}[\sinh(s)]^{-2} - 2s^{-4}\Big],
\end{aligned} \tag{2.49}$$

and hence the expected nontrivial integral. We note once more that

$$\Big[s\coth(s) + s^{2}[\sinh(s)]^{-2} - 2\Big] \underset{s\downarrow 0}{=} O\big(s^{4}\big), \tag{2.50}$$

rendering (2.34) well-defined. ◇

The following alternative (though, equivalent) approach to $\zeta(z)$ is perhaps a bit more streamlined.

Theorem 2.51 *Let* $n \in \mathbb{N}_0$, $0 < \mathrm{Re}(z) < 1$, *and*[1] $\mathrm{Re}(2n+2z) > 1$. *Then*[2]

$$\begin{aligned}
\zeta(2n+2z) &= \frac{(-1)^n \pi^{2(n+z)}}{\Gamma(1-z)\Gamma(n+z)} \int_0^{\infty} dt\, t^{-z} \frac{d^n}{dt^n}\Big[(2t)^{-1}\big[t^{1/2}\coth\big(t^{1/2}\big) - 1\big]\Big] \\
&= \frac{(-1)^n 2^{-n} \pi^{2(n+z)}}{\Gamma(1-z)\Gamma(n+z)} \int_0^{\infty} ds\, s^{1-2z} \bigg(\frac{1}{s}\frac{d}{ds}\bigg)^{n} \Big[s^{-2}[s\coth(s) - 1]\Big].
\end{aligned} \tag{2.52}$$

In addition to $\zeta(3)$ *in* (2.34) *one thus obtains similarly,*

$$\begin{aligned}
\zeta(5) = \frac{\pi^4}{3} \int_0^{\infty} ds\, s^{-6}\Big[&2s^3\coth(s)[\sinh(s)]^{-2} + 3s^2[\sinh(s)]^{-2} \\
&+ 3s\coth(s) - 8\Big],
\end{aligned} \tag{2.53}$$

$$\begin{aligned}
\zeta(7) = \frac{\pi^6}{15} \int_0^{\infty} ds\, s^{-8}\Big[&4s^4[\coth(s)]^2[\sinh(s)]^{-2} \\
&+ 2s^4[\sinh(s)]^{-4} + 12s^3\coth(s)[\sinh(s)]^{-2} \\
&+ 15s^2[\sinh(s)]^{-2} + 15s\coth(s) - 48\Big],
\end{aligned} \tag{2.54}$$

etc.

---

[1]The condition $\mathrm{Re}(2n+2z) > 1$ takes effect only for $n = 0$, that is, we assume $(1/2) < \mathrm{Re}(z) < 1$ if $n = 0$.

[2]The second formula is mentioned since it appears to be advantageous (cf. (2.32)–(2.34)) to substitute $t = s^2$ after one performs the n differentiations w.r.t. t in the 1st line of (2.52).

Proof Assume that $n \in \mathbb{N}_0$, $0 < \mathrm{Re}(z) < 1$, and $\mathrm{Re}(2n + 2z) > 1$. Then,

$$\begin{aligned}
&\int_0^\infty dt\, t^{-z} \frac{d^n}{dt^n}\Big[\mathrm{tr}_{L^2((0,1);dx)} \big((-\Delta_D + t I_{L^2((0,1);dx)})^{-1}\big)\Big] \\
&= \int_0^\infty dt\, t^{-z} \frac{d^n}{dt^n}\Big[(2t)^{-1}\big[t^{1/2}\coth\big(t^{1/2}\big) - 1\big]\Big] \\
&= \sum_{k\in\mathbb{N}} \int_0^\infty dt\, t^{-z} \frac{d^n}{dt^n}\Big[\big(\pi^2 k^2 + t\big)^{-1}\Big] \\
&= \sum_{k\in\mathbb{N}} \int_0^\infty dt\, \frac{t^{-z}(-1)^n n!}{\big(\pi^2 k^2 + t\big)^{n+1}} \\
&= (-1)^n n! \sum_{k\in\mathbb{N}} \big(\pi^2 k^2\big)^{-z-n} \int_0^\infty du\, \frac{u^{-z}}{(1+u)^{n+1}} \\
&= (-1)^n n! \pi^{-2(n+z)} \zeta(2(n+z)) \frac{\Gamma(1-z)\Gamma(n+z)}{\Gamma(n+1)} \\
&= (-1)^n \pi^{-2(n+z)} \zeta(2(n+z)) \Gamma(1-z)\Gamma(n+z),
\end{aligned} \tag{2.55}$$

resulting in (2.52). (The condition $(1/2) < \mathrm{Re}(z) < 1$ if $n = 0$ guarantees convergence of the sum over k in (2.55).)

Alternatively, one can attempt to analytically continue the equation

$$\zeta(z) = \sin(\pi z/2)\pi^{z-1} \int_0^\infty ds\, s^{-z-1}[s\coth(s) - 1], \quad \mathrm{Re}(z) \in (1,2), \tag{2.56}$$

to the region $\mathrm{Re}(z) \geq 2$. For this purpose we first introduce

$$\coth(z) = \frac{1}{z} + \sum_{k=1}^{\infty} \frac{2^{2k} B_{2k}}{(2k)!} z^{2k-1}, \quad z \in \mathbb{C},\ 0 < |z| < \pi, \tag{2.57}$$

$$\begin{aligned}
&\coth_0(z) = \frac{1}{z}, \quad z \in \mathbb{C}\backslash\{0\}, \\
&\coth_n(z) = \frac{1}{z} + \sum_{k=1}^{n} \frac{2^{2k} B_{2k}}{(2k)!} z^{2k-1}, \quad z \in \mathbb{C}\backslash\{0\},\ n \in \mathbb{N}.
\end{aligned} \tag{2.58}$$

Theorem 2.59 *Let $n \in \mathbb{N}_0$, then,*

$$\begin{aligned}
&\zeta(z) = \sin(\pi z/2)\pi^{z-1} \int_0^\infty ds\, s^{-z}[\coth(s) - \coth_n(s)], \\
&\qquad\qquad \mathrm{Re}(z) \in (\max(1, 2n), 2n+2).
\end{aligned} \tag{2.60}$$

In particular,

$$\begin{aligned}\zeta(3) &= -\pi^2 \int_0^\infty ds\, s^{-3}[\coth(s) - \coth_1(s)] \\ &= -\pi^2 \int_0^\infty ds\, s^{-3}[\coth(s) - (1/s) - (s/3)],\end{aligned} \tag{2.61}$$

$$\begin{aligned}\zeta(5) &= \pi^4 \int_0^\infty ds\, s^{-5}[\coth(s) - \coth_2(s)] \\ &= \pi^4 \int_0^\infty ds\, s^{-5}\big[\coth(s) - (1/s) - (s/3) + \big(s^3/45\big)\big],\end{aligned} \tag{2.62}$$

$$\begin{aligned}\zeta(7) &= -\pi^6 \int_0^\infty ds\, s^{-7}[\coth(s) - \coth_3(s)] \\ &= -\pi^6 \int_0^\infty ds\, s^{-7}\big[\coth(s) - (1/s) - (s/3) + \big(s^3/45\big) - \big(2s^5/945\big)\big],\end{aligned}$$

etc.

(2.63)

Proof When trying to analytically continue (2.56) to the right, one notices that it is the small s-behavior of the integrand that invalidates this representation. We therefore split the integral at some point $a > 0$ and write

$$\begin{aligned}\zeta(z) = \sin(\pi z/2)\pi^{z-1} &\int_a^\infty ds\, s^{-z-1}[s\coth(s) - 1] \\ + \sin(\pi z/2)\pi^{z-1} &\int_0^a ds\, s^{-z-1}[s\coth(s) - 1],\end{aligned} \tag{2.64}$$

where the first integral is well-defined for $\mathrm{Re}(z) > 1$, and the second for $\mathrm{Re}(z) < 2$. In order to analytically continue the second integral to the right, one writes

$$\begin{aligned}\int_0^a ds\, s^{-z-1}[s\coth(s) - 1] &= \int_0^a ds\, s^{-z-1}\left[s\coth(s) - 1 - \sum_{k=1}^n \frac{2^{2k}B_{2k}}{(2k)!}s^{2k}\right] \\ &\quad + \int_0^a ds\, s^{-z-1}\sum_{k=1}^n \frac{2^{2k}B_{2k}}{(2k)!}s^{2k} \\ &= \int_0^a ds\, s^{-z-1}\left[s\coth(s) - 1 - \sum_{k=1}^n \frac{2^{2k}B_{2k}}{(2k)!}s^{2k}\right]\end{aligned}$$

$$+\sum_{k=1}^{n} \frac{2^{2k} B_{2k}}{(2k)!} \int_0^a ds\, s^{-z-1+2k}$$

$$= \int_0^a ds\, s^{-z-1}\bigg[s\coth(s) - 1 - \sum_{k=1}^{n} \frac{2^{2k} B_{2k}}{(2k)!} s^{2k}\bigg] + \sum_{k=1}^{n} \frac{2^{2k} B_{2k}}{(2k)!} \frac{a^{2k-z}}{2k-z}, \tag{2.65}$$

valid for $1 < \mathrm{Re}(z) < 2n+2$, $z \notin \{2\ell\}_{1\leq \ell \leq n}$.

In summary, up to this point we have shown that for $1 < \mathrm{Re}(z) < 2n+2$, $z \notin \{2\ell\}_{1\leq \ell \leq n}$, and for $a > 0$, one has

$$\begin{aligned}\zeta(z) = \sin(\pi z/2)\pi^{z-1}\bigg\{ &\int_a^\infty ds\, s^{-z-1}[s\coth(s) - 1] \\ &+ \int_0^a ds\, s^{-z-1}\bigg[s\coth(s) - 1 - \sum_{k=1}^{n} \frac{2^{2k} B_{2k}}{(2k)!} s^{2k}\bigg] \\ &+ \sum_{k=1}^{n} \frac{2^{2k} B_{2k}}{(2k)!} \frac{a^{2k-z}}{2k-z}\bigg\}. \end{aligned} \tag{2.66}$$

Restricting z to $\mathrm{Re}(z) \in (\max(1, 2n), 2n+2)$ and performing the limit $a \to \infty$ in (2.66), observing that the first and third terms on the right-hand side of (2.66) vanish in the limit, proves (2.60).

One notes that for $z = 2n$ the first two lines in (2.66) as well as all terms $k \neq n$ vanish and one confirms Euler's celebrated formula

$$\zeta(2n) = \lim_{z\to 2n} \bigg[\sin(\pi z/2)\pi^{z-1} \frac{2^{2n} B_{2n}}{(2n)!} \frac{a^{2n-z}}{2n-z}\bigg] = \frac{2^{2n-1}|B_{2n}|\pi^{2n}}{(2n)!}, \quad n \in \mathbb{N}. \tag{2.67}$$

Finally, one can take these investigations one step further as follows. Introducing

$$F(z) = \ln\left(z^{-1/2}\sinh\left(z^{1/2}\right)\right) = \sum_{k=1}^{\infty} \frac{2^{2k} B_{2k}}{2k(2k)!} z^k, \quad z \in \mathbb{C},\ |z| < \pi, \tag{2.68}$$

$$F_n(z) = \sum_{k=1}^{n} \frac{2^{2k} B_{2k}}{2k(2k)!} z^k, \quad z \in \mathbb{C},\ n \in \mathbb{N}, \tag{2.69}$$

one can show the following result.

Theorem 2.70 *Let $n \in \mathbb{N}_0$, then,*

$$\zeta(z) = (z/2)\pi^{z-1}\sin(\pi z/2) \int_0^\infty dt\, t^{-z/2-1}\,[F(t) - F_n(t)], \tag{2.71}$$
$$\mathrm{Re}(z) \in (\max(1, 2n), 2n+2).$$

In particular,

$$\begin{aligned}\zeta(3) &= -3\pi^2 \int_0^\infty ds\, s^{-4}\big[F\big(s^2\big) - F_1\big(s^2\big)\big] \\ &= -3\pi^2 \int_0^\infty ds\, s^{-4}\big[\ln\big(s^{-1}\sinh(s)\big) - \big(s^2/6\big)\big],\end{aligned} \tag{2.72}$$

$$\begin{aligned}\zeta(5) &= 5\pi^4 \int_0^\infty ds\, s^{-6}\big[F\big(s^2\big) - F_2\big(s^2\big)\big] \\ &= 5\pi^4 \int_0^\infty ds\, s^{-6}\big[\ln\big(s^{-1}\sinh(s)\big) - \big(s^2/6\big) + \big(s^4/180\big)\big],\end{aligned} \tag{2.73}$$

$$\begin{aligned}\zeta(7) &= -7\pi^6 \int_0^\infty ds\, s^{-8}\big[F\big(s^2\big) - F_3\big(s^2\big)\big] \\ &= -7\pi^6 \int_0^\infty ds\, s^{-8}\big[\ln\big(s^{-1}\sinh(s)\big) - \big(s^2/6\big) + \big(s^4/180\big) - \big(s^6/2835\big)\big], \\ &\quad \textit{etc.}\end{aligned} \tag{2.74}$$

Proof The computation

$$\begin{aligned}F'(t) - F_n'(t) &= \frac{1}{2t}\big[t^{1/2}\coth\big(t^{1/2}\big) - 1\big] - \frac{1}{2}\sum_{k=1}^{n} \frac{2^{2k}B_{2k}}{(2k)!}\, t^{k-1} \\ &= \frac{1}{2t}\Bigg[t^{1/2}\coth\big(t^{1/2}\big) - \sum_{k=0}^{n} \frac{2^{2k}B_{2k}}{(2k)!}\, t^{k}\Bigg] \\ &= \frac{1}{2t}\big[t^{1/2}\coth\big(t^{1/2}\big) - t^{1/2}\coth_n\big(t^{1/2}\big)\big], \quad t \geq 0,\end{aligned} \tag{2.75}$$

and (2.60) then show

$$\begin{aligned}\zeta(z) &= \sin(\pi z/2)\pi^{z-1} \int_0^\infty ds\, s^{-z}[\coth(s) - \coth_n(s)] \\ &= \sin(\pi z/2)\pi^{z-1} \int_0^\infty dt\, t^{-z/2}[F'(t) - F_n'(t)] \\ &= (z/2)\sin(\pi z/2)\pi^{z-1} \int_0^\infty dt\, t^{-z/2-1}[F(t) - F_n(t)], \\ &\qquad \mathrm{Re}(z) \in (\max(1, 2n), 2n+2),\end{aligned} \tag{2.76}$$

after an integration by parts.

Acknowledgements We are indebted to the anonymous referee for kindly bringing references [24] and [27] to our attention.

Klaus Kirsten was supported by the Baylor University Summer Sabbatical and Research Leave Program.

Appendix: Basic Formulas for the Riemann ζ-Function

We present a number of formulas for $\zeta(z)$ and special values of $\zeta(\,\cdot\,)$. It goes without saying that no such collection can ever attempt at any degree of completeness, and certainly our compilation of formulas is no exception in this context.

Definition

$$\zeta(z) = \sum_{k\in\mathbb{N}} k^{-z}, \quad z \in \mathbb{C},\ \mathrm{Re}(z) > 1 \tag{A.1}$$

$$= \left[1 - 2^{-z}\right]^{-1} \sum_{k\in\mathbb{N}_0} (2k+1)^{-z}, \quad \mathrm{Re}(z) > 1, \quad [23, \text{p. } 19] \tag{A.2}$$

$$= \left[1 - 2^{1-z}\right]^{-1} \sum_{k\in\mathbb{N}} (-1)^{k+1} k^{-z}, \quad \mathrm{Re}(z) > 0, \quad [23, \text{p. } 19]. \tag{A.3}$$

Functional Equation

$$\zeta(z) = 2^z \pi^{z-1} \sin(\pi z/2) \Gamma(1-z) \zeta(1-z), \quad z \in \mathbb{C},\ \mathrm{Re}(z) < 0. \tag{A.4}$$

Alternative Formulas

$$\zeta(z) = \Gamma(z)^{-1} \int_0^\infty dt\, \frac{t^{z-1}}{e^t - 1}, \quad z \in \mathbb{C},\ \mathrm{Re}(z) > 1 \tag{A.5}$$

$$= \mu^z \Gamma(z)^{-1} \int_0^\infty dt\, \frac{t^{z-1}}{e^{\mu t} - 1}, \quad z \in \mathbb{C},\ \mathrm{Re}(z) > 1\ \mathrm{Re}(\mu) > 0, \quad [14, 3.4111] \tag{A.6}$$

$$= \Gamma(z)^{-1} [1 - 2^{1-z}]^{-1} \int_0^\infty dt\, \frac{t^{z-1}}{e^t + 1}, \quad z \in \mathbb{C},\ \mathrm{Re}(z) > 0 \tag{A.7}$$

$$= \mu^z \Gamma(z)^{-1} [1 - 2^{1-z}]^{-1} \int_0^\infty dt\, \frac{t^{z-1}}{e^{\mu t} + 1}, \quad z \in \mathbb{C},\ \mathrm{Re}(z) > 0\ \mathrm{Re}(\mu) > 0, \quad [14, 3.4113], \tag{A.8}$$

where

$$\Gamma(z) = \int_0^\infty dt\, t^{z-1} e^{-t}, \quad z \in \mathbb{C},\ \mathrm{Re}(z) > 0. \tag{A.9}$$

In addition,

$$\zeta(x) = \Gamma(x)^{-1} \int_0^1 \int_0^1 ds dt\, \frac{[\ln(st)]^{x-2}}{1 - st}, \quad x > 3, \quad [34] \tag{A.10}$$

$$= e^{i\pi(1-x)} \Gamma(x)^{-1} \int_0^1 dt\, \frac{\ln(t)^{x-1}}{1-t}, \quad x > 1,\ \text{Jensen (1895)},\ [14, 4.2714], \tag{A.11}$$

$$= \pi^{z/2} \Gamma(z/2)^{-1} \int_0^\infty dt\, t^{(z/2)-1} \sum_{k \in \mathbb{N}} e^{-k^2 \pi t}, \tag{A.12}$$

and

$$\zeta(z) = \pi^{z/2} \Gamma(z/2)^{-1} \sum_{k \in \mathbb{N}} \int_0^\infty dt\, t^{(z/2)-1} e^{-k^2 \pi t}, \quad z \in \mathbb{C},\ \mathrm{Re}(z) > 1, \quad [32] \tag{A.13}$$

$$= \frac{2^{z-1}}{z-1} - 2^z \int_0^\infty dt\, \frac{\sin(z \arctan(t))}{(1+t^2)^{z/2}(e^{\pi t}+1)}, \quad z \in \mathbb{C}\backslash\{1\}, \tag{A.14}$$

[14, 9.5134], [23, p. 21]

$$= \frac{2^{z-1}}{[1 - 2^{1-z}]} \int_0^\infty dt\, \frac{\cos(z \arctan(t))}{(1+t^2)^{z/2} \cosh(\pi t/2)}, \quad z \in \mathbb{C}\backslash\{1\}, \tag{A.15}$$

[23, p. 21]

$$= \frac{1}{2} + \frac{1}{z-1} + 2 \int_0^\infty dt\, \frac{\sin(z \arctan(t))}{(1+t^2)^{z/2}(e^{2\pi t}-1)}, \quad z \in \mathbb{C}\backslash\{1\}, \tag{A.16}$$

Jensen's formula (1895), [23, p. 21]

$$= a^z \frac{2^{z-1}}{[2^z - 1]} \Gamma(z)^{-1} \int_0^\infty dt\, \frac{t^{z-1}}{\sinh(at)}, \quad \mathrm{Re}(z) > 1,\ a > 0,\ [14, 3.5231] \tag{A.17}$$

$$= \Gamma(z+1)^{-1} 4^{-1} (2a)^{z+1} \int_0^\infty dt\, \frac{t^z}{[\sinh(at)]^2}, \quad \mathrm{Re}(z) > -1,\ \mathrm{Re}(a) > 0, \tag{A.18}$$

[14, 3.5271]

$$= \Gamma(z+1)^{-1}4^{-1}(2a)^{z+1}\big[1-2^{1-z}\big]^{-1}\int_0^\infty dt\,\frac{t^z}{[\cosh(at)]^2}, \tag{A.19}$$

$\mathrm{Re}(z) > -1,\ z \neq 1,\ \mathrm{Re}(a) > 0,$ [14, 3.5273]

$$= \Gamma(z+1)^{-1}\big[2-2^{2-z}\big]^{-1}\int_0^\infty dt\,\frac{t^z}{\cosh(t)+1}, \quad \mathrm{Re}(z) > 0,\ z \neq 1, \tag{A.20}$$

[14, 3.5316]

$$= 2^{-1} + \Gamma(z)^{-1}2^{z-1}\int_0^\infty dt\,t^{z-1}e^{-2t}\coth(t), \quad \mathrm{Re}(z) > 1, \quad [14, 3.5513] \tag{A.21}$$

$$= 2^{-1} + \Gamma(z)^{-1}2^{z-1}\int_0^\infty dt\,t^{z-1}e^{-2t}\coth(t), \quad \mathrm{Re}(z) > 1, \quad [14, 3.5513] \tag{A.22}$$

$$= \Gamma(z)^{-1}2^{z-1}\int_0^\infty dt\,t^{z-1}\frac{e^{-t}}{\sinh(t)}, \quad \mathrm{Re}(z) > 1, \quad [14, 3.5521] \tag{A.23}$$

$$= \Gamma(z)^{-1}2^{z-1}\big[1-2^{1-z}\big]^{-1}\int_0^\infty dt\,t^{z-1}\frac{e^{-t}}{\cosh(t)}, \quad \mathrm{Re}(z) > 0,\ z \neq 1,$$

[14, 3.5523] (A.24)

$$= 2^z\Gamma(z)^{-1}\int_0^1 dt\,[\ln(1/t)]^{z-1}\frac{t}{1-t^2}, \quad \mathrm{Re}(z) > 0, \quad [14, 4.27212] \tag{A.25}$$

$$= \Gamma(z+1)^{-1}\int_0^\infty dt\,\frac{t^z e^t}{\big[e^t-1\big]^2}, \quad \mathrm{Re}(z) > 1, \quad [23, p. 20] \tag{A.26}$$

$$= \Gamma(z+1)^{-1}\big[1-2^{1-z}\big]^{-1}\int_0^\infty dt\,\frac{t^z e^t}{\big[e^t+1\big]^2}, \quad \mathrm{Re}(z) > 0, \quad [23, p. 20] \tag{A.27}$$

$$= 2\sin(\pi z/2)\int_0^\infty dt\,\frac{t^{-z}}{e^{2\pi t}-1}, \quad \mathrm{Re}(z) < 0, \quad [22, p. 104] \tag{A.28}$$

$$= (2^z-1)^{-1}\frac{2^{z-1}z}{z-1} + 2(2^z-1)^{-1}\int_0^\infty dt\,\frac{\sin(z\arctan(2t))}{[(1/4)+t^2]^{z/2}}\,\frac{1}{e^{2\pi t}-1}, \tag{A.29}$$

$z \in \mathbb{C}\backslash\{1\},$ [30, p. 279].

Specific Values

$$\zeta(2n) = \frac{(-1)^{n+1}(2\pi)^{2n} B_{2n}}{2(2n)!}, \quad n \in \mathbb{N}_0, \tag{A.30}$$

where B_m are the Bernoulli numbers generated, for instance, by

$$\frac{w}{e^w - 1} = \sum_{m \in \mathbb{N}_0} B_m \frac{w^m}{m!}, \quad w \in \mathbb{C}, \ |w| < 2\pi, \tag{A.31}$$

in particular,

$$B_0 = 1, B_1 = -1/2, \ B_2 = 1/6, \ B_3 = 0, B_4 = -1/30, B_5 = 0, B_6 = 1/42, \text{etc.}, \tag{A.32}$$

$$B_{2k+1} = 0, \ k \in \mathbb{N}. \tag{A.33}$$

Moreover, one has the **generating functions** for $\zeta(2n)$,

$$-(\pi z/2)\cot(\pi z) = \sum_{n \in \mathbb{N}_0} \zeta(2n) z^{2n}, \quad |z| < 1, \ \zeta(0) = -1/2, \tag{A.34}$$

$$-(\pi z/2)\coth(\pi z) = \sum_{n \in \mathbb{N}_0} (-1)^n \zeta(2n) z^{2n}, \quad |z| < 1, \ \zeta(0) = -1/2, \tag{A.35}$$

and [32]

$$(n!/6)[\zeta(n-2) - 3\zeta(n-1) + 2\zeta(n)] = \int_0^\infty dt \, \frac{t^n e^t}{(e^t - 1)^4}, \quad n \in \mathbb{N}, \ n \geq 4. \tag{A.36}$$

Choosing $k = 2n$, $n \in \mathbb{N}$, even, employing (A.30) for $\zeta(2n)$, $\zeta(2n-2)$, yields a formula for $\zeta(2n-1)$. Moreover,

$$\zeta(2n+1) = \frac{1}{(2n)!} \int_0^\infty dt \, \frac{t^{2n}}{e^t - 1}, \quad n \in \mathbb{N} \tag{A.37}$$

$$= \frac{(-1)^{n+1}(2\pi)^{2n+1}}{2(2n+1)!} \int_0^1 dt \, B_{2n+1}(t)\cot(\pi t), \quad n \in \mathbb{N}, \quad [9], \tag{A.38}$$

where $B_m(\,\cdot\,)$ are the Bernoulli polynomials,

$$B_m(z) = \sum_{j=0}^{m} \binom{m}{j} B_j z^{m-j}, \quad t \in \mathbb{C}, \tag{A.39}$$

generated, for instance, by

$$\frac{we^{zw}}{e^w - 1} = \sum_{m\in\mathbb{N}_0} B_m(z)\frac{w^m}{m!}, \quad w \in \mathbb{C},\ |w| < 2\pi. \tag{A.40}$$

Explicitly,

$$\begin{aligned} &B_0(x) = 1,\ B_1(x) = x - (1/2),\ B_2(x) = x^2 - x + (1/6), \\ &B_3(x) = x^3 - (3/2)x^2 + (1/2)x,\ \text{etc.,} \end{aligned} \tag{A.41}$$

$$B_n(0) = B_n,\ n \in \mathbb{N}, \quad B_1(1) = -B_1 = 1/2,\ B_n(1) = B_n,\ n \in \mathbb{N}_0\backslash\{1\}, \tag{A.42}$$

$$B_n'(x) = nB_{n-1}(x), \quad n \in \mathbb{N},\ x \in \mathbb{R}. \tag{A.43}$$

In addition, for $n \in \mathbb{N}$,

$$\begin{aligned} \zeta(2n+1) &= \frac{a^2(2a)^{2n}}{[2^{-2n-1} - 1]}\frac{1}{(2n+1)!} \\ &\quad \times \int_0^\infty dt\, t^{2n+1}\frac{\cosh(at)}{[\sinh(at)]^2}, \quad a \neq 0, \quad [14, 3.5279] \end{aligned} \tag{A.44}$$

$$= \frac{2^{2n}}{[2^{2n} - 1]}[(2n)!]^{-1}\int_0^1 dt\, \frac{[\ln(t)]^{2n}}{1+t}, \quad [14, 4.2711] \tag{A.45}$$

$$= \frac{2^{2n+1}}{[2^{2n+1} - 1]}[(2n)!]^{-1}\int_0^1 dt\, \frac{[\ln(t)]^{2n}}{1-t^2}, \quad [14, 4.2711], \tag{A.46}$$

$$\zeta(n) = [(n-1)!]^{-1}\int_0^1 dt\, \frac{[\ln(1/t)]^{n-1}}{1-t}, \quad [14, 4.2729]. \tag{A.47}$$

Just for curiosity,

$$\zeta(3) = 1.2020569032\ldots. \tag{A.48}$$

Apery [1] proved in 1978 that $\zeta(3)$ is irrational (see also Beukers [2], van der Poorten [29], Zudilin [35], and [31], [33]).

Moreover,

$$\zeta(3) = \sum_{k\in\mathbb{N}} k^{-3} = \frac{8}{7}\sum_{k\in\mathbb{N}_0}(2k+1)^{-3} = \frac{4}{3}\sum_{k\in\mathbb{N}_0}(-1)^k(k+1)^{-3}, \quad [31] \tag{A.49}$$

$$= \frac{1}{2}\int_0^\infty dt\, \frac{t^2}{e^t - 1}, \quad [31] \tag{A.50}$$

$$= \frac{2}{3} \int_0^\infty dt \, \frac{t^2}{e^t + 1}, \quad [31] \tag{A.51}$$

$$= \frac{4}{7} \int_0^{\pi/2} dt \, t \ln([1/\cos(t)] + \tan(t)), \quad [31] \tag{A.52}$$

$$= \frac{8}{7} \bigg[\frac{\pi^2 \ln(2)}{4} + 2 \int_0^{\pi/2} dt \, t \ln(\sin(t)) \bigg], \quad [33] \tag{A.53}$$

$$= -\frac{1}{2} \int_0^1 \int_0^1 dx dy \, \frac{\ln(xy)}{1 - xy}, \quad [2] \tag{A.54}$$

$$= \int_0^1 \int_0^1 \int_0^1 dx dy dz \, \frac{1}{1 - xyz}, \quad [31] \tag{A.55}$$

$$= \pi \int_0^\infty dt \, \frac{\cos(2 \arctan(t))}{(1 + t^2)[\cosh(\pi t/2)]^2}, \quad [31] \tag{A.56}$$

$$= \frac{8\pi^2}{7} \int_0^1 dt \, \frac{t(t^4 - 4t^2 + 1) \ln(\ln(1/t))}{(1 + t^2)^4}, \quad [31] \tag{A.57}$$

$$= \frac{8\pi^2}{7} \int_1^\infty dt \, \frac{t(t^4 - 4t^2 + 1) \ln(\ln(t))}{(1 + t^2)^4}, \quad [31] \tag{A.58}$$

$$= 10 \int_0^{1/2} dt \, \frac{[\operatorname{arcsinh}(t)]^2}{t} \quad [12, \text{p. } 46] \tag{A.59}$$

$$= (2/7)\pi^2 \ln(2) + (4/7) \int_0^\pi dt \, t \ln(\sin(t/2)) \quad [12, \text{p. } 46] \tag{A.60}$$

$$= (2/7)\pi^2 \ln(2) - (8/7) \int_0^1 dt \, \frac{[\arcsin(t)]^2}{t} \quad [12, \text{p. } 46] \tag{A.61}$$

$$= (2/7)\pi^2 \ln(2) - (8/7) \int_0^{\pi/2} dt \, t^2 \cot(t) \quad [12, \text{p. } 46] \tag{A.62}$$

$$\zeta(3) = -\frac{2}{7}\pi^2 \ln(2) - \frac{16}{7} \int_0^1 dt \, \frac{\operatorname{arctanh}(t) \ln(t)}{t(1 - t^2)} \tag{A.63}$$

$$= -\frac{4}{3} \int_0^1 dt \, \frac{\ln(t) \ln(1 + t)}{t} \tag{A.64}$$

$$= -8 \int_0^1 dt \, \frac{\ln(t) \ln(1 + t)}{1 + t} \tag{A.65}$$

$$= \int_0^1 dt \, \frac{\ln(t) \ln(1 - t)}{1 - t} = \int_0^1 dt \, \frac{\ln(t) \ln(1 - t)}{t} \tag{A.66}$$

$$= \frac{1}{4}\pi^2 \ln(2) + \int_0^1 dt\, \frac{\ln(t)\ln(1+t)}{1-t} \tag{A.67}$$

$$= \frac{2}{13}\pi^2 \ln(2) + \frac{8}{13} \int_0^1 dt\, \frac{\ln(t)\ln(1-t)}{1+t} \tag{A.68}$$

$$= \frac{2}{7} \int_0^{\pi/2} dt\, \frac{t(\pi - t)}{\sin(t)}. \tag{A.69}$$

Formulas (A.63)–(A.69) were provided by Glasser and Ruehr and can be found in [21, Problem 80-13]. Finally, we also recall,

$$\zeta(3) = 1 + \int_0^\infty dt\, \frac{6t - 2t^3}{(1+t^2)^3} \frac{1}{e^{2\pi t} - 1}, \quad [17, \text{p. } 274] \tag{A.70}$$

$$= \frac{6}{7} + \frac{2}{7} \int_0^\infty dt\, \frac{\sin(3\arctan(2t))}{[(1/4) + t^2]^{3/2}} \frac{1}{e^{2\pi t} - 1} \tag{A.71}$$

$$= \frac{6}{7} + \frac{8}{7} \int_0^\infty dt\, \frac{\sin(3\arctan(t))}{(1+t^2)^{3/2}} \frac{1}{e^{\pi t} - 1} \tag{A.72}$$

$$= 2 - 8 \int_0^\infty dt\, \frac{\sin(3\arctan(t))}{(1+t^2)^{3/2}} \frac{1}{e^{\pi t} + 1} \tag{A.73}$$

$$= 1 + 2 \int_0^\infty dt\, \frac{\sin(3\arctan(t))}{(1+t^2)^{3/2}} \frac{1}{e^{2\pi t} - 1}. \tag{A.74}$$

Formulas (A.71)–(A.73) are due to Jensen (1895) and are special cases of results to be found in [30, p. 279] (cf. (A.16), (A.29)); finally, (A.74) is a consequence of (A.72) and (A.73).

For more on $\zeta(3)$ see also [12, p. 42–45].

For a wealth of additional formulas, going beyond what is recorded in this appendix, we also refer to [24] and [27].

References

1. R. Apéry, Irrationalité de $\zeta(2)$ et $\zeta(3)$. Astérisque **61**, 11–13 (1979)
2. F. Beukers, A note on the irrationality of $\zeta(2)$ and $\zeta(3)$. Bull. Lond. Math. Soc. **11**, 268–272 (1979)
3. M. Bordag, G.L. Klimchitskaya, U. Mohideen, V.M. Mostepanenko, *Advances in the Casimir Effect* (Oxford Science Publications, Oxford, 2009)
4. A.A. Bytsenko, G. Cognola, L. Vanzo, S. Zerbini, Quantum fields and extended objects in space-times with constant curvature spatial section. Phys. Rep. **266**, 1–126 (1996)
5. S. Coleman, *Aspects of Symmetry: Selected Lectures of Sidney Coleman* (Cambridge University Press, Cambridge, 1985)
6. L.A. Dikii, The zeta function of an ordinary differential equation on a finite interval. Izv. Akad. Nauk SSSR Ser. Mat. **19**(4), 187–200 (1955)
7. L.A. Dikii, Trace formulas for Sturm–Liouville differential operators. Am. Math. Soc. Transl. (2) **18**, 81–115 (1961)

8. J.S. Dowker, R. Critchley, Effective Lagrangian and energy momentum tensor in de Sitter space. Phys. Rev. **D13**, 3224–3232 (1976)
9. R.J. Dwilewicz, J. Mináč, Values of the Riemann zeta function at integers. MATerials MATemàtics **2009**(6), 26 pp
10. E. Elizalde, *Ten Physical Applications of Spectral Zeta Functions*. Lecture Notes in Physics, vol. 855 (Springer, Berlin, 2012)
11. G. Esposito, A.Y. Kamenshchik, G. Pollifrone, *Euclidean Quantum Gravity on Manifolds with Boundary*. Fundamental Theories of Physics, vol. 85 (Kluwer, Dordrecht, 1997)
12. S.R. Finch, *Mathematical Constants*. Encyclopedia of Mathematics and Its Applications, vol. 94 (Cambridge University Press, Cambridge, 2003)
13. F. Gesztesy, K. Kirsten, Effective Computation of Traces, Determinants, and ζ-Functions for Sturm–Liouville Operators. J. Funct. Anal. **276**, 520–562 (2019)
14. I.S. Gradshteyn, I.M. Ryzhik, *Table of Integrals, Series, and Products*, corrected and enlarged edition, prepared by A. Jeffrey, (Academic, San Diego, 1980)
15. M. Haase, *The Functional Calculus for Sectorial Operators*. Operator Theory: Advances and Applications, vol. 169 (Birkhäuser, Basel, 2006)
16. S.W. Hawking, Zeta function regularization of path integrals in curved space-time. Commun. Math. Phys. **55**, 133–148 (1977)
17. P. Henrici, *Applied and Computational Complex Analysis, Vol. I: Power Series–Integration–Conformal Mapping–Location of Zeros*, reprinted 1988 (Wiley, New York, 1974)
18. K. Kirsten, *Spectral Functions in Mathematics and Physics* (Chapman&Hall/CRC, Boca Raton, 2002)
19. K. Kirsten, A.J. McKane, Functional determinants by contour integration methods. Ann. Phys. **308**, 502–527 (2003)
20. K. Kirsten, A.J. McKane, Functional determinants for general Sturm–Liouville problems. J. Phys. A **37**, 4649–4670 (2004)
21. M.S. Klamkin (ed.), *Problems in Applied Mathematics: Selections from SIAM Review* (SIAM, Philadelphia, 1990)
22. E. Lindelöf, *Le Calcul des Résides et ses Applications a la Théorie des Fonctions* (Chelsea, New York, 1947)
23. W. Magnus, F. Oberhettinger, R.P. Soni, *Formulas and Theorems for the Special Functions of Mathematical Physics*. Grundlehren, vol. 52, 3rd edn. (Springer, Berlin, 1966)
24. M.S. Milgram, Integral and series representations of Riemann's zeta function and Dirichlet's eta function and a medley of related results. J. Math. **2013**, article ID 181724, 17p.
25. K.A. Milton, *The Casimir Effect: Physical Manifestations of Zero-Point Energy* (World Scientific, River Edge, 2001)
26. D.B. Ray, I.M. Singer, R-Torsion and the Laplacian on Riemannian Manifolds. Adv. Math. **7**, 145–210 (1971)
27. S.K. Sekatskii, *Novel integral representations of the Riemann zeta-function and Dirichlet eta-function, closed expressions for Laurent series expansions of powers of trigonometric functions and digamma function, and summation rules*, arXiv:1606.02150
28. H.M. Srivastava, J. Choi, *Series Associated with the Zeta and Related Functions* (Kluwer, Dordrecht, 2001)
29. A. van der Poorten, A proof that Euler missed .... Apéry's proof of the irrationality of $\zeta(3)$. An informal report. Math. Intell. **1**(4), 195–203 (1979)
30. E.T. Whittaker, G.N. Watson, *A Course of Modern Analysis*, 4th edn., reprinted 1986 (Cambridge University Press, Cambridge, 1927)
31. Wikipedia: Apery's constant, https://en.wikipedia.org/wiki/Apery's_constant
32. Wikipedia: Riemann zeta function, https://en.wikipedia.org/wiki/Riemann_zeta_function
33. WolframMathWorld: Apery's constant, http://mathworld.wolfram.com/AperysConstant.html
34. WolframMathWorld: Riemann Zeta Function, http://mathworld.wolfram.com/RiemannZetaFunction.html
35. W. Zudilin, *An elementary proof of Apéry's theorem*, arXiv:0202159

Real- and Complex-Energy Non-conserving Particle Number Pairing Solution

Rodolfo M. Id Betan

Abstract Many-body open quantum systems are characterized by the correlations between bound and scattering states. In contrast to a closed system (i.e., a very well bound many-body system), the continuous part of the energy spectrum has to be considered explicitly due to the proximity of the Fermi level to the continuum's threshold. In this work we show how to introduce these correlations through the continuum single-particle level density (CSPLD) in the pairing framework. By isolating the resonances of the system using an analytic continuation, we arrive at the Berggren (complex-energy) representation of the pairing solution.

Keywords Pairing · Continuum · Resonances · Berggren · Gamow

1 Introduction

The study of many-body systems requires the use of a single-particle representation in order to build a many-body basis. Within this basis one finds the many-body eigenvalues and eigenfunctions. If the system has a Fermi level that is close to the continuum's threshold, the correlations with the continuous part of the energy spectrum become important and have to be included. In this contribution, we present a basis that includes explicitly the continuous part of the energy spectrum of the pairing interaction. One main reason why we incorporate the continuum is that the expansion of the wave function has the correct asymptotic behavior required to describe loosely bound systems. In nuclear physics, the inclusion of the continuous part of the energy spectrum led to the development of the real energy continuum shell model [1, 2] and the complex-energy shell model [3, 4].

R. M. Id Betan (✉)
Instituto de Física Rosario (CONICET-UNR), Facultad de Ciencias Exactas, Ingeniería y Agrimensura (UNR), Santa Fe, Argentina
e-mail: idbetan@ifir-conicet.gov.ar

Instituto de Estudios Nucleares y Radiaciones Ionizantes (UNR), Ocampo y Esmeralda, Rosario, Santa Fe, Argentina

Ş. Kuru et al. (eds.), *Integrability, Supersymmetry and Coherent States*, CRM Series in Mathematical Physics, https://doi.org/10.1007/978-3-030-20087-9_9

The pairing correlation is a crucial part of the two-body residual interaction. Within the Hartree–Fock–Bogoliubov framework, the pairing effect on weakly bound nuclei was studied in Refs. [5–8]. Pairing correlations within the Bardeen–Cooper–Schrieffer (BCS) approximation were studied in Refs. [9–12]. In Ref. [13] the pairing with correlations in the continuum has been obtained using the continuum single-particle level density (CSPLD) [14]. This density was used by Mosel [15] to study hot nuclei. Fowler and Engelbrech [16] used the CSPLD to calculate the nuclear partition functions at high temperature. In Ref. [17], the CSPLD was used to calculate the contribution of unbound states within the nuclear Hartree–Fock approximation at finite temperature. Other implementations of the CSPLD in nuclear physics may be found in Refs. [18, 19].

The physical properties of nuclear systems with continuous spectrum have been studied using the CSPLD within the framework of the pairing solution in Refs. [20, 21], but without a proper proof. In the appendix of Ref. [22], a box representation has been used to justify the use of the CSPLD within the BCS framework, and in Ref. [23] the exact pairing solution within the Richardson formalism [24, 25].

The first goal of this contribution is to apply the CSPLD to the pairing solution without having to resort to box normalization, in an heuristic and plausible way, without mathematical rigor.

For very narrow resonances, i.e., for long-lived quasi-stationary systems, the CSPLD shows a very sharp peak around the resonant energy. Since the width of very narrow resonances may be many orders of magnitude smaller than the resonant energy, integration is extremely sensitive to the discretization used to approximate the integral by a quadrature. The second goal of this contribution is to avoid this drawback by doing an analytic continuation from the continuous spectrum to the complex-energy plane. This continuation separates the resonant contribution from the non-resonant one, and naturally leads to the Berggren representation [26].

In Sect. 2, we develop the non-conserving particle number pairing solution in systems with continuous spectrum, without appealing to the box normalization. In Sect. 3, the Berggren representation and the pole approximation are obtained by performing the analytic continuation of the BCS equations. In Sect. 4, we present our conclusions. In the Appendix we show how the discrete and the continuous gaps combine into a single pairing gap.

2 Continuum Real-Energy Representation

Let us start with a many-body Hamiltonian $H = \sum_i T_i + \sum_{j<i} V_{ij}$, and let us reduce it by using the mean-field approximation to obtain $H = H_{sp} + V$, where $H_{sp} = \sum_i h_i$ is the sum of single-particle Hamiltonians h_i and V is the residual interaction. We will assume that the single-particle Hamiltonians h_i have discrete ε_j and continuum ε eigenvalues. Their corresponding eigenfunctions generate the single-particle basis used to solve the many-body problem.

In the pairing framework the residual interaction is modeled by a potential V_P that correlates only one particle with its time reverse companion with strength G [27]. When the system contains bound and continuum states, the Hamiltonian reads

$$H = H_{sp} + V_P \tag{1}$$

$$H_{sp} = \sum_j \varepsilon_j \hat{n}_j + \int_0^\infty d\varepsilon \; \varepsilon \; \hat{n}(\varepsilon) \tag{2}$$

$$V_P = -G \; P^+ P \tag{3}$$

where

$$\hat{n}_j = \sum_m a^+_{jm} a_{jm} \,, \qquad \hat{n}(\varepsilon) = \sum_{\nu m} a^+_{\nu m}(\varepsilon) a_{\nu m}(\varepsilon) \tag{4}$$

and

$$P^+ = \sum_j A^+_j + \int_0^\infty d\varepsilon \; A^+(\varepsilon) \tag{5}$$

$$A^+_j = \sum_{m>0} a^+_{jm} a^+_{j\bar{m}} \qquad A^+(\varepsilon) = \sum_{\nu m>0} a^+_{\nu m}(\varepsilon) a^+_{\nu \bar{m}}(\varepsilon) \tag{6}$$

The index $j \equiv \{n, l, j\}$ labels the principal quantum number and the orbital and total angular momentum of the valence bound states of the mean-field. The index $\nu \equiv \{l, j\}$ labels the angular and total angular momentum of the continuum (scattering) states of real energy ε. The quantum number m is the projection of the total angular momentum j. The operators $a^\dagger_{jm}$ and a_{jm}, respectively, create and annihilate a valence bound state with real negative energy ε_{nlj}. They satisfy the usual anti-commutation relation $\{a_{jm}, a^+_{j'm'}\} = \delta_{jj'} \; \delta_{mm'}$. Similarly, the operators $a^\dagger_{\nu m}(\varepsilon)$ and $a_{\nu m}(\varepsilon)$ create and annihilate continuum states with real positive energy ε in the single-particle state $\{\nu, m\}$. They satisfy the anti-commutation relation normalized to the Dirac delta in energy, $\{a_{\nu m}(\varepsilon), a^+_{\nu' m'}(\varepsilon')\} = \delta_{\nu\nu'} \; \delta_{mm'} \; \delta(\varepsilon - \varepsilon')$. The dash on the quantum number m in Eq. (6) is a short-hand notation for the time reversed state, i.e., $a^+_{n\bar{m}} \equiv (-)^{j-m} a^+_{n,-m}$ for $n = j$ or $n = \nu(\varepsilon)$.

By introducing the operators $P^+_d = \sum_j A^+_j$ and $P^+_c = \int_0^\infty d\varepsilon \; A^+(\varepsilon)$ one can write the pairing interaction as $V_P = -G P^+_d P_d - G P^+_d P_c - G P^+_c P_d - G P^+_c P_c$. It is clear from this expression that the coupling between particles in different configurations is all the same. This is an unwanted feature of the formalism, since one expects that states in bound configurations to have stronger correlations than states in continuum configurations. We will show below that the strength of the coupling with states in the continuum is modulated by the continuum single-particle level density.

2.1 Canonical Transformation

The so-called BCS (Bardeen–Cooper–Schrieffer) [28] solution of the pairing Hamiltonian can be obtained by performing the Bogolyubov transformation [29] to quasiparticle operators α^+_{nm} in terms of unknown dimensionless coefficients u and v,

$$\alpha^+_{jm} = u_j a^+_{jm} - v_j a_{j\bar{m}} \tag{7}$$

$$\alpha^+_{\nu m}(\varepsilon) = u_\nu(\varepsilon) a^+_{\nu m}(\varepsilon) - v_\nu(\varepsilon) a^+_{\nu\bar{m}}(\varepsilon) \tag{8}$$

By inverting the above equation we end up with expressions such as

$$(u_n^2 + v_n^2) a^+_{nm} = u_n\, \alpha^+_{nm} + v_n\, \alpha_{n\bar{m}}\,. \tag{9}$$

This relation is equally valid for bound $n = j$ and continuum $n = \nu(\varepsilon)$ states. In order to get this equation, the anti-commutation relation has been used. Notice that the Dirac delta does not appear in the above expression for the continuum states, which justifies the election of the same normalization for bound and continuum states,

$$u_j^2 + v_j^2 = 1 \tag{10}$$

$$u_\nu^2(\varepsilon) + v_\nu^2(\varepsilon) = 1 \tag{11}$$

After solving the BCS equations, we will find that $u_\nu(\varepsilon)$ and $v_\nu(\varepsilon)$ do not depend on the quantum number ν.

As a consequence of the normalization (10) and (11), the quasi-particle operators satisfy the same anti-commutation relation (canonical transformation) as the particle operators

$$\{\alpha_{jm}, \alpha^+_{j'm'}\} = \delta_{jj'}\, \delta_{mm'} \tag{12}$$

$$\{\alpha_{\nu m}(\varepsilon), \alpha^+_{\nu' m'}(\varepsilon')\} = \delta_{\nu\nu'}\, \delta_{mm'}\, \delta(\varepsilon - \varepsilon') \tag{13}$$

2.2 Physical Argument for Introducing the CSPLD

In writing the particle number operator in terms of the quasi-particle operators, we found the singularity $\delta(\varepsilon - \varepsilon)$, which we proposed to avoid by introducing the CSPLD,

$$\begin{aligned} \hat{N} &= \sum_j \hat{n}_j + \int_0^\infty d\varepsilon\, \hat{n}(\varepsilon) \\ &= \sum_{jm} v_j^2 + \sum_{jm} (u_j^2 - v_j^2) \alpha^+_{jm} \alpha_{jm} + \sum_{jm} u_j v_j \left(\alpha^+_{jm} \alpha^+_{j\bar{m}} + hc \right) \end{aligned} \tag{14}$$

$$+\int d\varepsilon \sum_{\nu m} v_\nu^2(\varepsilon)\,\delta(\varepsilon-\varepsilon) + \int d\varepsilon \sum_{\nu m} \left[u_\nu^2(\varepsilon) - v_\nu^2(\varepsilon)\right] \alpha_{\nu m}^+(\varepsilon)\alpha_{\nu m}(\varepsilon)$$

$$+\int d\varepsilon \sum_{\nu m} u_\nu(\varepsilon)v_\nu(\varepsilon)\left[\alpha_{\nu m}^+(\varepsilon)\alpha_{\nu\bar{m}}^+(\varepsilon) + hc\right] \tag{15}$$

where hc denotes Hermitian conjugate.

Let us calculate the expectation of the particle number operator in the BCS vacuum wave function, which is defined so that it satisfies $\alpha_{jm}|BCS\rangle = 0$ and $\alpha_{\nu m}(\varepsilon)|BCS\rangle = 0$,

$$\langle BCS|\hat{N}|BCS\rangle = \sum_j \langle BCS|\hat{n}_j|BCS\rangle + \int d\varepsilon\ \langle BCS|\hat{n}(\varepsilon)|BCS\rangle \tag{16}$$

$$= \sum_{jm} v_j^2 + \int d\varepsilon \sum_{\nu m} v_\nu^2(\varepsilon)\,\delta(\varepsilon-\varepsilon) \tag{17}$$

Physically, this equation means that, on average, the particles will be distributed in a set of bound states $\{j, m\}$ with probability $\langle BCS|\hat{n}_j|BCS\rangle = (2j+1)v_j^2$ (where $(2j+1)$ is the degeneracy of the state j), and, less likely, in the continuum states $\nu(\varepsilon)$ with probability proportional to $\langle BCS|\hat{n}_\nu(\varepsilon)|BCS\rangle = (2j_\nu+1)v_\nu^2(\varepsilon)$. Where $\hat{n}_\nu(\varepsilon)$ comes from written $\hat{n}(\varepsilon) = \sum_\nu \hat{n}_\nu(\varepsilon)$.

Since the continuum states represent scattering states, they cannot hold any states unless there is a resonance in the system at the energy ε_ν. In such a situation the long-lived quasi-stationary state will have a measurable probability in the continuum state $\nu(\varepsilon)$ given by $\langle BCS|\hat{n}_\nu(\varepsilon)|BCS\rangle$, with a distribution peaked at some energy ε_ν, and then, the singular distribution $\delta(\varepsilon-\varepsilon)$ may be replaced by a non-singular distribution $g_\nu(\varepsilon)$. Using this ansatz, the singular delta in Eq. (15) is substituted by $g_\nu(\varepsilon)$, and then the average particle number reads

$$\langle BCS|\hat{N}|BCS\rangle = \sum_{jm} v_j^2 + \int d\varepsilon \sum_{\nu m} v_\nu^2(\varepsilon)\, g_\nu(\varepsilon) \tag{18}$$

The distribution $g_\nu(\varepsilon)$ represents the continuum single-particle level density, and it must be chosen according to the physical system. For example, in nuclear physics, we would use the Breit–Wigner (Lorentzian) distribution, and therefore $\langle BCS|\hat{n}_\nu(\varepsilon)|BCS\rangle = (2j_\nu+1)v_\nu^2(\varepsilon)\,g_\nu(\varepsilon)$ gives the probability that a nucleon populates the resonant state with quantum number ν and energy ε.

The replacement of $\delta(\varepsilon-\varepsilon)$ by the Breit–Wigner distribution may seem arbitrary. However, it has been found that, when we use the resonant (Gamow) state to describe the decay of a resonance, the ensuing decay energy spectrum is formally similar to what one obtains from the Fermi Golden Rule, but with the energy delta function replaced by the Breit–Wigner distribution [30, 31]. Hence, there are reasons to believe that such replacement can be justified in a more rigorous version of the formalism presented here.

2.3 BCS Solution

To solve the BCS equations [32], we need to build the BCS Hamiltonian $H_{BCS} = H - \lambda\hat{N}$ and write it down in terms of the quasi-particle operators of Eqs. (7) and (8). The parameter λ is the Fermi level, and it is fixed by the condition that the particle number average of Eq. (18) corresponds to the number of particles in the system, $\langle BCS|\hat{N}|BCS\rangle = N$.

The replacement of $\delta(\varepsilon - \varepsilon)$ by $g_\nu(\varepsilon)$ in Eq. (15) is equivalent to the substitution of the contraction $\widehat{a^+_{\nu m}(\varepsilon)a_{\nu m}}(\varepsilon) = v^2_\nu(\varepsilon)\delta(\varepsilon - \varepsilon)$ by

$$\widehat{a^+_{\nu m}(\varepsilon)a_{\nu m}}(\varepsilon) = v^2_\nu(\varepsilon)g_\nu(\varepsilon) \tag{19}$$

in the diagonal part of the H_{BCS} Hamiltonian. This is so because $\alpha^+\alpha = \widehat{\alpha^+\alpha} + \mathcal{N}(\alpha^+\alpha)$, $\mathcal{N}$ being the normal ordering operator according to the $|BCS\rangle$ vacuum. Applying the same approximation to the interaction part V_P of H_{BCS} leads to

$$\widehat{a^+_{\nu m}(\varepsilon)a^+_{\nu\bar{m}}}(\varepsilon) = (-)^{j-m}\, u_\nu(\varepsilon)\, v_\nu(\varepsilon) g_\nu(\varepsilon) \tag{20}$$

Applying the formalism boils down to finding the Fermi level λ and the pairing gap Δ using the gap equation and the particle number Eq. (18) for the number of particles of the system N (the calculation can be found in the Appendix),

$$\frac{4}{G} = \sum_{jm} \frac{1}{\sqrt{(\varepsilon_j - \lambda)^2 + \Delta^2}} + \sum_{\nu m} \int_0^\infty d\varepsilon\, \frac{g_\nu(\varepsilon)}{\sqrt{(\varepsilon - \lambda)^2 + \Delta^2}} \tag{21}$$

$$N = \sum_{jm} v_j^2 + \sum_{\nu m} \int_0^\infty d\varepsilon\, v^2(\varepsilon)\, g_\nu(\varepsilon) \tag{22}$$

where $v^2(\varepsilon)$ does not depend on the state ν (see Eq. (44) in the Appendix), and

$$v_j^2 = \frac{1}{2}\left(1 - \frac{\varepsilon_j - \lambda}{\sqrt{(\varepsilon_j - \lambda)^2 + \Delta^2}}\right) \tag{23}$$

$$v^2(\varepsilon) = \frac{1}{2}\left(1 - \frac{\varepsilon - \lambda}{\sqrt{(\varepsilon - \lambda)^2 + \Delta^2}}\right) \tag{24}$$

Notice that the inclusion of the continuum adds a contribution to the paring gap given by Δ_c (see the Appendix). However, the pairing gap does not appear as a new parameter in the BCS equations. Instead, it is combined with the usual (here called discrete) gap Δ_d into a single constant $\Delta = \Delta_d + \Delta_c$.

3 Continuum Complex-Energy Representation

The existence of resonances has been our motivation to include the single-particle level density. In the limit case that the resonances have a very long half-life (compared with the characteristic time of the system), they will show up as sharp peaks in $g_\nu(\varepsilon)$ for some particular quantum number ν. Such peaks can be parametrized using a Lorentzian distribution [33],

$$g_\nu(\varepsilon) = \begin{cases} \frac{1}{\pi} \frac{\Gamma_\nu/2}{(\varepsilon-\epsilon_\nu)^2-(\Gamma_\nu/2)^2} & \text{resonant states} \\ 0 & \text{non-resonant states} \end{cases} \tag{25}$$

3.1 *Berggren Representation*

By performing the analytic continuation of the BCS Eqs. (21) and (22) to the lower half of the complex-energy plane, and by applying the Cauchy theorem with the CSPL given in Eq. (25), we get

$$\frac{4}{G} = \sum_{jm} \frac{1}{\sqrt{(\varepsilon_j - \lambda)^2 + \Delta^2}} + \sum_{\nu m} \frac{1}{\sqrt{(\varepsilon_\nu - \lambda)^2 + \Delta^2}} + \int_\gamma d\varepsilon \, \frac{\sum_{\nu m} g_\nu(\varepsilon)}{\sqrt{(\varepsilon - \lambda)^2 + \Delta^2}} \tag{26}$$

$$N = \sum_{jm} v_j^2 + \sum_{\nu m} v_\nu^2 + \int_\gamma d\varepsilon \, v^2(\varepsilon) \left(\sum_{\nu m} g_\nu(\varepsilon) \right) \tag{27}$$

where $\varepsilon_\nu = \epsilon_\nu - i\,\frac{\Gamma_\nu}{2}$ and γ is a complex contour in the lower complex-energy plane that results from the deformation of the positive real-energy axis. The summation $\sum_\nu$ is done over the poles enclosed between the positive real-energy axis and the contour γ, while $\sum_m$ arises, as before, from the degeneracy of the state ν.

The first two terms of the BCS equations (26) and (27) correspond to the bound states. The second summations in Eqs. (26) and (27) correspond to the resonant (Gamow) states. The resonant states have complex energy and are solutions of the single-particle Hamiltonian h_i with purely outgoing boundary conditions. The single-particle representation formed by bound states ε_j, discrete complex-energy states ε_ν, and complex-energy scattering states $\varepsilon \in \gamma$ is called Berggren representation [26].

3.2 Pole Approximation

For very narrow resonances, the contribution from the complex contour (the third terms in Eqs. (26) and (27)) can be neglected. Within this approximation, we obtain a purely discrete representation for Eqs. (26) and (27) with each term being associated with the poles of the S-matrix,

$$\frac{2}{G} = \sum_n \frac{\Omega_n}{\sqrt{(\varepsilon_n - \lambda)^2 + \Delta^2}} \tag{28}$$

$$N = 2 \sum_n \Omega_n v_n^2 \tag{29}$$

where $n = \{j, \nu\}$, and $\Omega_n = (\sum_m 1)/2$, the pair degeneracy of the level n. This pole approximation makes $\frac{2}{G}$ and N complex, since now the last terms in Eqs. (26) and (27) are not present to cancel the complex-imaginary contribution from the complex resonant energies. However, if the imaginary parts of the resulting $\frac{2}{G}$ and N in Eqs. (26) and (27) are small compared to their real parts, one may consider that the pole approximation is acceptable. Otherwise, the results of the formalism cannot be accepted on physical grounds.

4 Conclusions

Correlations with the continuum part of the energy spectrum are important in open systems, e.g., in nuclei far from the stability line. Taking into account these correlations in many-body calculations is very hard, because the dimension rapidly increases with the number of particles and with the dimension of the model space. The pairing interaction, jointly with the continuum single-particle level density (CSPLD), overcomes these problems. In this work we have shown, in a heuristic manner, how to introduce the CSPLD in the non-conserving particle number solution of the pairing Hamiltonian. This model has been implemented in Ref. [13] to study many-nucleon systems close to the nuclear drip-line. Assuming that the resonances are sharp, we have provided a complex-energy representation of the model's solution. Our complex-energy representation of the solution has the advantage of isolating the resonant contribution from the non-resonant contribution of the continuum spectrum.

There are two ways in which this work can be expanded. First, although the ansatz used in this work seems plausible, it still needs a more rigorous mathematical approach. Second, in our approach the analytical continuation has been done in the same way as Berggren [26]. However, there are other ways of doing the analytic continuation (see, for example, review [34]) that would lead to expressions that are similar to, but different from, those in Eqs. (26)–(29). It would be interesting to obtain and compare them.

Appendix

In this section we will deduce the BCS gap equation. As argued in Sect. 2, we use the following expressions for the singular contractions:

$$a^{+}_{\nu m}(\varepsilon)\widehat{\,\,a^{+}_{\nu' m'}}(\varepsilon) = \delta_{\nu\nu'}\,\delta_{mm'}\,v^{2}_{\nu}(\varepsilon)\,g_{\nu}(\varepsilon) \tag{30}$$

$$a^{+}_{\nu m}(\varepsilon)\widehat{\,\,a^{+}_{\nu' \bar{m}'}}(\varepsilon) = \delta_{\nu\nu'}\,\delta_{m,-m'}\,(-)^{j-m}\,u_{\nu}(\varepsilon)\,v_{\nu}(\varepsilon)\,g_{\nu}(\varepsilon) \tag{31}$$

$$a_{\nu m}(\varepsilon)\widehat{\,\,a_{\nu' \bar{m}'}}(\varepsilon) = \delta_{\nu\nu'}\,\delta_{m,-m'}\,(-)^{j-m}\,u_{\nu}(\varepsilon)\,v_{\nu}(\varepsilon)\,g_{\nu}(\varepsilon) \tag{32}$$

where we have substituted $\delta(\varepsilon-\varepsilon)$ by $g_{\nu}(\varepsilon)$.

The diagonal part of the BCS Hamiltonian reads,

$$\begin{aligned}
H_{sp} - \lambda\hat{N} = &\sum_{jm}(\varepsilon_j-\lambda)v_j^2 + \sum_{\nu m}\int d\varepsilon(\varepsilon-\lambda)v_{\nu}^2(\varepsilon)g(\varepsilon)\\
&+\sum_{jm}(\varepsilon_j-\lambda)(u_j^2-v_j^2)\alpha^{+}_{jm}\alpha_{jm}\\
&+\sum_{\nu m}\int d\varepsilon(\varepsilon-\lambda)\left[u_{\nu}^2(\varepsilon)-v_{\nu}^2(\varepsilon)\right]\alpha^{+}_{\nu m}(\varepsilon)\alpha_{\nu m}(\varepsilon)\\
&+\sum_{jm}(\varepsilon_j-\lambda)(-)^{j-m}u_jv_j\left(\alpha^{+}_{jm}\alpha^{+}_{j\bar{m}}+hc\right)\\
&+\sum_{\nu m}\int d\varepsilon(\varepsilon-\lambda)(-)^{j_{\nu}-m}u_{\nu}(\varepsilon)v_{\nu}(\varepsilon)\left[\alpha^{+}_{\nu m}(\varepsilon)\alpha^{+}_{\nu\bar{m}}(\varepsilon)+hc\right]
\end{aligned} \tag{33}$$

For the pairing interaction we have

$$V_P = -GP_d^{+}P_d - GP_d^{+}P_c - GP_c^{+}P_d - GP_c^{+}P_c \tag{34}$$

where $P_d^{+} = \sum_j A_j^{+}$ and $P_c^{+} = \int_0^{\infty} d\varepsilon\; A^{+}(\varepsilon)$. Thus,

$$\begin{aligned}
-GP_d^{+}P_d = &-\frac{\Delta_d^2}{G} + 2\Delta_d\sum_{jm}u_jv_j\alpha^{+}_{jm}\alpha_{jm}\\
&-\frac{\Delta_d}{2}\sum_{jm}(-)^{j-m}(u_j^2-v_j^2)\left(\alpha^{+}_{jm}\alpha^{+}_{j\bar{m}}+hc\right)\\
&-G(\text{terms with four }\alpha)
\end{aligned} \tag{35}$$

$$
\begin{aligned}
- GP_c^+ P_c = & -\frac{\Delta_c^2}{G} + 2\Delta_c \sum_{\nu m} \int d\varepsilon u_\nu(\varepsilon) v_\nu(\varepsilon) \alpha_{\nu m}^+(\varepsilon) \alpha_{\nu m}(\varepsilon) \\
& -\frac{\Delta_c}{2} \sum_{\nu m} \int d\varepsilon (-)^{j_\nu - m} \left[u_\nu^2(\varepsilon) - v_\nu^2(\varepsilon) \right] \left[\alpha_{\nu m}^+(\varepsilon) \alpha_{\nu \bar{m}}^+(\varepsilon) + hc \right] \\
& -G(\text{terms with four } \alpha)
\end{aligned} \tag{36}
$$

$$
\begin{aligned}
-G(P_d^+ P_c + P_c^+ P_d) = & -2\frac{\Delta_d \Delta_c}{G} + 2\Delta_c \sum_{jm} u_j v_j \alpha_{jm}^+ \alpha_{jm} \\
& +2\Delta_d \sum_{\nu m} \int d\varepsilon u_\nu(\varepsilon) v_\nu(\varepsilon) \alpha_{\nu m}^+(\varepsilon) \alpha_{\nu m}(\varepsilon) \\
& -\frac{\Delta_d}{2} \sum_{\nu m} \int d\varepsilon (-)^{j_\nu - m} \left[u_\nu^2(\varepsilon) - v_\nu^2(\varepsilon)) \right] \left[\alpha_{\nu m}^+(\varepsilon) \alpha_{\nu \bar{m}}^+(\varepsilon) + hc \right] \\
& -\frac{\Delta_c}{2} \sum_{jm} (-)^{j-m} (u_j^2 - v_j^2) \left(\alpha_{jm}^+ \alpha_{j\bar{m}}^+ + hc \right) \\
& -G(\text{terms with four } \alpha)
\end{aligned} \tag{37}
$$

where we have introduced the *discrete* Δ_d and the *continuum* Δ_c gaps,

$$
\Delta_d = \frac{G}{2} \sum_{jm} u_j v_j \tag{38}
$$

$$
\Delta_c = \frac{G}{2} \sum_{\nu m} \int d\varepsilon u_\nu(\varepsilon) v_\nu(\varepsilon) g_\nu(\varepsilon) \tag{39}
$$

Then, the pairing interaction reads

$$
\begin{aligned}
V_P = & -\frac{\Delta^2}{G} + 2\Delta \sum_{jm} u_j v_j \alpha_{jm}^+ \alpha_{jm} + 2\Delta \sum_{\nu m} \int d\varepsilon u_\nu(\varepsilon) v_\nu(\varepsilon) \alpha_{\nu m}^+(\varepsilon) \alpha_{\nu m}(\varepsilon) \\
& -\frac{\Delta}{2} \sum_{jm} (-)^{j-m} (u_j^2 - v_j^2) \left(\alpha_{jm}^+ \alpha_{j\bar{m}}^+ + hc \right) \\
& -\frac{\Delta}{2} \sum_{\nu m} \int d\varepsilon (-)^{j_\nu - m} \left[u_\nu^2(\varepsilon) - v_\nu^2(\varepsilon)) \right] \left[\alpha_{\nu m}^+(\varepsilon) \alpha_{\nu \bar{m}}^+(\varepsilon) + hc \right] \\
& -G(\text{terms with four } \alpha)
\end{aligned} \tag{40}
$$

Notice that the unknown discrete and continuum gap do not appear separately but in the combination $\Delta = \Delta_d + \Delta_c$.

By adding Eqs. (33) and (40), we get the expression of the BCS Hamiltonian $H_{BCS} = H_{sp} + V_P - \lambda\hat{N}$ in terms of the quasi-particle operators. By eliminating the "dangerous" terms [29] that contain $\alpha^+_{nm}\alpha^+_{n\bar{m}} + hc$ with $n = j$ and $n = \nu(\varepsilon)$, we get the following two equations:

$$-\frac{\Delta}{2}(u_j^2 - v_j^2) + (\varepsilon_j - \lambda)u_j v_j = 0 \tag{41}$$

$$-\frac{\Delta}{2}\left[u_\nu^2(\varepsilon) - v_\nu^2(\varepsilon)\right] + (\varepsilon - \lambda)u_\nu(\varepsilon)v_\nu(\varepsilon) = 0 \tag{42}$$

which, together with the normalization conditions $u_j^2 + v_j^2 = 1$ and $u_\nu^2(\varepsilon) + v_\nu^2(\varepsilon) = 1$, yield

$$v_j^2 = \frac{1}{2}\left[1 - \frac{(\varepsilon_j - \lambda)}{\sqrt{(\varepsilon_j - \lambda)^2 + \Delta^2}}\right] \tag{43}$$

$$v_\nu^2(\varepsilon) = \frac{1}{2}\left[1 - \frac{(\varepsilon - \lambda)}{\sqrt{(\varepsilon - \lambda)^2 + \Delta^2}}\right] \tag{44}$$

Eq. (44) shows that the occupation probability in the continuum does not depend on the quantum number ν, i.e., $v_\nu(\varepsilon) = v(\varepsilon)$. The coefficients u_j^2 and $u^2(\varepsilon)$ are obtained from the normalization condition. Substituting Eqs. (43) and (44) into Eqs. (41) and (42) we get

$$u_j v_j = \frac{\Delta}{2\sqrt{(\varepsilon_j - \lambda)^2 + \Delta^2}} \tag{45}$$

$$u(\varepsilon)v(\varepsilon) = \frac{\Delta}{2\sqrt{(\varepsilon - \lambda)^2 + \Delta^2}} \tag{46}$$

which can be used to calculate the gap parameter,

$$\Delta = \Delta_d + \Delta_c \tag{47}$$

$$= \frac{G}{2}\sum_{jm} u_j v_j + \frac{G}{2}\sum_{\nu m}\int_0^\infty d\varepsilon\, u(\varepsilon)v(\varepsilon)g_\nu(\varepsilon) \tag{48}$$

The above equation reduces to the so-called gap equation,

$$\frac{4}{G} = \sum_{jm}\frac{1}{E_j} + \sum_{\nu m}\int_0^\infty d\varepsilon\, \frac{g_\nu(\varepsilon)}{E(\varepsilon)} \tag{49}$$

where we have introduced the quasi-particle energies, $E_j = \sqrt{(\varepsilon_j - \lambda)^2 + \Delta^2}$ and $E(\varepsilon) = \sqrt{(\varepsilon - \lambda)^2 + \Delta^2}$.

References

1. J. Okołowicz, M. Płoszajczak, I. Rotter, Phys. Rep. **374**, 271 (2003)
2. A. Volya, V. Zelevinsky, Phys. Rev. C **74**, 064314 (2006)
3. R. Id Betan, R.J. Liotta, N. Sandulescu, T. Vertse, Phys. Rev. Lett. **89**, 042501 (2002)
4. N. Michel, W. Nazarewicz, M. Płoszajczak, K. Bennaceur, Phys. Rev. Lett. **89**, 042502 (2002)
5. J. Dobaczewski, W. Nazarewicz, T.R. Werner, J.F. Berger, C.R. Chinn, J. Decharge, Phys. Rev. C **53**, 02809 (1996)
6. K. Bennaceur, J. Dobaczewski, M. Płoszajczak, Phys. Rev. C **60**, 034308 (1999)
7. S.A. Fayans, S.V. Tolokonnikov, D. Zawischa, Physics Letters B **491**, 245 (2000)
8. M. Grasso, N. Sandulescu, N. Van Giai, R.J. Liotta, Phys. Rev. C **64**, 064321 (2001)
9. N. Sandulescu, R.J. Liotta, R. Wyss, Phys. Lett. B **394**, 6 (1997)
10. N. Sandulescu, O. Civitarese, R.J. Liotta, Phys. Rev. C **61**, 044317 (2000)
11. N. Sandulescu, N. Van Giai, R.J. Liotta, Phys. Rev. C **61**, 061301(R) (2000)
12. A.T. Kruppa, P.H. Heenen, R.J. Liotta, Phys. Rev. C **63**, 044324 (2001)
13. R. M. Id Betan, Nucl. Phys. **879**, 14 (2012)
14. E. Beth, G. Uhlenbeck, Physica **4**, 915 (1937)
15. U. Mosel, Physcis Letters B **46**, 8 (1973).
16. W.A. Fowler, C.A. Engelbrech, Astrophys. J. **226**, 984 (1978).
17. P. Bonche, S. Levit, D. Vautherin, Nucl. Phys. A **427**, 278 (1984)
18. D.R. Dean, U. Mozel, Z. Phys. A **322**, 647 (1985)
19. R.J. Charity, L.G. Sobotka, Phys. Rev. C **71**, 024310 (2005)
20. R.M. Id Betan, Phys. Rev. C **85**, 064309 (2012)
21. R.M. Id Betan, C.E. Repetto, Nucl. Phys. A **960**, 131 (2017)
22. R.M. Id Betan, Nucl. Phys. A **959**, 147 (2017)
23. R. M. Id Betan, IOP Conf. Ser.: J. Phys.: Conf. Ser. **839**, 012003 (2017)
24. R.W. Richardson, Phys. Lett. **3**, 277 (1963)
25. R.W. Richardson, N. Sherman, Nucl. Phys. **52**, 221 (1964)
26. T. Berggren, Nucl. Phys. A **109**, 265 (1968)
27. A.M. Lane, *Nuclear Theory. Pairing Force Correlations and Collective Motion* (W. A. Benjamin, Inc., New York, 1964)
28. J. Bardeen, J.N. Cooper, J.R. Schrieffer, Phys. Rev. **106**, 162 (1957)
29. N.N. Bogolyubov, Sov. Phys. JETP **7**, 41 (1958)
30. R. de la Madrid, Nucl. Phys. A **940**, 297 (2015)
31. R.M. Id Betan, R. de la Madrid, Nucl. Phys. A **970** 398 (2018)
32. A.L. Fetter, J.D. Walecka, *Quantum Theory of Many-Particle Systems* (McGraw-Hill Book Company, New York, 1971)
33. V.I. Kukulin, V.M. Krasnolposky, J. Horacek, *Theory of Resonances* (Kluwer Academic Publishers, Dordrecht, 1988)
34. R. de la Madrid, G. Garcia-Calderon, J.G. Muga, Czech. J. Phys. **55**, 1141 (2005)

Jacobi Polynomials as $su(2,2)$ Unitary Irreducible Representation

Enrico Celeghini, Mariano A. del Olmo, and Miguel A. Velasco

Abstract An infinite-dimensional irreducible representation of $su(2,2)$ is explicitly constructed in terms of ladder operators for the Jacobi polynomials $J_n^{(\alpha,\beta)}(x)$ and the Wigner d_j-matrices where the integer and half-integer spins $j := n + (\alpha + \beta)/2$ are considered together. The 15 generators of this irreducible representation are realized in terms of zero or first order differential operators and the algebraic and analytical structure of operators of physical interest discussed.

Keywords Jacobi polynomials · Lie algebras · Irreducible representations · Wigner matrices · Operators on special functions

PACS 02.20.Sv, 02.30Gp, 03.65Db

Mathematics Subject Classification (2000) 17B15, 17B81, 33C45

1 Introduction

The classification of the functions that can be defined "special," where "special" means something more than "useful," is an open problem [1].

E. Celeghini
Dpto di Fisica, Università di Firenze and INFN–Sezione di Firenze, Firenze, Italy

Dpto de Física Teórica and IMUVA, Univ. de Valladolid, Valladolid, Spain
e-mail: celeghini@fi.infn.it

M. A. del Olmo (✉)
Dpto de Física Teórica and IMUVA, Univ. de Valladolid, Valladolid, Spain
e-mail: marianoantonio.olmo@uva.es

M. A. Velasco
Departamento de Física Teórica, Atómica y Óptica, Universidad de Valladolid, Valladolid, Spain

Ş. Kuru et al. (eds.), *Integrability, Supersymmetry and Coherent States*, CRM Series in Mathematical Physics, https://doi.org/10.1007/978-3-030-20087-9_10

The actual main line of work for a possible unified theory of special functions is the Askey scheme that is based on the analytical theory of linear differential equations [2–4].

A possible scheme, different from the Askey one, seems to emerge in these last years by means of a generalization of the classical special functions, principally related to the introduction of d-orthogonal polynomials by means of difference equations, q-polynomials, and exceptional polynomials [5–15].

We follow here a point of view closely related to a field of mathematics seemingly quite far from special functions: Lie algebras. It is an idea first introduced by Wigner [16] and Talman [17] and later developed mainly by Miller [18] and Vilenkin and Klimyk [19–21].

However, our approach starts from well-established concepts, the "old style" orthogonal polynomials and looks for possible connections with the "old style" Lie group theory. Thus in this paper, as Jacobi polynomials have three parameters we simply attempt to relate them with a Lie algebra of rank three.

While other researches are focused on the general relations between special functions and Lie algebras we consider a further step connecting special functions and irreducible representations (IR) of Lie algebras. This restriction of the Lie counterpart that has quite more properties of the abstract algebra gives a lot of additional information on the special functions [22, 23].

Starting from the seminal work by Truesdell [24], where a sub-class of special functions was defined by means of a set of formal properties, we propose indeed a possible definition of a fundamental sub-class of special functions that we call "algebraic special functions" (ASF).

These ASF are related to the hypergeometric functions but they are constructed from the following algebraic assumptions:

1. A set of differential recurrence relations exists on these ASF that can be associated with a set of operators that span a Lie algebra.
2. These ASF support a characteristic IR of this algebra.
3. A vector space can be constructed on these ASF where the ladder operators have all the appropriate properties for realizing this IR of the associated Lie algebra.
4. The differential equations that define the ASF are related to the diagonal elements of the universal enveloping algebra (UEA) and, in particular, to the Casimir invariants of the whole algebra and subalgebras.

From these assumptions, we have that:

1. The exponential maps of the algebra define the associated group and allow to obtain from the ASF other different sets of functions. If the transformation is unitary, another algebraically equivalent basis of the space is thus obtained. When the transformations are not unitary, as in the case of coherent states, sets with different properties are found (like overcomplete sets).
2. The vector space of the operators acting on the L^2-space of functions is isomorphic to the UEA built on the algebra.

The starting point of our work has been the paradigmatic example of Hermite functions that are a basis on the Hilbert space of the square integrable functions defined on the configuration space $\mathbb{R}$. As it is well known from the algebraic discussion of the harmonic oscillator, besides the continuous basis $\{|x\rangle\}_{x\in\mathbb{R}}$ determined by the configuration space, a discrete basis $\{|n\rangle\}_{n\in\mathbb{N}}$—related to the Weyl–Heisenberg algebra $h(1)$—can be introduced such that Hermite functions are the transition matrix elements from one basis to the other.

In previous papers we have presented the direct connection between some special functions and specific IRs of Lie algebras in cases where the Lie structure was smaller [25–28].

In this paper we discuss in detail the symmetries of the Jacobi functions introduced in [29]. The fact that a $su(2, 2)$ symmetry exists inside the hypergeometric functions ${}_2F_1$ [30, 31] is, of course, the starting point of our discussion.

This is a further confirmation of the line introduced in [25–27] in terms of the Jacobi polynomials that satisfy the required conditions 1–4 and thus deserves an additional analysis to that presented in [29]. As shown there, Jacobi polynomials indeed can be associated with well-defined "algebraic Jacobi functions" (AJF) that satisfy the preceding assumptions.

The AJF support an IR of $su(2, 2)$ (a real form of A_3) a Lie algebra of rank 3 related to the three parameters, $\{n, \alpha, \beta\}$, of the Jacobi polynomials $J_n^{(\alpha,\beta)}(x)$ and, alternatively, to the three parameters $\{j, m, q\}$ of the AJF. These two triplets of parameters are indeed belonging to the Cartan subalgebra of $su(2, 2)$.

The procedure consists in starting from well-known orthogonality conditions of the Jacobi polynomials and defines the orthonormal AJF. The recurrence relations of the Jacobi polynomials are then rewritten by means of differential operators acting on the AJF as ladder operators, whose explicit action remembers the operators $J_\pm$ of the $su(2)$ representation. In this way we obtain twelve non-diagonal operators that together with three Cartan (diagonal) operators close the Lie algebra $su(2, 2)$ in a well-defined IR of AJF. All this analysis can also be transferred to the d_j-Wigner matrices [32].

From the Lie algebra point of view for both, AJF and Wigner d_j–matrices, the relevant algebraic chains are $su(2, 2) \supset su(2) \otimes su(2) \supset su(2)$ to consider together integer and half-integer spin j and $su(2, 2) \supset su(1, 1)$ to describe separately bosons and fermions.

The paper is organized as follows. Section 2 is devoted to recall the main properties of the AJF relevant for our discussion and their relations with the Wigner d_j-matrices. In Sect. 3 we study the symmetries of the AJF that keep invariant the principal parameter j changing only m and/or q. We thus construct the ladder operators that determine a $su(2) \oplus su(2)$ algebra and allow to build up the irreducible representations defined by the same Casimir invariant of both $su(2)$, i.e., $su_j(2) \otimes su_j(2)$. In Sect. 4 we construct four new sets of ladder operators that change the three parameters j, m, and q adding to all of them $\pm 1/2$. Each of these sets generates a $su(1, 1)$ algebra to which ∞-many IRs of $su(1, 1)$—supported by the AJF and the d_j-matrices—are associated. In Sect. 5 we show that the ladder operators, obtained in the previous sections, span all together a $su(2,2)$ algebra and that both AJF and Wigner d_j-matrices are a basis of the IR of $su(2,2)$ (that is characterized by the eigenvalue $-3/2$ of the quadratic Casimir of $su(2,2)$). Finally some conclusions and comments are included.

2 Algebraic Jacobi Functions and Their Structure

The Jacobi polynomial of degree $n \in \mathbb{N}$, $J_n^{(\alpha,\beta)}(x)$, is defined in terms of the hypergeometric functions ${}_2F_1$ [33–35] by

$$J_n^{(\alpha,\beta)}(x) = \frac{(\alpha+1)_n}{n!} \; {}_2F_1\left[-n, 1+\alpha+\beta+n; \alpha+1; \frac{1-x}{2}\right], \tag{1}$$

where $(a)_n := a\,(a+1)\cdots(a+n-1)$ is the Pochhammer symbol.

Now we include an x-depending factor related to the integration measure of the Jacobi polynomials and we define—alternatively to $\{n,\alpha,\beta\}$—three other parameters $\{j,m,q\}$:

$$j := n + \frac{\alpha+\beta}{2}, \qquad m := \frac{\alpha+\beta}{2}, \qquad q := \frac{\alpha-\beta}{2},$$

such that

$$n = j - m, \qquad \alpha = m + q, \qquad \beta = m - q .$$

In order to obtain an algebra representation, as we will prove later, we have to impose the following restrictions for $\{j,m,q\}$:

$$j \geq |m|, \qquad j \geq |q|, \qquad 2j \in \mathbb{N}, \quad j-m \in \mathbb{N}, \quad j-q \in \mathbb{N}, \tag{2}$$

thus $\{j,m,q\}$ are all together integers or half-integers. The conditions (2) rewritten in terms of the original parameters $\{n,\alpha,\beta\}$ exhibit that they are all integers satisfying

$$n \in \mathbb{N}, \qquad \alpha,\beta \in \mathbb{Z}, \qquad \alpha \geq -n, \qquad \beta \geq -n, \qquad \alpha+\beta \geq -n.$$

We thus define

$$\hat{\mathcal{J}}_j^{m,q}(x) := \sqrt{\frac{\Gamma(j+m+1)\,\Gamma(j-m+1)}{\Gamma(j+q+1)\,\Gamma(j-q+1)}}$$

$$\times \left(\frac{1-x}{2}\right)^{\frac{m+q}{2}} \left(\frac{1+x}{2}\right)^{\frac{m-q}{2}} J_{j-m}^{(m+q,m-q)}(x) . \tag{3}$$

Note that usually the Jacobi polynomials $J_n^{(\alpha,\beta)}(x)$ are defined for $\alpha > -1$ and $\beta > -1$ ($\alpha,\ \beta \in \mathbb{R}$) in such a way that a unique weight function $w(x)$ allows their normalization. However (see also [36, p. 49]) we have to change such restrictions since the normalization inside the functions and their algebraic properties requires Eq. (2). So, in addition to integer or half-integer conditions, we have to restrict to $j \geq |m|$ in Eq. (3) ($\hat{\mathcal{J}}_j^{m,q}(x) = 0$ when $|q| > j \in \mathbb{N}/2$). This can be obtained assuming

$$\mathcal{J}_j^{m,q}(x) := \lim_{\varepsilon \to 0} \hat{\mathcal{J}}_{j+\varepsilon}^{m,q}(x)$$

indeed

$$\mathcal{J}_j^{m,q}(x) = \begin{cases} \hat{\mathcal{J}}_j^{m,q}(x)\ \forall \{j, m, q\} & \text{verifying all conditions (2)} \\ 0 & \text{otherwise} \end{cases}. \tag{4}$$

In conclusion, the basic objects of this paper that we call "algebraic Jacobi functions" (AJF) have the final form (4).

The AJF (4) reveal additional symmetries hidden inside the Jacobi polynomials. Indeed we have

$$\begin{aligned} \mathcal{J}_j^{m,q}(x) &= \mathcal{J}_j^{q,m}(x), \\ \mathcal{J}_j^{m,q}(x) &= (-1)^{j-m}\, \mathcal{J}_j^{m,-q}(-x), \\ \mathcal{J}_j^{m,q}(x) &= (-1)^{j-q}\, \mathcal{J}_j^{-m,q}(-x), \\ \mathcal{J}_j^{m,q}(x) &= (-1)^{m+q}\, \mathcal{J}_j^{-m,-q}(x)\,. \end{aligned} \tag{5}$$

The proof of these properties is straightforward. The first one can be proved taking into account the following property of the Jacobi polynomials for integer coefficients (n, α, β) [36]:

$$J_n^{\alpha,\beta}(x) = \frac{(n+\alpha)!\,(n+\beta)!}{n!\,(n+\alpha+\beta)!} \left(\frac{x+1}{2}\right)^{-\beta} J_{n+\beta}^{\alpha,-\beta}(x)\,,$$

while the second relation can be derived from the well-known symmetry of the Jacobi polynomials [33]

$$J_n^{(\alpha,\beta)}(x) = (-1)^n\, J_n^{(\beta,\alpha)}(-x), \tag{6}$$

and the last two properties can be proved using the first two ones.

The AJF for m and q fixed verify the orthonormality relation

$$\int_{-1}^{1} \mathcal{J}_j^{m,q}(x)\,(j+1/2)\,\mathcal{J}_{j'}^{m,q}(x)\,dx = \delta_{j\,j'} \tag{7}$$

as well as the completeness relation

$$\sum_{j=\sup(|m|,|q|)}^{\infty} \mathcal{J}_j^{m,q}(x)\ (j+1/2)\ \mathcal{J}_j^{m,q}(y) = \delta(x-y). \tag{8}$$

Both relations are similar to those of the Legendre polynomials [25] and the associated Legendre polynomials [26]: all are orthonormal up to the factor $j+1/2$. These relations allow us to state that $\{\mathcal{J}_j^{m,q}(x);\ m,q \text{ fixed}\}_{j=\sup(|m|,|q|)}^{\infty}$ is a basis in the space of square integrable functions defined in $\mathbb{E}=[-1,1]$. Considering

$$\mathbb{E}\times\mathbb{Z}\times\mathbb{Z}/2 := \bigcup_{m-q\in\mathbb{Z}}\ \bigcup_{q\in\mathbb{Z}/2} \mathbb{E}_{m,q}\,,$$

where $\mathbb{E}_{m,q}$ is the configuration space $\mathbb{E}=[-1,1]$ with m and q fixed and $\mathbb{Z}\times\mathbb{Z}/2$ is related to the set of pairs (m,q) with m and q both integer or half-integer, then $\{\mathcal{J}_j^{m,q}(x)\}$ is a basis of $L^2(\mathbb{E},\mathbb{Z},\mathbb{Z}/2)$ [29].

The Jacobi equation

$$E_n^{(\alpha,\beta)}\, J_n^{(\alpha,\beta)}(x) = 0\,,$$

where

$$E_n^{(\alpha,\beta)} \equiv (1-x^2)\frac{d^2}{dx^2} - ((\alpha+\beta+2)x + (\alpha-\beta))\,\frac{d}{dx} + n(n+\alpha+\beta+1)\,,$$

rewritten in terms of these new functions $\mathcal{J}_j^{m,q}(x)$ and of the new parameters $\{j,m,q\}$ becomes

$$\mathcal{E}_j^{m,q}\ \mathcal{J}_j^{m,q}(x) = 0\,, \tag{9}$$

with

$$\mathcal{E}_j^{m,q} \equiv -\left(1-x^2\right)\frac{d^2}{dx^2} + 2x\,\frac{d}{dx} + \frac{2\,m\,q\,x + m^2 + q^2}{1-x^2} - j(j+1)\,, \tag{10}$$

where the symmetry under the interchange between m and q is evident.

It is worth noticing that the AJFs (4), with the substitution $x=\cos\beta$ with $0\le\beta\le\pi$, are essentially the Wigner d_j rotation matrices [32, 36]

$$d^j(\beta)_q^m = \sqrt{\frac{(j+m)!(j-m)!}{(j+q)!(j-q)!}}\left(\sin\frac{\beta}{2}\right)^{q-m}\left(\cos\frac{\beta}{2}\right)^{m+q} J_{j-m}^{(m-q,m+q)}(\cos\beta)$$

that verify the conditions (2). The explicit relation between them is

$$d^j(\beta)_q^m = \mathcal{J}_j^{m,-q}(\cos\beta). \tag{11}$$

Equation (5) are equivalent to the well-known relations among the $d^j(\beta)^m_q$, for instance,

$$d^j(\beta)^q_m = (-1)^{q-m}\, d^j(\beta)^m_q .$$

The starting point for finding the algebra representation of the AJF is now the construction of the rising/lowering differential applications [18] that change the labels $\{j, m, q\}$ of the AJF by 0 or 1/2. The fundamental limitation of the analytical approach [16–21] is that the indices are considered as parameters that, in iterated applications, must be introduced by hand. This problem has been solved in [25] where a consistent vector space framework was introduced to allow the iterated use of recurrence formulas by means of operators of which the parameters involved are eigenvalues.

Indeed—in order to realize the needed operator structure on the set $\{\mathcal{J}_j^{m,q}(x)\}$—we introduce not only the operators X and D_x of the configuration space :

$$X\, f(x) = x\, f(x), \qquad D_x\, f(x) = f'(x),$$

but also three other operators J, M, and Q such that

$$(J,\ M,\ Q)\ :\ \mathcal{J}_j^{m,q}(x) \to\ (j,\ m,\ q)\, \mathcal{J}_j^{m,q}(x), \tag{12}$$

that are diagonal on the AJF and, thus, belong—in the algebraic scheme—to the Cartan subalgebra.

3 Algebra Representations for $\Delta j = 0$

We start from the differential–difference applications verified by the Jacobi functions (a complete list of which can be found in Refs. [33–35]). The procedure is laborious, so that, we only sketch the simplest case with $\Delta j = 0$, related to $su(2)$ and well known for the d_j in terms of the angle [37].

Let us start from the operators that change the values of m only. The relations [33]

$$\frac{d}{dx} J_n^{(\alpha,\beta)}(x) = \frac{1}{2}(n+\alpha+\beta+1)\, J_{n-1}^{(\alpha+1,\beta+1)}(x)\ ,$$

$$\frac{d}{dx}\Big[(1-x)^{\alpha}(1+x)^{\beta} J_n^{(\alpha,\beta)}(x)\Big] = -2(n+1)(1-x)^{\alpha-1}(1+x)^{\beta-1} J_{n+1}^{(\alpha-1,\beta-1)}(x)$$

allow us to define the operators

$$A_{\pm} := \pm\sqrt{1-X^2}\, D_x + \frac{1}{\sqrt{1-X^2}}\,(XM+Q), \tag{13}$$

that act on the algebraic Jacobi functions $\mathcal{J}_j^{m,q}(x)$ as

$$A_\pm\, \mathcal{J}_j^{m,q}(x) = \sqrt{(j \mp m)\,(j \pm m + 1)}\;\, \mathcal{J}_j^{m\pm 1,\,q}(x). \tag{14}$$

The operators (13) are a generalization for $Q \neq 0$ of the operators $J_\pm$ introduced in [26] for the associated Legendre functions related to the AJF with $q = 0$. Indeed Eq. (14) that are independent from q coincide with Eqs. (2.11) and (2.12) of Ref. [26].

Defining now $A_3 := M$ and taking into account the action of the operators $A_\pm$ and A_3 on the AJFs, Eqs. (14) and (12), it is easy to check that $A_\pm$ and A_3 close a $su(2)$ algebra that commutes with J and Q, denoted in the following by $su_A(2)$:

$$[A_3, A_\pm] = \pm A_\pm \qquad [A_+, A_-] = 2A_3.$$

Thus, the AJFs $\{\mathcal{J}_j^{m,q}(x)\}$, with j and q fixed such that $2j \in \mathbb{N}$, $j - m \in \mathbb{N}$ and $-j \leq m \leq j$, support the $(2j+1)$-dimensional IR of the Lie algebra $su_A(2)$ independent from the value of q.

Similarly to [26], starting from the differential realization (13) of the $A_\pm$ operators, the Jacobi differential equation (9) is shown to be equivalent to the Casimir equation of $su_A(2)$

$$[\mathcal{C}_A - J(J+1)]\; \mathcal{J}_j^{m,q}(x) \equiv \left[A_3^2 + \frac{1}{2}\{A_+, A_-\} - J(J+1)\right] \mathcal{J}_j^{m,q}(x) = 0\,.$$

Indeed, this equation reproduces the operatorial form of (9), i.e., it gives

$$\mathcal{E}_J^{M,Q} \equiv -(1-X^2)D_x^2 + 2XD_x + \frac{1}{1-X^2}(2XMQ + M^2 + Q^2) - J(J+1)\,. \tag{15}$$

On the other hand, we can make use of the factorization method [38–40], relating second order differential equations to product of first order ladder operators in such a way that the application of the first operator modifies the values of the parameters of the second one. Taking into account this fact, iterated application of (13) gives the two equations

$$\begin{aligned} &[A_+\,A_- - (J+M)\,(J-M+1)]\;\, \mathcal{J}_j^{m,q}(x) = 0\,, \\ &[A_-\,A_+ - (J-M)\,(J+M+1)]\;\, \mathcal{J}_j^{m,q}(x) = 0\,, \end{aligned} \tag{16}$$

that reproduce again the operator form of the Jacobi equation (9). These are particular cases of a general property: the defining Jacobi equation can be recovered applying to $\mathcal{J}_j^{m,q}$ the Casimir operator of any involved algebra and subalgebra as well as any diagonal product of ladder operators.

Now, using the symmetry under the interchange of the labels m and q of the AJF (see first relation of (5)), we construct the algebra of operators that changes q

leaving j and m unchanged. From $A_\pm$ two new operators $B_\pm$ are thus defined

$$B_\pm := \pm\sqrt{1-X^2}\,D_x + \frac{1}{\sqrt{1-X^2}}\,(XQ+M)\,, \tag{17}$$

and their action on the AJF is

$$B_\pm\,\mathcal{J}_j^{m,q}(x) = \sqrt{(j\mp q)\,(j\pm q+1)}\;\mathcal{J}_l^{m,q\pm 1}(x). \tag{18}$$

Obviously also the operators $B_\pm$ and $B_3 := Q$ close a $su(2)$ algebra we denote $su_B(2)$

$$[B_3, B_\pm] = \pm B_\pm \qquad [B_+, B_-] = 2B_3,$$

and the AJFs $\{\mathcal{J}_j^{m,q}(x)\}$, with j and m fixed such that $2j\in\mathbb{N}$, $j-q\in\mathbb{N}$ and $-j\le q\le j$, close the $(2j+1)$-dimensional IR of the Lie algebra $su(2)_B$ independent from the value of m.

Again we can recover the Jacobi equation (9) from the Casimir, $\mathcal{C}_B$, of $su_B(2)$

$$[\mathcal{C}_B - J(J+1)]\,\mathcal{J}_j^{m,q}(x) = \left[B_3^2 + \frac{1}{2}\{B_+, B_-\} - J(J+1)\right]\mathcal{J}_j^{m,q}(x) = 0\,.$$

A more complex algebraic scheme appears in common applications of the operators $A_\pm$ and $B_\pm$. As the operators $\{A_\pm, A_3\}$ commute with $\{B_\pm, B_3\}$, the algebraic structure is the direct sum of the two Lie algebras

$$su_A(2)\oplus su_B(2).$$

A new symmetry of the AJFs emerges in the space of $\mathcal{J}_j^{m,q}(x)$ when only j is fixed. Both for $\{j, m, q\}$, integer or half-integer (see Eqs. (14), (18) and (12)) we have the IR of the algebra $su(2)\oplus su(2)$

$$su_j(2)\oplus su_j(2)\,.$$

So that the AJFs $\{\mathcal{J}_j^{m,q}(x)\}$ for fixed j and $-j\le m\le j$, $-j\le q\le j$ determine the IR with $\mathcal{C}_A = \mathcal{C}_B = j(j+1)$. From (13) and (17), taking into account that always the operators M and Q have been written at the right of X and D_x, it can be shown that $A_\pm^\dagger = A_\mp$, $B_\pm^\dagger = B_\mp$ and the representation would be unitary with a suitable inner product. In Fig. 1 the action of the operators $A_\pm$, $B_\pm$ on the parameters $\{j, m, q\}$ that label the AJFs corresponds to the plane $\Delta j = 0$.

In conclusion, $\{\mathcal{J}_j^{m,q}(x)\}$ with j fixed is the basis of an IR of $su(2)\oplus su(2)$ of dimension $(2j+1)^2$ symmetrical under the interchange of A with B.

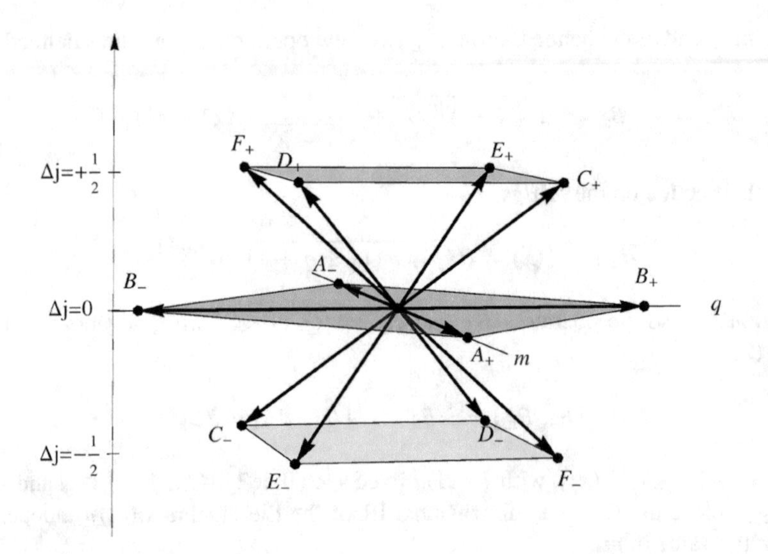

Fig. 1 Root diagram of $su(2,2)$. The coordinates displayed on the planes correspond to the pairs $\{m, q\}$, while the parameter Δj is represented in the vertical axis. The Cartan elements at the origin are not included

4 Other Ladder Operators Acting on AJF and $su(1,1)$ Representations

As we mentioned before there are many differential–difference relations between the Jacobi polynomials for different values of the parameters [33, 34]. Starting from them we construct a $su(2,2)$ representation supported by the AJF. The Lie algebra $su(2,2)$ has fifteen infinitesimal generators, where three of them are Cartan generators (for instance, J, M, and Q). As the four generators that commute with J (i.e., $A_\pm$ and $B_\pm$) have been introduced in the preceding paragraph, we have to construct eight non-diagonal operators more. They are

$$\begin{aligned}
C_\pm &:= \pm \frac{(1+X)\sqrt{1-X}}{\sqrt{2}}\, D_x - \frac{1}{\sqrt{2(1-X)}} \Big(X\,(J+\tfrac{1}{2}\pm\tfrac{1}{2}) - (J+\tfrac{1}{2}\pm\tfrac{1}{2}+M+Q)\Big),\\
D_\pm &:= \mp \frac{(1-X)\sqrt{1+X}}{\sqrt{2}}\, D_x + \frac{1}{\sqrt{2\,(1+X)}} \Big(X(J+\tfrac{1}{2}\pm\tfrac{1}{2}) + (J+\tfrac{1}{2}\pm\tfrac{1}{2}+M-Q)\Big),\\
E_\pm &:= \mp \frac{(1-X)\sqrt{1+X}}{\sqrt{2}}\, D_x + \frac{1}{\sqrt{2\,(1+X)}} \Big(X(J+\tfrac{1}{2}\pm\tfrac{1}{2}) + (J+\tfrac{1}{2}\pm\tfrac{1}{2}-M+Q)\Big),\\
F_\pm &:= \mp \frac{(1+X)\sqrt{1-X}}{\sqrt{2}}\, D_x + \frac{1}{\sqrt{2\,(1-X)}} \Big(X(J+\tfrac{1}{2}\pm\tfrac{1}{2}) - (J+\tfrac{1}{2}\pm\tfrac{1}{2}-M-Q)\Big).
\end{aligned} \tag{19}$$

All these differential operators act on the space $\{\mathcal{J}_j^{m,q}\}$ for $\{j, m, q\}$ integer and half-integer such that $j \geq |m|, |q|$. The explicit form of their action is

$$\begin{aligned}
C_\pm\, \mathcal{J}_j^{m,q}(x) &= \sqrt{\left(j+m+\tfrac{1}{2}\pm\tfrac{1}{2}\right)\left(j+q+\tfrac{1}{2}\pm\tfrac{1}{2}\right)}\; \mathcal{J}_{j\pm 1/2}^{m\pm 1/2,\, q\pm 1/2}(x),\\
D_\pm\, \mathcal{J}_j^{m,q}(x) &= \sqrt{\left(j+m+\tfrac{1}{2}\pm\tfrac{1}{2}\right)\left(j-q+\tfrac{1}{2}\pm\tfrac{1}{2}\right)}\; \mathcal{J}_{j\pm 1/2}^{m\pm 1/2,\, q\mp 1/2}(x)\\
E_\pm\, \mathcal{J}_j^{m,q}(x) &= \sqrt{\left(j-m+\tfrac{1}{2}\pm\tfrac{1}{2}\right)\left(j+q+\tfrac{1}{2}\pm\tfrac{1}{2}\right)},\; \mathcal{J}_{j\pm 1/2}^{m\mp 1/2,\, q\pm 1/2}(x),\\
F_\pm\, \mathcal{J}_j^{m,q}(x) &= \sqrt{\left(j-m+\tfrac{1}{2}\pm\tfrac{1}{2}\right)\left(j-q+\tfrac{1}{2}\pm\tfrac{1}{2}\right)}\; \mathcal{J}_{j\pm 1/2}^{m\mp 1/2,\, q\mp 1/2}(x).
\end{aligned} \tag{20}$$

From (19) or (20) we have

$$C_\pm^\dagger = C_\mp, \qquad D_\pm^\dagger = D_\mp, \qquad E_\pm^\dagger = E_\mp, \qquad F_\pm^\dagger = F_\mp,$$

i.e., all these rising/lowering operators could have the hermiticity properties required by the representation to be unitary. The operators (19) change all parameters by $\pm 1/2$, so that in Fig. 1 they correspond to the planes $\Delta j = \pm 1/2$. In [29] also quadratic forms of operators (19) that change the parameters in $(\pm 1, 0)$ instead of $\pm 1/2$ have been considered.

From Eq. (19) it is easily stated that

$$\begin{aligned}
D_\pm(X, D_x, M, Q) &= C_\pm(-X, -D_x, M, -Q),\\
E_\pm(X, D_x, M, Q) &= C_\pm(-X, -D_x, -M, Q),\\
F_\pm(X, D_x, M, Q) &= -C_\pm(X, D_x, -M, -Q).
\end{aligned} \tag{21}$$

Thus, because of the Weyl symmetry of the roots, we limit ourselves to discuss the operators $C_\pm$. Taking thus into account their action on the Jacobi functions we get

$$[C_+, C_-] = -2C_3, \qquad [C_3, C_\pm] = \pm C_\pm \tag{22}$$

where

$$C_3 := J + \frac{1}{2}(M+Q) + \frac{1}{2}. \tag{23}$$

Hence $\{C_\pm, C_3\}$ close a $su(1,1)$ algebra we can denote $su_C(1,1)$.

As in the cases of the operators $A_\pm$ and $B_\pm$, we obtain the Jacobi differential equation from the Casimir $\mathcal{C}_C$ of $su_C(1,1)$, written in terms of (19) and (23),

$$\mathcal{C}_C\, \mathcal{J}_j^{m,q}(x) \equiv \left[C_3^2 - \frac{1}{2}\{C_+, C_-\}\right] \mathcal{J}_j^{m,q}(x) = \frac{1}{4}\left[(m+q)^2 - 1\right] \mathcal{J}_j^{m,q}(x).$$

Indeed

$$\left[\mathcal{C}_C - \frac{1}{4}(M+Q)^2 + \frac{1}{4}\right] \mathcal{J}_j^{m,q}(x)$$
$$\equiv \left[C_3^2 - \tfrac{1}{2}\{C_+, C_-\} - \tfrac{1}{4}(M+Q)^2 + 1/4\right] \mathcal{J}_j^{m,q}(x) = 0 \tag{24}$$

allows us to recover the Jacobi equation (9). Analogously the same result derives from eqs.

$$[C_+ C_- - (J+M)(J+Q)]\ \mathcal{J}_j^{m,q}(x) = 0,$$
$$[C_- C_+ - (J+1+M)(J+1+Q)]\ \mathcal{J}_j^{m,q}(x) = 0, \tag{25}$$

obtained by the factorization method.

From (24) we see that since $(m+q) = 0, \pm 1, \pm 2, \pm 3, \cdots$ the unitary IRs of $su(1,1)$ with $\mathcal{C}_C = (m+q)^2/4 - 1/4 = -1/4, 0, 3/4, 2, 15/4, \cdots$ are obtained. Hence, the set of AJF supports infinite unitary IRs of the discrete series of $su_C(1,1)$ [41].

Similar results can be found for the other ladder operators $D\pm$, $E\pm$, $F\pm$, up to an eventual multiplicative factor, with the substitutions (21) in all Eqs. (22)–(25).

5 The AJF Representation of $su(2,2)$

To obtain the root system of the simple Lie algebra A_3 (that has $su(2,2)$ as one of its real forms) we have only simply to add to Fig. 1 the three points in the origin corresponding to the elements J, M, and Q of the Cartan subalgebra.

The commutators of the generators $A_\pm$, $B_\pm$, $C_\pm$, $D_\pm$, $E_\pm$, $F_\pm$, J, M, Q are

$$[J, A_\pm] = 0, \quad [J, M] = 0, \quad [J, B_\pm] = 0, \quad [J, Q] = 0,$$

$$[J, C_\pm] = \pm\tfrac{C_\pm}{2}, \quad [J, D_\pm] = \pm\tfrac{D_\pm}{2}, \quad [J, E_\pm] = \pm\tfrac{E_\pm}{2}, \quad [J, F_\pm] = \pm\tfrac{F_\pm}{2},$$

$$[M, B_\pm] = 0, \quad [M, Q] = 0,$$

$$[M, C_\pm] = \pm\tfrac{C_\pm}{2}, \quad [M, D_\pm] = \pm\tfrac{D_\pm}{2}, \quad [M, E_\pm] = \mp\tfrac{E_\pm}{2}, \quad [M, F_\pm] = \mp\tfrac{F_\pm}{2},$$

$$[Q, A_\pm] = 0,$$

$$[Q, C_\pm] = \pm\tfrac{C_\pm}{2}, \quad [Q, D_\pm] = \mp\tfrac{D_\pm}{2}, \quad [Q, E_\pm] = \pm\tfrac{E_\pm}{2}, \quad [Q, F_\pm] = \mp\tfrac{F_\pm}{2},$$

$$[A_+, A_-] = 2A_3, \quad [A_3, A_\pm] = \pm A_\pm, \quad (A_3 = M),$$

$$[B_+, B_-] = 2B_3, \quad [B_3, B_\pm] = \pm B_\pm, \quad (B_3 = Q),$$

$$[C_+, C_-] = -2C_3, \quad [C_3, C_\pm] = \pm C_\pm, \quad (C_3 = J + \tfrac{1}{2}(M + Q) + \tfrac{1}{2}),$$

$$[D_+, D_-] = -2D_3, \quad [D_3, D_\pm] = \pm D_\pm, \quad (D_3 = J + \tfrac{1}{2}(M - Q) + \tfrac{1}{2}),$$

$$[E_+, E_-] = -2E_3, \quad [E_3, E_\pm] = \pm E_\pm, \quad (E_3 = J + \tfrac{1}{2}(-M + Q) + \tfrac{1}{2}),$$

$$[F_+, F_-] = -2F_3, \quad [F_3, F_\pm] = \pm F_\pm, \quad (F_3 = J - \tfrac{1}{2}(M + Q) + \tfrac{1}{2}),$$

$$[A_\pm, B_\pm] = 0, \quad [A_\pm, B_\mp] = 0,$$

$$[A_\pm, C_\pm] = 0, \quad [A_\pm, C_\mp] = \pm E_\mp, \quad [A_\pm, D_\pm] = 0, \quad [A_\pm, D_\mp] = \mp F_\mp,$$

$$[A_\pm, E_\pm] = \pm C_\pm, \quad [A_\pm, E_\mp] = 0, \quad [A_\pm, F_\pm] = D_\pm, \quad [A_\pm, F_\mp] = 0,$$

$$[B_\pm, C_\pm] = 0, \quad [B_\pm, C_\mp] = \mp D_\mp, \quad [B_\pm, D_\pm] = \pm C_\pm, \quad [B_\pm, D_\mp] = 0,$$

$$[B_\pm, E_\pm] = 0, \quad [B_\pm, E_\mp] = \mp F_\mp, \quad [B_\pm, F_\pm] = \pm E_\pm, \quad [B_\pm, F_\mp] = 0,$$

$$[C_\pm, D_\pm] = 0, \quad [C_\pm, D_\mp] = \mp B_\pm, \quad [C_\pm, E_\pm] = 0, \quad [C_\pm, E_\mp] = \mp A_\pm,$$

$$[C_\pm, F_\pm] = 0, \quad [C_\pm, F_\mp] = 0,$$

$$[D_\pm, E_\pm] = 0, \quad [D_\pm, E_\mp] = 0, \quad [D_\pm, F_\pm] = 0, \quad [D_\pm, F_\mp] = \mp A_\pm,$$

$$[E_\pm, F_\pm] = 0, \quad [E_\pm, F_\mp] = \mp B_\pm.$$

The quadratic Casimir of $su(2,2)$ has the form

$$\begin{aligned}
\mathcal{C}_{su(2,2)} &= \tfrac{1}{2}\left(\{A_+, A_-\} + \{B_+,\ B_-\} - \{C_+, C_-\} - \{D_+, D_-\} - \{E_+, E_-\}\right. \\
&\quad \left. -\{F_+, F_-\}\right) + \tfrac{1}{2}\left(A_3^2 + B_3^2 + C_3^2 + D_3^2 + E_3^2 + F_3^2\right) \\
&= \tfrac{1}{2}\left(\{A_+, A_-\} + \{B_+,\ B_-\} - \{C_+, C_-\} - \{D_+, D_-\} - \{E_+, E_-\}\right. \\
&\quad \left. -\{F_+, F_-\}\right) + 2J(J+1) + M^2 + Q^2 + \tfrac{1}{2}\,,
\end{aligned}$$

that, applied on the $\{\mathcal{J}_j^{m,q}(x)\}$, gives

$$\mathcal{C}_{su(2,2)}\, \mathcal{J}_j^{m,q}(x) = -\frac{3}{2}\, \mathcal{J}_j^{m,q}(x)\,. \tag{26}$$

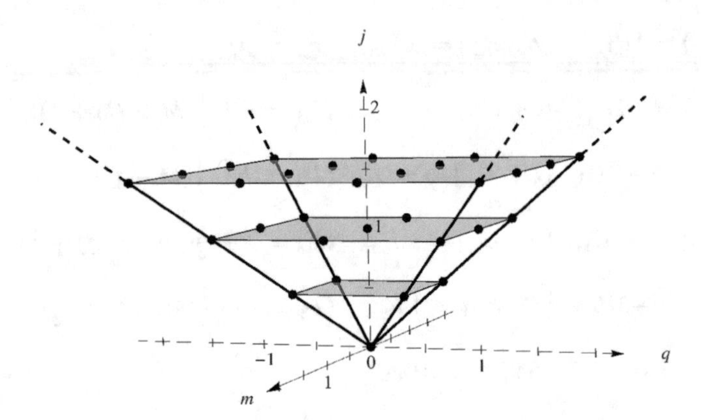

Fig. 2 IR of $su(2,2)$ supported by the AJF $\mathcal{J}_l^{m,q}(x)$ represented by the black points. The horizontal planes correspond to IR of $su_A(2) \oplus su_B(2)$

The relation (26) shows that the infinite-dimensional IR of $su(2,2)$ generated by $\{\mathcal{J}_j^{m,q}(x)\}$ contains all $j = 0, 1/2, 1, \ldots,$. From it and taking into account the differential realization of the operators involved, (12), (13), (17), and (19), we recover again the Jacobi equation (9) that, as in the previous sections, can be obtained also from the Casimir of any subalgebra of $su(2,2)$ as well as from any diagonal product of ladder operators.

In this IR of $su(2,2)$ the integer and half-integer values of $\{j, m, q\}$ are put all together (see Fig. 2). The symmetries of the AJF, where integer and half-integer values of $\{j, m, q\}$ belong to different IRs, have been considered in [29].

6 Resume and Conclusions

The Jacobi polynomials and the d_j-matrices look to be more general examples of the properties described in [25–29] for special functions. This suggests that the following properties could be assumed for a possible classification of the ASF, a relevant subset of generic special functions:

1. ASF are a basis of $L^2(\mathbb{F})$, the space of integrable functions defined on an appropriate space $\mathbb{F}$.
2. ASF are a basis of an IR of a Lie algebra $\mathcal{G}$.
3. All the diagonal elements of the UEA[$\mathcal{G}$] can be written in terms of the fundamental second order differential equation determined by the quadratic Casimir of $\mathcal{G}$.
4. All the non-diagonal elements of the UEA[$\mathcal{G}$] can be written as first order differential operators.
5. Every basis of $L^2(\mathbb{F})$ can be obtained applying an element of the Lie group G to the ASF.

6. Every operator acting on $L^2(\mathbb{F})$ belongs to UEA[$\mathcal{G}$].

Returning now to the particular case of the AJF the previous remarks become:

1. AJF are a basis of an IR of the Lie algebra $su(2, 2)$.
2. All the diagonal elements of the UEA[$su(2, 2)$] can be obtained from Eq. (9).
3. All the non-diagonal elements of the UEA[$su(2, 2)$] can be written as first order differential operators.
4. The set of AJF $\{\mathcal{J}_j^{m,q}(x)\}$ is a basis in $L^2(\mathbb{E}, \mathbb{Z}, \mathbb{Z}/2)$, where $\mathbb{E} = [-1, 1]$.
5. Every basis of $L^2(\mathbb{E}, \mathbb{Z}, \mathbb{Z}/2)$ can be obtained under the action of $SU(2, 2)$ on the set of AJF, i.e., it can be written as $\{g\, \mathcal{J}_j^{m,q}(x)\}$ where $g \in SU(2, 2)$.
6. Every operator acting on $L^2(\mathbb{E}, \mathbb{Z}, \mathbb{Z}/2)$ belongs to the UEA[$su(2, 2)$].

As a final point we recall the connection between the IR of $SU(2)$,

$$D_j(\alpha, \beta, \gamma)_m^{m'} = e^{-i\alpha m'}\, d_j(\beta)_m^{m'}\, e^{-i\gamma m}\,,$$

where α, β, γ are the Euler angles [37], the Wigner d_j-matrices, and the Jacobi polynomials $P_{j-m'}^{m'-m, m'+m}$. This implies that all the results of this paper can be extended to $\{D_j(\alpha, \beta, \gamma)_m^{m'}\}$ that have similar properties of the $\{\mathcal{J}_j^{m,q}(x)\}$ and are a basis of the square integrable functions defined in the space $\{\alpha, \beta, \gamma\}$.

Acknowledgements This research is supported in part by the Ministerio de Economía y Competitividad of Spain under grant MTM2014-57129-C2-1-P and the Junta de Castilla y León (Projects VA057U16, VA137G18, and BU229P18).

References

1. M. Berry, Why are special functions special? Phys. Today **54**, 11 (2001)
2. G.E. Andrews, R. Askey, R. Roy, *Special Functions* (Cambridge University Press, Cambridge, 1999)
3. G. Heckman, H. Schlichtkrull, *Harmonic Analysis and Special Functions on Symmetric Spaces* (Academic, New York, 1994)
4. R. Koekoek, P.A. Lesky, R.F. Swarttouw, *Hypergeometric Orthogonal Polynomials and Their q-Analogues* (Springer, Berlin, 2010) (and references therein)
5. E.I. Jafarov, J. Van der Jeugt, A finite oscillator model related to $\mathfrak{sl}(2|1)$. J. Phys. A: Math. Theor. **45** 275301 (2012). Discrete series representations for $\mathfrak{sl}(2|1)$, Meixner polynomials and oscillator models, J. Phys. A: Math. Theor. **45**, 485201 (2012); The oscillator model for the Lie superalgebra $\mathfrak{sh}(2|2)$ and Charlier polynomials, J. Math. Phys. **54**, 103506 (2013)
6. J. Van der Jeugt, Finite oscillator models described by the Lie superalgebra $sl(2|1)$, in *Symmetries and Groups in Contemporary Physics*, ed. by C. Bai, J.-P. Gazeau, M.-L. Ge. Nankai Series in Pure, Applied Mathematics and Theoretical Physics, vol. 11 (World Scientific, Singapore, 2013), p. 301
7. L. Vinet, A. Zhedanov, A "missing" family of classical orthogonal polynomials. J. Phys. A: Math. Theor. **44**, 085201 (2011)

8. V.X. Genest, L. Vinet, A. Zhedanov, d-Orthogonal polynomials and $su(2)$. J. Math. Anal. Appl. **390**, 472 (2012)
9. A. Zaghouani, Some basic d-orthogonal polynomials sets. Georgian Math. J. **12**, 583 (2005)
10. I. Lamiri, A. Ouni, d-Orthogonality of Humbert and Jacobi type polynomials. J. Math. Anal. Appl. **341**, 24 (2008)
11. P. Basseilhac, X. Martin, L. Vinet, A. Zhedanov, Little and big q-Jacobi polynomials and the Askey-Wilson algebra (2018). arXiv:1806.02656v2
12. R. Floreanini, L. Vinet, Quantum algebras and q-special functions. Ann. Phys. **221**, 53 (1993); On the quantum group and quantum algebra approach to q-special functions, Lett. Math. Phys. **27**, 179 (1993)
13. T.H. Koornwinder, q-special functions, a tutorial, in *Representations of Lie Groups and Quantum Groups*, ed. by V. Baldoni, M.A. Picardello (Longman Scientific and Technical, New York, 1994), pp. 46–128
14. D. Gómez-Ullate, N. Kamran, R. Milson, An extended class of orthogonal polynomials defined by a Sturm-Liouville problem. J. Math. Anal. Appl. **359**, 352 (2009).
15. A.J. Durán, Constructing bispectral dual Hahn polynomials. J. Approx. Theory **189**, 1–28 (2015)
16. E.P. Wigner, *The Application of Group Theory to the Special Functions of Mathematical Physics* (Princeton University Press, Princeton, 1955)
17. J.D. Talman, *Special Functions: A Group Theoretic Approach* (Benjamin, New York, 1968)
18. W. Miller Jr., *Lie Theory and Special Functions* (Academic, New York, 1968)
19. N.J. Vilenkin, *Special Functions and the Theory of Group Representations* (American Mathematical Society, Providence, 1968)
20. N.J. Vilenkin, A.U. Klimyk, *Representation of Lie Groups and Special Functions*, vols. 1–3 (Kluwer, Dordrecht, 1991–1993) (and references therein)
21. N.J. Vilenkin, A.U. Klimyk, *Representation of Lie Groups and Special Functions: Recent Advances* (Kluwer, Dordrecht, 1995)
22. E. Celeghini, M. Gadella, M.A. del Olmo, $SU(2)$, associated Laguerre polynomials and rigged Hilbert spaces, in *Quantum Theory and Symmetries with Lie Theory and Its Applications in Physics*, vol. 2, ed. by V. Dobrev, pp. 373–383 in Springer Proceedings in Mathematics & Statistics, vol. 255 (Springer, Singapore, 2018)
23. T.H. Koornwinder, Representation of $SU(2)$ and Jacobi polynomials (2016). arXiv:1606.08189 [math.CA]
24. C. Truesdell, *An Essay Toward a Unified Theory of Special Functions*. Annals of Mathematical Studies, vol. 18 (Princeton University Press, Princeton, 1949)
25. E. Celeghini, M.A. del Olmo, Coherent orthogonal polynomials. Ann. Phys. **335**, 78 (2013)
26. E. Celeghini, M.A. del Olmo, Algebraic special functions and $SO(3, 2)$. Ann. Phys. **333**, 90 (2013)
27. E. Celeghini, M. Gadella, M.A. del Olmo, Spherical harmonics and rigged Hilbert spaces. J. Math. Phys. **59**, 053502 (2018)
28. E. Celeghini, M.A. del Olmo, Group theoretical aspects of $L^2(\mathbb{R}^+)$, $L^2(\mathbb{R}^2)$ and associated Laguerre polynomials in *Physical and Mathematical Aspects of Symmetries*, ed. by S. Duarte, J.P. Gazeau, et al., (Springer, New York, 2018), pp. 133–138
29. E. Celeghini, M.A. del Olmo, M.A. Velasco, Lie groups, algebraic special functions and Jacobi polynomials. J. Phys.: Conf. Ser. **597**, 012023 (2015)
30. W. Miller, Jr., *Symmetry and Separation of Variables* (Addison-Wesley, Reading, 1977)
31. G. Lauricella, Sulle funzioni ipergeometriche a pui variable. Rend. Circ. Mat. Palermo **7**, 111 (1893)
32. E. Wigner, Einige Folgerungen aus der Schrödingerschen Theorie für die Termstrukturen. Z. Phys. **43**, 624 (1927)
33. F.W.J. Olver, D.W. Lozier, R.F. Boisvert, C.W. Clark, *NIST Handbook of Mathematical Functions* (Cambridge University Press, New York, 2010)
34. Y.L. Luke, *The Special Functions and Their Approximations*, vol.1 (Academic Press, San Diego, 1969), pp. 275–276

35. M. Abramowitz, I. Stegun, *Handbook of Mathematical Functions with Formulas, Graphs, and Mathematical Tables* (Dover, San Diego, 1972)
36. L.C. Biedenharn, J.D. Louck, *Angular Momentum in Quantum Mechanics* (Addison-Wesley, Reading, 1981)
37. W.-K. Tung, *Group Theory in Physics* (World Scientific, Singapore, 1985)
38. E. Schrödinger, Further studies on solving eigenvalue problems by factorization. Proc. Roy. Irish Acad. **A46**, 183 (1940/1941); The Factorization of the Hypergeometric Equation, Proc. Roy. Irish Acad. **A47**, 53 (1941)
39. L. Infeld, T.E. Hull, The factorization method. Rev. Mod. Phys. **23**, 21 (1951)
40. D. Fernández, J. Negro, M.A. del Olmo, Group approach to the factorization of the radial oscillator equation. Ann. Phys. **252**, 386 (1996)
41. V. Bargmann, Irreducible unitary representations of the Lorentz group. Ann. Math. **48**, 368 (1947)

Infinite Square-Well, Trigonometric Pöschl-Teller and Other Potential Wells with a Moving Barrier

Alonso Contreras-Astorga and Véronique Hussin

Abstract Using mainly two techniques, a point transformation and a time dependent supersymmetry, we construct in sequence several quantum infinite potential wells with a moving barrier. We depart from the well-known system of a one-dimensional particle in a box. With a point transformation, an infinite square-well potential with a moving barrier is generated. Using time dependent supersymmetry, the latter leads to a trigonometric Pöschl-Teller potential with a moving barrier. Finally, a confluent time dependent supersymmetry transformation is implemented to generate new infinite potential wells, all of them with a moving barrier. For all systems, solutions of the corresponding time dependent Schrödinger equation fulfilling boundary conditions are presented in a closed form.

Keywords Infinite square-well potential · Pöschl-Teller potential · Supersymmetry · Point transformation · Moving barrier

1 Introduction

There are physical problems where the boundary conditions of the underlying equation can move. Examples of them are the so-called Stefan problems, where temperature as a function of position and time on a system of water and ice has to be found, the interface water–ice imposes a boundary condition that changes its position with time [1]. Another example was drafted by Fermi, he theorized

A. Contreras-Astorga (✉)
Cátedras CONACYT—Departamento de Física, Cinvestav, Ciudad de México, Mexico

Department of Physics, Indiana University Northwest, Gary, IN, USA
e-mail: alonso.contreras.astorga@gmail.com; alonso.contreras@conacyt.mx

V. Hussin
Centre de Recherches Mathématiques & Département de Mathématiques et de Statistique, Université de Montréal, Montréal, QC, Canada
e-mail: veronique.hussin@umontreal.ca

Ş. Kuru et al. (eds.), *Integrability, Supersymmetry and Coherent States*, CRM Series in Mathematical Physics, https://doi.org/10.1007/978-3-030-20087-9_11

the origin of cosmic radiation as particles accelerated by collisions with a moving magnetic field [2], this problem was later studied by Ulam [3] in a classical framework where the statistical properties of particles in a box with oscillating infinite barriers were analyzed numerically. In this paper, we are interested in systems ruled by the time dependent Schrödinger equation. In particular, we show how different quantum systems with a moving boundary condition and their solutions can be generated using basically two tools, a point transformation and a time dependent supersymmetry.

The point transformation we use was introduced in [4, 5] where the authors mapped solutions between two time dependent Schrödinger equations with different potentials. This transformation can be used, for example, to map solutions of the harmonic oscillator to solutions of the free particle system.

On the other hand, the supersymmetry technique or SUSY helps as well to map solutions between two Schrödinger equations, but in this case the potentials share properties like asymptotic behavior or a similar discrete spectrum in case of time independent potentials. A first version links two time independent one-dimensional Schrödinger equations [6, 7]. In this article we use the time dependent version that links two time dependent Schrödinger equations [8, 9]. The involved potentials are referred as SUSY partners and if the link is made through a first-order differential operator, often called intertwining operator, the technique is known as 1-SUSY. Examples of time dependent SUSY partners of the harmonic oscillator can be found in [10, 11].

Finkel et al. [12] showed that the time independent SUSY technique and the time dependent version were related by the previously mentioned point transformation.

The structure of this article is as follows. The quantum particle in a box is revised in Sect. 2. In Sect. 3, we use a point transformation to generate an infinite square-well potential with a moving barrier. A brief review of time dependent SUSY is given in Sect. 4 and it is applied to the infinite square-well potential with a moving barrier to generate the exactly solvable system of a Pöschl-Teller potential with a moving barrier. In Sect. 5, we apply for the second time a supersymmetric transformation to the infinite square-well potential to obtain a biparametric family of infinite potential wells with a moving barrier. Exact solutions of the time dependent Schrödinger equation for each potential are given in the corresponding section. We finish this article with our conclusion.

2 Quantum Infinite Square-Well Potential

The quantum particle in a one-dimensional infinite square-well potential or particle in a box is a common example of an exact solvable model in textbooks, see, for example, [13–15]. It represents a particle trapped in the interval $0 < y < L$ with impenetrable barriers, placed at zero and L, and inside that one-dimensional box the particle is free to move. The corresponding time independent Schrödinger equation is

$$\frac{d^2}{dy^2}\psi(y) + \frac{2m}{\hbar^2}\left(E - \widetilde{V}_0\right)\psi(y) = 0, \tag{1}$$

where E is a real parameter representing the energy of the particle, m is its mass, and $\hbar$ is Planck's constant. Through this article we will use units where $m = 1/2$ and $\hbar = 1$. The one-dimensional infinite potential well $\widetilde{V}_0(y)$ is

$$\widetilde{V}_0(y) = \begin{cases} 0, & 0 < y < L, \\ \infty, & \text{otherwise}, \end{cases} \tag{2}$$

where L is a positive real constant. The solution of this eigenvalue problem is well known, eigenfunctions and eigenvalues are given by

$$\psi_n(y) = \sqrt{\frac{2}{L}}\sin\left(\frac{n\pi y}{L}\right), \qquad E_n = \left(\frac{n\pi}{L}\right)^2, \qquad n = 1, 2, 3, \ldots. \tag{3}$$

Functions $\psi_n(y)$ satisfy the boundary conditions $\psi_n(0) = \psi_n(L) = 0$. We will use this system to construct a variety of infinite potential wells where one of the barriers is moving.

3 From the Particle in a Box to the Infinite Square-Well Potential with a Moving Barrier

In this section we will use a point transformation in order to obtain from the stationary potential (2) an infinite square-well potential with a moving barrier. First we introduce a general point transformation [4, 5, 12, 16] and then we apply it on the particle in a box system. The simplified notation for the transformation was introduced in [16].

3.1 Point Transformation

Consider a one-dimensional time independent Schrödinger equation in the spatial variable y as

$$\frac{d^2}{dy^2}\psi(y) + \left(E - \widetilde{V}_0\right)\psi(y) = 0 \tag{4}$$

where $\widetilde{V}_0 = \widetilde{V}_0(y)$ and a solution ψ are known. Now let us take arbitrary functions $A = A(t)$ and $B = B(t)$ and let the variable y be defined in terms of a temporal parameter t and a new spatial variable x as:

$$y(x,t) = x\exp\left[4\int A(t)dt\right] + 2\int B(t)\exp\left[4\int A(t)dt\right]dt \tag{5}$$

then the function

$$\phi(x,t) = \psi(y(x,t)) \exp\left\{-i\left[A(t)x^2 + B(t)x + E\int \exp\left[8\int A(t)dt\right]dt + \int\left[2iA(t) + B^2(t)\right]dt\right]\right\}, \tag{6}$$

is solution of the equation

$$i\frac{\partial}{\partial t}\phi(x,t) + \frac{\partial^2}{\partial x^2}\phi(x,t) - V_0(x,t)\phi(x,t) = 0. \tag{7}$$

The last equation is a time dependent Schrödinger equation where the potential is given by

$$V_0(x,t) = \widetilde{V}_0(y(x,t)) \exp\left[8\int A(t)dt\right] + \left[\frac{d}{dt}A(t) - 4A^2(t)\right]x^2 + \left[\frac{d}{dt}B(t) - 4A(t)B(t)\right]x. \tag{8}$$

3.2 *Infinite Square-Well Potential with a Moving Barrier*

We can use the point transformation to obtain an infinite square-well potential with a moving barrier. Without considering for this moment boundary conditions of the problem, we will transform the potential $\widetilde{V}_0(y) = 0$ into $V_0(x,t) = 0$. Apparently we are mapping a potential to itself but it will not be the case once we incorporate the boundary conditions. To make this transformation, functions $A(t)$ and $B(t)$ such that $V_0(x,t) = \widetilde{V}_0(y) = 0$ in (8) need to be found. By setting $\widetilde{V}_0(y) = 0$ and $V_0(x,t) = 0$ in (8) we get

$$0 = \left[\frac{d}{dt}A(t) - 4A^2(t)\right]x^2 + \left[\frac{d}{dt}B(t) - 4A(t)B(t)\right]x. \tag{9}$$

Coefficients of the previous polynomial in x give us a system of coupled differential equations that can be solved:

$$\begin{aligned} &\frac{d}{dt}A(t) - 4A^2(t) = 0, && \Rightarrow \quad A(t) = -\frac{1}{4t + c_1}; \\ &\frac{d}{dt}B(t) - 4A(t)B(t) = 0, && \Rightarrow \quad B(t) = \frac{c_2}{4t + c_1}; \end{aligned} \tag{10}$$

where c_1 and c_2 are real constants. Once these two functions are known, the change of variable defined in (5) can be evaluated,

$$y(x,t) = \frac{2x - c_2}{2(4t + c_1)}. \tag{11}$$

At this point, we can discuss boundary conditions of the potential $V_0(x,t)$. The barriers of the potential (2) of the initial problem are located at $y_1 = 0$ and at $y_2 = L$, using the change of variable (11) the new barriers will then be placed at $x_1 = c_2/2$ and $x_2 = 4Lt + c_1L + c_2/2$, respectively. Thus, the potential $V_0(x,t)$ is

$$V_0(x,t) = \begin{cases} 0, & x_1 < x < \ell(t), \\ \infty, & \text{otherwise}, \end{cases} \tag{12}$$

where

$$\ell(t) = 4Lt + c_1L + c_2/2. \tag{13}$$

This potential is an infinite square-well potential with a moving barrier. The meaning of the constants c_1 and c_2 can be extracted directly from the position of the boundaries of this potential. Indeed, the position of the fixed barrier is $c_2/2$, while the moving barrier is located at $\ell(t)$ and it is moving with a constant velocity $4L$. At $t_0 = -c_1/4$ we have an ill-defined problem (a particle in a box of length zero), so we should avoid this singularity, moreover, this time t_0 separates two problems: one of a contracting box and one where the potential well is expanding as time increases.

Finally, solutions of the time dependent Schrödinger Eq. (7) where $V_0(x,t)$ is (12) can be constructed using (3), (6), (10), and (11):

$$\phi_n(x,t) = \sqrt{\frac{2}{L(4t + c_1)}} \sin\left\{\frac{n\pi}{L}\left[\frac{2x - c_2}{2(4t + c_1)}\right]\right\} \exp\left[\frac{i}{4t + c_1}x^2 - \frac{ic_2}{4t + c_1}x\right]$$
$$\times \exp\left[i\left(\frac{n\pi}{L}\right)^2 \frac{1}{4(4t + c_1)} + \frac{i\,c_2^2}{4(4t + c_1)}\right]. \tag{14}$$

We will fix the constant $c_2 = 0$ so that one barrier is always at zero. We take as well $c_1 = 1$, then the moving barrier will be at L when $t = 0$. The singularity of the problem will be located at $t = -1/4$. For this selection of constant the functions A, B in (10) and the change of variable y in (11) simplify to

$$A(t) = -\frac{1}{4t + 1}, \quad B(t) = 0, \quad y(x,t) = \frac{x}{4t + 1}; \tag{15}$$

the potential reads

$$V_0(x,t) = \begin{cases} 0, & 0 < x < \ell(t), \quad \text{where} \quad \ell(t) = L(4t + 1) \\ \infty, & \text{otherwise}, \end{cases} \tag{16}$$

and the solutions of the time dependent Schrödinger equation can be written as

$$\phi_n(x,t) = \sqrt{\frac{2}{\ell}} \sin\left(\frac{n\pi}{\ell}x\right) \exp\left\{ i\frac{L}{\ell}\left[x^2 + \left(\frac{n\pi}{2L}\right)^2\right]\right\}. \tag{17}$$

Note that $\phi_n(0,t) = \phi_n(\ell,t) = 0$, satisfying the required boundary conditions of the physical problem. For this specific selection of the constants c_1 and c_2 the domain of the time variable for the contracting box is $(-\infty, -1/4)$ and for the expanding well is $(-1/4, \infty)$. Functions (17) are normalized $\int_0^\ell |\phi_n|^2 dx = 1$ at any given time. They form a complete orthogonal set at any fixed time and the expectation value of the energy $\langle E\rangle_{\phi_n}(t) = \int_0^\ell \phi_n^*(-\partial_x^2\phi_n)dx = (n\pi/\ell)^2$. System (16) and its solutions (17) were also discussed in [17–21]. In Fig. 1 three different probability densities were plotted at four different times, see (17): in blue $|\phi_1|^2$, in purple $|\phi_2|^2$, and in yellow $|\phi_3|^2$; for the times $t = 1/4,\ 1/2,\ 3/4,\ 1$ and the parameter $L = 1$. Since the times we used are greater than $t_0 = -1/4$, the plotted potential represents an expanding box.

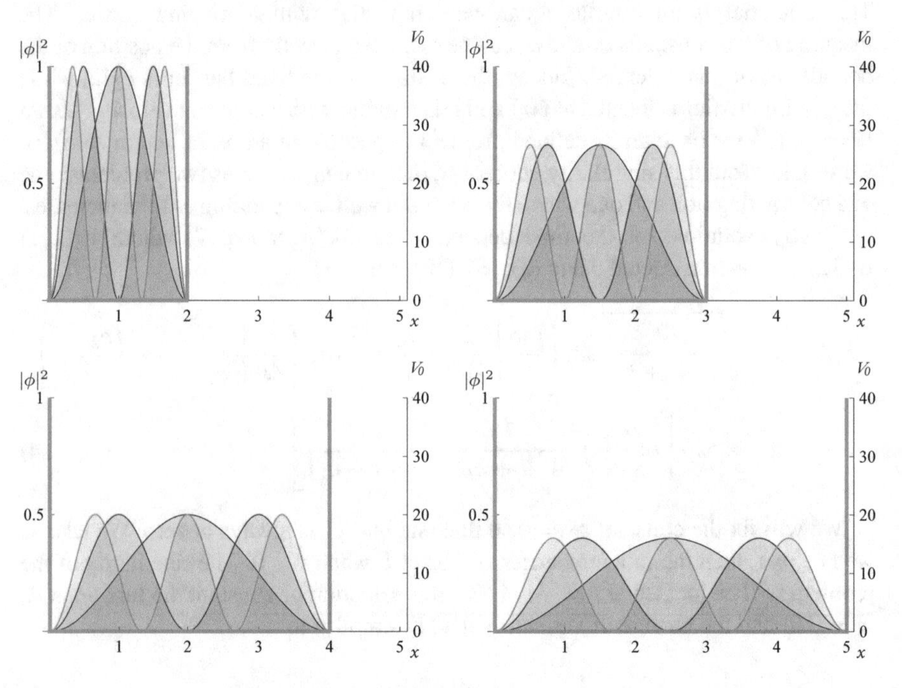

Fig. 1 Infinite square-well potential with a moving barrier, see (16). Plot of the probability densities, see (17), $|\phi_1|^2$ (blue), $|\phi_2|^2$ (purple), $|\phi_3|^2$ (yellow), at four different times: $t = 1/4$ (top left), $t = 1/2$ (top right), $t = 3/4$ (bottom left), $t = 1$ (bottom right), for the parameter $L = 1$

4 From the Infinite Well Potential with a Moving Barrier to a Pöschl-Teller Potential with a Moving Barrier

In this section we introduce our second tool, a time dependent SUSY transformation introduced in [8, 9], the notation is adopted from [10]. Then, we apply it to the infinite well potential with a moving barrier to generate a Pöschl-Teller potential with a moving barrier.

4.1 Time Dependent Supersymmetric Quantum Mechanics

We start out with a time dependent Schrödinger Eq. (7) where the potential V_0 is a real known function. Next, we propose the existence of an operator $\mathcal{L}_1$ intertwining two Schrödinger operators

$$S_1\mathcal{L}_1 = \mathcal{L}_1 S_0, \tag{18}$$

where the Schrödinger operators are defined as $S_j = i\partial_t + \partial_x^2 - V_j$, $j = 0, 1$. Now, if $\mathcal{L}_1$ is a differential operator of the form $\mathcal{L}_1 = A_1\left(-\partial_x + \frac{u_x}{u}\right)$ where $A_1 = A_1(t)$, $u = u(x,t)$ and the subindex in u_x represents partial derivation with respect to x, then the intertwining relationship (18) and the form of the Schrödinger operators impose the conditions:

$$V_1 = V_0 + i(\ln A_1)_t - 2(\ln u)_{xx}, \qquad i\partial_t u + u_{xx} - V_0 u = c(t), \tag{19}$$

where $c(t)$ is an integration function. This function $c(t)$ can be absorbed in the potential term, and it will be reflected in the solution u of the Schrödinger equation as a time dependent phase, in this work it will be set $c(t) = 0$. Note that now u satisfies $S_0 u = 0$, i.e., it is a solution of the initial system. Furthermore, it can be seen from (19) that in order to avoid new singularities in V_1, the functions A_1 and u must not vanish.

The potential V_1 in (19) is in general a complex function. Since we are interested in a Hermitian operator S_1, we must ask that the imaginary part of V_1 vanishes, $\mathrm{Im}(V_1) = 0$. Taking (19) and considering V_0 as a real function, A_1 and u must also satisfy $i\left(\ln|A_1|^2\right)_t = 2\left(\ln(u/u^*)\right)_{xx}$, since the left-hand side depends only on time we can say that

$$\frac{\partial^3}{\partial x^3}\ln\left(\frac{u}{u^*}\right) = 0 \tag{20}$$

is a reality condition to generate a Hermitian operator S_1, and then $|A_1|$ is fixed to

$$|A_1| = \exp\left\{2\int \mathrm{Im}\left[\frac{\partial^2}{\partial x^2}\ln u(x,t)\right]dt\right\}. \tag{21}$$

If this condition is inserted into (19) along with $A_1 = |A_1|$, then the expression of the new potential simplifies to

$$V_1 = V_0 - \frac{\partial^2}{\partial x^2} \ln |u|^2. \tag{22}$$

The intertwining relation (18) ensures that if ϕ solves the equation $S_0\phi = 0$, then $\chi = \mathcal{L}_1\phi$ solves $S_1\chi = 0$. Direct substitution shows that there is an extra function $\chi_\epsilon = 1/A_1u^*$, often called missing state, that also solves $S_1\chi_\epsilon = 0$. Consult [8–11] for more details on this technique.

4.2 *Trigonometric Pöschl-Teller Potential with a Moving Barrier*

In order to apply a 1-SUSY transformation to the time dependent potential defined in (16), we need to select a transformation function $u(x,t)$ fulfilling three conditions: (i) $u(x,t)$ must satisfy the time dependent Schrödinger equation $S_0u = 0$, (ii) $u(x,t) \neq 0$ to avoid new singularities inside the domain of the potential, and (iii) $\partial_x^3 \ln(u/u^*) = 0$ to generate a Hermitian potential V_1. One function satisfying all three conditions is $\phi_1(x,t)$ in (17), thus, we will use it as transformation function:

$$u(x,t) = \sqrt{\frac{2}{\ell}} \sin\left(\frac{\pi}{\ell}x\right) \exp\left\{ i\frac{L}{\ell}\left[x^2 + \left(\frac{\pi}{2L}\right)^2\right]\right\}, \qquad \ell = L(4t+1). \tag{23}$$

Then, we need to calculate the function A_1, see (21), and the intertwining operator $\mathcal{L}_1 = A_1\left(-\partial_x + \frac{u_x}{u}\right)$:

$$A_1(t) = 4t+1, \qquad \mathcal{L}_1 = -(4t+1)\frac{\partial}{\partial_x} + i2x + \frac{\pi}{L}\cot\left(\frac{\pi x}{\ell}\right). \tag{24}$$

The 1-SUSY partner V_1 of (16) can be obtained directly from (22) as

$$V_1(x,t) = \begin{cases} 2\left(\frac{\pi}{\ell}\right)^2 \csc^2\left(\frac{\pi x}{\ell}\right), & 0 < x < \ell(t), \\ \infty, & \text{otherwise}, \end{cases} \tag{25}$$

it coincides with a trigonometric Pöschl-Teller potential at any fixed time [22], and it is a particular case of a class of potentials found in [19]. Solutions of the time dependent Schrödinger equation for this potential can be obtained applying the operator $\mathcal{L}_1$ onto solutions ϕ_n, see (17) and (24):

$$\chi_n(x,t) = \mathcal{L}_1\phi_n(x,t) \tag{26}$$
$$= \frac{\pi}{L}\sqrt{\frac{2}{\ell}}\left[\cot\left(\frac{\pi x}{\ell}\right)\sin\left(\frac{n\pi x}{\ell}\right) - n\cos\left(\frac{n\pi x}{\ell}\right)\right]\exp\left\{i\frac{L}{\ell}\left[x^2 + \left(\frac{n\pi}{2L}\right)^2\right]\right\},$$

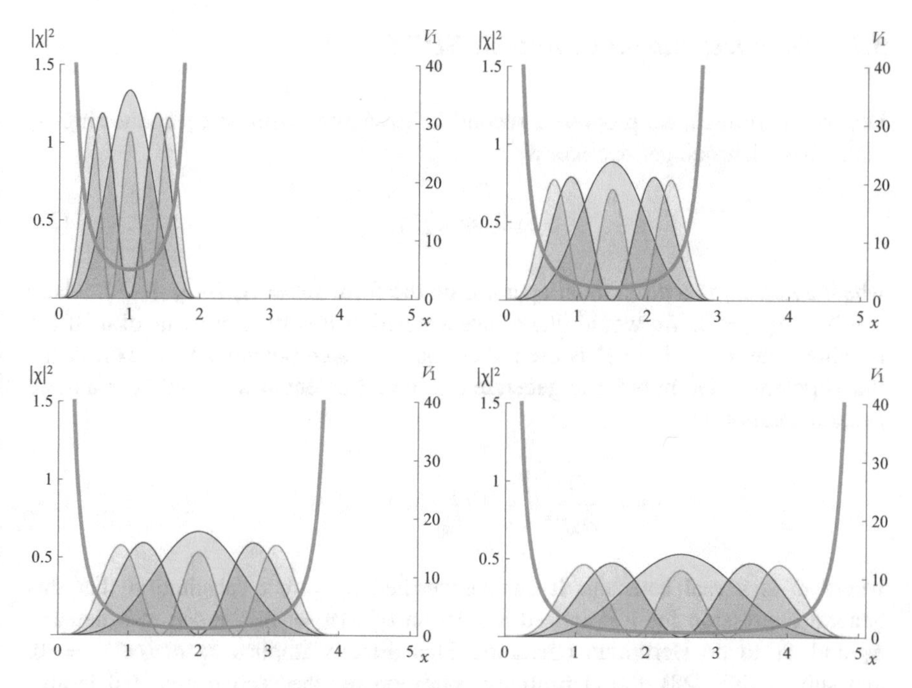

Fig. 2 In gray a trigonometric Pöschl-Teller potential with a moving barrier, see (25). Moreover, normalized probability densities are also plotted, see (26), $|\chi_2|^2$ (blue), $|\chi_3|^2$ (purple), $|\chi_4|^2$ (yellow), at four different times: $t = 1/4$ (top left), $t = 1/2$ (top right), $t = 3/4$ (bottom left), $t = 1$ (bottom right), for the parameter $L = 1$

where $n = 2, 3, 4, \cdots$. In this problem $\lim_{x\to 0} \chi_n(x,t) = \lim_{x\to \ell} \chi_n(x,t) = 0$. There is no square integrable missing state χ_ϵ. In Fig. 2 the Pöschl-Teller potential with a moving barrier and the normalized probability densities corresponding to χ_2, χ_3, and χ_4 are shown at four different times $t = 1/4,\ 1/2,\ 3/4,\ 1$, for the parameter $L = 1$.

5 Confluent SUSY Partners: More Potentials with a Moving Barrier

The 1-SUSY technique introduced in Sect. 4 has a constraint, the transformation function u must never vanish. To underpass this restriction a second iteration can be performed, the particular iteration we will use is known as confluent SUSY, see [10] for the time dependent version and [23] for the time independent case. This technique will be applied again to V_0 and will generate a new family of infinite potential wells with a moving barrier. A different approach to get around the mentioned restriction was followed in [19], leading to different potentials than the ones constructed in this section.

5.1 Time Dependent Confluent SUSY

Departing from S_1 we propose a second intertwining operator $\mathcal{L}_2$ connecting S_1 with a new Schrödinger operator S_2

$$S_2\mathcal{L}_2 = \mathcal{L}_2 S_1, \tag{27}$$

where again $\mathcal{L}_2$ is a differential operator on the form $\mathcal{L}_2 = A_2\left(-\partial_x + \frac{v_x}{v}\right)$, where v solves $S_1 v = 0$. We would like to use a function v written in terms of u. If the missing state $\chi_\epsilon = 1/A_1 u^*$ is used, then the generated potential V_2 is exactly the initial potential V_0. In order to generate a different potential we should use a more general solution v:

$$v = \frac{1}{A_1 u^*}\left(\omega + \int_{x_0}^{x} |u(s,t)|^2 ds\right), \tag{28}$$

where ω is a real constant. It can be verified by direct substitution that this general expression for v is indeed a solution of $S_1 v = 0$. We can also demand S_2 and S_1 to be Hermitian operators. This directly implies $\partial_x^3 \ln(v/v^*) = 0$, and substituting (28), the Hermiticity condition for the second potential is also $\partial_x^3 \ln(u/u^*) = 0$. Analogous to (21), since ω was chosen real, A_2 can be fixed as $A_2 = A_1$. Under these considerations the new potential is given by

$$V_2 = V_1 - 2\partial_{xx} \ln|v| = V_0 - 2\partial_{xx} \ln\left(\omega + \int_{x_0}^{x} |u(s,t)|^2 ds\right). \tag{29}$$

From the intertwining relation (27) and using the functions χ_n (solving $S_1\chi = 0$), we can see that functions $\xi_n = \mathcal{L}_2\chi_n = \mathcal{L}_2\mathcal{L}_1\phi_n$ will solve the equation $S_2\xi = 0$. Finally, a missing solution can be found as

$$\xi_\epsilon = \frac{1}{A_2 v^*} = \frac{u}{\omega + \int_{x_0}^{x} |u(s,t)|^2 ds}. \tag{30}$$

Confluent and 1-SUSY techniques present similarities. Indeed, both use only one transformation function u fulfilling $S_0 u = 0$, both require $\partial_x^3 \ln(u/u^*) = 0$ to generate Hermitian potentials, but the regularity condition is different. In 1-SUSY u must be nodeless, for confluent SUSY the transformation function satisfies a more relaxed condition: $\int_{x_0}^{x} |u(s,t)|^2 ds \neq -\omega$, this last condition could be met, for example, by any square integrable solution.

5.2 More Potentials with a Moving Barrier

Departing from the infinite well potential with a moving barrier in (16), we can notice that solutions $\phi_n(x,t)$ (see (17)), when $n \geq 2$ cannot be used as transformation function for a 1-SUSY transformation because they have at least one zero in the interval $(0,\ell)$. With the confluent SUSY algorithm presented in this section we can surpass this restriction.

By selecting $u(x,t) = \phi_m(x,t)$ (see (17)), where $m \in \mathbb{N}$ is a fixed number, a confluent SUSY partner of the infinite square-well potential with a moving barrier can be constructed. First we need to find the function $A_2 = A_1 = |A_1|$, see (21), and the intertwining operators $\mathcal{L}_1 = A_1(-\partial_x + u_x/u)$ and $\mathcal{L}_2 = A_1(-\partial_x + v_x/v)$:

$$
\begin{aligned}
A_2(t) &= A_1(t) = 4t+1,\\
\mathcal{L}_1 &= -(4t+1)\frac{\partial}{\partial_x} + i2x + \frac{m\pi}{L}\cot\left(\frac{m\pi x}{\ell}\right),\\
\mathcal{L}_2 &= -(4t+1)\frac{\partial}{\partial_x} + i2x - \frac{m\pi}{L}\cot\left(\frac{m\pi x}{\ell}\right)\\
&\quad + \frac{4m\pi\ell\sin^2\left(\frac{m\pi x}{\ell}\right)}{2m\pi L(x+\ell\omega) - L\ell\sin\left(\frac{2m\pi x}{\ell}\right)},
\end{aligned}
\tag{31}
$$

where we fixed $x_0 = 0$ in the definition of v, see (28), (29), and (30). Then, using (29) an expression for the potential V_2 can be obtained:

$$
V_2(x,t) = \begin{cases} \dfrac{32\left(\frac{m\pi}{\ell}\right)^2\sin\left(\frac{m\pi x}{\ell}\right)\left[\sin\left(\frac{m\pi x}{\ell}\right) - \frac{m\pi}{\ell}\cos\left(\frac{m\pi x}{\ell}\right)(x+\omega\ell)\right]}{\left[\sin\left(\frac{2m\pi x}{\ell}\right) - 2\frac{m\pi}{\ell}(x+\omega\ell)\right]^2}, & 0 < x < \ell(t),\\ \infty, & \text{otherwise}, \end{cases}
\tag{32}
$$

where $\omega \in (\infty,-1] \cup [0,\infty)$ is a constant introduced by confluent algorithm.

Solutions $\xi_n(x,t)$ for these potentials can as well be found with help of intertwining operators $\mathcal{L}_1$ and $\mathcal{L}_2$, see (17) and (31)), when $n \neq m$:

$$
\begin{aligned}
\xi_n(x,t) &= \mathcal{L}_2\mathcal{L}_1\phi_n\\
&= \left(\frac{\pi}{L}\right)^2\sqrt{\frac{2}{\ell}}\sin\left(\frac{n\pi}{\ell}x\right)\exp\left\{i\frac{L}{\ell}\left[x^2 + \left(\frac{n\pi}{2L}\right)^2\right]\right\}\\
&\quad\times\frac{\left[(m^2+n^2)\sin\left(\frac{2m\pi x}{\ell}\right) + \frac{2m\pi}{\ell}(m^2-n^2)(x+\omega\ell)\right] - 4mn\ell\cot\left(\frac{n\pi x}{\ell}\right)\sin^2\left(\frac{m\pi x}{\ell}\right)}{\frac{2m\pi}{\ell}(x+\omega\ell) - \sin\left(\frac{2m\pi x}{\ell}\right)}.
\end{aligned}
\tag{33}
$$

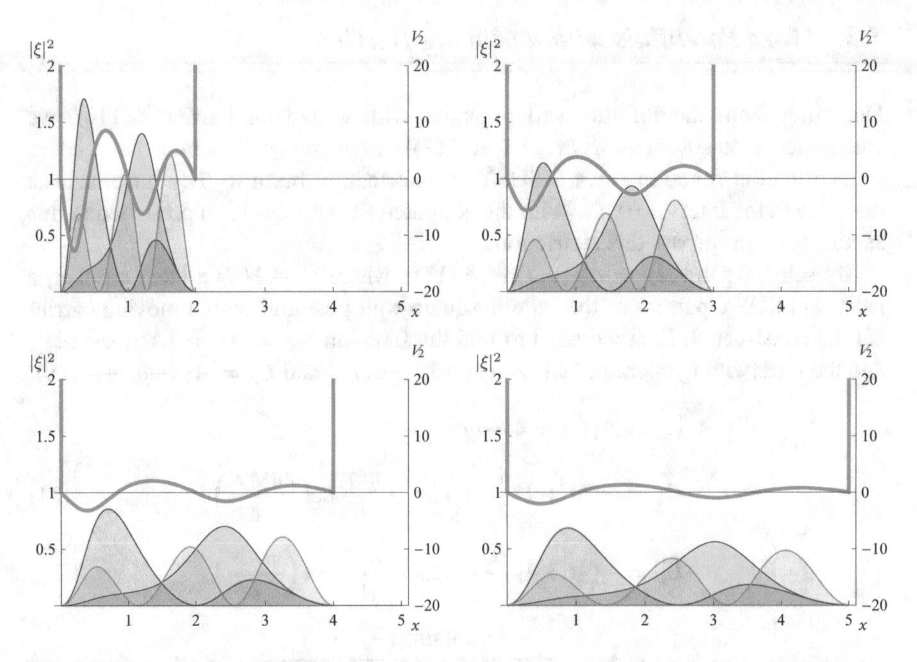

Fig. 3 In gray a confluent SUSY partner of the infinite square-well potential with a moving barrier, see (32). Moreover, normalized probability densities are also plotted, see (33) and (34), $|\xi_1|^2$ (blue), $|\xi_\epsilon|^2$ (purple), $|\xi_3|^2$ (yellow), at four different times: $t = 1/4$ (top left), $t = 1/2$ (top right), $t = 3/4$ (bottom left), $t = 1$ (bottom right), for the parameters $L = 1$, $\omega = 0.4$ and $m = 2$

If $n = m$, then the corresponding solution is the missing state $\xi_\epsilon(x, t)$ (see (30)):

$$\xi_\epsilon(x,t) = \sqrt{\frac{2}{\ell}} \sin\left(\frac{m\pi}{\ell}x\right) \\ \times \exp\left\{i\frac{L}{\ell}\left[x^2 + \left(\frac{m\pi}{2L}\right)^2\right]\right\} \frac{2m\pi}{\frac{2m\pi}{\ell}(x+\ell\omega) - \sin\left(\frac{2m\pi x}{\ell}\right)}. \tag{34}$$

Solutions ξ_n satisfy $\lim_{x\to 0}\xi_n(x,t) = \lim_{x\to\ell}\xi_n(x,t) = 0$ whereas ξ_ϵ fulfills $\lim_{x\to 0}\xi_\epsilon(x,t) = \lim_{x\to\ell}\xi_\epsilon(x,t) = 0$ only if $\omega \neq -1, 0$. When ω is set equal to -1 or zero, the function ξ_ϵ is not square integrable. The confluent SUSY partner of the infinite square-well potential with fixed barriers $\widetilde{V}(y)$, see (2), is reported in [24, 25], the potentials $V_2(x,t)$ are a dynamic version of them.

In Fig. 3 the potential V_2 in (32) is illustrated, the parameters used where $L = 1$, $\omega = 0.4$, and $m = 2$. This system has sharp edges at $x = 0$ and $x = \ell$. Normalized probability densities of three solutions are shown as well, corresponding to ξ_1, ξ_ϵ, and ξ_3. For the special cases $\omega = -1$ and $\omega = 0$ one of the edges of V_2 is smooth while the other is sharp, as can be seen in Figs. 4 and 5, moreover these special situations do not present a square integrable missing state ξ_ϵ.

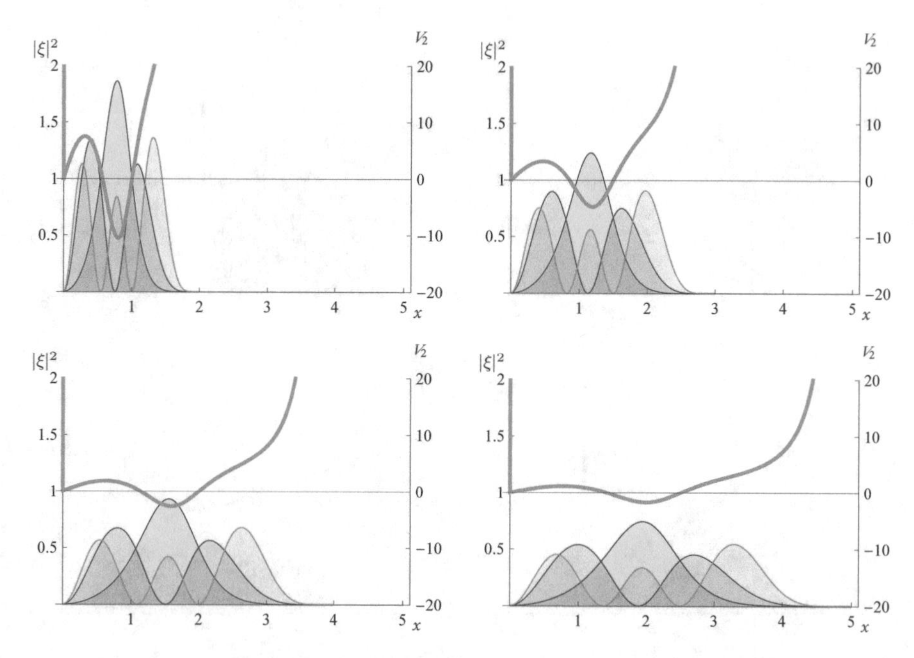

Fig. 4 In gray a confluent SUSY partner of the infinite square-well potential with a moving barrier, see (32). Moreover, normalized probability densities are also plotted, see (33), $|\xi_1|^2$ (blue), $|\xi_3|^2$ (purple), $|\xi_4|^2$ (yellow), at four different times: $t = 1/4$ (top left), $t = 1/2$ (top right), $t = 3/4$ (bottom left), $t = 1$ (bottom right), for the parameters $L = 1$, $\omega = -1$ and $m = 2$

6 Conclusions

In this article we showed how to generate different infinite potential wells with a moving boundary condition. Through a series of transformations, we obtained the infinite square-well potential, a trigonometric Pöschl-Teller potential, and the confluent SUSY partners of the infinite square-well potential, where one of the barriers is fixed and the other is moving with a constant velocity. For all these systems, exact solutions of the time dependent Schrödinger equations fulfilling the moving boundary conditions were given in a closed form.

As a continuation of the present work, it would be interesting to study different sets of coherent and squeezed states for the constructed systems and the calculation of relevant physical quantities and mathematical properties of such states.

Acknowledgements This work has been supported in part by research grants from Natural sciences and engineering research council of Canada (NSERC). ACA would like to thank the Centre de Recherches Mathématiques for kind hospitality.

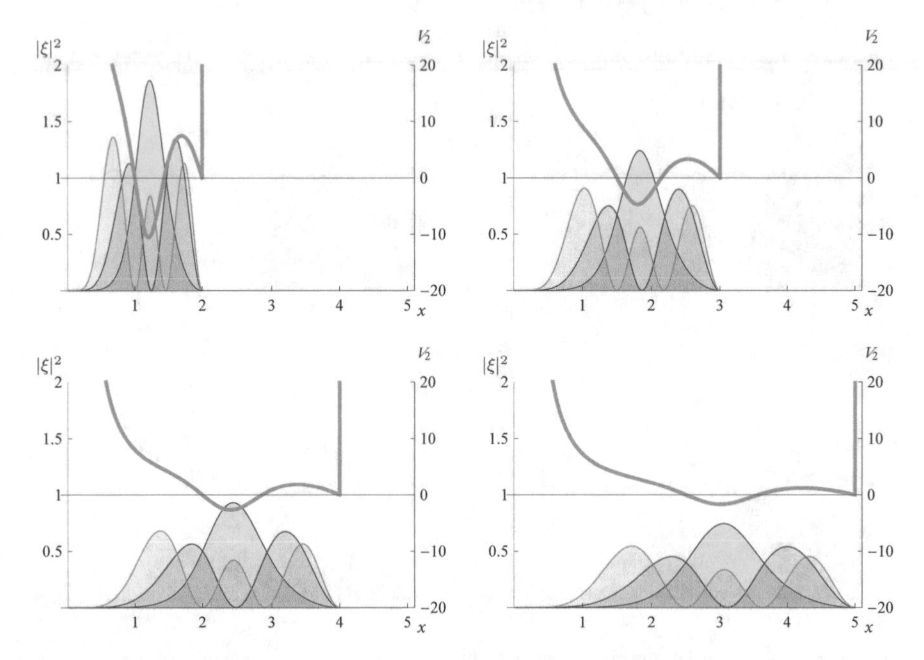

Fig. 5 In gray a confluent SUSY partner of the infinite square-well potential with a moving barrier, see (32). Moreover, normalized probability densities are also plotted, see (33), $|\xi_1|^2$ (blue), $|\xi_3|^2$ (purple), $|\xi_4|^2$ (yellow), at four different times: $t = 1/4$ (top left), $t = 1/2$ (top right), $t = 3/4$ (bottom left), $t = 1$ (bottom right), for the parameters $L = 1$, $\omega = 0$ and $m = 2$

References

1. J. Crank, *Free and Moving Boundary Problems* (Clarendon, Oxford, 1984)
2. E. Fermi, On the origin of the cosmic radiation. Phys. Rev. **75**, 1169–1174 (1949)
3. S.M. Ulam, On some statistical properties of dynamical systems, in *Proceedings Fourth Berkeley Symposium on Mathematical Statistics and Problem*, vol. 3 (University of California Press, Berkeley, CA, 1961), pp. 315–320
4. J.R. Ray, Exact solutions to the time-dependent Schrödinger equation. Phys. Rev. A **26**, 729–733 (1982)
5. G.W. Bluman, On mapping linear partial differential equations to constant coefficient equations. SIAM J. Appl. Math. **43**, 1259–1273 (1983)
6. F. Cooper, A. Khare, U. Sukhatme, Supersymmetry and quantum mechanics. Phys. Rep. **251**, 267–385 (1995)
7. D.J. Fernández C., Supersymmetric quantum mechanics. AIP Conf. Proc. **1287**, 3–36 (2010)
8. V.B. Matveev, M.A. Salle, *Darboux Transformations and Solitons* (Springer, Berlin, 1991)
9. V.G. Bagrov, B.F. Samsonov, L.A. Shekoyan, Darboux transformation for the nonsteady Schrödinger equation. Russ. Phys. J. **38**, 706–712 (1995)
10. A. Contreras-Astorga, A time-dependent anharmonic oscillator. IOP Conf. Series J. Phys. Conf. Ser. **839**, 012019 (2017)
11. K. Zelaya, O. Rosas-Ortiz, Exactly solvable time-dependent oscillator-like potentials generated by Darboux transformations. IOP Conf. Series J. Phys. Conf. Ser. **839**, 012018 (2017)
12. F. Finkel, A. González-López, N. Kamran, M.A. Rodríguez, On form-preserving transformations for the time-dependent Schrödinger equation. J. Math. Phys. **40**, 3268–3274 (1999)

13. C. Cohen-Tannoudji, B. Diu, F. Laloe, *Quantum Mechanics*, vol. I (Wiley/Hermann, Paris, 1977)
14. R. Shankar, *Principles of Quantum Mechanics* (Plenum, New York, 1994)
15. L.D. Landau, E.M. Lifshitz, *Quantum Mechanics Non-Relativistic Theory* (Pergamon, Exeter, 1991)
16. A. Schulze-Halberg, B. Roy, Time dependent potentials associated with exceptional orthogonal polynomials. J. Math. Phys. **55**, 123506 (2014)
17. S.W. Doescher, M.H. Rice, Infinite square-well potential with a moving wall. Am. J. Phys. **37**, 1246–1249 (1969)
18. D.N. Pinder, The contracting square quantum well. Am. J. Phys. **58**, 54–58 (1990)
19. T.K. Jana, P. Roy, A class of exactly solvable Schrödinger equation with moving boundary condition. Phys. Lett. A **372**, 2368–2373 (2008)
20. M.L. Glasser, J. Mateo, J. Negro, L.M. Nieto, Quantum infinite square well with an oscillating wall. Chaos, Solitons Fractals **41**, 2067–2074 (2009)
21. O. Fojón, M. Gadella, L.P. Lara, The quantum square well with moving boundaries: a numerical analysis. Comput. Math. Appl. **59**, 964–976 (2010)
22. A. Contreras-Astorga, D.J. Fernández C., Supersymmetric partners of the trigonometric Pöschl–teller potentials. J. Phys. A Math. Theor. **41**, 475303 (2008)
23. D.J. Fernández C., E. Salinas-Hernández, The confluent algorithm in second-order supersymmetric quantum mechanics. J. Phys. A Math. Theor. **36**, 2537–2543 (2003)
24. D.J. Fernández, V. Hussin, O. Rosas-Ortiz, Coherent states for Hamiltonians generated by supersymmetry. J. Phys. A Math. Theor. **40**, 6491 (2007)
25. M.-A. Fiset, V. Hussin, Supersymmetric infinite wells and coherent states. J. Phys. Conf. Ser. **624**, 012016 (2015)

Variational Method Applied to Schrödinger-Like Equation

Elso Drigo Filho, Regina M. Ricotta, and Natália F. Ribeiro

Abstract In this work we propose to adapt the variational method to analyze a specific equation derived from a statistical model for the DNA molecule. The referred equation is a Schrödinger-like equation with an additional position-dependent function multiplying its second order derivative term. The use of the adapted variational approach is shown to be a suitable technique for the calculation of the ground state for two similar potential problems. In the first problem the additional function and the potential have an exponential position-dependence while for the second the additional function has a quadratic position-dependence and the potential has a quadratic and inverse quadratic position-dependence.

Keywords Variational method · Schrödinger-like equation · Position-dependent mass · Ground state solution · Non-exact potential · Energy-dependent potential

1 Introduction

In this work we address the problem of solving analytically a linear second order differential equation that is formally equal to a Schrödinger equation with an additional function multiplying the second order derivative term. To do this we developed an adaptation of the usual variational method of quantum mechanics [1–6].

E. Drigo Filho
Instituto de Biociências, Letras e Ciências Exatas, IBILCE-UNESP, São José do Rio Preto, SP, Brazil
e-mail: elso@ibilce.unesp.br

R. M. Ricotta (✉)
Faculdade de Tecnologia de São Paulo, FATEC/SP-CEETEPS, São Paulo, SP, Brazil
e-mail: regina@fatecsp.br

N. F. Ribeiro
Centro Universitário do Norte Paulista, UNORP, São José do Rio Preto, SP, Brazil

Ş. Kuru et al. (eds.), *Integrability, Supersymmetry and Coherent States*, CRM Series in Mathematical Physics, https://doi.org/10.1007/978-3-030-20087-9_12

The main motivation to develop this methodology comes from the Peyrard–Bishop–Dauxois (PBD) model for the DNA molecule [7]. In this model a non-harmonic interaction is used to simulate the stacking interaction. The transfer integral formalism [8] used in this approach to deal the thermodynamical properties of the model leaves the system dominated by the ground state in the thermodynamic limit. It is described by a non-exactly solvable Schrödinger-like equation that has an additional function multiplying the second order derivative term that depends on the position, [7]

$$-\frac{1}{m(x)}\frac{d^2\Psi(x)}{dx^2}+V(x)\Psi(x)=E\Psi(x), \tag{1}$$

where the function $m(x)$ has position-dependence and is related with the stacking interaction, $V(x)$ is the potential and represents the H-bond interaction, E is the energy eigenvalue of the system, and $\Psi(x)$ is the eigenfunction. The functions $m(x)$ and $V(x)$ also depend on the parameters of the model, in particular the temperature [7]. In the thermodynamic limit, the ground state eigenfunction solution of Eq. (1) can be used to determine the thermodynamic properties of the studied system. For example, the phase transition curve can be calculated through the evaluation of mean stretching of the variable, $\langle x\rangle$, in terms of the temperature by the expression:

$$\langle x\rangle=\int_{-\infty}^{+\infty}x|\Psi(x)|^2dx\ . \tag{2}$$

In general, it is not possible to determine its exact analytical solutions from Eq. (1) and results from computational simulations are used to analyze the process [9].

We introduce a semi-analytical method based on the variational method to determine the ground state solution of Eq. (1). We interpret here that the function $m(x)$ plays a role similar to that of the mass that depends on the position. Thus, we call this function as a position-dependent mass. The physical system is a classical statistical one and we are only interested in its ground state solution. However, the equation in question shows similarity to the Schrödinger equation with position-dependent mass [10, 11].

It is opportune to emphasize that there are several problems related to the description of quantum systems with position-dependent mass as, for instance, the ordering ambiguity. It is also important to note that the operator on Eq. (1) is not formally self-adjoint, which could allow the existence of non-real eigenvalues E. In this context, the mathematical structure of the quantum Hamiltonian depends on the so-called von Roos ambiguity parameters [10, 11]. However, although we analyze a Schrödinger-like equation, we are not analyzing a quantum system. The Roos parameters are fixed and we are only interested in the ground state solution of the equation. Thus there is no ambiguity and that is why the ordering ambiguity is not discussed here.

The applicability of the proposed methodology is tested for two particular potential problems. In the first problem the additional function and the potential have an exponential position-dependence. This problem has a solution in the literature [10], which enabled the comparison of the results obtained through the variational method and the analytical solution.

In the second one the additional function has a quadratic position-dependence and the potential has a quadratic and inverse quadratic position-dependence. The resulting equation does not present exact analytical solution. The adapted variational method suggested here is used to determine its ground state energy eigenvalue.

2 The Adapted Variational Method

We propose to adapt the variational method to obtain the approximate analytical ground state solution of Eq. (1). Usually, in quantum mechanics, this approach consists in the choice of a function $\Psi_\mu(x)$, the trial function that depends on set of parameters μ. It is used to compute the energy eigenvalue. The variational principle guarantees that the mean energy obtained from this trial function is always an upper limit of the real ground state energy of the system (E_0). The mean value of energy is equal to E_0 if the trial function is the exact solution of the Schrödinger equation. The variational parameters in the trial eigenfunction are varied until the expectation value of the energy is minimum.

Equation (1) can be rewritten if we multiply all the terms of the equation by $m(x)$. This procedure leads to the equation,

$$-\frac{d^2\Psi(x)}{dx^2} + \underbrace{m(x)(V(x) - E)}_{V_{eff}-\epsilon_f}\Psi(x) = 0 \tag{3}$$

The term $m(x)(V(x) - E)\Psi(x)$ can be separated in a term depending on the position and a constant term, namely $-\epsilon_f$. Thus, Eq. (3) is rewritten as the following effective equation:

$$H_{eff}\Psi(x) = \epsilon_f \Psi(x) \tag{4}$$

where H_{eff} is the effective Hamiltonian, given by

$$H_{eff} = -\frac{d^2}{dx^2} + V_{eff}, \tag{5}$$

which defines the effective potential V_{eff}, given by

$$V_{eff} - \epsilon_f = m(x)(V(x) - E). \tag{6}$$

The variational method should be adapted to solve Eq. (3) since the energy eigenvalue of the Schrödinger-like equation (1) is inserted in the effective potential V_{eff}. We notice that Eq. (5) with V_{eff} defined in (6) has similar structure of the Schrödinger problems with energy-dependent potentials [12, 13]. The term that originally is interpreted as the energy eigenvalue is now a fixed constant ϵ_f. Thus, the adopted procedure is to vary the energy eigenvalue E, and consequently to vary the effective potential in order to compute the mean energy, as usual in the variational method,

$$\langle H_{eff} \rangle = \frac{\int_V \Psi_\mu^* H_{eff} \Psi_\mu dV}{\int_V \Psi_\mu^* \Psi_\mu dV} = \epsilon(E, \mu), \tag{7}$$

where H_{eff} is the effective Hamiltonian given by (5), Ψ_μ is the trial function, and μ is the variational parameter. As indicated in (7), the value of the mean energy $\epsilon(E, \mu)$ depends on the energy eigenvalue E.

After integrating (7) the variational parameter is determined by the minimization of $\epsilon(E, \mu)$ with respect to this parameter. After the minimization process, the energy eigenvalue is determined when the calculated $\epsilon(E, \mu)$, Eq. (7), becomes equal to the fixed ϵ_f. This value can be found, for instance, through a graphic representation of $\epsilon(E, \mu)$ versus E. The correct energy eigenvalue E is obtained from the intersection point where $\epsilon(E, \mu) = \epsilon_f$.

In what follows we illustrate the approach with two different examples to show the applicability of this technique. We restrict the approach to one variational parameter; however, it can be extended to a larger number of parameters.

2.1 Mass and Potential with Exponential Position-Dependence

In this case we analyze Eq. (1) with the mass and the potential with an exponential position-dependence, given by

$$m(x) = m_0 e^{-cx} , \qquad V(x) = V_0 e^{-cx} - \gamma e^{cx}. \tag{8}$$

The equivalent effective Schrödinger equation to be solved is

$$-\frac{1}{m_0}\frac{d^2\Psi(x)}{dx^2} + (V_0 e^{-2cx} - E e^{-cx})\Psi(x) = \gamma \Psi(x). \tag{9}$$

Adopting $V_0 = 1$ and $\gamma = -0.25$, we realize that Eq. (9) allows exact/analytical solution [10]; it is formally similar to the Schrödinger equation of a particle with constant mass (m_0) under the influence of the Morse potential. Notice, however, that it is in a different format, since the energy eigenvalue (E) of the system is inserted in the potential. Moreover, the resulted equation is an equation of eigenfunctions with the constant eigenvalue γ. For this reason, the usual variational method is not applicable and should be restructured. In this case, it is necessary to vary the energy E inserted in the potential, until obtaining the fixed value of $\gamma = -0.25$. The chosen trial function is the wave function of the Morse potential [2]:

$$\Psi_\mu \propto e^{\frac{a_1}{c}e^{-cx}} e^{-b_1 x}, \tag{10}$$

with $a_1 = -\sqrt{\bar{V}_0}$, $b_1 = -\frac{c}{2} + \frac{\mu}{2\sqrt{\bar{V}_0}}$, $\bar{V}_0 = \frac{2m_0}{\hbar^2} V_0$ and μ is the variational parameter. With this trial function (10) the mean value of H_{eff} obtained from (9) is determined by numerical integration of (7); the minimization is achieved graphically by plotting the mean value as a function of the variational parameter μ. Thus, the variational parameter that minimizes the energy γ is found as the global minimum of this curve. Figure 1 shows the graphic representation of different values of the eigenvalues, E, against the mean value of H_{eff}. It can be observed that the pursued value of $\gamma = -0.25$ is obtained when the energy inserted in the effective potential is $E = 2$. This energy, therefore, is the energy of the ground state for the original problem. The solution submitted in reference [10] shows that, using the equivalent parameters adopted in this work ($\hbar^2 = 2m_0 = 1$, $c = 1$, $V_0 = 1$ e $\gamma = -0.25$), the lowest energy calculated analytically is $E_0 = 2$, coinciding with the result found with the adapted variational method, as proposed here. From this result, we acquire confidence that the variational method can be adapted to obtain solutions of the Schrödinger-like equation (1).

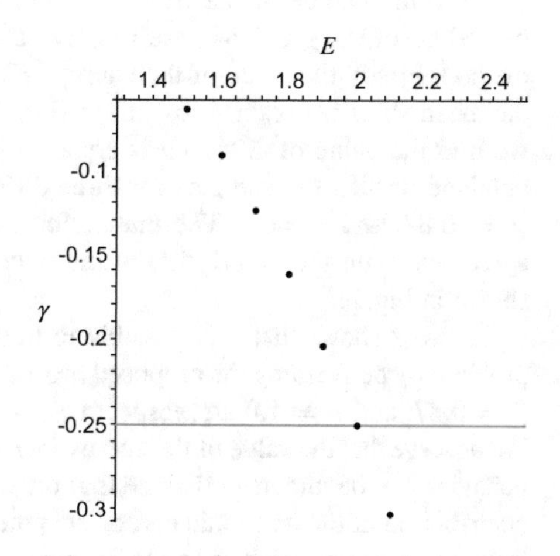

Fig. 1 Graphic representation of γ, the mean value of H_{eff} obtained with the variational method, as a function of the energy E for $m(x)$ and $V(x)$ given by Eq. (8). The continuum line represents the constant value $\gamma = -0.25$. The discrete points are the minimum mean values for the Hamiltonian H_{eff} obtained by using the trial function, Eq. (10), for different values of E

2.2 Mass with Quadratic Position-Dependence and Potential with Quadratic and Inverse Quadratic Position-Dependence

The second case used to evaluate the viability of the adapted variational method sets the mass and potential as given by

$$m(x) = m_0 x^2 , \quad V(x) = -\frac{\gamma}{x^2} + x^2. \tag{11}$$

For this case, Eq. (3) is rewritten as,

$$-\frac{d^2\Psi(x)}{dx^2} + (-Ex^2 + x^4)\Psi(x) = \gamma\Psi(x). \tag{12}$$

As previously mentioned, the position-dependent mass Schrödinger equation (1) leads to an equation with the energy inserted in the effective potential $V_{eff} = -Ex^2 + x^4$ and the eigenvalue is given by a fixed constant, γ. To apply the adapted variational method to this problem, the trial eigenfunction to be used is taken from the results presented in [14], based on the solution of a Schrödinger equation for a potential similar to the one used in Eq. (12), where the supersymmetric quantum mechanics technique was applied; it is given by

$$\Psi_\mu \propto e^{-\frac{\gamma}{2}x^2 - \frac{\mu}{4}x^4}, \tag{13}$$

where μ is the variational parameter.

From the choice of the trial wave function, it is possible to follow the same procedure of the previous case to solve the Eq. (12) with the adapted variational method. Firstly the value of the energy (E) inserted in the potential is varied until the mean value is obtained, by integrating (7) and minimizing it with respect to μ, we find the value of E which is equal to the fixed constant γ. To exemplify the obtained results, the energies for three different values of γ were fixed, $\gamma = 0.5$, $\gamma = 0.87$, and $\gamma = 1$. The integration and the minimization was performed in agreement with the description in the previous problem. The obtained results are shown in Fig. 2.

Figure 2 shows that it is possible to find the energy of the ground state for the problem by performing the proposed procedures. The obtained energies for $\gamma = 0.5$, $\gamma = 0.87$, and $\gamma = 1.0$ are, respectively, $E = 1.3376$, $E = 0.5$, and $E = 0.1647$. We observe that the value of the energy increases when the value of γ increases. This behavior can be understood since that the decrease of the value of E decreases the contribution of the harmonic part of the potential. This effect causes the potential to become narrower which causes an increase in the eigenvalue of γ. Once the problem does not have an analytical exact solution, the use of the adapted variational method enabled us to analyze the problem by using an approximated solution.

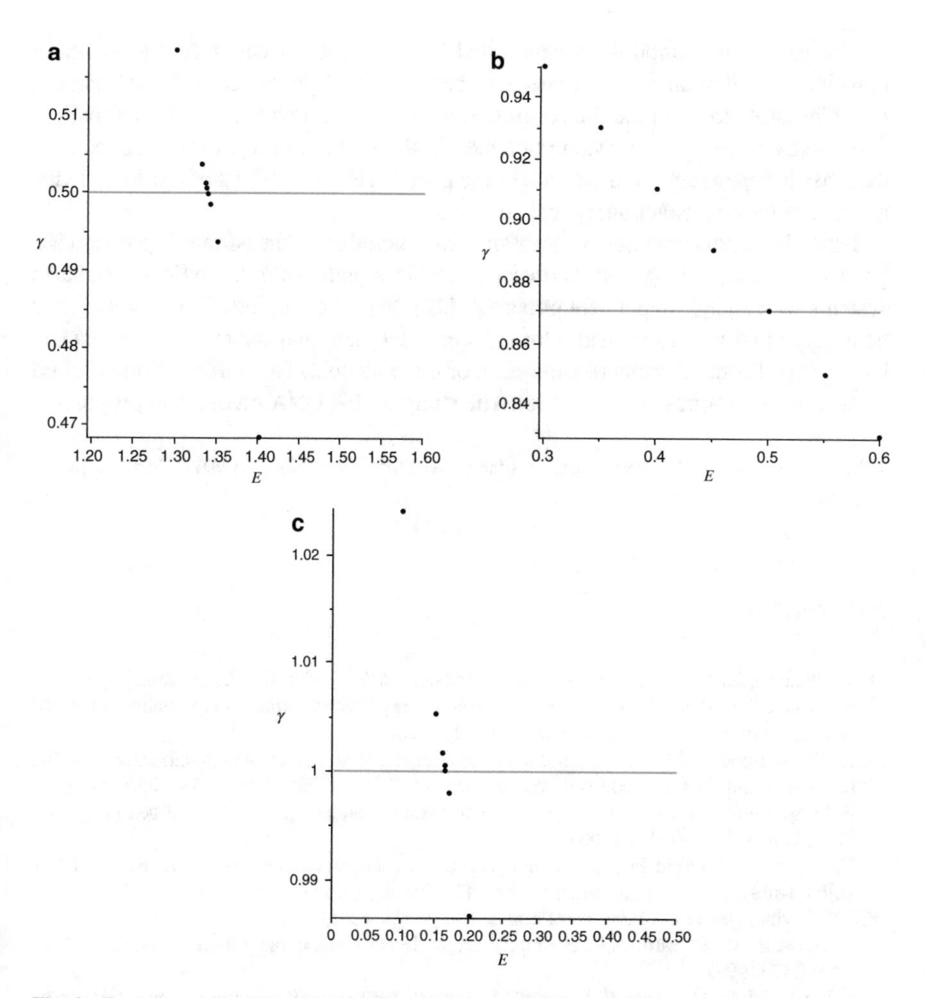

Fig. 2 Graphics of the eigenvalue obtained with the variational method, $\epsilon(E, \mu)$, as a function of the energy E for $m(x)$ and $V(x)$ given by Eq. (11). The continuum line represents the constant value γ. The points are the minimum mean values for the Hamiltonian H_{eff} obtained by using the trial function, Eq. (13), for different values of E. (**a**) $\gamma = 0.5$, (**b**) $\gamma = 0.87$, and (**c**) $\gamma = 1.0$

3 Conclusions

The main novelty of this paper is to present a semi-analytical method, the adapted variational method to solve a second order differential equation that looks like a Schrödinger equation with position-dependent mass, Eq. (1). The motivation to study this type of equation comes from a statistical mechanics model to describe the thermal and dynamic properties of DNA. The variational method, although it is commonly used in the context of quantum mechanics to study the Schrödinger equation, has not been yet applied to solve similar problems as the ones analyzed here.

The proposed methodology was tested in two particular cases. In the first case, in which the potential and the mass have exponential dependence with the position, it was possible to compare the result obtained with the adapted variational method with the exact energy eigenvalue obtained in the literature [10]. In the second case the mass is dependent on the square of the position and to the best of our knowledge it does not have an exact/analytical solution.

From the results obtained, it is possible to conclude that the adopted approach is a good alternative to study systems described by an equation similar to Eq. (1) and also systems with energy-dependent potentials [12]. In particular, the PBD model [7] can be analyzed by using the methodology suggested here and quantitative results can be obtained for the thermal denaturation of the molecule. This information is useful for testing and improving the model. The study of this DNA model is in progress.

Acknowledgement EDF would like to thank FAPESP (Proc. No. 2017/01757-9) for partial support.

References

1. J.J. Sakurai, *Modern Quantum Mechanics Revised Edition* (Addison-Wesley, Reading, 994)
2. E. Drigo Filho, R.M. Ricotta, Morse potential energy spectra through the variational method and supersymmetry. Phys. Lett. A **269**, 269–276 (2000)
3. E. Drigo Filho, R.M. Ricotta, Induced variational method from supersymmetric quantum mechanics and the screened Coulomb potential. Mod. Phys. Lett. A **15**, 1253–1259 (2000)
4. E. Drigo Filho, R.M. Ricotta, Supersymmetric variational energies of 3d confined potentials. Phys. Lett. A **320**, 95–102 (2003)
5. N.F. Ribeiro, E. Drigo Filho, Thermodynamics of a Peyrard Bishop one-dimensional lattice with on-site hump potential. Braz. J. Phys. **41**, 195–200 (2011)
6. I.N. Levine, *Quantum Chemistry* (Prentice Hall, Englewood, 2013)
7. T. Dauxois, M. Peyrard, A.R. Bishop, Entropy-driven DNA denaturation. Phys. Rev. E **47**, R44–R47 (1993)
8. D.J. Scalapino, M. Sears, R.A. Ferrel, Statistical mechanics of one-dimensional Ginzburg–Landau fields. Phys. Rev. B **6**, 3409–3416 (1972)
9. M. Peyrard, Nonlinear dynamics and statistical physics of DNA. Nonlinearity **17**, R1–R40 (2004)
10. A. de Souza Dutra, C.A.S. de Almeida, Exact solvability of potentials with spatially dependent effective masses. Phys. Lett. A **275**, 25–30 (2000)
11. O. Mustafa, S. Habib Mazharimousavi, Ordering ambiguity revisited via position dependent mass pseudo-momentum operators. Int. J. Theor. Phys. **46**(7), 1786–1796 (2007)
12. J. Garcia-Martinez, J. Garcia-Ravelo, J.J. Peña, A. Schulze-Halberg, Exactly solvable energy-dependent potentials. Phys. Lett. A **373**, 3619–3623 (2009)
13. A. Schulze-Halberg, O. Yesiltas, Generalized Schrödinger equations with energy-dependent potentials: formalism and applications. J. Math. Phys. **59**, 113503 (2018)
14. G.R.P. Borges, E. Drigo Filho, R.M. Ricotta. Phys. A **389**, 3892–3899 (2010)

The Lippmann–Schwinger Formula and One Dimensional Models with Dirac Delta Interactions

Fatih Erman, Manuel Gadella, and Haydar Uncu

Abstract We show how a proper use of the Lippmann–Schwinger equation simplifies the calculations to obtain scattering states for one dimensional systems perturbed by N Dirac delta equations. Here, we consider two situations. In the former, attractive Dirac deltas perturbed the free one dimensional Schrödinger Hamiltonian. We obtain explicit expressions for scattering and Gamow states. For completeness, we show that the method to obtain bound states use comparable formulas, although not based on the Lippmann–Schwinger equation. Then, the attractive N deltas perturbed the one dimensional Salpeter equation. We also obtain explicit expressions for the scattering wave functions. Here, we need regularisation techniques that we implement via heat kernel regularisation.

Keywords Scattering states · Schrödinger and Salpeter one dimensional Hamiltonians · Contact perturbations · Gamow wave functions · Lippmann–Schwinger equation

1 Introduction

One of the more used tools in order to understand quantum mechanics are the solvable models, in particular those which are one dimensional due to their simplicity [1–4]. The more often studied among these models is the free particle

F. Erman
Department of Mathematics, Izmir Institute of Technology, Urla, Izmir, Turkey
e-mail: fatih.erman@gmail.com

M. Gadella (✉)
Departamento de Física Teórica, Atómica y Óptica and IMUVA, Universidad de Valladolid, Valladolid, Spain
e-mail: manuelgadella1@gmail.com

H. Uncu
Department of Physics, Adnan Menderes University, Aydın, Turkey
e-mail: huncu@adu.edu.tr

Ş. Kuru et al. (eds.), *Integrability, Supersymmetry and Coherent States*, CRM Series in Mathematical Physics, https://doi.org/10.1007/978-3-030-20087-9_13

Schrödinger Hamiltonian decorated with Dirac delta interactions. Relativistic one dimensional approaches for the free particle Hamiltonian, such as those named after Salpeter or Dirac, have also been perturbed with contact interactions of delta type [5–7]. The purpose of the present article is to give a brief review of the recent work by the authors including the perturbation by N Dirac deltas of the one dimensional Schrödinger and Salpeter free Hamiltonians [6, 8, 9].

From the physics point of view, point potentials may represent interactions which are very localised in the space and strong and have a vast amount of applications for modelling real physical systems. A well-known model using Dirac delta potentials in non-relativistic quantum mechanics is the so-called Kronig–Penney model [10], and it is actually a reference model in describing the band gap structure of metals in solid state physics [11]. In addition, Dirac delta interactions in one or more dimensions serve as simple pedagogical toy models for the understanding of several quantum non-trivial concepts [12–19].

From the mathematical point of view, contact potentials are the result of the theory of self-adjoint extensions of symmetric operators with equal deficiency indices. In general, there are two methods to obtain these extensions. One is by defining some matching conditions at the nodes (points that support the contact potentials). Other uses the construction of the resolvent operator and often requires a renormalisation due to possible divergences in the construction of the resolvent of the self-adjoint extension. Still a third method relies on a theorem of von Neumann that characterises all self-adjoint extensions of a symmetric operator with equal deficiency indices, although this one has been less used.

We also want to show how the Lippmann–Schwinger formula is useful for this purpose as a simplifying computational tool. Here, we shall use the simplest form of this equation which acquires mathematical sense on Gelfand triplets. The Lippmann–Schwinger formula gives an equation satisfied by the incoming and outgoing plane waves after a scattering process due to a potential V. It has the following form:

$$|k^{\pm}\rangle = |k\rangle - R_0(E_k \pm i0)\, V\, |k^{\pm}\rangle\,, \tag{1}$$

where $|k^{\pm}\rangle$ refers to the full scattered incoming $(+)$ and outgoing $(-)$ plane waves, $|k\rangle$ is the free plane wave, V the potential and $R_0(E_k \pm i0)$ is the free resolvent, also called the Green operator. Since it is a function of the complex variable z, $R(z)$, and has a branch cut at the spectrum of the free Hamiltonian (usually $\mathbb{R}^+ \equiv [0, \infty)$), we denote by $R_0(E_k \pm i0)$ the upper and lower limits of $R(z)$ as the imaginary part of z goes to zero. Here, $E_k = (\hbar^2 k^2)/2m$.

This paper contains three more sections. In Sect. 2, we briefly discuss the consequences of adding N Dirac delta perturbations to the one dimensional free Schrödinger Hamiltonian. In Sect. 3, we do the same with the one dimensional Salpeter Hamiltonian. The analysis of bound states is particularly relevant in both cases. We finish our discussion with the concluding remarks.

2 One Dimensional Schrödinger Hamiltonian with N Dirac Delta Interactions

The objective of this section is to study the one dimensional Schrödinger Hamiltonian $H_0 = \frac{p^2}{2m}$ perturbed by N Dirac deltas located at some points in the real axis. This study includes the search for bound states, scattering coefficients and resonances provided they exist. As is well known, this perturbed Hamiltonian has the form

$$H := \frac{p^2}{2m} - \sum_{i=1}^{N} \lambda_i \, \delta(x - a_i) , \qquad V := -\sum_{i=1}^{N} \lambda_i \, \delta(x - a_i) , \tag{2}$$

where λ_i and $i = 1, 2, \ldots, N$, $i = 1, 2, \ldots, N$ are *positive* real numbers. The a_i show the points supporting the deltas and are called *nodes*. Each of the $-\lambda_i$, with $\lambda_i > 0$, is the intensity of the delta located at a_i for all value of i. These coefficients are chosen to be negative if we want to have bound states. The Schrödinger equation produced by (2) is

$$-\frac{\hbar^2}{2m} \frac{d^2\psi(x)}{dx^2} - \sum_{i=1}^{N} \lambda_i \, \delta(x - a_i) \, \psi(x) = E\psi(x) . \tag{3}$$

It is interesting to rewrite the interaction V in such a way that the calculations with the aid of the Lippmann–Schwinger equation become easy. For simplicity, let us assume that we have only one first. Then, the potential is $V = \lambda \, \delta(x - a)$ and the wave function is $\psi(x) = \langle x|\psi\rangle$ [20–24]. In this notation, $(V\psi)(x) = \langle x|V\psi\rangle$ and $\langle x|a\rangle = \delta(x - a)$. Thus,

$$(V\psi)(x) = \lambda \, \delta(x - a) \, \psi(a) . \tag{4}$$

Next, we note that the potential can be written as $V = \lambda \, |a\rangle\langle a|$, since then,

$$\langle x|V\psi\rangle = \lambda \, \langle x|a\rangle\langle a|\psi\rangle = \lambda \, \delta(x - a) \, \psi(a) = (V\psi)(x) . \tag{5}$$

The generalisation of the expression for the potential V in the case of having N nodes is the following:

$$V = -\sum_{i=1}^{N} \lambda_i \, |a_i\rangle\langle a_i| . \tag{6}$$

This is the desired expression. Let us clarify the vectors $|x\rangle$ for any real number x are the generalised eigenvalues of the position (multiplication) operator in one dimension with eigenvalue x. As is well known, these vectors do not belong to the

Hilbert space on which the multiplication operator acts, but instead to an extension of it endowed with a weak topology. We do not want to enter in these kind of details here, see [21–24]. Vectors $|a_i\rangle$ are precisely of this type with $x = a_i$.

The first objective is the search for scattering states. We are introducing the procedure in the sequel, although we shall skip some steps in order to reach the final result as straightforward as possible. Details may be found in [8, 9]. Let us use (6) in the Lippmann–Schwinger equation (1) and multiply the result from the left by the bra $\langle x|$. We have

$$\langle x|k^{\pm}\rangle = \langle x|k\rangle + \sum_{j=1}^{N} \lambda_j \, \langle x|G_0(E_k \pm i0)|a_j\rangle\langle a_j|k^{\pm}\rangle \,. \tag{7}$$

For convenience, we shall use the notation $G_0(x, y; E_k \pm i0) := \langle x|G_0(E_k \pm i0)|y\rangle$ in the sequel. Also, we recall that $\langle x|k\rangle$ is the free plane wave and $\psi_k^{\pm}(x) := \langle x|k^{\pm}\rangle$ the perturbed plane wave in the coordinate representation. In consequence, (7) can be written as (Henceforth we shall consider the sign plus in (7) only, for simplicity. Similar results would be obtained with the other choice.)

$$\psi_k^{+}(x) = e^{ikx} + \sum_{j=1}^{N} \lambda_j \, G_0(x, a_j; E_k + i0) \, \psi_k^{+}(a_j) \,, \tag{8}$$

The goal is now to obtain the explicit form of $\psi^{+}(x)$, for which we have to find the explicit form of the terms under the sum in (8). First, let as choose as values of x in (8) the $\{a_j\}$. We obtain the following linear system of N equations for N indeterminates:

$$\begin{aligned} e^{ika_i} &= \psi^{+}(a_i)\,[1 - \lambda_i \, G_0(a_i, a_i; E_k + i0)] \\ &\quad - \sum_{j \neq i}^{N} \lambda_j \, G_0(a_i, a_j; E_k + i0) \, \psi^{+}(a_j) \,, \qquad i = 1, 2, \ldots, N \,. \end{aligned} \tag{9}$$

This system can be rewritten in matrix form. If $\Phi \equiv \{\Phi_{ij}\}$ is the $N \times N$ matrix with matrix elements

$$\Phi_{ij}(E_k + i0) = \begin{cases} 1 - \lambda_i \, G_0(a_i, a_i; E_k + i0) \text{ if } i = j \,, \\ \\ \lambda_j \, G_0(a_i, a_j; E_k + i0) \quad \text{if } i \neq j \,. \end{cases} \tag{10}$$

Then, Eqs. (9) take the form,

$$\sum_{j=1}^{N} \Phi_{ij}(E_k + i0) \, \psi_k^{+}(a_j) = e^{ika_j} \,, \qquad j = 1, 2, \ldots, N \,, \tag{11}$$

with solution,

$$\psi_k^+(a_j) = \sum_{j=1}^{N} \left[\Phi^{-1}(E_k + i0)\right]_{ij} e^{ika_j}\,, \tag{12}$$

where Φ^{-1} is the inverse of the matrix Φ. In consequence, the final form of (8) is

$$\psi_k^+(x) = e^{ikx} + \sum_{j=1}^{N} \lambda_j\, G_0\left(x, a_j; E_k + i0\right) \left[\Phi^{-1}(E_k + i0)\right]_{ij} e^{ika_j}\,. \tag{13}$$

Then, we have to find the Green function $G_0(x, a_j; E_k + i0)$. We do not intend to describe the procedure here, which is explained in detail in [9]. Once we have obtained this Green function, using (10), we finally get all matrix elements of Φ. The final results are

$$G_0\left(x, a_j; E_k + i0\right) = \frac{im}{\hbar^2 k}\, e^{k|x-a_i|} \tag{14}$$

and

$$\Phi_{ij}(E_k + i0) = \begin{cases} 1 - \dfrac{im\lambda_i}{\hbar^2 k} & \text{if} \quad i = j\,, \\[2ex] -\sqrt{\lambda_i\,\lambda_j}\, \dfrac{im}{\hbar^2 k}\, e^{ik|a_i - a_j|} & \text{if} \quad i \neq j\,. \end{cases} \tag{15}$$

Then, we have determined all the perturbed plane waves $\psi_k^+(x)$. For $\psi_k^-(x)$, we follow a similar procedure. Always recall that $E_k = (\hbar^2 k^2)/2m$.

2.1 *Search for Bound States*

So far, we have found the scattering states corresponding to the total (or perturbed) Hamiltonian, for which we have used the Lippmann–Schwinger equation as main tool. Next, we search for the possible existence of bound states, where the search could be carried out with similar tools to those used in the precedent discussion.

We proceed as follows: Let us use the simplified notation $|f_i\rangle := \sqrt{\lambda_i}\,|a_i\rangle$, so that the total Hamiltonian (2) may be written as

$$H = \frac{p^2}{2m} - \sum_{i=1}^{N} |f_i\rangle\langle f_i|\,. \tag{16}$$

The corresponding Schrödinger equation reads

$$\left\langle x \left| \frac{p^2}{2m} \right| \psi \right\rangle - \sum_{i=1}^{N} \langle x|f_i\rangle \langle f_i|\psi\rangle = E \, \langle x|\psi\rangle \, . \tag{17}$$

Bound states correspond to solutions of (17) with negative E and square integrable wave function $\psi(x) \equiv \langle x|\psi\rangle$.

Next, insert the completeness relation $1 = \frac{1}{2\pi\hbar} \int |p\rangle\langle p| \, dp$ in front of $|\psi\rangle$ and $|f_i\rangle$. Define $\widetilde{\psi}(p) := \langle p|\psi\rangle$, which is indeed the Fourier transform of $\langle x|\psi\rangle$, and write $\phi(a_i) := \langle f_i|\psi\rangle = \sqrt{\lambda_i}\,\langle a_i|\psi\rangle = \sqrt{\lambda_i}\,\psi(a_i)$. Recall that $\langle x|p\rangle = e^{\frac{i}{\hbar}px}$. Then, (17) becomes

$$\int_{-\infty}^{\infty} \frac{dp}{2\pi\hbar}\, e^{\frac{i}{\hbar}px}\, \widetilde{\psi}(p) \left(\frac{p^2}{2m} - E \right) = \sum_{i=1}^{N} \sqrt{\lambda_i} \int_{-\infty}^{\infty} \frac{dp}{2\pi\hbar}\, e^{\frac{i}{\hbar}p(x-a_i)}\, \phi(a_i) \, . \tag{18}$$

From (18) and the properties of the Fourier transform, we have that

$$\widetilde{\psi}(p) = \sum_{i=1}^{N} \sqrt{\lambda_i}\, \frac{e^{-\frac{i}{\hbar}pa_i}}{\dfrac{p^2}{2m} - E}\, \phi(a_i) \, . \tag{19}$$

But $\widetilde{\psi}(p)$ is the Fourier transform of the solution $\psi(x)$ of the Schrödinger equation (17). Let us use this idea to conclude that (take $x = a_i$)

$$\psi(a_i) = \sum_{i=1}^{N} \sqrt{\lambda_i} \int_{-\infty}^{\infty} \frac{dp}{2\pi\hbar}\, \frac{e^{-\frac{i}{\hbar}pa_i}}{\dfrac{p^2}{2m} - E}\, \phi(a_i) \, . \tag{20}$$

Multiply both sides in (20) by $\sqrt{\lambda_i}$ and recalling that $\phi(a_i) = \sqrt{\lambda_i}\,\psi(a_i)$, we arrive to an equation of the form:

$$\sum_{j=1}^{N} \Phi_{ij}(E)\, \phi(a_j) = 0 \, . \tag{21}$$

Find details in [8]. It is beyond a mere coincidence that the matrix elements $\Phi \equiv \{\Phi_{ij}(E)\}$ are identical to those of (15) with the replacement $k = \sqrt{2m|E|}$, so that [8]

$$\Phi_{ij}(E) = \begin{cases} 1 - \dfrac{m\lambda_i}{\hbar\sqrt{2m|E|}} & \text{if} \quad i = j \, , \\[2ex] -\dfrac{m\sqrt{\lambda_i\lambda_j}}{\hbar\sqrt{2m|E|}} \exp\left(-\sqrt{2m|E|}\,|a_i - a_j|/\hbar\right) & \text{if} \quad i \neq j \, . \end{cases} \tag{22}$$

Since Eq. (21) has come directly from (17), it is a necessary condition for the existence of solutions of (17) with the desired properties. This equation has nontrivial solutions $\{\phi(a_j)\}$ if and only if $\det \Phi(E) = 0$. Therefore, the bound states energies are solutions of the transcendental equation $\det \Phi(E) = 0$.[1]

For a systematic calculation of the bound states, let us consider the following eigenvalue problem:

$$\Phi(E)A(E) = \omega(E)A(E)\,, \tag{23}$$

where $\omega(E)$ are the eigenvalues of the $N \times N$ matrix $\Phi(E)$ and $A(E)$ their corresponding eigenvectors. Equations (21) and (23) coincide if and only if $\omega(E) = 0$ and then, the bound states energies have to be the solutions of the transcendental equation $\omega(E) = 0$ its eigenvectors being those with components equal to $\phi(a_j)$. If we assume no degeneracy, the wave function corresponding to the energy value E_i with eigenvector $A(E) \equiv (\phi(a_1), \dots, \phi(a_N))$ takes the form (19) with $E = E_i$. In the coordinate representation, the wave function is just its Fourier transform. For further comments, see [6, 8].

2.2 *Resonances and Gamow States*

The Lippmann–Schwinger equation is also useful for the construction of Gamow states, which are vector states for resonances. In a resonant scattering process [25] produced by a Hamiltonian pair, say $\{H_0, H\}$, where H_0 is a *free* Hamiltonian and $H = H_0 + V$, where V is the interaction, the Gamow vectors, $\psi^{\pm}$, for a resonance with energy E_R and inverse of the mean life given by Γ are two eigenvectors of H with respective eigenvalues $E_R \pm \Gamma/2$, i.e., $H\psi^{\pm} = (E_R \mp \Gamma/2)\,\psi^{\pm}$ [25]. This property shows that the Gamow vector ψ^{+} decays exponentially as $t \longmapsto \infty$ (and ψ^{-} decays exponentially as $t \longmapsto -\infty$, they are time reversal of each other). This situation produces two problems, one from the point of view of physics and the other from the point of view of mathematics.

Although exponential decay for simple quantum unstable systems has been detected for essentially for all values of time, deviations for these exponential law have been detected for very short or very large times [26, 27]. Since these deviations certainly occur under these conditions only, they are very difficult to be detected. For most values of time, exponential decay serves as an excellent approximation. This is why Gamow vectors are useful as good approximations of decaying states.

A self-adjoint operator on Hilbert space, as is the case of the Hamiltonian H, cannot have complex eigenvalues with corresponding eigenvectors in this Hilbert space. Thus, Gamow vectors are well-defined objects on some extensions of Hilbert spaces called rigged Hilbert spaces [25, 28–30].

---

[1] As a matter of fact, this also follows because $\Phi(E)$ appears in the denominator of the resolvent of the total Hamiltonian H.

Let us briefly sketch the use of (1) to obtain an explicit expression of the Gamow vectors as eigenvectors of H with eigenvalue $E_R \pm \Gamma/2$. Details may be found in [9, 31]. If we multiply Eq. (1) to the right by the bra $\langle\psi|$, we obtain a complex function on the variable k. With adequate choices of the space of bras, this results on meromorphic functions of complex variable defined at least on a half plane [29, 30]. Let us assume that this is the case and omit the bra $\langle\psi|$. Then, if we define k_R as

$$z_R := E_R - \Gamma/2 = \frac{k_R^2 \hbar^2}{2m}, \tag{24}$$

we may consider the analytic extension of (1) to the value of k given by k_R,

$$|k_R^+\rangle = |k_R\rangle - G_0(z_R)\, V\, |k_R^+\rangle. \tag{25}$$

It is important to remark that z_R is a pole of the Green function corresponding to the total Hamiltonian H, but not of the free Hamiltonian H_0, just by the characterisation of resonances using the resolvent [32]. Then, $G_0(z_R)$ is well defined and so is $|k_R^+\rangle$, which has the property [9, 31, 33]

$$H|k_R^+\rangle = z_R\, |k_R^+\rangle. \tag{26}$$

Thus, $|k_R^+\rangle$ is one of the Gamow vectors with resonance pole z_R (the other can be obtained exactly in the same way, just replacing z_R by its complex conjugate z_R^* and taking the minus sign in (1). This Gamow vector in the coordinate representation is $\psi_R^+(x) := \langle x|k_R^+\rangle$, so that

$$(H\psi_R^+)(x) = \langle x|H|k_R^+\rangle = z_R\, \langle x|k_R^+\rangle = z_R\, \psi_R^+(x). \tag{27}$$

Now, let us go back to the N Dirac deltas interaction and, consequently, take in (25) the form of the potential given by $V = -\sum_{i=1}^N \lambda_i\, |a_i\rangle\langle a_i|$. Multiply the result of this operation to the right by the bra $\langle x|$ and divide k_R into real and imaginary parts, $k_R = k_r - ik_I$. We have that $\langle x|k_R\rangle = e^{ik_R x} = e^{ik_r x}\, e^{-ik_I x}$ and

$$\begin{aligned}
\psi_k^+(x) &= \langle x|k_R^+\rangle = \langle x|k_R\rangle + \sum_{i=1}^N \lambda_i\, \langle x|G_0(z_R)|a_i\rangle\langle a_i|k_R^+\rangle \\
&= e^{ik_r x}\, e^{k_I x} + \sum_{i=1}^N \lambda_i\, G_0(x, a_i; z_R)\, \psi_R^+(a_i) = e^{ik_r x}\, e^{k_I x} \\
&+ \sum_{i=1}^N \lambda_i \sum_{j=1}^N \frac{im\sqrt{\lambda_i\, \lambda_j}}{\hbar^2 (k_r - ik_I)} \left[e^{i(k_r - ik_I)\,|x-a_i|}\, \Phi^{-1}(z_R) \right]_{ij} e^{i(k_r - ik_I) a_j}.
\end{aligned} \tag{28}$$

A similar result can be obtained for the Gamow wave function $\psi^-(x)$. In principle, both Gamow functions will be equally suitable to play the role of wave function for the resonance state. The only technical difference is that one represents the time reversal of the other [30]. Observe that $\psi_k^+(x) \longmapsto \infty$ as $x \longmapsto \infty$. Gamow wave functions cannot be normalised in the usual sense of square integrable normalisation, but in sharp contrast with the plane waves (Dirac kets) which are not normalisable although bounded, Gamow functions show an exponential growing at the spatial infinite. This behaviour has been often called the exponential catastrophe. This is not such a problem with a proper interpretation of the Gamow wave function in terms of generalised functions in a suitable rigged Hilbert space. Still, this exponential behaviour creates some particular problems such as the difficulties arisen in order to fix a proper definition of averages of observables in Gamow states [34, 35].

3 One Dimensional Salpeter Hamiltonian with N Deltas

The one dimensional Salpeter Hamiltonian decorated with N Dirac deltas has the following form ($c = 1$):

$$H := \sqrt{p^2 + m^2} - \sum_{i=1}^{N} \lambda_i \, \delta(x - a_i) \,, \qquad H_0 := \sqrt{p^2 + m^2} \,. \tag{29}$$

Here, H_0 is the free Salpeter Hamiltonian. The definition of a self-adjoint version for H in (29) is not as simple as is in the Schrödinger case, where it is sufficient to impose correct matching conditions at the nodes. This self-adjoint version is usually determined by a proper choice of the resolvent operator of H, which should be obtained from the resolvent operator of H_0 by the Krein formula. However, this procedure leads to divergences in our case, so that a regularisation procedure is in order here [5, 6]. We have chosen heat kernel regularisation for several reasons discussed in [6]. Let us sketch briefly the procedure. First of all, we write the Hamiltonian H as in (29) as

$$H = \sqrt{p^2 + m^2} - \sum_{i=1}^{N} \lambda_i \, |a_i\rangle\langle a_i| \,, \tag{30}$$

exactly as we did for the cases studied in the previous section. The next step is to write an ϵ-regularised version of (30) as

$$H_\epsilon = \sqrt{p^2 + m^2} - \sum_{i=1}^{N} \lambda_i(\epsilon) \, |a_i^\epsilon\rangle\langle a_i^\epsilon| \,, \tag{31}$$

where the new kets $|a_i^\epsilon\rangle$ are defined in such a way that $\langle x|a_i^\epsilon\rangle := K_{\epsilon/2}(x, a_i)$, where the function $K_t(x, y)$ is the so-called heat kernel, which is the fundamental solution of the heat equation of the form:

$$\sqrt{p^2+m^2}\, K_t(x, y) = -\frac{\partial\, K_t(x, y)}{\partial\, t}\,, \tag{32}$$

and the weights $\lambda(\epsilon)$ are also chosen as functions of the parameter ϵ, such that $\lim_{\epsilon\to 0^+} \lambda_i(\epsilon) \longmapsto \lambda_i$, $i = 1, 2, \ldots, N$. The interest of this choice for $\langle x|a_i^\epsilon\rangle$ comes after the limiting property $\langle x|a_i^\epsilon\rangle \longmapsto \langle x|a_i\rangle = \delta(x - a_i)$ as $\epsilon \longmapsto 0^+$.

Now, we go back to the Lippmann–Schwinger equation (1), where in the present case $E_k = \sqrt{p^2+m^2}$ and V is as in (31). This gives

$$|k^\pm(\epsilon)\rangle = |k\rangle + \sum_{j=1}^{N} \lambda_j(\epsilon)\, R_0(E_k \pm i0)\, |a_j^\epsilon\rangle\langle a_j^\epsilon|k^\pm\rangle\,. \tag{33}$$

Let us choose the plus sign in (33) and use for brevity the following notation: $|f_i^\epsilon\rangle := \sqrt{\lambda_i(\epsilon)}\, |a_i^\epsilon\rangle$. Then, we choose one subindex i and isolate the corresponding term in (33):

$$\begin{aligned} |k^+(\epsilon)\rangle = |k\rangle &+ R_0(E_k + i0)\, |f_i^\epsilon\rangle\langle f_i^\epsilon|k^+(\epsilon)\rangle \\ &+ \sum_{j\neq i}^{N} R_0(E_k + i0)\, |f_j^\epsilon\rangle\langle f_j^\epsilon|k^+(\epsilon)\rangle\,, \end{aligned} \tag{34}$$

before multiplying (34) to the left by the ket $\langle f_i^\epsilon|$. This gives

$$\begin{aligned} &\left[1 - \langle f_i^\epsilon|\, R_0(E_k + i0)\, |f_i^\epsilon\rangle\right] \langle f_i^\epsilon|k^+(\epsilon)\rangle \\ &- \sum_{i\neq j}^{N} \left[\langle f_i^\epsilon|\, R_0(E_k + i0)\, |f_i^\epsilon\rangle\right] \langle f_i^\epsilon|k^+(\epsilon)\rangle = \langle f_i^\epsilon|k\rangle\,, \end{aligned} \tag{35}$$

expression valid for $i = 1, 2, \ldots, N$. This may be written in the matrix form as

$$\sum_{j=1}^{N} T_{ij}(\epsilon, E_k + i0)\, \langle f_j^\epsilon|k^+(\epsilon)\rangle = \langle f_i^\epsilon|k\rangle\,, \qquad j = 1, 2, \ldots, N\,, \tag{36}$$

with

$$T_{ij}(\epsilon, E_k + i0) = \begin{cases} 1 - \langle f_i^\epsilon|\, R_0(E_k + i0)\, |f_i^\epsilon\rangle & \text{if} \quad i = j\,, \\ -\langle f_i^\epsilon|\, R_0(E_k + i0)\, |f_j^\epsilon\rangle & \text{if} \quad i \neq j\,. \end{cases} \tag{37}$$

Therefore, we may write the solution of (36) as

$$\left\langle f_i^\epsilon | k^+(\epsilon) \right\rangle = \sum_{j=1}^{N} \left[T^{-1}(\epsilon, E_k + i0) \right]_{ij} \left\langle f_j^\epsilon | k \right\rangle . \tag{38}$$

We use (38) in (35) and, then, multiply the result to the left by the bra $\langle x|$. This gives

$$\begin{aligned} \psi^+(\epsilon, x) &:= \langle x | k^+(\epsilon) \rangle \\ &= \langle x | k \rangle + \sum_{i,j=1}^{N} \langle x | \, R_0(E_k + i0) \, | f_i^\epsilon \rangle \, [T^{-1}(\epsilon, E_k + i0)]_{ij} \, \langle f_j^\epsilon | k \rangle \\ &= e^{ikx} + \sum_{i,j=1}^{N} \langle x | \, R_0(E_k + i0) \, | a_i^\epsilon \rangle \, [\Phi^{-1}(\epsilon, E_k + i0)]_{ij} \, \langle a_j^\epsilon | k \rangle \, , \end{aligned} \tag{39}$$

with

$$\Phi_{ij}(\epsilon, E_k + i0) = \begin{cases} \frac{1}{\lambda_i(\epsilon)} - \langle a_i^\epsilon | \, R_0(E_k + i0) \, | a_i^\epsilon \rangle & \text{if} \quad i = j \, , \\ -\langle a_i^\epsilon | \, R_0(E_k + i0) \, | a_j^\epsilon \rangle & \text{if} \quad i \neq j \, . \end{cases} \tag{40}$$

The next step is to take the limit $\epsilon \longmapsto 0$, for which we need a determination of the functions $\lambda_i(\epsilon)$ for all values of $i = 1, 2, \ldots, N$. This has been motivated and determined in Section II in [6] and is

$$\frac{1}{\lambda_i(\epsilon)} = \frac{1}{\lambda_i(M_i)} + \int_0^\infty dt \, K_{t+\epsilon}(a_i, a_i) \, e^{tM_i} \, , \tag{41}$$

where $K_t(x, y)$ is the heat kernel and M_i is an unphysical renormalisation scale that is chosen to be the energy of the bound state E_B^i corresponding to the bound state of the i-th delta [6]. This gives in the limit $\epsilon \longmapsto 0$,

$$\psi_k^+(x) = e^{ikx} + \sum_{i,j=1}^{N} \langle x | \, R_0(E_k + i0) \, | a_i \rangle \, [\Phi^{-1}(E_k + i0)]_{ij} \, e^{ika_j} \, . \tag{42}$$

Here,

$$\langle x | \, R_0(E_k + i0) \, | a_i \rangle = \frac{i\sqrt{k^2 + m^2}}{k} \, e^{ik|x - a_i|} + \frac{1}{\pi} \int_m^\infty d\mu \, e^{-\mu|x - a_i|} \, \frac{\sqrt{\mu^2 - m^2}}{\mu^2 + k^2}$$

and

$$\Phi_{ij}(E_k + i0) =$$

$$= \begin{cases} -\frac{1}{\lambda(E_\lambda, E_B^i)} - \frac{i E_k}{\sqrt{E_k^2 - m^2}} & \text{if } i = j\,, \\ -\frac{i E_k}{\sqrt{E_k^2 - m^2}} e^{i\sqrt{E_k^2 - m^2}\,|x - a_j|} - \frac{1}{\pi}\int_m^\infty d\mu\, e^{-\mu|x - a_i|} \frac{\sqrt{\mu^2 - m^2}}{\mu^2 + E_k^2 - m^2} & \text{if } i \neq j\,, \end{cases}$$

where

$$\frac{1}{\lambda\left(E_\lambda, E_B^i\right)}$$

$$= \frac{-E_k}{\pi\sqrt{E_k^2 - m^2}} \operatorname{arctanh}\left(\frac{\sqrt{E_k^2 - m^2}}{E_k}\right) - \frac{E_B^i}{\pi\sqrt{m^2 - \left(E_B^i\right)^2}}\left(\frac{\pi}{2} + \arcsin\frac{E_B^i}{m}\right),$$

where E_B^i has been defined before and $\mu := \min_i E_B^i$. The conclusion is that the Lippmann–Schwinger equation gives in a rather straightforward manner the exact form of the scattering states in a rather cumbersome situation as the one discussed along the present section. Explicit expressions for transmission and reflection coefficients can be also derived from the above expressions.

4 Concluding Remarks

The Lippmann–Schwinger equation is a useful tool that permits to obtain explicit forms for the scattering states produced by some potential. When this potential is a finite set of Dirac delta interactions, one may find explicit expressions for these scattering states. We have shown that this is the case when perturbing the free Schrödinger one dimensional and the Salpeter Hamiltonians with N attractive deltas. In the first case, we have also shown that the Lippmann–Schwinger equation gives explicit expressions for Gamow wave functions which are the wave function for the purely exponential decay part of resonance states. The discussion on the search for bound states for the Schrödinger case includes similar methods.

The one dimensional Salpeter Hamiltonian with N attractive deltas is much more complicated as it requires of a regularisation procedure that we implement with the use of the heat kernel for the pseudo-differential operator $\sqrt{-d^2/dx^2 + m^2}$. In this case, we also obtain the exact form of the scattering states.

Acknowledgements We dedicate this paper to Professor Véronique Hussin for her contributions to science and her friendship. The present work has been fully financed by TUBITAK from Turkey under the "2221 - Visiting Scientist Fellowship Programme". We are very grateful to TUBITAK for this support. We also acknowledge Osman Teoman Turgut for clarifying discussions and his interest in the present research. This work was also sponsored by the Ministerio de Economía y Competitividad of Spain (Project No. MTM2014-57129-C2-1-P with EU-FEDER support) and the Junta de Castilla y León (Projects VA057U16, VA137G18 and BU229P18).

References

1. S. Albeverio, F. Gesztesy, R. Høeg-Krohn, H. Holden, *Solvable Models in Quantum Mechanics* (AMS Chelsea Series, Providence RI, 2004)
2. Y.N. Demkov, V.N. Ostrovskii, *Zero-range Potentials and Their Applications in Atomic Physics* (Plenum, New York, 1988)
3. M. Belloni, R.W. Robinett, The infinite well and Dirac delta function potentials as pedagogical, mathematical and physical models in quantum mechanics. Phys. Rep. **540**, 25–122 (2014)
4. S. Albeverio, P. Kurasov, *Singular Perturbations of Differential Operators Solvable Schrödinger-type Operators* (Cambridge University Press, Cambridge, 2000)
5. M.H. Al-Hashimi, A.M. Shalaby, U.-J.Wiese, Asymptotic freedom, dimensional transmutation, and an infrared conformal fixed point for the δ-function potential in one-dimensional relativistic quantum mechanics. Phys. Rev. D **89**, 125023 (2014)
6. F. Erman, M. Gadella, H. Uncu, One-dimensional semirelativistic Hamiltonian with multiple Dirac delta potentials. Phys. Rev. D **95**, 045004 (2017)
7. M. Calçada, J.T. Lunardi, L.A. Manzoni, W. Monteiro, Distributional approach to point interactions in one-dimensional quantum mechanics. Front. Phys. **2**, 23 (2014)
8. F. Erman, M. Gadella, S. Tunalı, H. Uncu, A singular one-dimensional bound state problem and its degeneracies. Eur. Phys. J. Plus **132**, 352 (2017)
9. F. Erman, M. Gadella, H. Uncu, On scattering from the one dimensional multiple Dirac delta potentials. Eur. J. Phys. **39**, 035403 (2018)
10. R. de L. Kronig, W.G. Penney, Quantum mechanics of electrons in crystal lattices. Proc. R. Soc. A **130**, 499 (1931)
11. C. Kittel, *Introduction to Solid State Physics* 8th edn. (Wiley, New York, 2005)
12. I.R. Lapidus, Resonance scattering from a double δ-function potential. Am. J. Phys. **50**, 663–664 (1982)
13. P. Senn, Threshold anomalies in one dimensional scattering. Am. J. Phys. **56**, 916–921 (1988)
14. P.R. Berman, Transmission resonances and Bloch states for a periodic array of delta function potentials. Am. J. Phys. **81**, 190–201 (2013)
15. S.H. Patil, Quadrupolar, triple δ-function potential in one dimension. Eur. J. Phys. 629–640 (2009)
16. V.E. Barlette, M.M. Leite, S.K. Adhikari, Integral equations of scattering in one dimension. Am. J. Phys. **69**, 1010–1013 (2001)
17. D. Lessie, J. Spadaro, One dimensional multiple scattering in quantum mechanics. Am. J. Phys. **54**, 909–913 (1986)
18. J.J. Alvarez, M. Gadella, L.M. Nieto, A study of resonances in a one dimensional model with singular Hamiltonian and mass jump. Int. J. Theor. Phys. **50**, 2161–2169 (2011)
19. J.J. Alvarez, M. Gadella, L.P. Lara, F.H. Maldonado-Villamizar, Unstable quantum oscillator with point interactions: Maverick resonances, antibound states and other surprises. Phys. Lett. A **377**, 2510–2519 (2013)
20. A. Bohm, in *The Rigged Hilbert Space and Quantum Mechanics*. Springer Lecture Notes in Physics, vol. 78 (Springer, New York, 1978)
21. J.E. Roberts, Rigged Hilbert spaces in quantum mechanics. Commun. Math. Phys. **3**, 98–119 (1966)

22. J.P. Antoine, Dirac formalism and symmetry problems in quantum mechanics. I. General formalism. J. Math. Phys. **10**, 53–69 (1969)
23. O. Melsheimer, Rigged Hilbert space formalism as an extended mathematical formalism for quantum systems. J. Math. Phys. **15**, 902–916 (1974)
24. M. Gadella, F. Gómez, On the mathematical basis of the Dirac formulation of quantum mechanics. Int. J. Theor. Phys. **42**, 2225–2254 (2003)
25. A. Bohm, *Quantum Mechanics. Foundations and Applications* (Springer, Berlin, New York, 2002)
26. M.C. Fischer, B. Gutiérrez-Medina, M.G. Raizen, Observation of the quantum Zeno and anti-Zeno effects in an unstable system. Phys. Rev. Lett. **87**, 40402 (2001)
27. C. Rothe, S.L. Hintschich, A.P. Monkman, Violation of the exponential-decay law at long times. Phys. Rev. Lett. **96**, 163601 (2006)
28. A. Bohm, Resonance poles and Gamow vectors in the rigged Hilbert space formulation of quantum mechanics. J. Math. Phys. **22** (12), 2813–2823 (1981)
29. A. Bohm, M. Gadella, in *Dirac Kets, Gamow Vectors and Gelfand Triplets*. Springer Lecture Notes in Physics, vol. 348 (Springer, Berlin, 1989)
30. O. Civitarese, M. Gadella, Physical and mathematical aspects of Gamow states. Phys. Rep. **396**, 41–113 (2004)
31. O. Civitarese, M. Gadella, Gamow states as solutions of a modified Lippmann–Schwinger equation. Int. J. Mod. Phys. E **25**, 1650075 (2016)
32. M. Reed, B. Simon, *Analysis of Operators* (Academic, New York, 1978), p. 55
33. M. Gadella, F. Gómez, The Lippmann–Schwinger equations in the rigged Hilbert space. J. Phys. A: Math. Gen. **35**, 8505–8511 (2002)
34. T. Berggren, Expectation value of an operator in a resonant state. Phys. Lett. B **373**, 1–4 (1996)
35. O. Civitarese, M. Gadella, R. Id Betan, On the mean value of the energy for resonance states. Nucl. Phys. A **660**, 255–266 (1999)

Hermite Coherent States for Quadratic Refractive Index Optical Media

Zulema Gress and Sara Cruz y Cruz

Abstract Ladder and shift operators are determined for the set of Hermite–Gaussian modes associated with an optical medium with quadratic refractive index profile. These operators allow to establish irreducible representations of the $su(1,1)$ and $su(2)$ algebras. Glauber coherent states, as well as $su(1,1)$ and $su(2)$ generalized coherent states, were constructed as solutions of differential equations admitting separation of variables. The dynamics of these coherent states along the optical axis is also evaluated.

Keywords Hermite–Gaussian modes · Ladder operators · Coherent states · Self-focusing media · Paraxial beams · Ermakov equation

1 Introduction

The problem of addressing the construction, analysis, and possible applications of coherent states has been a very important issue in quantum physics. Yet, the concept of coherent state has also been considered within the framework of classical optics in the context of the quantum mechanics-classical optics analogy [1–3]. This analogy, based on the formal equivalence between the paraxial Helmholtz equation and the time-dependent Schrödinger equation [4–6], enables, for instance, the use of operatorial methods for the description of light propagation phenomena [7, 8]. In particular, diverse families of coherent states as linear combinations of

Z. Gress
Universidad Autónoma del Estado de Hidalgo, Ciudad del Conocimiento, Hidalgo, Mexico

S. Cruz y Cruz (✉)
Instituto Politécnico Nacional, UPIITA, Ciudad de México, Mexico
e-mail: sgcruzc@ipn.mx

Ş. Kuru et al. (eds.), *Integrability, Supersymmetry and Coherent States*, CRM Series in Mathematical Physics, https://doi.org/10.1007/978-3-030-20087-9_14

paraxial beams can be constructed from a pure algebraic point of view in the Barut–Girardello as well as in the Perelomov approaches (see the details in [9]).

In this work we deal with the construction of Glauber, $SU(1,1)$ and $SU(2)$ coherent states for the set of Hermite–Gaussian modes in an optical medium with a quadratic refractive index profile, as solutions of differential equations admitting separation of variables. This approach leads us to wave-packet type expressions for these coherent states instead of the usual expressions defined algebraically as linear combinations of paraxial modes.

With this end we organize our work as follows. In Sect. 2 we briefly discuss on the Hermite–Gaussian modes and the functions featuring their behavior. In Sect. 3 we present the construction of ladder operators for the Hermite–Gaussian modes and their relation to the canonical variables of position and direction of propagation. This relation will be important in the statement of the differential equations defining the coherent states. Next, in Sect. 4, we construct the Hermite–Glauber coherent states as eigenfunctions of the annihilation operators. Sections 5 and 6 contain, respectively, the construction of Hermite-$su(1,1)$ and Hermite-$su(2)$ coherent states in the Perelomov approach. Finally, we summarize our results in Sect. 7.

2 The Hermite–Gaussian Beams

We are interested in optical media with quadratic refractive index profiles of the form

$$n^2(r) = n_0^2\left(1 - \Omega^2 r^2\right), \tag{1}$$

where $r^2 = x^2 + y^2$, n_0 is called reference or bulk refractive index, and Ω is a constant that defines the confining properties of the medium. In the paraxial approximation it is assumed that $\Omega^2 r^2 \ll 1$ so that the refractive index is nearly constant in regions close to the optical axis (that is chosen as the z-axis). The paraxial Helmholtz equation takes the form of a time-dependent Schrödinger equation

$$\frac{i}{k_0}\frac{\partial}{\partial z}U(\mathbf{r},z) = \left[-\frac{1}{2k_0^2 n_0}\left(\frac{\partial^2}{\partial x^2} + \frac{\partial^2}{\partial y^2}\right) + \frac{n_0}{2}\Omega^2 r^2\right]U(\mathbf{r},z) = \mathcal{H}U(\mathbf{r},z), \tag{2}$$

where k_0 is the wave number in free space, $\mathbf{r} = (x, y)$ is the transversal radial vector, and the function $U(\mathbf{r},z)$ is the amplitude of the electric field. It is well known that this equation has wave-packet type solutions of the form [9]

$$U_{n,m}(\mathbf{r},z)=\frac{1}{\sqrt{\pi 2^{n+m}n!m!}}\frac{\sqrt{2}}{w(z)}e^{-i(n+m+1)\chi(z)}e^{i\frac{k_0n_0r^2}{2R(z)}}e^{-\frac{r^2}{w^2(z)}}$$
$$\times H_n\left(\frac{\sqrt{2}}{w(z)}x\right)H_m\left(\frac{\sqrt{2}}{w(z)}y\right), \tag{3}$$

where $H_p(x)$ stands for the Hermite polynomial of degree p, and the functions $\chi(z)$, $R(z)$, and $w(z)$ are given by

$$w(z)=w_0\left[\cos^2(\Omega z)+\frac{1}{(\Omega z_R)^2}\sin^2(\Omega z)\right]^{1/2}, \quad w_0=w(0), \quad z_R=\frac{k_0n_0w_0^2}{2}, \tag{4}$$

$$\frac{1}{R(z)}=\frac{d}{dz}\ln w(z), \quad \chi(z)=\frac{2}{k_0n_0}\int^z\frac{dt}{w^2(z)}. \tag{5}$$

These modes are the Hermite–Gaussian beams associated with a parabolic medium and the functions $w(z)$, $R(z)$, $\chi(z)$ are the width, the radius of curvature, and the Gouy phase shift of the beam (see [9]). The constants w_0 and z_R are called, respectively, the beam waist or waist radius and the Rayleigh range. In the limit $\Omega\to 0$ we recover the well- known beam width, radius of curvature, and Gouy phase shift of a Hermite–Gaussian mode in a homogeneous medium [10]

$$w(z)=w_0\sqrt{1+\left(\frac{z}{z_R}\right)^2}, \quad R(z)=z\left[1+\left(\frac{z_R}{z}\right)^2\right], \quad \chi(z)=\arctan\left(\frac{z}{z_R}\right). \tag{6}$$

In this case, the constants w_0 and z_R correspond to the minimum width of the beam and to the distance from the focus along the optical axis at which the beam doubles its transversal area. Due to this fact these parameters are considered, respectively, the transversal and longitudinal characteristic lengths of the system. In the general case $\Omega>0$, the beam width is an oscillating function of period $\frac{\pi}{\Omega}$ due to the confining properties of the medium.

2.1 *Stationary Hermite–Gaussian Modes*

Another interesting case occurs whenever $\frac{1}{z_R}=\Omega$, i.e., when the beam waist w_0 and Ω are related through

$$w_0=\sqrt{\frac{2}{k_0n_0\Omega}}.$$

In this case the divergent nature of light is balanced with the focalization properties of the medium resulting in a beam that propagates with a constant width w_0. A simple inspection to (5) leads to the conclusion that, in this limit, the Hermite–Gaussian modes become plane waves, as their radii of curvature becomes infinite and their Gouy phase shifts $\chi(z) = \Omega z = \frac{z}{z_R}$ are proportional to z. The corresponding amplitudes

$$U^s_{n,m}(\mathbf{r}, z) = \frac{1}{\sqrt{\pi 2^{n+m} n! m!}} \frac{\sqrt{2}}{w_0} e^{-i(n+m+1)\frac{z}{z_R}} e^{-\frac{r^2}{w_0^2}} H_n\left(\frac{\sqrt{2}}{w_0} x\right) H_m\left(\frac{\sqrt{2}}{w_0} y\right), \tag{7}$$

turn into (stationary) eigenmodes of the operator $\mathcal{H}$ with eigenvalues $\frac{\Omega}{k_0}(n+m+1)$ in complete analogy to the quantum harmonic oscillator potential. In this context, it is worthwhile to note that, as the function $w(z)$ in (4) fulfills the Ermakov equation (check [9])

$$\frac{d^2 w}{dz^2} + \Omega^2 w = \frac{w_0^4}{z_R^2 w^3},$$

the Hermite–Gaussian modes (3) can be obtained from the stationary ones (7) through the quantum Arnold–Ermakov transformation (see, e.g., [11–13]).

3 Ladder Operators for the Hermite–Gaussian Modes

The Hermite–Gaussian modes (3) can be written in the form

$$U_{n,m}(\mathbf{r}, z) = \frac{\sqrt{2}}{w(z)} e^{-i\beta_{n,m}\chi(z)} e^{i\frac{k_0 n_0 r^2}{2R(z)}} \Psi_{n,m}(u(x,z), v(y,z)), \tag{8}$$

where

$$u(x,z) = \frac{\sqrt{2}}{w(z)} x, \quad v(x,z) = \frac{\sqrt{2}}{w(z)} y,$$

$\beta_{n,m} = n + m + 1$, and the functions $\Psi_{n,m}(u, v)$ are the square integrable solutions to the eigenvalue problem [9]

$$H\Psi(u,v) = \beta\Psi(u,v), \tag{9}$$

with

$$H = -\frac{1}{2}\left(\frac{\partial^2}{\partial u^2} + \frac{\partial^2}{\partial v^2}\right) + \frac{1}{2}\left(u^2 + v^2\right). \tag{10}$$

The form of this operator suggests that we may write

$$\Psi_{n,m}(u, v) = \frac{1}{\sqrt{n!m!}} \left(a^+\right)^n \left(b^+\right)^m \Psi_{0,0}(u, v), \quad \Psi_{0,0}(u, v) = \frac{1}{\sqrt{\pi}} e^{-\frac{1}{2}(u^2+v^2)}, \tag{11}$$

where $a^\pm$ and $b^\pm$ are the ladder operators of the harmonic oscillator

$$a^\pm = \frac{1}{\sqrt{2}}\left(\mp\frac{\partial}{\partial u} + u\right), \quad b^\pm = \frac{1}{\sqrt{2}}\left(\mp\frac{\partial}{\partial v} + v\right), \tag{12}$$

fulfilling the boson algebra

$$\left[a^-, a^+\right] = \left[b^-, b^+\right] = \mathbb{I}, \qquad \left[a^\pm, b^\pm\right] = 0. \tag{13}$$

The substitution of (11) into (8) leads us to

$$U_{n,m}(\mathbf{r}, z) = \frac{1}{\sqrt{n!m!}} e^{-i(n+m)\chi(z)} e^{i\frac{k_0 n_0 r^2}{2R(z)}} \left(a^+\right)^n \left(b^+\right)^m e^{-i\frac{k_0 n_0 r^2}{2R(z)}} U_{0,0}(\mathbf{r}, z),$$

which, in turn, allows us to write

$$U_{n,m}(\mathbf{r}, z) = \frac{1}{\sqrt{n!m!}} \left(A^+\right)^n \left(B^+\right)^m U_{0,0}(\mathbf{r}, z), \tag{14}$$

with the identification of the non-autonomous invariant operators [9]

$$\begin{aligned} A^+ &= e^{-i\chi(z)} e^{i\frac{k_0 n_0 x^2}{2R(z)}} a^+ e^{-i\frac{k_0 n_0 x^2}{2R(z)}}, & A^- &= \left(A^+\right)^\dagger, \\ B^+ &= e^{-i\chi(z)} e^{i\frac{k_0 n_0 y^2}{2R(z)}} b^+ e^{-i\frac{k_0 n_0 y^2}{2R(z)}}, & B^- &= \left(B^+\right)^\dagger. \end{aligned} \tag{15}$$

The algebraic structure of the operators $A^\pm$, $B^\pm$ is inherited from that of $a^\pm$ and $b^\pm$. If $N_x = A^+A^-$ and $N_y = B^+B^-$, we have

$$\left[A^-, A^+\right] = \mathbb{I}, \quad \left[B^-, B^+\right] = \mathbb{I}, \quad \left[A^\pm, B^\pm\right] = 0, \tag{16}$$

$$\left[N_x, A^\pm\right] = \pm A^\pm, \qquad \left[N_y, B^\pm\right] = \pm B^\pm. \tag{17}$$

Also, the action of $A^{\pm}$, $B^{\pm}$ on the Hermite–Gaussian beams can be readily stated

$$\begin{aligned} &A^{+}U_{n,m}(\mathbf{r},z)=\sqrt{n+1}U_{n+1,m}(\mathbf{r},z), \quad A^{-}U_{n,m}(\mathbf{r},z)=\sqrt{n}U_{n-1,m}(\mathbf{r},z),\\ &B^{+}U_{n,m}(\mathbf{r},z)=\sqrt{m+1}U_{n,m+1}(\mathbf{r},z), \quad B^{-}U_{n,m}(\mathbf{r},z)=\sqrt{m}U_{n,m-1}(\mathbf{r},z),\\ &N_{x}U_{n,m}(\mathbf{r},z)=nU_{n,m}(\mathbf{r},z), \quad N_{y}U_{n,m}(\mathbf{r},z)=mU_{n,m}(\mathbf{r},z). \end{aligned} \tag{18}$$

It can be also shown that, in terms of the canonical variables of position $\mathbf{r}$ and propagation direction $\mathbf{p}=-\frac{i}{k_0}\left(\frac{\partial}{\partial x},\frac{\partial}{\partial y}\right)$

$$\begin{aligned} A^{+}&=-ie^{-i\chi}w\left[\frac{k_0p_x}{2}-\bar{S}x\right], \quad A^{-}=ie^{i\chi}w\left[\frac{k_0p_x}{2}-Sx\right],\\ B^{+}&=-ie^{-i\chi}w\left[\frac{k_0p_y}{2}-\bar{S}y\right], \quad B^{-}=ie^{i\chi}w\left[\frac{k_0p_y}{2}-Sy\right], \end{aligned} \tag{19}$$

where the bar stands for complex conjugation and

$$S(z)=\frac{k_0n_0}{2R(z)}+\frac{i}{w^2(z)}.$$

At this stage, it is important to stress that the form of the Hermite–Gaussian modes (8), as well as that of the operators $A^{\pm}$, and $B^{\pm}$, depends strongly on the beam width $w(z)$. This is due to the fact that, according to Eqs. (3)–(5), this function encodes all the information about the beam propagation along the longitudinal axis. In turn, the behavior of $w(z)$ is determined by the choice of its initial value w_0. This means that, for a fixed value of Ω defining a particular optical medium, the form of the Hermite–Gaussian modes and the corresponding ladder operators will be intrinsically dependent on the parameter w_0.

To close this section, let us note that the paraxial Helmholtz equation (2) defines the z-variation of the field amplitude $U(\mathbf{r},z)$ through

$$U(\mathbf{r},z)=\mathcal{W}(z)U(\mathbf{r},0), \tag{20}$$

where

$$\mathcal{W}(z)=e^{-ik_0\mathcal{H}z}, \quad \text{with} \quad \mathcal{H}=\frac{\mathbf{p}^2}{2n_0}+\frac{n_0}{2}\Omega^2\mathbf{r}^2,$$

can be interpreted as the *evolution operator* of the electromagnetic modes. Note also that

$$U_{n,m}(\mathbf{r},0)=\frac{\sqrt{2}}{w_0}\,\Psi_{n,m}(u(x,0),v(y,0))$$

is an eigenfunction of the operator H for the fixed value $z = 0$. In this way, combining (20) and (8) we have

$$\mathcal{W}(z)U_{n,m}(\mathbf{r},0) = \frac{\sqrt{2}}{w(z)} e^{-i\beta_{n,m}\chi(z)} e^{i\frac{k_0 n_0 r^2}{2R(z)}} \Psi_{n,m}\left(u(x,z), v(y,z)\right).$$

This means that if $\Psi_{n,m}(u(x,0), v(y,0))$ is an eigenfunction of H at $z = 0$, corresponding to the eigenvalue $\beta_{n,m}$, we have

$$\mathcal{W}(z)\Psi_{n,m}(u(x,0), v(y,0)) = \frac{w_0}{w(z)} e^{-i\beta_{n,m}\chi(z)} e^{i\frac{k_0 n_0 r^2}{2R(z)}} \Psi_{n,m}\left(u(x,z), v(y,z)\right). \tag{21}$$

4 Hermite–Glauber Coherent States

The Glauber coherent states $U_{\alpha_x,\alpha_y}(\mathbf{r},z)$ of the Hermite type are constructed as eigenstates of the annihilation operators A^- and B^-. Thus, they must fulfill the differential equations

$$A^- U_{\alpha_x,\alpha_y}(\mathbf{r},z) = \frac{1}{2} w e^{i\chi}\left[\frac{\partial}{\partial x} - 2iSx\right] U_{\alpha_x,\alpha_y}(\mathbf{r},z) = \alpha_x U_{\alpha_x,\alpha_y}(\mathbf{r},z),$$

$$B^- U_{\alpha_x,\alpha_y}(\mathbf{r},z) = \frac{1}{2} w e^{i\chi}\left[\frac{\partial}{\partial y} - 2iSy\right] U_{\alpha_x,\alpha_y}(\mathbf{r},z) = \alpha_y U_{\alpha_x,\alpha_y}(\mathbf{r},z). \tag{22}$$

The normalized solution is given by

$$U_{\alpha_x,\alpha_y}(\mathbf{r},z) = \frac{1}{\sqrt{\pi}} \frac{\sqrt{2}}{w(z)} \exp\left\{\frac{1}{2}\left[e^{-2i\chi(z)}\left(\alpha_x^2 + \alpha_y^2\right) - \left(|\alpha_x|^2 + |\alpha_y|^2\right)\right]\right\}$$
$$\times e^{i\frac{k_0 n_0 r^2}{2R(z)}} \exp\left[-\left(\frac{x}{w(z)} - e^{-i\chi(z)}\alpha_x\right)^2 - \left(\frac{y}{w(z)} - e^{-i\chi(z)}\alpha_y\right)^2\right]. \tag{23}$$

In Fig. 1 we present some plots concerning Hermite–Glauber coherent states (23). They are Gaussian wave-packets (left) that remain Gaussian as they propagate along the optical axis with oscillating width, given by (4), of period π/Ω (center). The center of the wave-packet follows an elliptic trajectory given by the parametric equations (see Fig. 1 (right))

$$x(z) = w(z)\mathrm{Re}\left(e^{-i\chi(z)}\alpha_x\right), \quad y(z) = w(z)\mathrm{Re}\left(e^{-i\chi(z)}\alpha_y\right).$$

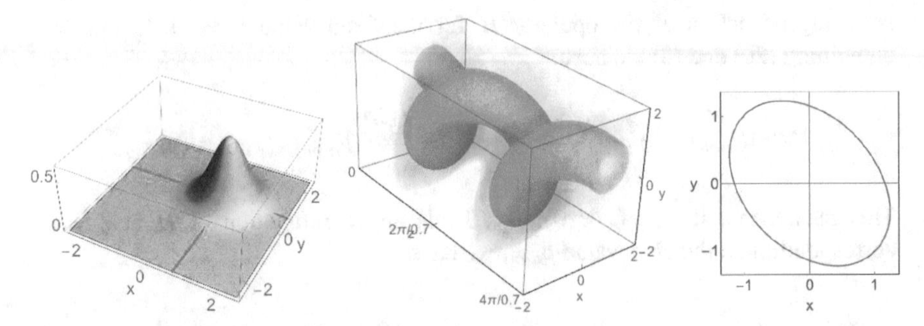

Fig. 1 Hermite–Glauber coherent state for $\alpha_x = \frac{1}{\sqrt{2}}(1+i)$, $\alpha_y = \frac{1}{\sqrt{2}}(1-i)$, and $z_R = 0.7/\Omega$: (Left) Intensity distribution in the xy-plane. (Center) Dynamics of the coherent beam as it propagates along the z-axis. (Right) The trajectory of the center of the wave-packet in the transversal plane. The transversal variables x, y are measured in units of w_0 while the longitudinal one z is measured in units of z_R

From the geometrical optics point of view, this trajectory would correspond to the projection on the transversal plane of the path that follows a light ray that propagates in this medium. The corresponding components of the propagation direction **p** are defined by

$$p_x(z) = \frac{w(z)}{k_0}\mathrm{Re}\left(e^{-i\chi(z)}\bar{S}(z)\alpha_x\right), \quad p_y(z) = \frac{w(z)}{k_0}\mathrm{Re}\left(e^{-i\chi(z)}\bar{S}(z)\alpha_y\right).$$

5 Hermite-$su(1,1)$ Coherent States

It is well known that the generators of the $su(1,1)$ and $su(2)$ algebras can be realized as higher order compositions of single mode boson operators [14–18]. In the non-degenerate case these realizations correspond to the Schwinger representation of those algebras [19]. Indeed the operators

$$\mathcal{K}_- = B^- A^-, \qquad \mathcal{K}_+ = B^+ A^+, \qquad \mathcal{K}_0 = \frac{1}{2}H, \tag{24}$$

fulfill the commutation rules of the $su(1,1)$ algebra

$$[\mathcal{K}_-, \mathcal{K}_+] = 2\mathcal{K}_0, \qquad [\mathcal{K}_0, \mathcal{K}_\pm] = \pm\mathcal{K}_\pm. \tag{25}$$

From its definition, it is clear that the operator $\mathcal{K}_-$ ($\mathcal{K}_+$) lows (raises) both labels n, m of the Hermite–Gaussian modes, meaning that, under the action of these operators the Hermite–Gaussian modes are transformed in such a way that the number $|n-m|$ is left invariant. Hence, the set of Hermite–Gaussian modes can be classified into hierarchies defined by the number $N = |n-m|$. The complete space of modes is, thus, decomposed as the direct sum of subspaces $\mathcal{H}_{\pm}^{N}$ which are the span of each N-hierarchy and correspond to a particular representation of the $su(1,1)$ algebra. Here the $\pm$ signs stand for the cases $n-m>0$ and $n-m<0$, respectively.

On the other hand, it is well known that the $su(1,1)$ coherent states of the Perelomov type can be obtained as eigenvectors of the operator [14]

$$(\mathbf{n}\mathcal{K}) = n_0\mathcal{K}_0 - n_1\mathcal{K}_1 - n_2\mathcal{K}_2, \tag{26}$$

where $\mathbf{n}=(n_0,n_1,n_2)$ is a vector fulfilling $\mathbf{n}^2=n_0^2-n_1^2-n_2^2=1$, $n_0>0$, that can be parametrized as

$$(n_0,n_1,n_2) = (\cosh\tau_0, \sinh\tau_0\cos\phi_0, \sinh\tau_0\sin\phi_0),$$

and the operators $\mathcal{K}_1,\mathcal{K}_2$ are the quadratures of the $su(1,1)$ algebra given by

$$\mathcal{K}_1 = \frac{1}{2}\left(\mathcal{K}_+ + \mathcal{K}_-\right), \qquad \mathcal{K}_2 = \frac{1}{2i}\left(\mathcal{K}_+ - \mathcal{K}_-\right). \tag{27}$$

Hence, the construction of Perelomov coherent states $U_\xi^N(\mathbf{r},z)$ involves the solution of two eigenvalue equations. The first one will select the corresponding N-subspace from the space of modes. For the choice $n-m>0$ (the case $n-m<0$ can be constructed by interchanging the roles of n and m) we have

$$\left(N_x-N_y\right)U_\xi^N(\mathbf{r},z) = NU_\xi^N(\mathbf{r},z). \tag{28}$$

The second equation will define the coherent state in this particular subspace:

$$(\mathbf{n}\mathcal{K})\,U_\xi^N(\mathbf{r},z) = kU_\xi^N(\mathbf{r},z), \tag{29}$$

where k is the Bargmann parameter given by $k=\frac{1}{2}(n-m+1)$. The complex number ξ labeling the coherent state is fixed by the parameters τ_0 and ϕ_0 in the form

$$\frac{\xi}{|\xi|} = e^{-i\phi_0}, \qquad \tanh|\xi| = \tanh\frac{\tau_0}{2}.$$

In the coordinate representation the operators $\hat{N}=N_x-N_y$, $\mathcal{K}_0$, $\mathcal{K}_1$, and $\mathcal{K}_2$ have the forms

$$\hat{N} = \frac{w^2}{4}\left[-\left(\frac{\partial^2}{\partial x^2}-\frac{\partial^2}{\partial y^2}\right)+2i\mathrm{Re}(S)\left(x\frac{\partial}{\partial x}-y\frac{\partial}{\partial y}\right)+4|S|^2\left(x^2-y^2\right)\right],$$

$$\mathcal{K}_0 = \frac{w^2}{8}\left[-\left(\frac{\partial^2}{\partial x^2}+\frac{\partial^2}{\partial y^2}\right)+2i\mathrm{Re}(S)\left(x\frac{\partial}{\partial x}+y\frac{\partial}{\partial y}\right)+4|S|^2\left(x^2+y^2\right)\right]$$
$$+\frac{i}{2}w^2\mathrm{Re}(S),$$

$$\mathcal{K}_1 = \frac{w^2}{4}\left[\cos 2\chi \frac{\partial^2}{\partial x \partial y} - 2i\mathrm{Re}\left(e^{2i\chi} S\right)\left(y\frac{\partial}{\partial x} + x\frac{\partial}{\partial y}\right) - 4\mathrm{Re}\left(e^{2i\chi} S^2\right) xy\right],$$

$$\mathcal{K}_2 = \frac{w^2}{4}\left[-\sin 2\chi \frac{\partial^2}{\partial x \partial y} + 2i\mathrm{Im}\left(e^{2i\chi} S\right)\left(y\frac{\partial}{\partial x} + x\frac{\partial}{\partial y}\right) + 4\mathrm{Im}\left(e^{2i\chi} S^2\right) xy\right].$$

From these expressions it is clear that the operators $\hat{N}$ and $(\mathbf{nK})$ will contain first as well as second order symmetries for an arbitrary value of z. However, for $z = 0$ the operators $\hat{N}$, $\mathcal{K}_0$, and $\mathcal{K}_1$ only contain second order symmetries, as $\chi(0) = 0$ and $S(0) = \frac{i}{w_0^2}$. In this case the set of Eqs. (28)–(29) can be solved using separation of variables by setting $\tau_0 = 0$. As the coherent states will not depend on the parameter ϕ_0 we may choose $\phi_0 = 0$ without loss of generality. For these choices of the parameter we have $\xi = 0$. The set (28)–(29) now have the form

$$\hat{N} U_0^N(\mathbf{r}, 0) = -\frac{w_0^2}{4}\left[\frac{\partial^2}{\partial x^2} - \frac{\partial^2}{\partial y^2} + \frac{4}{w_0^4}\left(x^2 - y^2\right)\right] U_0^N(\mathbf{r}, 0) = N U_0^N(\mathbf{r}, 0),$$

$$\begin{aligned}\mathcal{K}_0 U_0^N(\mathbf{r}, 0) &= -\frac{w_0^2}{8}\left[\frac{\partial^2}{\partial x^2} + \frac{\partial^2}{\partial y^2} + \frac{4}{w_0^4}\left(x^2 + y^2\right)\right] U_0^N(\mathbf{r}, 0) \\ &= \frac{1}{2}(N+1)\, U_0^N(\mathbf{r}, 0).\end{aligned}$$

The normalized solution fulfilling both equations reads

$$U_0^N(\mathbf{r}, 0) = \frac{1}{\sqrt{\pi 2^N N!}} \frac{\sqrt{2}}{w_0} e^{-\frac{r^2}{w_0^2}} H_N\left(\frac{\sqrt{2}}{w_0} x\right). \tag{30}$$

Now, in order to determine the evolution of this coherent state in the longitudinal coordinate, consider the fact that $U_0^N(\mathbf{r}, 0)$ is an eigenfunction of H at $z = 0$ with eigenvalue $\beta = N + 1$, therefore

$$U_0^N(\mathbf{r}, z) = \frac{1}{\sqrt{\pi 2^N N!}} \frac{\sqrt{2}}{w(z)} e^{-i(N+1)\chi(z)} e^{i\frac{k_0 n_0 r^2}{2R(z)}} e^{-\frac{r^2}{w^2(z)}} H_N\left(\frac{\sqrt{2}}{w(z)} x\right). \tag{31}$$

In Fig. 2 we present some plots showing the behavior of this coherent state for $N = 4$ and $z_R = 0.5/\Omega$. The intensity distribution exhibits the rectangular symmetry characteristic of the Hermite–Gaussian modes, being N the number of nodes in the x direction. As z grows, the width of the beam oscillates with a period defined by the value of z_R.

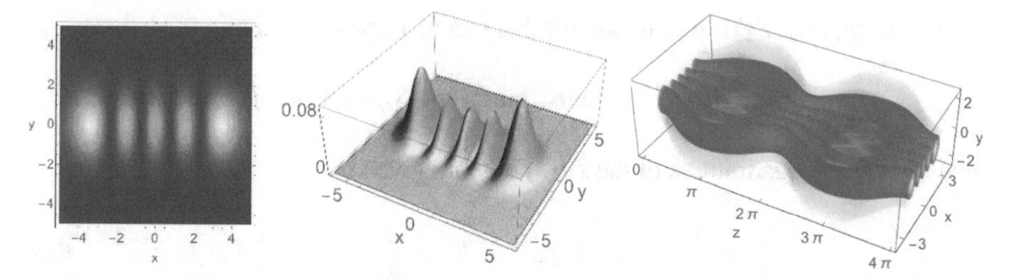

Fig. 2 Hermite-$su(1,1)$ coherent state for $N = 4$ and $z_R = 0.5/\Omega$: (Left, Center) Intensity distribution as a function of x and y. (Right) Dynamics of the coherent beam as it propagates along the z-axis. The width of the beam oscillates with a period $2\pi z_R$. The transversal variables x, y are measured in units of w_0 while the longitudinal one z is measured in units of z_R

6 Hermite-$su(2)$ Coherent States

Now let us consider the second order operators

$$\mathcal{J}_- = A^- B^+, \qquad \mathcal{J}_+ = A^+ B^-, \qquad \mathcal{J}_0 = \frac{1}{2}\left(N_x - N_y\right), \tag{32}$$

fulfilling the $su(2)$ commutation relations

$$[\mathcal{J}_-, \mathcal{J}_+] = -2\mathcal{J}_0, \qquad [\mathcal{J}_0, \mathcal{J}_\pm] = \pm\mathcal{J}_\pm. \tag{33}$$

In this case it is possible to see that the operators $\mathcal{J}_\pm$ induce transformations on the Hermite–Gaussian modes leaving the number $n + m$ invariant. The Hermite–Gaussian modes are classified now into hierarchies defined by the number $j = \frac{1}{2}(n+m)$. This means that the space of modes can be decomposed as the direct sum of subspaces $\mathcal{H}^j$ spanned by the corresponding j-hierarchies.

Following the Perelomov approach, we construct the $su(2)$ coherent states $U^j_\xi(\mathbf{r}, z)$ as the simultaneous solutions of two equations: the first one selects a particular j-subspace

$$HU^j_\xi(\mathbf{r}, z) = (2j+1)U^j_\xi(\mathbf{r}, z), \tag{34}$$

while the second one will define the coherent state

$$(\mathbf{n}\cdot\mathbf{J})U^j_\xi(\mathbf{r}, z) = -jU^j_\xi(\mathbf{r}, z), \tag{35}$$

where

$$(\mathbf{n}\cdot\mathbf{J}) = n_0\mathcal{J}_0 + n_1\mathcal{J}_1 + n_2\mathcal{J}_2,$$

with $\mathbf{n} = (n_0, n_1, n_2)$ a unitary vector that can be parametrized as

$$(n_0, n_1, n_2) = (\cos\theta_0, \sin\theta_0 \cos\phi_0, \sin\theta_0 \sin\phi_0)\,,$$

and $\mathcal{J}_1, \mathcal{J}_2$ the quadratures of the $su(2)$ algebra given by

$$\mathcal{J}_1 = \frac{1}{2}\left(\mathcal{J}_+ + \mathcal{J}_-\right), \qquad \mathcal{J}_2 = \frac{1}{2i}\left(\mathcal{J}_+ - \mathcal{J}_-\right). \tag{36}$$

The complex parameter ξ labeling the coherent state is now defined in terms of θ_0 and ϕ_0 by

$$\frac{\xi}{|\xi|} = e^{-i\phi_0}, \qquad \tan|\xi| = \tan\frac{\theta_0}{2}.$$

The expressions for the quadratures in the coordinate representation are

$$\mathcal{J}_1 = \frac{w^2}{4}\left[-\frac{\partial^2}{\partial x \partial y} + 2i\mathrm{Re}(S)\left(y\frac{\partial}{\partial x} + x\frac{\partial}{\partial y}\right) + 4|S|^2 xy\right], \tag{37}$$

$$\mathcal{J}_2 = -\frac{i}{2}\left(x\frac{\partial}{\partial y} - y\frac{\partial}{\partial x}\right). \tag{38}$$

In this case it is possible to see that for $z = 0$ and $\phi_0 = 0, \pi$ the differential operator $\mathbf{n} \cdot \mathbf{J}$ contains only second order symmetries, while for $z = 0$, $\theta_0 = \frac{\pi}{2}$ and $\phi_0 = \pm\frac{\pi}{2}$ it only contains first order symmetries.

In the first case, for the choice $\phi_0 = 0$, we have $\xi = \frac{\theta_0}{2}$ and the set of Eqs. (34)–(35) reduces to

$$-\frac{w_0^2}{4}\left[\frac{\partial^2}{\partial x^2} + \frac{\partial^2}{\partial y^2} - \frac{4}{w_0^4}\left(x^2 + y^2\right)\right] U_\xi^j(\mathbf{r}, 0) = (2j+1)\, U_\xi^j(\mathbf{r}, 0),$$

$$-\frac{w_0^2}{8}\left[\cos\theta_0\left(\frac{\partial^2}{\partial x^2} - \frac{\partial^2}{\partial y^2}\right) + 2\sin\theta_0 \frac{\partial^2}{\partial x \partial y}\right] U_\xi^j(\mathbf{r}, 0)$$
$$+\frac{1}{2w_0^2}\left[\cos\theta_0\left(x^2 - y^2\right) + 2\sin\theta_0 xy\right] U_\xi^j(\mathbf{r}, 0) = -jU_\xi^j(\mathbf{r}, 0).$$

The normalized solution reads

$$U_\xi^j(\mathbf{r}, 0) = \frac{1}{\sqrt{\pi 2^{2j}(2j)!}} \frac{\sqrt{2}}{w_0} e^{-\frac{r^2}{w_0^2}} H_{2j}\left(\frac{\sqrt{2}}{w_0}\left[y\cos\frac{\theta_0}{2} - x\sin\frac{\theta_0}{2}\right]\right). \tag{39}$$

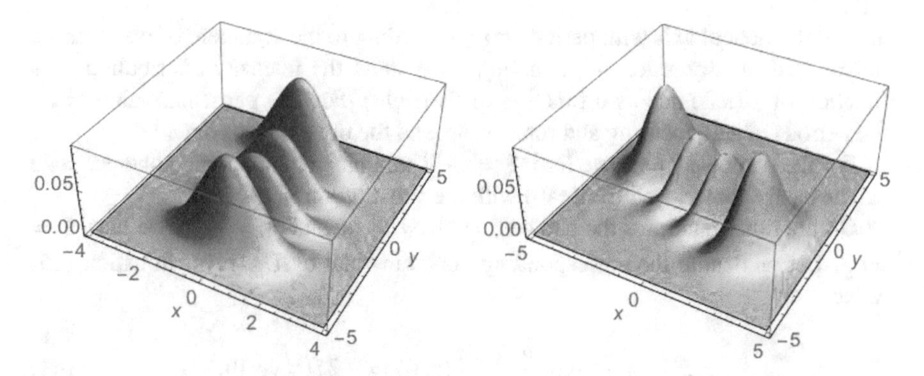

Fig. 3 Hermite-$su(2)$ coherent state for $j = \frac{3}{2}$ and $z_R = 0.5/\Omega$: Intensity distribution as a function of x and y for $\theta_0 = 0$ (left) and $\theta_0 = \frac{3\pi}{4}$ (right). As a function of θ_0 this distribution rotates with a period $4\pi z_R$ around the optical axis. The transversal variables x, y are measured in units of w_0

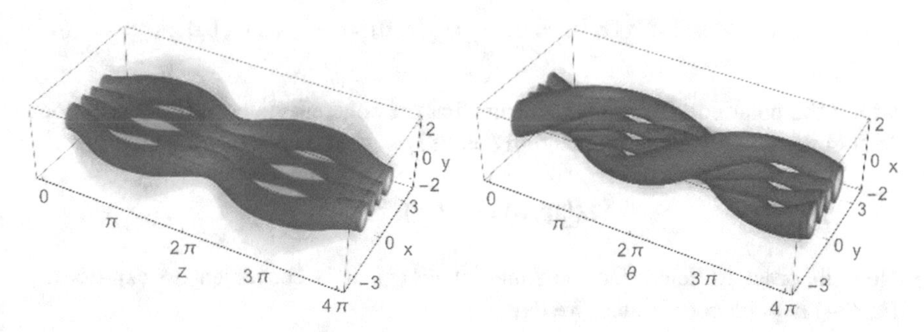

Fig. 4 Hermite-$su(2)$ coherent state for $j = \frac{3}{2}$ and $z_R = 0.5/\Omega$: Intensity distribution as a function of x, y, z (left) and x, y, θ_0 (right). For this value of z_R, the oscillation period of the beam width coincides with the rotation period of this distribution with respect to θ_0. The transversal variables x, y are measured in units of w_0 while the longitudinal one z is measured in units of z_R

As this function is an eigenstate of H for $z = 0$ with eigenvalue $\beta = 2j + 1$, its dynamics in z can be readily obtained:

$$U_\xi^j(\mathbf{r}, z) = \frac{1}{\sqrt{\pi 2^{2j}(2j)!}} \frac{\sqrt{2}}{w(z)} e^{-i(2j+1)\chi(z)} e^{i\frac{k_0 n_0 r^2}{2R(z)}} e^{-\frac{r^2}{w^2(z)}} \times$$
$$\times H_{2j}\left(\frac{\sqrt{2}}{w(z)}\left[y \cos\frac{\theta_0}{2} - x \sin\frac{\theta_0}{2}\right]\right). \tag{40}$$

In Figs. 3 and 4 we show the intensity distributions associated with these states for $z_R = 0.5/\Omega$. The coherent state exhibits the typical rectangular symmetry of Hermite–Gaussian states, where the parameter j defines the number of nodes of the field intensity. The variation of the parameter θ_0 induces rotations of the distribution around the optical axis with period $4\pi z_R$ according to the argument of the Hermite polynomial in expression (40). In Fig. 4 we show the intensity distribution as a function of z (left) and as a function of θ_0 (right). For this particular value of z_R the periods of self-focusing and rotation around the optical axis coincide.

For the choice $\phi_0 = \pi$ we have $\xi = -\frac{\theta_0}{2}$ and the corresponding coherent state can be easily constructed from (40) with the substitution $\theta_0 \to -\theta_0$.

On the other hand, for the choice $z = 0$, $\theta_0 = \frac{\pi}{2}$, and $\phi_0 = \mp\frac{\pi}{2}$, we have $\xi = \pm i\frac{\pi}{4}$. Let us denote the corresponding coherent state by $U^j_\pm(\mathbf{r}, 0)$. Equation (35) reduces to

$$\pm i\left(x\frac{\partial}{\partial y} - y\frac{\partial}{\partial x}\right)U^j_\pm(\mathbf{r}, 0) = -2jU^j_\pm(\mathbf{r}, 0). \tag{41}$$

This equation can be identified with the eigenvalue equation for the orbital angular momentum in the z-direction

$$\mp k_0 L_z U^j_\pm(\mathbf{r}, 0) = \pm i\frac{\partial}{\partial\theta}U^j_\pm(\mathbf{r}, 0) = -2jU^j_\pm(\mathbf{r}, 0), \tag{42}$$

with θ the polar coordinate. This means that the coherent state $U^j_\pm(\mathbf{r}, 0)$ is a mode of definite orbital angular momentum $\ell = \pm 2j$ so that it can be written as

$$U^j_\pm(\mathbf{r}, 0) = \psi^j(r)e^{\pm i2j\theta}.$$

Now, in order to determine the function $\psi^j(r)$, it is convenient to express the Eq. (34) in polar coordinates. We have

$$-\frac{w_0^2}{4}\left(\frac{d^2}{dr^2} + \frac{1}{r}\frac{d}{dr} - \frac{4j^2}{r^2}\right)\psi^j(r) + \frac{1}{w_0^2}r^2\psi^j(r) = (2j+1)\psi^j(r).$$

This equation can be transformed into a confluent hypergeometric one [20, 21]. Indeed, with the substitution

$$\psi^j(r) = \varrho^j e^{-\frac{\varrho}{2}}\phi(\varrho), \qquad \varrho = \frac{2}{w_0^2}r^2,$$

we get

$$\left[\varrho\,\frac{d^2}{d\varrho^2} + (2j+1-\varrho)\,\frac{d}{d\varrho}\right]\phi(\varrho) = 0.$$

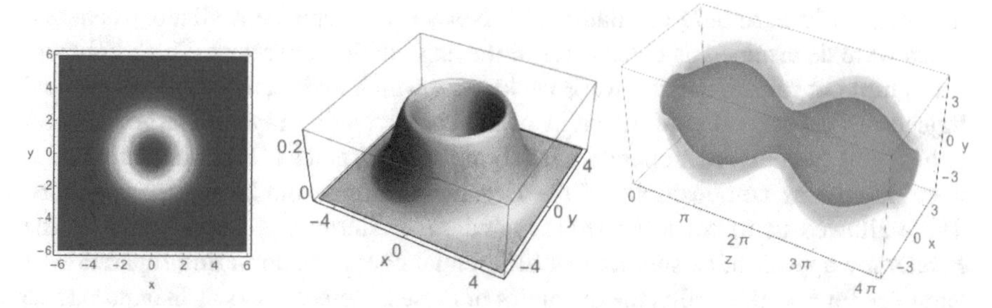

Fig. 5 Hermite-$su(2)$ coherent state for $2j = 3$ and $z_R = 0.5/\Omega$: (Left, Center) Intensity distribution as a function of x and y. (Right) Dynamics of the coherent beam as it propagates along the z-axis. The width of the beam oscillates with a period $2\pi z_R$. The transversal variables x, y are measured in units of w_0 while the longitudinal one z is measured in units of z_R

The solution to this equation that leads to a well-behaved function $\psi^j(r)$ is a constant, and, hence, the normalized coherent state has the form

$$U^j_{\pm}(\mathbf{r}, 0) = \frac{1}{\sqrt{\pi(2j)!}} \frac{\sqrt{2}}{w_0} \left(\frac{\sqrt{2}}{w_0} r\right)^j e^{-\frac{r^2}{w_0^2}} e^{\mp i2j\theta}, \tag{43}$$

while its evolved version reads

$$U^j_{\pm}(\mathbf{r}, z) = \frac{1}{\sqrt{\pi(2j)!}} \frac{\sqrt{2}}{w(z)} \left(\frac{\sqrt{2}}{w(z)} r\right)^j e^{-i(2j+1)\chi(z)} e^{i\frac{k_o n_0 r^2}{2R(z)}} e^{-\frac{r^2}{w_0^2}} e^{\mp i2j\theta}. \tag{44}$$

As these coherent states are eigenvectors of the orbital angular momentum operator, they exhibit axial symmetry. Indeed, they correspond to the Laguerre–Gaussian modes of degree $p = 0$ and order $\ell = \pm 2j$ (see [10]). In Fig. 5 we show the behavior of the intensity distribution for $j = \frac{3}{2}$ as a function of the transversal coordinates, as well as its evolution along the optical axis.

7 Summary

We addressed the construction of ladder operators for the set of Hermite–Gaussian modes associated with an optical medium with quadratic refractive index. The case of the homogeneous medium is recovered in the limit as the parabolicity parameter of the medium $\Omega \to 0$. In this limit the ladder operators coincide with the creation and annihilation operators for the Hermite–Gaussian modes in free space [1], with the creation and annihilation operators for the harmonic states of the free particle [11], and with the invariant generalized ladder operators of the parametric harmonic oscillator [22]. Next, the z-dependent Glauber coherent states were determined as eigenstates of the annihilation operators. These coherent states turn out to be Gaussian wave-packets for which their centers follow classical trajectories associated with the paths of the corresponding rays in the geometrical optics limit. Second order operators satisfying $su(1,1)$ and $su(2)$ commutation rules were defined as compositions of the Hermite–Gaussian modes ladder operators. These allowed us to construct $su(1,1)$ and $su(2)$ Hermite coherent states, in the Perelomov approach, as solutions of differential equations admitting separation of variables for $z = 0$. Finally, the dynamics of these coherent states as they propagate along the optical axis were determined by the means of the corresponding evolution operator $\mathcal{W}(z)$.

Acknowledgements The financial support of CONACyT, Mexico (Project A1-S-24569 and grant 257292 for ZG), Instituto Politécnico Nacional, Mexico (Project SIP20180377), the Spanish MINECO (Pro. MTM2014-57129-C2-1-P), and Junta de Castilla y León, Spain (VA137G18) is acknowledged. The authors are indebted to Prof. J. Negro for enlightening comments and to the anonymous referee for valuable suggestions. Z. Gress is grateful to the Valladolid University for kind hospitality.

References

1. G. Nienhuis, L. Allen, Paraxial wave optics and harmonic oscillators. Phys. Rev. A **48**, 656 (1993)
2. S.G. Krivoshlykov, N.I. Petrov, I.N. Sisakyan, Correlated coherent states and propagation of arbitrary Gaussian beams in longitudinally homogeneous quadratic media exhibiting absorption or amplification. Sov. J. Quantum Electron. **16**, 933 (1986)
3. N.I. Petrov, Macroscopic quantum effects for classical light. Phys. Rev. A **90**, 043814 (2014)
4. D. Stoler, Operator methods in physical optics. J. Opt. Soc. Am. **71**, 334 (1981)
5. M.A.M. Marte, S. Stenholm, Paraxial light and atom optics: the optical Schrodinger equation and beyond. Phys. Rev. A **56**, 2940 (1997)
6. S. Cruz y Cruz, O. Rosas-Ortiz, Leaky modes of waveguides as a classical optics analogy of quantum resonances. Adv. Math. Phys. **2015**, 281472 (2015)
7. D. Gloge, D. Marcuse, Formal quantum theory of light rays. J. Opt. Soc. Am. **59**, 1629 (1969)
8. G. Nienhuis, J. Visser, Angular momentum and vortices in paraxial beams. J. Opt. A: Pure Appl. Opt. **6**, S248 (2004)
9. S. Cruz y Cruz, Z. Gress, Group approach to the paraxial propagation of Hermite–Gaussian modes in a parabolic medium. Ann. Phys. **383**, 257 (2017)
10. A.E. Siegman, *Lasers* (University Science Books, Mill Valley, CA, 1986)
11. J. Guerrero, F.F. López-Ruiz, V. Aldaya, F. Cossío, Harmonic states for the free particle. J. Phys. A: Math. Theor. **44**, 445307 (2011)
12. J. Guerrero, F. F. López-Ruiz, The quantum Arnold transformation and the Ermakov–Pinney equation. Phys. Scr. **87** 038105 (2013)
13. J. Guerrero, F.F. López-Ruiz, On the Lewis–Riesenfeld (Dodonov–Man'ko) invariant method. Phys. Scr. **90** 074046 (2015)
14. A. Perelomov, *Generalized Coherent States and their Applications* (Springer, Berlin, 1986)
15. R.R. Puri, $SU(m, n)$ coherent states in the bosonic representation and their generation in optical parametric processes. Phys. Rev. A **50**, 5309 (1994)

16. P. Shanta, S. Chaturvedi, V. Srinivasan, G.S. Agarwal, C.L. Mehta, Unified approach to multiphoton coherent states. Phys. Rev. Lett. **72**, 1447 (1994)
17. R.R. Puri, G.S. Agarwal, $SU(1, 1)$ coherent states defined via a minimum-uncertainty product and an equality of quadrature variances. Phys. Rev.A **53**, 1786 (1996)
18. I. Dhand, B.C. Sanders, H. de Guise, Algorithms for $SU(n)$ boson realizations and $\mathcal{D}$-functions. J. Math. Phys. **56**, 111705 (2015)
19. J. Schwinger, *Quantum Theory of Angular Momentum* (Academic, New York, 1965), pp. 229–279
20. M. Abramowitz, I. Stegun, *Handbook of Mathematical Functions with Formulas, Graphs and Mathematical Tables* (Dover, Washington, DC, 1970)
21. J. Negro, L.M. Nieto, O. Rosas-Ortiz, Confluent hypergeometric equations and related solvable potentials in quantum mechanics. J. Math. Phys. **41**,7964 (2000)
22. O. Castaños, D. Schuch, O. Rosas-Ortiz, Generalized coherent states for time-dependent and nonlinear Hamiltonian operators via complex Riccati equations. J. Phys. A: Math. Theor. **46**, 075304 (2013)

Analysis of $\mathbb{C}P^{N-1}$ Sigma Models via Soliton Surfaces

Piotr P. Goldstein and Alfred M. Grundland

Abstract In this paper we present results obtained from the study of an invariant formulation of completely integrable $\mathbb{C}P^{N-1}$ Euclidean sigma models in two dimensions defined on the Riemann sphere, having finite actions. Surfaces connected with the $\mathbb{C}P^{N-1}$ models, invariant recurrence relations linking the successive projection operators, and immersion functions of the surfaces are discussed in detail. We show that immersion functions of 2D-surfaces associated with the $\mathbb{C}P^{N-1}$ model are contained in 2D-spheres in the $\mathfrak{su}(N)$ algebra. Making use of the fact that the immersion functions of the surfaces satisfy the same Euler–Lagrange equations as the original projector variables, we derive surfaces induced by surfaces and prove that the stacked surfaces coincide with each other, which demonstrates the idempotency of the recurrent procedure. We also demonstrate that the $\mathbb{C}P^{N-1}$ model equations admit larger classes of solutions than the ones corresponding to rank-1 Hermitian projectors. This fact allows us to generalize the Weierstrass formula for the immersion of 2D-surfaces in the $\mathfrak{su}(N)$ algebra and to show that, in general, these surfaces cannot be conformally parametrized. Finally, we consider the connection between the structure of the projective formalism and the possibility of spin representations of the $\mathfrak{su}(2)$ algebra in quantum mechanics.

Keywords Spin matrices · Sigma models · Soliton surfaces · Integrable systems · Weierstrass formula for immersion

P. P. Goldstein
Theoretical Physics Division, National Centre for Nuclear Research, Warsaw, Poland
e-mail: piotr.goldstein@ncbj.gov.pl

A. M. Grundland (✉)
Centre de Recherches Mathématiques, Université de Montréal, Montréal, QC, Canada
e-mail: grundlan@crm.umontreal.ca

Ş. Kuru et al. (eds.), *Integrability, Supersymmetry and Coherent States*, CRM Series in Mathematical Physics, https://doi.org/10.1007/978-3-030-20087-9_15

1 Introduction

Integrable models and their continuous deformations under various types of dynamics have produced a great deal of interest and activity in several branches of mathematics, physics, and biology. Soliton surfaces associated with integrable models, and with the $\mathbb{C}P^{N-1}$ sigma model in particular, have been shown to play an essential role in many problems with physical applications (see, e.g., [2, 11, 16, 17, 19, 20]). The possibility of using a rank-1 projector formalism associated with the $\mathbb{C}P^{N-1}$ model to construct soliton surfaces has yielded many new results concerning the intrinsic geometric properties of such surfaces (see, e.g., [1]). In this vein, it has recently proved fruitful to extend such characterizations of soliton surfaces via their immersion functions in Lie algebras, based on projectors of rank higher than one. The construction of such surfaces related to the completely integrable $\mathbb{C}P^{N-1}$ sigma model in two dimensions has been accomplished by representing the equation of motion for the model as a conservation law which in turn provides a closed differential for the surface. This is the so-called generalized Weierstrass formula for immersion [15]. The results obtained [7, 9, 10, 12–14, 18] proved to be fruitful from the point of view of constructing multileaf soliton surfaces immersed in Lie algebras. This paper contains a survey of these results for the immersion of soliton surfaces, particularly as applied to the integrable $\mathbb{C}P^{N-1}$ sigma model, and its link to the quantum mechanics of spin representations of the $\mathfrak{su}(2)$ algebra.

2 The Projector Formalism and Solitons Obtained via $\mathbb{C}P^{N-1}$ Sigma Models

The description of the $\mathbb{C}P^{N-1}$ model in terms of projection operators and the properties of the orthogonal projection matrices P (mapping onto one-dimensional subspaces of $\mathbb{C}^N$) can be summarized as follows.

The $\mathbb{C}P^{N-1}$ models are defined by their action integral and can be defined in terms of the homogeneous variables (see, e.g., [25] and the references therein)

$$\mathbb{C} \supseteq \Omega \ni \xi = \xi^1 + i\xi^2 \mapsto f = (f^0, f^1, \ldots, f^{N-1}) \in \mathbb{C}^N \setminus \{0\}. \tag{2.1}$$

The action integral for a $\mathbb{C}P^{N-1}$ model on a Riemann surface $\mathcal{R}$ having domain in $\Omega \subset \mathbb{C}$ can be written in terms of the homogeneous variables

$$\mathcal{A}(f) = \frac{1}{4}\int_{\Omega} \frac{1}{f^\dagger f}\left(\partial_+ f^\dagger P \partial_- f + \partial_- f^\dagger P \partial_+ f\right) d\xi d\bar{\xi}, \tag{2.2}$$

where P is a rank-1 Hermitian projector

$$P = \frac{f \otimes f^\dagger}{f^\dagger f}, \qquad P^2 = P, \quad P^\dagger = P, \quad \operatorname{tr} P = 1, \tag{2.3}$$

and the complex derivatives ∂_+ and ∂_- with respect to ξ and $\bar{\xi}$ are given by

$$\partial_+ = \frac{1}{2}\left(\partial_1 - \mathrm{i}\,\partial_2\right), \quad \partial_- = \frac{1}{2}\left(\partial_1 + \mathrm{i}\,\partial_2\right), \quad \partial_1 = \frac{\partial}{\partial \xi^1}, \quad \partial_2 = \frac{\partial}{\partial \xi^2}. \tag{2.4}$$

In this formulation the action integral (2.2) can be expressed in a more compact way in terms of the projectors P, which are explicitly scaling-invariant, namely

$$\mathcal{A}(P) = \int_\Omega \operatorname{tr}\left(\partial_+ P \cdot \partial_- P\right) d\xi d\bar{\xi}, \tag{2.5}$$

and its extremum is subject to the algebraic constraints (2.3).

The Euler–Lagrange (EL) equations corresponding to the action integral (2.5) with these constraints take the simple form [7, 13]

$$[\partial_+ \partial_- P, P] = \varnothing, \tag{2.6}$$

or equivalently can be written as the conservation law (CL)

$$\partial_+ [\partial_- P, P] + \partial_- [\partial_+ P, P] = \varnothing. \tag{2.7}$$

In what follows we assume that the model (2.7) is defined on the Riemann sphere $S^2 = \mathbb{C} \cup \{\infty\}$ and that its action functional (2.5) is finite. According to [5, 6, 21], for finite action integrals, all rank-1 projectors $P(\xi, \bar{\xi})$ as well as the corresponding homogeneous vectors

$$f_0 = \left(f_0^0, f_0^1, \ldots, f_0^{N-1}\right) \in \mathbb{C}^N \setminus \{0\} \tag{2.8}$$

can be obtained by acting on the holomorphic (or antiholomorphic) solution f_0 (or f_{N-1}) with raising (creation) and lowering (annihilation) operators. In terms of f, the recurrence relations are given by [5, 6]

$$f_{k+1} = P_+(f_k) = (\mathbb{I}_N - P_k)\,\partial_+ f_k, \quad f_{k-1} = P_-(f_k) = (\mathbb{I}_N - P_k)\,\partial_- f_k, \qquad P_\pm^0 = \mathbb{I}_N, \quad P_\pm^N f_k = 0, \quad k = 0, 1, \ldots, N-1, \tag{2.9}$$

where P_+ is the creation operator and P_- is the annihilation operator.

The equations satisfied by f_k, $k = 0, \ldots N-1$ are invariant under multiplication of the f's by any scalar function of $(\xi, \bar{\xi})$. Therefore "holomorphic" (or "antiholomorphic") means that there exists a solution (2.8) whose all components are holomorphic (or similarly f_{N-1} with all antiholomorphic components). The holomorphic or antiholomorphic property of f can easily be checked by dividing all components of f by the first one f_0^0 (or f_{N-1}^0) and thus setting the first component equal to 1.

In terms of the projectors P_k the raising and lowering operators are given by Goldstein and Grundland [7]

$$\begin{aligned} P_{\pm 1} &= \Pi_{\pm}(P), \\ \Pi_{+}(P) &= \begin{cases} \frac{\partial_+ P P \partial_- P}{\mathrm{tr}(\partial_+ P P \partial_- P)} & \text{for } \partial_+ P P \partial_- P \neq \varnothing, \\ \varnothing & \text{for } \partial_+ P P \partial_- P = \varnothing, \end{cases} \\ \Pi_{-}(P) &= \begin{cases} \frac{\partial_- P P \partial_+ P}{\mathrm{tr}(\partial_- P P \partial_+ P)} & \text{for } \partial_- P P \partial_+ P \neq \varnothing, \\ \varnothing & \text{for } \partial_- P P \partial_+ P = \varnothing, \end{cases} \end{aligned} \tag{2.10}$$

where P stands for one of the projectors $\{P_0, P_1, \ldots, P_{N-1}\}$. Note that Eq. (2.10) are nonlinear and the objects on which they act have to remain normalized to retain their projective character.

The set of N rank-1 projectors $\{P_0, P_1, \ldots, P_{N-1}\}$ satisfies the orthogonality and completeness relations

$$\begin{aligned} P_j P_k &= \delta_{jk} P_j \quad \text{(no summation)}, \\ \sum_{j=0}^{N-1} P_j &= \mathbb{I}_N. \end{aligned} \tag{2.11}$$

The first vector f_0 is holomorphic and the last one f_{N-1} is antiholomorphic [5]

$$\partial_- f_0 = 0, \qquad \partial_+ f_{N-1} = 0. \tag{2.12}$$

Therefore the annihilation operator acting on f_0 and the creation operator acting on f_{N-1} yield zero. Thus the sequence of solutions in the $\mathbb{C}P^{N-1}$ model consists of N vectors f_k or N rank-1 projectors P_k, $0 \leq k \leq N-1$. The procedure given in [5, 6], for $N > 2$, allows us to construct three classes of solutions: holomorphic f_0, antiholomorphic f_{N-1}, and mixed solutions f_k, $1 \leq k \leq N-2$.

Note that if the target space f of P is a direction of a holomorphic or an antiholomorphic function $f(\xi)$ or $f(\bar{\xi})$ (i.e., if it depends on only one of the two independent variables ξ or $\bar{\xi}$) then we obtain an analogue of the first Frenet formula [8],

$$\begin{aligned} \partial_+ P P &= \frac{1}{f^\dagger f} (\mathbb{I}_N - P)\, \partial_+ f \otimes f^\dagger, \\ P \partial_- P &= \frac{1}{f^\dagger f} f \otimes \partial_- f^\dagger\, (\mathbb{I}_N - P)\,. \end{aligned} \tag{2.13}$$

For a given set of N rank-1 projector solutions P_k of the EL equations (2.7) written as conservation laws

$$\partial_+ [\partial_- P_k, P_k] + \partial_- [\partial_+ P_k, P_k] = \varnothing, \tag{2.14}$$

the generalized Weierstrass formula for the immersion (GWFI) of two-dimensional surfaces is defined by the contour integral [12, 15] of the $\mathfrak{su}(N)$ matrix-valued 1-form

$$X_k(\xi, \bar{\xi}) = \mathrm{i} \int_{\gamma_k} \left(- [\partial_+ P_k, P_k]\, d\xi + [\partial_- P_k, P_k]\, d\bar{\xi} \right) \in \mathfrak{su}(N) \simeq \mathbb{R}^{N^2-1}, \tag{2.15}$$

where γ_k is a trajectory in $\mathbb{C}$. The CLs ensure that the contour integral is locally independent on the trajectory γ_k and can be explicitly integrated [12]

$$\begin{aligned} X_k(\xi, \bar{\xi}) &= -\mathrm{i} \left(P_k + 2 \sum_{j=0}^{k-1} P_j \right) + \mathrm{i}\, c_k \mathbb{I}_N \in \mathfrak{su}(N), \\ c_k &= \frac{1}{N} (1 + 2k)\,. \end{aligned} \tag{2.16}$$

For each index k, the projectors P_k satisfy the eigenvalue equations

$$(X_k - \mathrm{i} \lambda_k \mathbb{I}_N)\, P_k = \varnothing \tag{2.17}$$

with

$$\lambda_k = \begin{cases} c_{k-2} & j < k, \\ c_{k-1} & j = k, \\ c_k & j > k. \end{cases} \tag{2.18}$$

Note that

$$\left[X_k, P_j \right] = 0, \qquad 0 \leq k,\ j \leq N - 1, \tag{2.19}$$

whence the immersion functions X_k span a Cartan subalgebra of $\mathfrak{su}(N)$

$$\left[X_k, X_j\right] = 0, \qquad 0 \leq k,\ j \leq N-1. \tag{2.20}$$

The matrix-valued immersion functions $X_k(\xi, \bar{\xi})$ satisfy the following minimal polynomial identities [7]:

1. For any mixed solution of the EL equations (2.14), $1 \leq k \leq N-2$, the minimal polynomial is cubic

$$\begin{aligned} &[X_k - \mathrm{i}\, c_k \mathbb{I}_N]\,[X_k - \mathrm{i}(c_k - 1)\mathbb{I}_N]\,[X_k - \mathrm{i}(c_k - 2)\mathbb{I}_N] = \varnothing, \\ &1 \leq k \leq N-2. \end{aligned} \tag{2.21}$$

2. For any holomorphic ($k = 0$) or antiholomorphic ($k = N-1$) solutions of EL equations (2.14), the minimal polynomial is quadratic

$$\begin{aligned} k =&0 : [X_0 - \mathrm{i}\, c_0 \mathbb{I}_N]\,[X_0 - \mathrm{i}(c_0 - 1)\mathbb{I}_N] = \varnothing,\ c_0 + c_N = 2, \\ k =&N-1 : [X_{N-1} + \mathrm{i}\, c_0 \mathbb{I}_N]\,[X_{N-1} + \mathrm{i}(c_0 - 1)\mathbb{I}_N] = \varnothing. \end{aligned} \tag{2.22}$$

For the sake of uniformity, the inner product is defined by

$$(A, B) = -\frac{1}{2}\mathrm{tr}\,(A \cdot B)\,, \qquad A, B \in \mathfrak{su}(N). \tag{2.23}$$

The quadratic expressions for the immersion functions X_k are given by

$$(X_k, X_k) = \frac{1}{2}\,[N c_k\,(2 - c_k) - 1] = const. \tag{2.24}$$

This means that the surfaces described by (2.21) and (2.22) are submanifolds of the compact sphere with the radius $\left[\frac{N}{2} c_k\,(2 - c_k) - \frac{1}{2}\right]^{1/2}$ immersed in $\mathbb{R}^{N^2-1} \simeq \mathfrak{su}(N)$.

The projectors P_k fulfill the completeness relation (2.11) which implies in turn that the immersion functions X_k satisfy the linear relation

$$\sum_{k=0}^{N-1} (-1)^k\, X_k = \varnothing. \tag{2.25}$$

We can reconstruct all projectors P_k using the immersion functions X_k and the unit matrix $\mathbb{I}_N$. The inverse formulae are given by Goldstein and Grundland [7]

$$P_k = X_k^2 - 2\,\mathrm{i}\,(c_k - 1)\, X_k - c_k\,(c_k - 2)\,\mathbb{I}_N, \tag{2.26}$$

but these formulae are nonlinear.

For the $\mathbb{C}P^{N-1}$ models, the spectral problem is closely related to the immersion functions of 2D soliton surfaces. The $\mathbb{C}P^{N-1}$ models with finite action integrals (2.5) are completely integrable. The linear spectral problem (LSP) associated with the CL (2.7) is given by Zakharov and Mikhailov [24]

$$\begin{aligned} \partial_+ \Phi_k &= U_k^1 \Phi_k = \frac{2}{1+\lambda} [\partial_+ P_k, P_k] \Phi_k, \\ \partial_- \Phi_k &= U_k^2 \Phi_k = \frac{2}{1-\lambda} [\partial_- P_k, P_k] \Phi_k, \end{aligned} \tag{2.27}$$

where $\lambda \in \mathbb{C}$ is the spectral parameter. An explicit solution of the LSP (2.6) for which the wavefunctions Φ_k tend to the unit matrix $\mathbb{I}_N$ as $\lambda \to \infty$ [6]

$$\begin{aligned} \Phi_k &= \mathbb{I}_N + \frac{4\lambda}{(1-\lambda)^2} \sum_{j=0}^{k-1} P_j - \frac{2}{1-\lambda} P_k \in SU(N), \ \lambda = \mathrm{i}t, \ t \in \mathbb{R}, \\ \Phi_k^{-1} &= \mathbb{I}_N - \frac{4\lambda}{(1+\lambda)^2} \sum_{j=0}^{k-1} P_j - \frac{2}{1+\lambda} P_k, \end{aligned} \tag{2.28}$$

and the zero-curvature condition (ZCC) takes the form

$$\partial_- U_k^1 - \partial_+ U_k^2 + \left[U_k^1, U_k^2\right] \approx [\partial_+ \partial_- P_k, P_k] = 0. \tag{2.29}$$

The relation between the wavefunctions Φ_k and the immersion functions X_k is given by the Sym-Tafel (ST) formulas [23] yielding soliton surfaces

$$\begin{aligned} X_k^{ST}\left(\xi, \bar{\xi}, \lambda\right) &= -\mathrm{i}\, \Phi_k^{-1} \frac{\partial}{\partial \lambda} \Phi_k + \mathrm{i}\, c_k \mathbb{I}_N \\ &= \frac{-2\mathrm{i}}{(1-\lambda)^2} \left[P_k + 2 \sum_{j=0}^{k-1} P_j - c_k \mathbb{I}_N \right] \in \mathfrak{su}(N). \end{aligned} \tag{2.30}$$

These formulae are identical to the generalized Weierstrass immersion functions X_k given in the expression (2.16), up to a multiplicative constant. Therefore in conformal coordinates we obtain, as a result, a sequence of surfaces X_k whose structural equations are identical to the EL equations (2.6) for the $\mathbb{C}P^{N-1}$ model

$$[\partial_+ \partial_- X_k, X_k] = 0, \ X_k^\dagger = -X_k \in \mathfrak{su}(N), \ 0 \le k \le N-1 \tag{2.31}$$

and thus they are the soliton surfaces for the $\mathbb{C}P^{N-1}$ model. The corresponding creation and annihilation operators for the immersion functions X_k were found in

[18]. They are defined by

$$X_{\pm 1} = \mathcal{X}_{\pm}(P), \tag{2.32}$$

and their explicit forms are [18]

$$\mathcal{X}_{\pm}(X) = \begin{cases} \frac{\partial_{\pm} X X \partial_{\mp} X}{\mathrm{tr}(\partial_{\pm} X X \partial_{\mp} X)} & \partial_{\pm} X X \partial_{\mp} X \neq \varnothing, \\ \varnothing & \partial_{\pm} X X \partial_{\mp} X = \varnothing, \end{cases} \tag{2.33}$$

where X stands for one of the immersion functions $\{X_0, X_1, \ldots, X_{N-1}\}$. Note that X is orthogonal to $\mathcal{X}_{\pm}(X)$, i.e.,

$$(X, \mathcal{X}_{\pm}(X)) = 0. \tag{2.34}$$

Note that in view of the linear dependence of the immersion function X_k, i.e., Eq. (2.25), the ST formula (2.30) leads us to the following differential constraint on the wavefunctions Φ_k:

$$\sum_{j=0}^{N-1} (-1)^j X_j^{ST} = \left(1 - \lambda^2\right) \partial_\lambda \ln \prod_{j=0}^{N-1} \Phi_j^{(-1)^j} = 0.$$

This implies that the expression

$$\prod_{2l<N} \Phi_{2l} \prod_{2l+1<N} \Phi_{2l+1}^{-1}$$

is independent of the spectral parameter λ, but it may depend on the coordinates $\xi, \bar{\xi} \in \mathbb{C}$.

Let us now explore certain geometrical aspects of surfaces immersed in the $\mathfrak{su}(N)$ algebra. The complex tangent vectors are

$$\partial_{\pm} X_k = -\mathrm{i}\, \partial_{\pm} P_k - 2\mathrm{i} \sum_{j=0}^{k-1} \partial_{\pm} P_j, \tag{2.35}$$

and the corresponding metric tensors defined on these surfaces X_k are conformally parametrized

$$\begin{aligned} (g_k)_{\pm\pm} &= -\frac{1}{2} \mathrm{tr}\,(\partial_{\pm} P_k \cdot \partial_{\pm} P_k) = 0, \\ (g_k)_{\pm\mp} &= -\frac{1}{2} \mathrm{tr}\,(\partial_{+} P_k \cdot \partial_{-} P_k) = \frac{1}{2}\left[N c_k\,(2 - c_k) + 1\right]. \end{aligned} \tag{2.36}$$

This gives us the following expressions for the first and the second fundamental forms [7]

$$
\begin{aligned}
I_k &= \operatorname{tr}(\partial_+ P_k \cdot \partial_- P_k)\, d\xi d\bar{\xi}, \\
I\!I_k &= -\operatorname{tr}(\partial_+ P_k \cdot \partial_- P_k)\, \partial_+ \left(\frac{[\partial_+ P_k, P_k]}{\operatorname{tr}(\partial_+ P_k \cdot \partial_- P_k)} \right) d\xi^2 + 2\mathrm{i}\, [\partial_- P_k, \partial_+ P_k]\, d\xi d\bar{\xi} \\
&\quad + \operatorname{tr}(\partial_+ P_k \cdot \partial_- P_k)\, \partial_- \left(\frac{[\partial_- P_k, P_k]}{\operatorname{tr}(\partial_+ P_k \cdot \partial_- P_k)} \right) d\bar{\xi}^2.
\end{aligned}
\tag{2.37}
$$

The Gaussian curvature K_k and the mean curvatures $\mathcal{H}_k$ (written in matrix form) take the forms

$$
\begin{aligned}
K_k &= \frac{-\partial_+ \partial_- \ln |\operatorname{tr}(\partial_+ P_k \cdot \partial_- P_k)|}{\operatorname{tr}(\partial_+ P_k \cdot \partial_- P_k)}, \\
\mathcal{H}_k &= \frac{-4\mathrm{i}}{\operatorname{tr}(\partial_+ P_k \cdot \partial_- P_k)} [\partial_+ P_k, \partial_- P_k] \in \mathfrak{su}(N),
\end{aligned}
\tag{2.38}
$$

where

$$
\operatorname{tr}(\mathcal{H}_k) = 0, \qquad (\mathcal{H}_k, \partial_\pm X_k) = 0. \tag{2.39}
$$

This result allows us to compute the Willmore functionals [7]

$$
W_k = \iint_{S^2} \operatorname{tr}([\partial_+ P_k, \partial_- P_k])^2\, d\xi^1 d\xi^2, \tag{2.40}
$$

where the integration is over the whole Riemann sphere $S^2 \cong \mathbb{C} \cup \{\infty\}$. The topological charges Q_k associated with these surfaces are [7]

$$
Q_k = -\frac{2}{\pi} \iint_{S^2} \operatorname{tr}(P_k \cdot [\partial_+ P_k, \partial_- P_k])\, d\xi^1 d\xi^2. \tag{2.41}
$$

The Euler–Poincaré characters associated with these surfaces are given by Goldstein and Grundland [7]

$$
\Delta_k = -\frac{1}{\pi} \iint_{S^2} \partial_+ \partial_- \ln |\operatorname{tr}(\partial_+ P_k \cdot \partial_- P_k)|\, d\xi^1 d\xi^2. \tag{2.42}
$$

The integrals (2.40)–(2.42) exist and provide global characterization of these surfaces. The fact that all soliton surfaces X_k possess the same value of the Euler–Poincaré character, equal to 2, and positive Gaussian curvature $K_k > 0$ means that all surfaces are homeomorphic to spheres.

It is easy to prove [8] that for $k \neq l$ the surfaces X_k and X_l do not have common points, with the exception of X_0 and X_1 in the $\mathbb{C}P^1$ model, where according to (2.25) X_0 coincides with X_1.

Proof Indeed, let X_k coincide with X_l at some point $\left(\xi, \bar{\xi}\right) \in \mathbb{C}$. In view of (2.16), subtracting X_l from X_k for $l > k$, we get

$$P_l - P_k + 2\sum_{j=k}^{l-1} P_j - \frac{2}{N}(l-k)\,\mathbb{I}_N = \varnothing. \tag{2.43}$$

This relation implies that

$$\begin{aligned} \left(1 - \frac{2}{N}(l-k)\right) P_k &= \varnothing, \\ \left(1 - \frac{1}{N}(l-k)\right) P_{l-1} &= \varnothing, \text{ for } k < l-1, \\ \left(1 - \frac{2}{N}\right) P_k &= \varnothing, \text{ for } k = l-1. \end{aligned} \tag{2.44}$$

Equations (2.44) can only be satisfied when $N = 2, l = 1, k = 0$. Hence $X_1 = X_0$. This equality holds for all points due to the fact that (2.25) holds.

The interesting case of surfaces having constant Gaussian curvature was discussed in detail by Delisle, Hussin, and Zakrzewski, also for higher-rank Grassmannians, in [3, 4].

3 Stack of Conformally Parametrized Surfaces

The immersion functions X_k of the 2D-soliton surfaces satisfy the same EL equations (2.31) as the original rank-1 projectors P_k (2.6). This suggests a possibility of further construction of surfaces induced by surfaces, etc., up to a whole stack of surfaces. An unexpected result comes from the fact that the surfaces Y_k over the surfaces X_k prove to be identical to the original surfaces X_k up to a factor of (-1) if we require that the induced surfaces be elements of the $\mathfrak{su}(N)$ algebra. The surfaces over surfaces are defined by the contour integrals

$$Y_k\left(\xi, \bar{\xi}\right) = \mathrm{i}\int_{\gamma_k} \left(-\left[\partial_+ X_k, X_k\right] d\xi + \left[\partial_- X_k, X_k\right] d\bar{\xi}\right) \in \mathfrak{su}(N). \tag{3.1}$$

The integrals in (3.1) are also contour-independent because the EL equations (2.31) written in terms of the $\mathfrak{su}(N)$-valued immersion functions X_k can be written as the CLs

$$\partial_+\left[\partial_- X_k, X_k\right] + \partial_-\left[\partial_+ X_k, X_k\right] = 0. \tag{3.2}$$

The complex tangent vectors $\partial_+ Y_k$ and $\partial_- Y_k$ are obviously

$$\partial_+ Y_k = -\,\mathrm{i}\,[\partial_+ X_k, X_k], \qquad \partial_- Y_k = \mathrm{i}\,[\partial_- X_k, X_k]. \tag{3.3}$$

We now derive surfaces obtained from rank-1 projectors induced by surfaces and show that the surfaces within the stack of surfaces coincide with each other, which demonstrates the idempotency of the recurrence procedure.

Proposition 1 ([9]) *Let the $\mathbb{C}P^{N-1}$ model be defined on the Riemann sphere and have a finite action functional. Then the surfaces Y_k over the surfaces X_k defined by the contour integral (expressed in terms of the rank-1 projectors P_k and the identity matrix)* (3.1) *are identical to the initial surface* (2.16) *from which the surfaces were derived, up to a factor of* (-1).

Outline of the proof. By direct computation of the complex tangent vectors of $\partial_+ Y_k$ and $\partial_- Y_k$ from the projector properties we obtain

$$\partial_+ Y_k = -\partial_+ X_k, \qquad \partial_- Y_k = -\partial_- X_k,$$

which implies

$$Y_k = -X_k \tag{3.4}$$

if we require that both Y_k and X_k be elements of the algebra $\mathfrak{su}(N)$. This means that the process of building the stack in which each next step is a surface over the previous step becomes idempotent.

4 Higher-Rank Projectors and Superposition Formula for Immersions

The EL equations (2.6) with the property $P^2 = P$ admit a larger class of solutions than the rank-1 Hermitian projectors P_k. We show that any linear combination of rank-1 projectors P_k satisfying the EL equations is itself a projector which satisfies the EL equations.

Proposition 2 ([9]) *Let P be any linear combination of rank-1 Hermitian projectors P_k where not all λ_k are necessarily zero*

$$P = \sum_{k=0}^{N-1} \lambda_k P_k, \qquad \lambda_k \in \mathbb{R}, \tag{4.1}$$

which have been obtained from P_0 by raising operators (2.10) *and each P_k satisfies the EL equations* (2.6). *Then P also satisfies the EL equations* (2.6). *If in addition,* *for all indices* $i, j \in \{0, \dots, N-1\}$ *we have*

$$\lambda_i = 0 \quad or \quad \lambda_j = 1, \tag{4.2}$$

then P *is a higher-rank projector satisfying the idempotency condition*

$$P^2 = P. \tag{4.3}$$

In this case P *maps the* $\mathbb{C}^N$ *space onto* $\mathbb{C}^m$*, where*

$$m = \sum_{i=0}^{N-1} \lambda_i.$$

Proof The proof is based on the Clebsch–Gordan decompositions. Namely, if P_k satisfies the EL equations (2.6), then the second mixed derivatives of P_k can be represented as a combination of at most three rank-1 neighboring projectors [9]

$$\partial_+\partial_- P_k = \hat{\alpha}_k P_{k-1} + \left(\hat{\alpha}_k + \check{\alpha}_k\right) P_k + \check{\alpha}_k P_{k+1}, \tag{4.4}$$

where the coefficients

$$\begin{aligned} &\hat{\alpha}_k = \mathrm{tr}\,(\partial_- P_k P_k \partial_+ P_k)\,, \qquad \check{\alpha}_k = \mathrm{tr}\,(\partial_+ P_k P_k \partial_- P_k)\,, \\ &\hat{\alpha}_k + \check{\alpha}_k = \mathrm{tr}\,(\partial_+ P_k \partial_- P_k)\,. \end{aligned} \tag{4.5}$$

are real-valued functions. The rank-1 projectors P_k are mutually orthogonal and they commute with each other. This implies that $\partial_+\partial_- P$ commutes with P, i.e.,

$$\begin{aligned} [\partial_+\partial_- P, P] &= \left[\partial_+\partial_- \sum_{i=0}^{N-1} \lambda_i P_i, \sum_{j=0}^{N-1} \lambda_j P_j\right] \\ &= \sum_{i,j=0}^{N-1} \lambda_i \lambda_j \left\{\check{\alpha}_i \left[P_{i-1}, P_j\right] + \hat{\alpha}_i \left[P_{i+1}, P_j\right] - \left(\check{\alpha}_i + \hat{\alpha}_j\right)\left[P_i, P_j\right]\right\} \\ &= 0. \end{aligned} \tag{4.6}$$

The linear combination of rank-1 projectors P_k is a projector of higher rank. If P satisfies the condition (4.3), then this implies that for all $i, j, \in \{0, \dots, N-1\}$, the coefficients λ_i satisfy (4.2).

Note that the coefficients $\hat{\alpha}_k$ and $\check{\alpha}_k$ have physical and geometrical interpretations. The coefficients $(\hat{\alpha}_k + \check{\alpha}_k)$ constitute the Lagrangian density in the action functional (2.5). Moreover, we have shown that

$$\mathrm{tr}\,(\partial_+ X_k \partial_- X_k) = -\mathrm{tr}\,(\partial_+ P_k \partial_- P_k)\,, \quad 0 \le k \le N-1,$$

which is also the Lagrange density for the 2D-surface immersion functions. It is also the non-diagonal element of the metric tensor $(g_k)_{\pm\mp}$ on the surface X_k (see Eq. (2.36)), which is also an element of the area of the surface X_k. In this way $\hat{\alpha}_k + \check{\alpha}_k$ determines the metric properties of the surfaces and hence all surfaces of the stack.

Note that the inverse theorem is not true. For $r > 1$, there exist decompositions of rank-r projectors which satisfy the EL equations (2.6) into rank-1 projectors which do not satisfy them. For example, the 2×2 identity matrix $\mathbb{I}_2$ obviously satisfies the EL equations (2.6), whereas it may be decomposed into $P_1 + (\mathbb{I}_2 - P_1)$ where P_1 can be any rank-1 projector function including those which do not satisfy the EL equations (2.6).

We now show that any linear combination of immersion functions X_k satisfying the EL equations is itself an immersion function which satisfies the EL equations.

Proposition 3 ([9]) *Let a matrix function X be a linear combination of immersion functions X_k of 2D-surfaces in the $\mathfrak{su}(N)$ algebra*

$$X = \sum_{k=0}^{N-1} \lambda_k X_k, \tag{4.7}$$

where λ_k are complex-valued constants. Let the X_k satisfy the EL equations (2.6). *Then X also satisfies the EL equations $[\partial_+\partial_- X, X] = 0$. If, in addition, all λ_k are real-valued constants, then the immersion function of the multileaf X is also an element of the $\mathfrak{su}(N)$ algebra.*

Proof The proof is straightforward, based on the fact that the 2D-surface immersion functions X_k are linear combinations of rank-1 projectors P_k and the identity matrix $\mathbb{I}_N$ (see Eq. (2.16)). Hence, $\partial_+\partial_- X$ commute with X. Note that if all λ_k are 0 or 1 then the surface is represented by the multileaf immersion function X which is a union of the 2D-surfaces X_k represented by those X_k for which $\lambda_k = 1$, which are immersed in the $N^2 - 1$ dimensional $\mathfrak{su}(N)$ algebra.

For $m > k$, the immersion functions X_k and X_m make a constant angle ϕ_{km} between them which does not depend on the choice of the P_0 solution of the EL equations (2.6), nor on the coordinates $(\xi, \bar{\xi}) \in \mathbb{C}$

$$\cos\phi_{km} = \frac{c_k(2 - c_m)}{\left\{\left[c_k(2 - c_k) - \frac{1}{N}\right]\left[c_m(2 - c_m) - \frac{1}{N}\right]\right\}^{1/2}} \in (0, 1), \tag{4.8}$$

since

$$(X_k, X_m) = \frac{N}{2} c_k(2 - c_m)$$

holds and (X_k, X_k) is given by (2.24). The fact that the angles ϕ_{km} between X_k and X_m remain the same does not imply anything about the mutual inclination of the surfaces (whose angle is not well defined). It means that given N, k, and m, we can simultaneously plot both surfaces $X_k, X_m \in \mathfrak{su}(N)$ with a compass whose legs are at a constant angle dependent on these three integers only. Obviously, the angle between the legs $\phi_{km} \in [0, \pi]$ is uniquely defined by its cosine. For example, in the $\mathbb{C}P^3$ case the cosines of ϕ_{km} for $N = 4$, $k = 0, 1, 2$, and $m = 1, 2, 3$ are given by

$k\backslash m$	1	2	3
0	$5/\sqrt{33}$	$\sqrt{3/11}$	$1/3$
1		$9/11$	$\sqrt{3/11}$
2			$5/\sqrt{33}$

A question arises: does the idempotency stated in Proposition 2 hold for the surfaces corresponding to higher-rank projectors?

Let P be a projector of rank $r > 1$ satisfying the EL equations (2.6) which have the projective property $P^2 = P$. Let the immersion function X be obtained from P by the contour integral (2.15) and the "surface over the surface." Will this Y prove to be the same immersion function as X up to a constant factor? The answer is negative, even if the projector P is a linear combination of rank-1 projectors (4.1) where the coefficients λ_k are 0 or 1. Direct computation shows that the complex tangent vectors $\partial_+ Y$ and $\partial_- Y$ have a nonzero remainder. Namely, we have

$$\partial_+ Y = -\sum_{k=0}^{N-1} \lambda^2 \partial_+ X_k + 2\mathrm{i} \sum_{l=0}^{N-1} \sum_{k=0}^{N-1} \lambda_k \lambda_l \left\{ \partial_+ P_k P_k - P_l \partial_+ P_l \right.$$
$$\left. + (\partial_+ P_k + \partial_+ P_l) \sum_{j=k}^{l-1} P_j + \sum_{j=k}^{l-1} P_j (\partial_+ P_k + \partial_+ P_l) \right\}. \tag{4.9}$$

and its respective complex conjugate. The first sum in (4.9) implies that $Y = -X$ provided that the λ_k are 0 or 1 but the remainder in (4.9) may be nonzero. The same holds for $\partial_- Y$. This means that the idempotency in the stack of surfaces does not hold if we start from projectors of rank-$r > 1$.

5 Spin Matrices Obtained via $\mathfrak{su}(2)$ Projective Structures

The matrices S which represent spins or isospins generate subalgebras of $\mathfrak{su}(2)$ in the subspace $\mathbb{C}^2$, which is linked with the $\mathbb{C}P^1$ model ($N = 2$) and can be expressed in terms of linear combination of rank-1 projectors [10]

$$S = s_0 P_0 + s_1 P_1, \quad S^\dagger = S, \quad s_k = k - \frac{1}{2}, \ k = 0, 1. \tag{5.1}$$

The eigenvalues of the matrix S representing particles of spin are $-\frac{1}{2}$ and $\frac{1}{2}$.

Proposition 4 *The matrices S which represent spins are conjugate to*

$$\left(-\frac{\mathrm{i}}{2}\right)(X_0 + X_1) = -\mathrm{i}\,X_0. \tag{5.2}$$

This means that every spin field is conjugate, up to a factor of $\left(-\frac{\mathrm{i}}{2}\right)$ to the surface X_0. Expressing X_0 in terms of the rank-1 projectors P_k, we get (5.1). Note that the expression on the right-hand side is conjugate to any $\mathfrak{su}(2)$ matrix whose eigenvalues are given by $s_0 = -\frac{1}{2}$ and $s_1 = \frac{1}{2}$.

Proposition 5 *There is a one-to-one correspondence of the surface* $-\mathrm{i}\,X_0$ *immersed in* $\mathfrak{su}(2)$ *algebra with the quantum spin fields, provided that the Killing form which determines the length of the spin matrix S is defined as*

$$(S, S) = \frac{3}{2}\mathrm{tr}\,(S \cdot S)\,. \tag{5.3}$$

Remark 1 The absence of the minus sign in the expression (5.3), in contrast to the classical Killing form (2.23), arises from the fact that the spin matrices S are Hermitian while elements of the $\mathfrak{su}(2)$ algebra are anti-Hermitian.

Proof The spin matrices S have to satisfy the two conditions:

1. The eigenvalues of a spin z-component (or those of the projection of the spin in any direction) assume the half integer values $s_0 = -\frac{1}{2}$ and $s_1 = \frac{1}{2}$.
2. The length of a spin field S has to be [17, 22]

$$[s(s+1)]^{1/2} = \frac{\sqrt{3}}{2}. \tag{5.4}$$

On the other hand, the square of the length has to be determined as a multiplier of the scalar product, i.e., the Killing form

$$s(s+1) = c\,\mathrm{tr}\,(S \cdot S)\,,$$

where c is a real constant. From (5.1), we get

$$s(s+1) = c\,\mathrm{tr}\left(\left[\sum_{k=0}^{1}\left(k - \frac{1}{2}\right)P_k\right]^2\right).$$

Substituting $s = \frac{1}{2}$ and making use of the orthogonality relation (2.11) for rank-1 projectors, we obtain $c = \frac{3}{2}$.

6 Final Remarks

The technique presented above for constructing an increasing number of surfaces associated with $\mathbb{C}P^{N-1}$ sigma models on Euclidean spaces provides us with an effective tool for finding them without requiring any additional consideration, but proceeding directly from the given $\mathbb{C}P^{N-1}$ model.

In the next stage of this research, it would be worthwhile to identify the general solution of the EL equations (2.6), a result which, up to now, remains an open problem. It would also be interesting to extend this technique to different types of Grassmannian manifolds, such as

$$Gr_{p,q}(\mathbb{C}) = SU(p+q)/\left[S\left(U(p) \times U(q)\right)\right],$$

and also their symplectic analogue defined over the quaternionic field, e.g.,

$$Gr_{p,q}(\mathbb{H}) = Sp(p+q)/\left[Sp(p) \times Sp(q)\right]$$

in the compact case. Their important common properties are that the EL equations can be written in terms of higher-rank projectors. These equations share many properties such as an infinite number of local or nonlocal conserved quantities, infinite-dimensional symmetry algebras, Hamiltonian structures, complete integrability, and the existence of multisoliton solutions. Investigating these manifolds can provide us with much more diverse types of surfaces than those discussed here. It would also be pertinent to further develop these models via the $SU(2)$ coherent states approach [14]. This task will be undertaken in a future work.

References

1. D. Babelon, D. Bernard, M. Talon, *Introduction to Classical Integrable Systems* (Cambridge University Press, Cambridge, 2004)
2. A. Davydov, *Solitons in Molecular Systems* (Kluwer, New York, 1999)
3. L. Delisle, V. Hussin, W. Zakrzewski, Constant curvature solutions of Grassmannian sigma models: (1) Holomorphic solutions. J. Geom. Phys. **66**, 24–36 (2013)
4. L. Delisle, V. Hussin, W. Zakrzewski, Geometry of surfaces associated to Grassmannian sigma models. J. Phys.: Conf. Ser. **597**, 012029 (2015)
5. A. Din, W. Zakrzewski, General classical solutions of the $\mathbb{C}P^{N-1}$ model. Nucl. Phys. B **174**, 397–403 (1989)
6. A. Din, Z. Horvath, W. Zakrzewski, The Riemann-Hilbert problem and finite action $\mathbb{C}P^{N-1}$ solutions. Nucl. Phys. B **233**, 269–299 (1984)
7. P.P. Goldstein, A.M. Grundland, Invariant recurrence relations for $\mathbb{C}P^{N-1}$ models. J. Phys. A: Math. Gen. **43**, 265206 (2010)
8. P. Goldstein, A.M. Grundland, On the surfaces associated with $\mathbb{C}P^{N-1}$ models. J. Phys. Conf. Ser. **284**, 012031 (2011)
9. P. Goldstein, A.M. Grundland, On a stack of surfaces obtained from the $\mathbb{C}P^{N-1}$ sigma models. J. Phys. A: Math. Gen. **51**, 095201 (2018)

10. P.P. Goldstein, A.M. Grundland, S. Post, Soliton surfaces associated with sigma models; differential and algebraic aspects. J. Phys. A: Math. Theor. **45**, 395208 (2012)
11. D.G. Gross, T. Piran, S. Weinberg, *Two Dimensional Quantum Gravity and Random Surfaces* (World Scientific, Singapore, 1992)
12. A.M. Grundland, I. Yurdusen, On analytic description of two-dimensional surfaces associated with $\mathbb{C}P^{N-1}$ sigma model. J. Phys A: Math. Gen. **42**, 172001 (2009)
13. A.M. Grundland, A. Strasburger, W.J. Zakrzewski, Surfaces immersed in $\mathfrak{su}(N+1)$ lie algebras obtained from the $\mathbb{C}P^N$ sigma models. J. Phys. A: Math. Gen. **39**, 9187–9213 (2006)
14. A.M. Grundland, A. Strasburger, D. Dziewa-Dawidczyk, $\mathbb{C}P^N$ sigma models via the $SU(2)$ coherent states approach, in *Conference Proceedings on 50th Seminar Sophus Lie, Banach Center Publications*, vol. 113 (Institute Polish Academy of Sciences, Warsaw, 2017), 23 pp.
15. B. Konopelchenko, Induced surfaces and their integrable dynamics. Stud. Appl. Math. **96**, 9–51 (1996)
16. G. Landolfi, On the Canham-Helfrich membrane model. J. Phys. A: Math. Theor. **36**, 4699–4715 (2003)
17. N. Manton, P. Sutcliffe, *Topological Solitons* (Cambridge University Press, Cambridge, 2007)
18. S. Post, A.M. Grundland, Analysis of $\mathbb{C}P^{N-1}$ sigma models via projective structures. Nonlinearity **25**, 1–36 (2012)
19. R. Rajaraman, $\mathbb{C}P^n$ solitons in quantum Hall systems. Eur. J. Phys. B **28**, 157–162 (2002)
20. S. Safran, *Statistical Thermodynamics of Surfaces, Interfaces, and Membranes* (Addison-Wesley, Boston, 1994)
21. R. Sasaki, General class of solutions of the complex Grassmannian and $\mathbb{C}P^{N-1}$ sigma models. Phys. Lett. B **130**, 69–72 (1983)
22. L.I. Schiff, *Quantum Mechanics* (McGraw-Hill, New York, 1949), pp. 69–76, 216–228
23. A. Sym, Soliton surfaces. Lett. Nuovo Cimento **33**, 394–400 (1982), which also mentions J. Tafel's contribution
24. V.E. Zakharov, A.V. Mikhailov, Relativistically invariant two-dimensional models of field theory which are integrable by means of the inverse scattering problem method. J. Exp. Theor. Phys. **47**, 1017–1049 (1979)
25. W. Zakrzewski, *Low Dimensional Sigma Models* (Hilger, Bristol, 1989), pp. 46–74

On the Equivalence Between Type I Liouville Dynamical Systems in the Plane and the Sphere

Miguel A. González León, Juan Mateos Guilarte, and Marina de la Torre Mayado

Abstract Separable Hamiltonian systems either in sphero-conical coordinates on an S^2 sphere or in elliptic coordinates on a $\mathbb{R}^2$ plane are described in a unified way. A back and forth route connecting these Liouville Type I separable systems is unveiled. It is shown how the gnomonic projection and its inverse map allow us to pass from a Liouville Type I separable system with a spherical configuration space to its Liouville Type I partners where the configuration space is a plane and back. Several selected spherical separable systems and their planar cousins are discussed in a classical context.

Keywords Separation of variables · Sphero-conical coordinates · Elliptic coordinates · Liouville dynamical systems · Trajectory isomorphism

1 Introduction

Hamiltonian systems in $\mathbb{R}^2$ that admit separation of variables were completely determined by Liouville [1] and Morera [2], and can be classified, see [3], in four different types according to the system of coordinates where the separability is manifested: elliptic, polar, parabolic, and Cartesian, respectively. Thus Type I Liouville systems in $\mathbb{R}^2$ are defined by natural Hamiltonians: $H = K + \mathcal{U}$, $K = \frac{m}{2}\left((\frac{dx_1}{dt})^2 + (\frac{dx_2}{dt})^2\right)$, that are separable in elliptic coordinates [3].

M. A. González León (✉)
Departamento de Matemática Aplicada and IUFFyM, Universidad de Salamanca, Salamanca, Spain
e-mail: magleon@usal.es

J. Mateos Guilarte · M. de la Torre Mayado
Departamento de Física Fundamental and IUFFyM, Universidad de Salamanca, Salamanca, Spain
e-mail: guilarte@usal.es; marina@usal.es

Ş. Kuru et al. (eds.), *Integrability, Supersymmetry and Coherent States*, CRM Series in Mathematical Physics, https://doi.org/10.1007/978-3-030-20087-9_16

In this work we shall establish an isomorphism between this kind of mechanical systems and Liouville systems in S^2 that are separable in sphero-conical coordinates that, correspondingly, we shall call Type I Liouville systems in S^2.

This isomorphism will be constructed by mapping the configuration space S^2 by means of two gnomonic projections from the two S^2-hemispheres into two $\mathbb{R}^2$ planes, together with a redefinition of the physical time and the application of a linear transformation in the projecting planes. This procedure is a generalization of the method used in [4], where the orbits of the two fixed center problem on S^2 [5, 6] were determined by inverting these transformations. The inspiration was taken from the work of Borisov and Mamaev [7], based itself on the ideas of Albouy [8] and Albouy and Stuchi [9]. The main novelty of [4] was the simultaneous consideration of two gnomonic projections in order to study the complete set of orbits, identifying each trajectory crossing the equator of S^2 with the conjunction of two planar unbounded orbits: one of the two attractive center problem and another corresponding to the system of the two associated repulsive centers.

The idea of projecting dynamical systems in constant positive curvature surfaces to planar ones goes back to Appell [10, 11] in the nineteenth century. Higgs and Leemon [12] have studied the Kepler problem in S^2 using this kind of techniques. In modern times projective dynamics has been developed by Albouy [8, 13–15] and Albouy and Stuchi [9], see also [16] for a detailed historical review and references on problems defined in spaces of constant curvature.

The structure of this paper is as follows: In Sect. 2 the gnomonic projections will be constructed. Section 3 is devoted to describe the properties of Liouville type I systems, both in S^2 and $\mathbb{R}^2$. In Sect. 4 the isomorphism is established. Section 5 contains several selected examples and finally some comments and future perspectives are showed in the final section.

2 Gnomonic Projections from S^2 to $\mathbb{R}^2$

Let us consider the S^2 sphere embedded in $\mathbb{R}^3$, i.e., $(X, Y, Z) \in \mathbb{R}^3$, such that $X^2 + Y^2 + Z^2 = R^2$. Standard spherical coordinates in S^2:

$$X = R \sin\theta \cos\varphi \,, \quad Y = R \sin\theta \sin\varphi \,, \quad Z = R \cos\theta$$

$\theta \in [0, \pi]$, $\varphi \in [0, 2\pi)$, allow us to write the metric tensor in TS^2 (i.e., the restriction of Euclidean metric in $T\mathbb{R}^3$ to the sphere) in standard form:

$$ds^2 = R^2 \left(d\theta^2 + \sin^2\theta \, d\varphi^2\right)$$

The gnomonic projections from the North/South hemispheres: $S^2_+ = \{(X, Y, Z) \in S^2 / Z > 0\}$, $S^2_- = \{(X, Y, Z) \in S^2 / Z < 0\}$, to the $\mathbb{R}^2$ plane, with respect to the

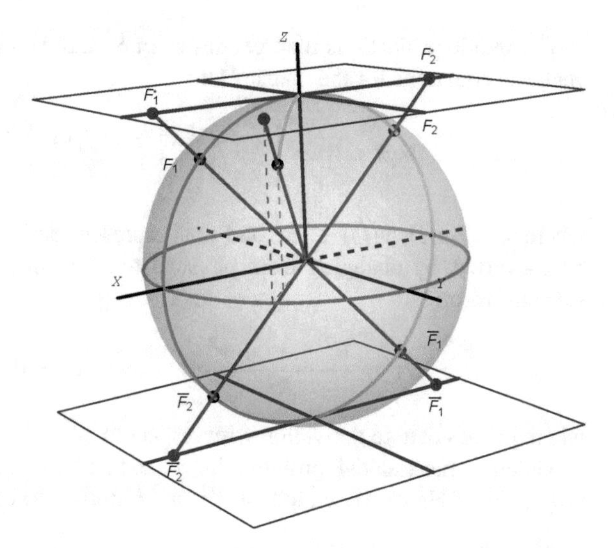

Fig. 1 Gnomonic projections Π_+ and Π_-

points $(0, 0, \pm R)$, see Fig. 1, are defined by the change of variables

$$\Pi_\pm : S^2_\pm \longrightarrow \mathbb{R}^2 \quad \Rightarrow \quad \begin{cases} x = \frac{R}{Z} X = R \tan\theta \cos\varphi \\ \\ y = \frac{R}{Z} Y = R \tan\theta \sin\varphi \end{cases} , \quad \varphi \in [0, 2\pi)$$

where $\theta \in \left[0, \frac{\pi}{2}\right)$ in the case of Π_+ and $\theta \in \left(\frac{\pi}{2}, \pi\right]$ for Π_-. The inverse maps $\Pi_\pm^{-1} : \mathbb{R}^2 \longrightarrow S^2_\pm$, read:

$$X = \frac{Rx}{\sqrt{R^2 + x^2 + y^2}} , \ Y = \frac{Ry}{\sqrt{R^2 + x^2 + y^2}} , \ Z = \frac{\pm R^2}{\sqrt{R^2 + x^2 + y^2}}$$

The projections $\Pi_\pm$ define two copies of the Riemannian manifold $(\mathbb{R}^2, g)$ where the metric tensor g in each copy is given by:

$$ds^2 = \frac{R^2}{(R^2 + x^2 + y^2)^2} \left((R^2 + y^2) dx^2 - 2xy\, dx\, dy + (R^2 + x^2) dy^2 \right) \qquad (1)$$

with associated Christoffel symbols: $\Gamma^1_{22} = \Gamma^2_{11} = 0$,

$$\Gamma^1_{11} = 2\Gamma^2_{12} = 2\Gamma^2_{21} = \frac{-2x}{R^2 + x^2 + y^2} , \quad \Gamma^2_{22} = 2\Gamma^1_{12} = 2\Gamma^1_{21} = \frac{-2y}{R^2 + x^2 + y^2} .$$

Gnomonic projections map geodesics in S^2 into straight lines in $\mathbb{R}^2$. In fact the geodesic equations for the metric (1):

$$\nabla_{\dot{\mathbf{x}}}\dot{\mathbf{x}} = 0 \Rightarrow \begin{cases} \ddot{x} + \Gamma^1_{11}\dot{x}^2 + 2\Gamma^1_{12}\dot{x}\dot{y} + \Gamma^1_{22}\dot{y}^2 = 0 \\ \ddot{y} + \Gamma^2_{11}\dot{x}^2 + 2\Gamma^2_{12}\dot{x}\dot{y} + \Gamma^2_{22}\dot{y}^2 = 0 \end{cases}$$

where $\mathbf{x} \equiv (x(t), y(t)) \in \mathbb{R}^2$ and dots represent derivative with respect to t, can be converted, by changing from physical to local (or *projected*) time, into trivial standard form:

$$d\tau = \frac{R^2 + x^2 + y^2}{R^2} dt \quad \Rightarrow \quad x'' = 0\,, \quad y'' = 0$$

where primes denote derivation with respect to τ.

Given a mechanical problem in S^2 defined by a potential function $\mathcal{U}$, the projection of Newton equations in S^2_+ or S^2_- to $(\mathbb{R}^2, g)$ can be written as:

$$\nabla_{\dot{\mathbf{x}}}\dot{\mathbf{x}} = -\mathrm{grad}\,\mathcal{U}(\mathbf{x}) \tag{2}$$

where $\mathbf{x} \equiv (x, y)$, and covariant derivatives and the gradient are associated with the g metric (1). Changing to projected time, Eq. (2) will be written as:

$$\mathbf{x}'' = -\mathrm{grad}\,\mathcal{U}(\mathbf{x}) \Rightarrow \begin{cases} x'' = -g^{11}\frac{\partial \mathcal{U}}{\partial x} - g^{12}\frac{\partial \mathcal{U}}{\partial y} \\ y'' = -g^{21}\frac{\partial \mathcal{U}}{\partial x} - g^{22}\frac{\partial \mathcal{U}}{\partial y} \end{cases} \tag{3}$$

where g^{ij} denote the components of g^{-1}, the inverse of the metric g.

We now pose the following question: Is it possible to understand Eq. (3) as Newton equations for a mechanical system in the Euclidean $\mathbb{R}^2$ plane with time τ?, in other words: Would it exists a function $\mathcal{V}(x_1, x_2)$ such that equations

$$x_1'' = -\frac{\partial \mathcal{V}}{\partial x_1}, \quad x_2'' = -\frac{\partial \mathcal{V}}{\partial x_2} \tag{4}$$

are equivalent to (3)?

The answer was given by Albouy [8] and developed explicitly by Borisov and Mamaev [7] for the case of the Killing problem restricted to the North hemisphere, i.e., the problem of two Kepler centers in S^2_+. The equivalence (trajectory isomorphism) was achieved in this concrete case via the linear transformation $x_1 = x$, $x_2 = \frac{1}{\sigma}y$, for an adequate value of σ parameter, in Eq. (3). Moreover, in [7] this isomorphism was extended to other mechanical systems and in general to systems admitting separation of variables in sphero-conical coordinates in S^2_+. In [4] the equivalence for the Killing problem was applied to the complete sphere considering the two projections Π_+ and Π_- simultaneously. A delicate point is the gluing of the inverse projections at the equator of the sphere. Orbits crossing the equator have to be described by the differentiable gluing of two pieces coming from unbounded orbits in each of the two planes, respectively.

In this work, following [7], we shall show that these results are valid for the class of Type I Liouville systems in the whole S^2, i.e., separable system in sphero-conical coordinates in S^2, that are transformed by gnomonic projections and the linear transformation into Liouville systems of Type I in $\mathbb{R}^2$ (separable in elliptic coordinates) with respect to the "non-physical" time τ.

3 Liouville Type I Systems in S^2 and $\mathbb{R}^2$

We shall refer to Hamilton–Jacobi separable spherical systems in sphero-conical coordinate as Liouville dynamical systems of Type I in S^2, in analogy with the planar case for elliptic coordinates, see [3],

Sphero-conical coordinates $U \in (\bar{\sigma}, 1)$, $V \in (-\bar{\sigma}, \bar{\sigma})$ describe points in an S^2-sphere by means of the geodesic distances $R\theta_1$ and $R\theta_2$ from the particle position to two fixed points that we choose without loosing generality as: $F_1 = (R\bar{\sigma}, 0, R\sigma)$, $F_2 = (-R\bar{\sigma}, 0, R\sigma)$, $\sigma = \cos\theta_f$, $\bar{\sigma} = \sin\theta_f$, see Fig. 2, in the form:

$$\theta_1 = \arccos(\sigma\cos\theta + \bar{\sigma}\sin\theta\cos\varphi)\,, \quad \theta_2 = \arccos(\sigma\cos\theta - \bar{\sigma}\sin\theta\cos\varphi)$$

Sphero-conical coordinates are thus defined by replicating on the sphere the "gardener" construction which allowed Euler to define elliptic coordinates in $\mathbb{R}^2$:

$$U = \sin\frac{\theta_1 + \theta_2}{2}\,, \quad V = \sin\frac{\theta_2 - \theta_1}{2}$$

and the change of coordinates is the following:

$$X = \frac{R}{\bar{\sigma}}UV\,, \quad Y^2 = \frac{R^2}{\sigma^2\bar{\sigma}^2}(U^2 - \bar{\sigma}^2)(\bar{\sigma}^2 - V^2)\,, \quad Z^2 = \frac{R^2}{\sigma^2}(1 - U^2)(1 - V^2)\,.$$

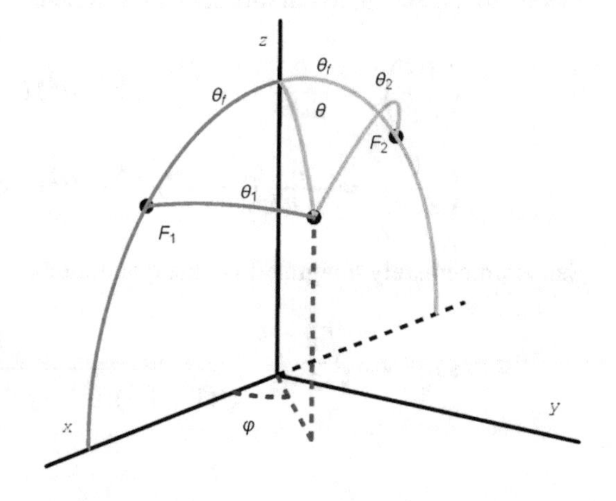

Fig. 2 Position of the particle in the sphere relative to two fixed points or foci

The kinetic energy of a particle moving on one S^2 sphere as configuration space expressed in sphero-conical coordinates reads

$$K = \frac{mR^2}{2}\left[\frac{U^2 - V^2}{(1 - U^2)(U^2 - \bar{\sigma}^2)}\left(\frac{dU}{dt}\right)^2 + \frac{U^2 - V^2}{(1 - V^2)(\bar{\sigma}^2 - V^2)}\left(\frac{dV}{dt}\right)^2\right],$$

where t is the physical time and we stress that K is singular in the equator, i.e., in the circle: $Z = 0 \equiv U = 1$. Changing from physical to local time, $d\varsigma = \frac{dt}{U^2 - V^2}$, the kinetic energy is rewritten as:

$$K = \frac{mR^2}{2(U^2 - V^2)}\left[\frac{1}{(1 - U^2)(U^2 - \bar{\sigma}^2)}\left(\frac{dU}{d\varsigma}\right)^2 + \frac{1}{(1 - V^2)(\bar{\sigma}^2 - V^2)}\left(\frac{dV}{d\varsigma}\right)^2\right].$$

We define a natural dynamical system as Liouville of Type I in S^2 if the potential energy is a function of the form:

$$\mathcal{U}(U, V) = \frac{1}{U^2 - V^2}\left(F(U) + G(V)\right). \tag{5}$$

This kind of potentials where the functions $F(U)$ and $G(V)$ are regular enough give rise to motion equations which are separable in the U and V evolutions.

Systems of this type are automatically completely integrable. The first integral of motion, the mechanical energy $E = K + \mathcal{U}$, leads to the separated expressions:

$$-\frac{mR^2\left(\frac{dV}{d\varsigma}\right)^2}{2(1 - V^2)(\bar{\sigma}^2 - V^2)} - G(V) - EV^2 = \frac{mR^2\left(\frac{dU}{d\varsigma}\right)^2}{2(1 - U^2)(U^2 - \bar{\sigma}^2)} + F(U) - EU^2$$

which necessarily must be equal to a constant $-\Omega$, a second invariant in involution with the energy. Rearranging these expressions we finally reduce the equations of motion to the uncoupled first-order ODEs system:

$$\left(\frac{dU}{d\varsigma}\right)^2 = \frac{2}{mR^2}(1 - U^2)(U^2 - \bar{\sigma}^2)\left(-\Omega + E\,U^2 - F(U)\right) \tag{6}$$

$$\left(\frac{dV}{d\varsigma}\right)^2 = \frac{2}{mR^2}(1 - V^2)(\bar{\sigma}^2 - V^2)\left(\Omega - E\,V^2 - G(V)\right). \tag{7}$$

that is immediately integrated via the quadratures:

$$\varsigma - \varsigma_0 = \pm R\sqrt{\frac{m}{2}}\int_{\bar{\sigma}}^{U}\frac{d\tilde{U}}{\sqrt{(1 - \tilde{U}^2)(\tilde{U}^2 - \bar{\sigma}^2)\left(-\Omega + E\,\tilde{U}^2 - F(\tilde{U})\right)}} \tag{8}$$

$$\varsigma - \varsigma_0 = \pm R\sqrt{\frac{m}{2}}\int_{-\bar{\sigma}}^{V} \frac{d\tilde{V}}{\sqrt{(1-\tilde{V}^2)(\bar{\sigma}^2-\tilde{V}^2)\,(\Omega - E\,\tilde{V}^2 - G(\tilde{V}))}}. \tag{9}$$

and the orbits are found by inversion, if possible, of these integrals. The physical time can be recovered by integration of the expression:

$$t = \int_{\varsigma_0}^{\varsigma} (U(\bar{\varsigma})^2 - V(\bar{\varsigma})^2)d\bar{\varsigma}$$

Liouville Type I systems in $\mathbb{R}^2$ are separable in elliptic coordinates [3]. Recall that Euler elliptic coordinates in $\mathbb{R}^2$ relative to the foci: $f_1 = (a, 0)$, $f_2 = (-a, 0)$ are defined as half the sum and half the difference of the distances from the particle position to the foci:

$$u = \frac{r_1 + r_2}{2a}\,,\ v = \frac{r_2 - r_1}{2a}\,;\ r_1 = \sqrt{(x_1 - a)^2 + x_2^2}\,,\ r_2 = \sqrt{(x_1 + a)^2 + x_2^2}. \tag{10}$$

The new coordinates vary in the intervals: $-1 < v < 1$, $1 < u < \infty$. In terms of these coordinates the particle position is defined to be

$$x_1 = auv\,, \quad x_2^2 = a^2\,(u^2 - 1)(1 - v^2), \tag{11}$$

implying a two-to-one map from $\mathbb{R}^2$ to the infinite "rectangle": $(-1, 1) \times (1, +\infty)$. The kinetic energy with respect to the local time $d\zeta = \frac{dt}{u^2 - v^2}$ in this coordinate system reads

$$K = \frac{ma^2}{2(u^2 - v^2)}\left(\frac{1}{u^2 - 1}\left(\frac{du}{d\zeta}\right)^2 + \frac{1}{1 - v^2}\left(\frac{dv}{d\zeta}\right)^2\right)$$

and the potential provides a Liouville Type I system in $\mathbb{R}^2$ if is of the form

$$\mathcal{V}(u, v) = \frac{1}{u^2 - v^2}\,(f(u) + g(v))\;, \tag{12}$$

for arbitrary but sufficiently regular functions $f(u)$ and $g(v)$. A standard separability process leads to the uncoupled first-order ODEs:

$$\left(\frac{du}{d\zeta}\right)^2 = \frac{2}{ma^2}\,(u^2 - 1)\,(-\lambda + h\,u^2 - f(u)) \tag{13}$$

$$\left(\frac{dv}{d\zeta}\right)^2 = \frac{2}{ma^2}\,(1 - v^2)\,(\lambda - h\,v^2 - g(v)) \tag{14}$$

depending on the energy h and the second constant of motion λ.

4 Trajectory Isomorphism Between Liouville Type I Systems in S^2 and $\mathbb{R}^2$

The gnomonic projection Π_+ from S^2_+ to $\mathbb{R}^2$ allows us to write the Cartesian coordinates (x, y) in terms of the sphero-conical ones:

$$x = \frac{R\sigma}{\bar{\sigma}} \frac{UV}{\sqrt{1-U^2}\sqrt{1-V^2}}, \quad y^2 = \frac{R^2}{\bar{\sigma}^2} \frac{(U^2-\bar{\sigma}^2)(\bar{\sigma}^2-V^2)}{(1-U^2)(1-V^2)}$$

The re-scaling $x_1 = x$, $x_2 = \frac{y}{\sigma}$ permits us to re-write these expressions in terms of Euler elliptic coordinates in the form (11) for $a = \frac{R\bar{\sigma}}{\sigma}$. Note that with this choice we have that $\Pi_+(F_1) = (a, 0) = f_1$ and $\Pi_+(F_2) = (-a, 0) = f_2$. Thus the $\mathbb{R}^2$ plane with coordinates (x_1, x_2) is equivalently described in terms of the sphero-conical coordinates or in the elliptic form (11) via the identifications:

$$u = \frac{\sigma U}{\bar{\sigma}\sqrt{1-U^2}}, \quad v = \frac{\sigma V}{\bar{\sigma}\sqrt{1-V^2}} \tag{15}$$

It is interesting to remark that Eq. (15) do not explicitly depend on the radius R of the sphere that only appears in the inter-foci distance a. This is thus a way to link sphero-conical and elliptic coordinates completely different to the well-known limiting process that identifies elliptic coordinates in the plane as the limiting case of sphero-conical coordinates in the sphere as the radius tends to infinity.

Let us consider a Liouville Type I system in S^2 with potential (5) and the corresponding separated first-order equations (6) and (7) with respect to the local time ς in S^2. The chain of changes from this local time ς to the elliptic local time ζ, via going back to the physical time t, changing to the projected time τ and finally defining ζ in terms of τ: $d\zeta = \frac{d\tau}{u^2-v^2}$, can be simply summarized in the form:

$$d\zeta = \bar{\sigma}\, d\varsigma$$

Using this expression and the identification (15) one easily realizes that Eqs. (6) and (7) are equivalent to Eqs. (13) and (14), and reciprocally, if the respective constants of motion are related through the equation:

$$h = \frac{E - \Omega}{\sigma^2}, \quad \lambda = \frac{\Omega}{\bar{\sigma}}.$$

and the potential energy in $\mathbb{R}^2$ is obtained from the potential energy in S^2, and vice versa, via the identities:

$$f(u) = \frac{\bar{\sigma}^2 u^2 + \sigma^2}{\sigma^2 \bar{\sigma}^2} F(U(u)), \quad g(v) = \frac{\bar{\sigma}^2 v^2 + \sigma^2}{\sigma^2 \bar{\sigma}^2} G(V(v)) \tag{16}$$

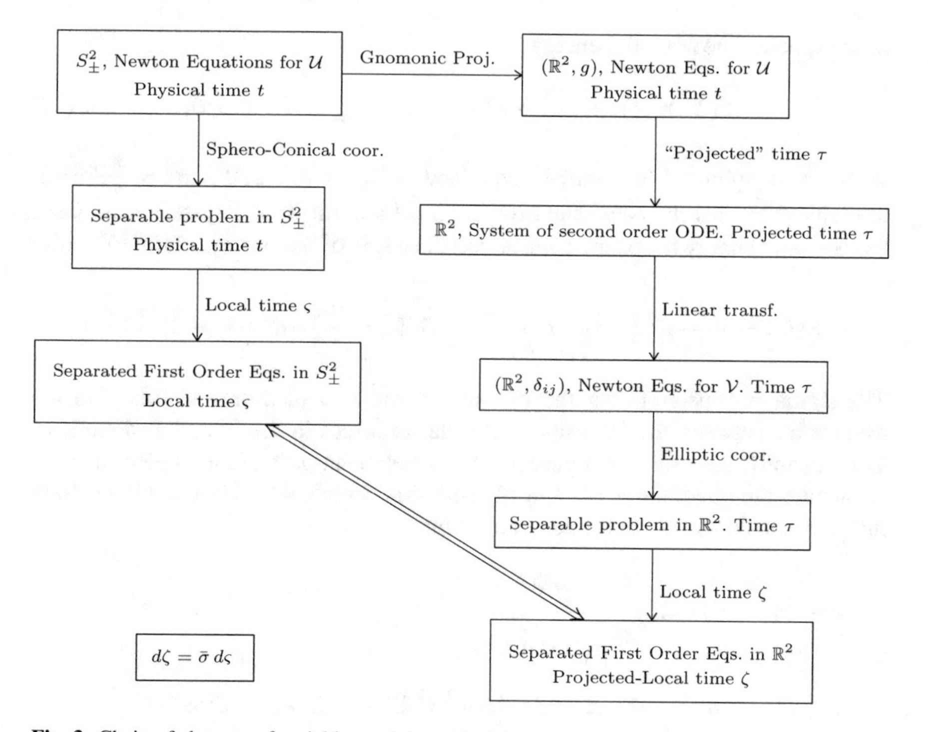

Fig. 3 Chain of changes of variables and time schedules

It is thus established the prescription to pass back and forth between a Liouville Type I separable systems in S^2_+ with a given physical time t and a Liouville Type I system in $\mathbb{R}^2$ with respect to a "non-physical" time τ.

An analogous procedure relative to the projection Π_- can be developed for S^2_-, and thus Newton equations on $S^2_\pm$ are equivalent to Newton equations (4) in the Euclidean planes. The orbits of a system with S^2 as configuration space require the determination of the orbits of two planar systems to be completely described in this projected picture.

In order to clarify the relationship between local times needed for separability in S^2 and $\mathbb{R}^2$ we include in Fig. 3 a diagram showing all the changes of time schedules.

5 Gallery of Selected Examples

5.1 The Neumann System

The Neumann system [17] consists of a particle constrained to move in an S^2 sphere of radius R subjected to maximally anisotropic linear attraction towards the center of the sphere. The potential energy is

$$\mathcal{U}(X,Y,Z) = aX^2 + bY^2 + cZ^2 , \quad a > b > c > 0, \tag{17}$$

where the couplings a, b, c may be redefined as $\frac{m\omega^2}{2} = a - c$, $0 < \sigma^2 = \frac{b-c}{a-c} < 1$, to easily show that the Neumann problem is a Liouville Type I system in S^2 since the potential energy in sphero-conical coordinates is of the standard form (5) with:

$$F(U) = \frac{m\omega^2}{2} R^2 \left(U^2 - \bar{\sigma}^2\right) U^2 , \quad G(V) = \frac{m\omega^2}{2} R^2 \left(\bar{\sigma}^2 - V^2\right) V^2 .$$

The sigma parameter fixing the position of the foci measures in this case the asymmetry between the intensity of the elastic forces in the X and Y directions. Consequently, the orbits of a particle in the Neumann problem are determined by evaluating the quadratures (8) and (9) with this choice of $F(U)$ and $G(V)$. Both integrals can be written in the compact form:

$$\varsigma - \varsigma_0 = \pm\sqrt{m} R \int_{\mathcal{X}_0}^{\mathcal{X}} \frac{d\tilde{\mathcal{X}}}{\sqrt{P_5(\tilde{\mathcal{X}})}} \tag{18}$$

$$P_5(\mathcal{X}) = -\mathcal{X}(1-\mathcal{X})(\mathcal{X}-\bar{\sigma}^2)\left(m\omega^2 R^2 \mathcal{X}^2 - (2E - m\omega^2 R^2 \bar{\sigma}^2)\mathcal{X} + 2\Omega\right)$$

where a new integration variable: $\mathcal{X}$ has been introduced: $U = \mathcal{X}^2$ for quadrature (8) and $V = \mathcal{X}^2$ for (9). Equation (18) is a hyperelliptic integral of genus 2, and obviously to obtain explicit expressions for these orbits requires the use of rank 2 theta functions, see [18, 19].

Taking into account the symmetry of the Neumann potential (17), the corresponding planar potentials $\mathcal{V}(x_1, x_2)$ will have identical expressions in both $\Pi_+(S^2_+)$ and $\Pi_-(S^2_-)$ planes. Applying (16) we obtain potential (12) with:

$$f(u) = \frac{m\omega^2 R^2 \sigma^2}{2} \frac{u^2(u^2-1)}{\bar{\sigma}^2 u^2 + \sigma^2} , \quad g(v) = \frac{m\omega^2 R^2 \sigma^2}{2} \frac{v^2(1-v^2)}{\bar{\sigma}^2 v^2 + \sigma^2}$$

that in Cartesian coordinates corresponds to the potential function:

$$\mathcal{V}(x_1, x_2) = \frac{m\omega^2 R^2}{2} \left(\frac{x_1^2 + \sigma^2 x_2^2}{R^2 + x_1^2 + \sigma^2 x_2^2} \right) \tag{19}$$

Thus orbits for (17) lying in S^2_+ or S^2_- are in one-to-one correspondence with bounded orbits of the planar system (19), see Fig. 4, whereas orbits that cross the equator have to be recovered from the projected pictures as the gluing of unbounded orbits of the two planar copies.

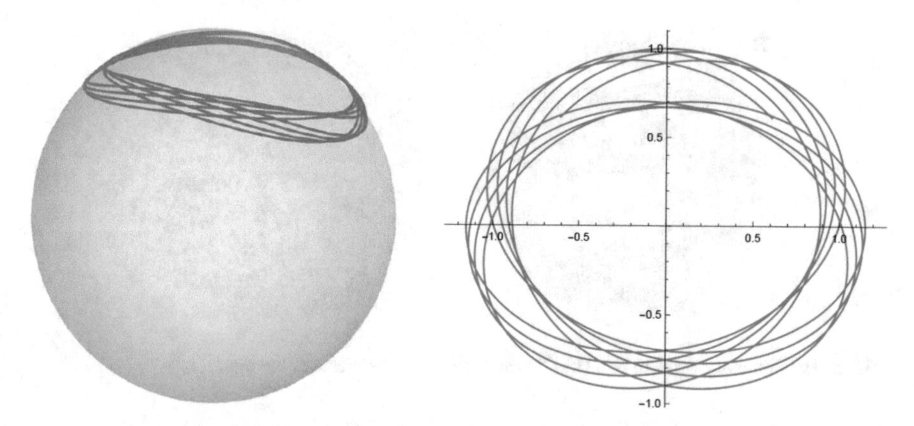

Fig. 4 An orbit of the Neumann problem and its corresponding planar orbit

5.2 *The Killing System*

In the Killing system [5, 6], see also [4] and the references therein, one massive particle is forced to move on an S^2-sphere of radius R under the action of a gravitational field created by two (e.g., attractive, $\gamma_1 > 0, \gamma_2 > 0$) Keplerian centers. Fixing the centers in the above defined F_1 and F_2 points the potential energy reads

$$\mathcal{U}(\theta_1, \theta_2) = -\frac{\gamma_1}{R}\text{cotan}\,\theta_1 - \frac{\gamma_2}{R}\text{cotan}\,\theta_2,$$

and thus the test mass feels the presence of two attractive centers in the North hemisphere and two (repulsive) ones located at the antipodal points in the South hemisphere with identical strengths. In sphero-conical coordinates the potential energy is written in two different expressions depending on the hemisphere that it is considered. In both cases $\mathcal{U}_\pm(U, V)$ is of Liouville Type I in S^2 form (5), with:

$$F_\pm(U) = \mp\frac{\gamma_1 + \gamma_2}{R}U\sqrt{1 - U^2}, \quad G(V) = -\frac{\gamma_1 - \gamma_2}{R}V\sqrt{1 - V^2}.$$

Applying the general procedure explained above the dynamics in the S^2_+ hemisphere can be described by the planar potential:

$$\mathcal{V}_+(x_1, x_2) = -\frac{\gamma_1}{\sigma^2\sqrt{(x_1 - \frac{R\bar{\sigma}}{\sigma})^2 + x_2^2}} - \frac{\gamma_2}{\sigma^2\sqrt{(x_1 + \frac{R\bar{\sigma}}{\sigma})^2 + x_2^2}}$$

that corresponds to the problem of two attractive centers in $\mathbb{R}^2$. In a parallel way, the problem in the South hemisphere is orbitally equivalent to the planar problem:

$$\mathcal{V}_-(x_1, x_2) = \frac{\gamma_2}{\sigma^2\sqrt{(x_1 - \frac{R\bar{\sigma}}{\sigma})^2 + x_2^2}} + \frac{\gamma_1}{\sigma^2\sqrt{(x_1 + \frac{R\bar{\sigma}}{\sigma})^2 + x_2^2}},$$

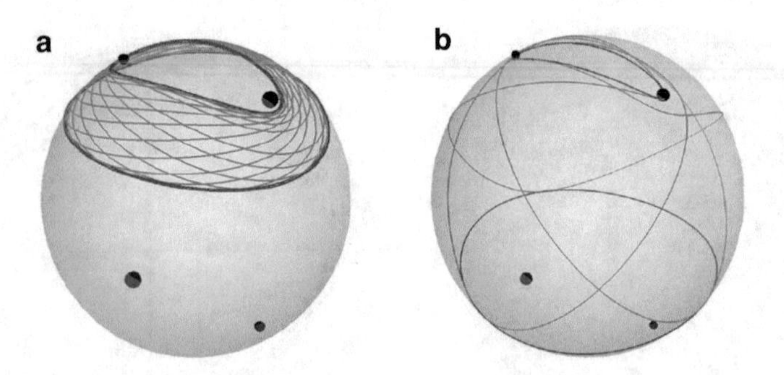

Fig. 5 (**a**) Planetary orbit in S^2_+. (**b**) Closed orbit in S^2 that crosses the equator

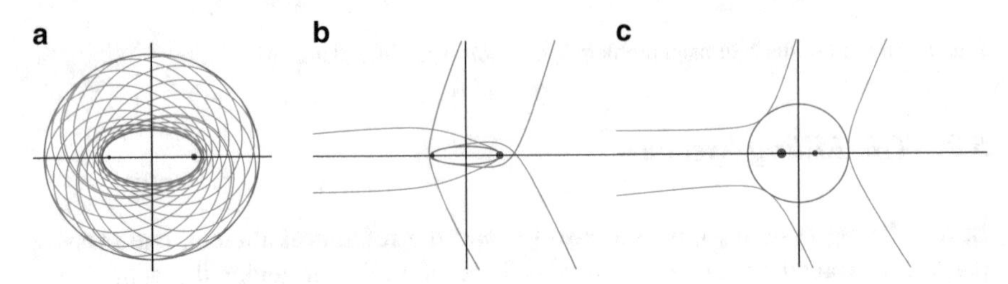

Fig. 6 (**a**) Planetary orbit projected in $\mathbb{R}^2$. (**b** and **c**) Projections of the closed orbit on $\Pi_+(S^2_+)$ and $\Pi_-(S^2_-)$, respectively

i.e., the planar potential of two repulsive centers where the roles of the points $(\pm\frac{R\bar{\sigma}}{\sigma}, 0)$, and thus the strengths of the centers in modulus, are interchanged with respect to the attractive potential $\mathcal{V}_+(x_1, x_2)$ in $\Pi_+(S^2_+)$.

In [4] a complete analysis of the different types of orbits for this problem is performed, including the integration and inversion of the involved elliptic integrals that lead to explicit expressions in terms of Jacobi elliptic functions for all the available regimes in the bifurcation diagram. Two examples of planetary type orbits are represented in Fig. 5. In Fig. 6 we can see their corresponding orbits in the projected planar systems.

5.3 *Inverse Gnomonic Projection of the Garnier System from* $\mathbb{R}^2$ *to* S^2

The Garnier system [3, 20] corresponds to a planar anharmonic oscillator which is isotropic in the quartic power of the distance to the center but anisotropic in the quadratic term. Using non-dimensional coordinates and couplings the potential energy is defined to be:

$$\mathcal{V}(x_1, x_2) = \frac{1}{2}\left(x_1^2 + x_2^2 - 1\right)^2 + \frac{a^2}{2}x_2^2, \qquad 0 < a < 1.$$

Changing to Euler elliptic variables (11) it is easily seen that it is a Liouville Type I system in $\mathbb{R}^2$ since the potential energy takes the form (12) where

$$f(u) = \frac{a^4}{2}(u^2-1)\left(u^2-\frac{1}{a^2}\right)^2, \quad g(v) = \frac{a^4}{2}(1-v^2)\left(v^2-\frac{1}{a^2}\right)^2.$$

The quadratures are thus

$$\varsigma - \varsigma_0 = \mp a\int_1^u \frac{dz}{\sqrt{(z^2-1)P_6(z)}}, \quad \varsigma - \varsigma_0 = \mp a\int_{-1}^v \frac{dz}{\sqrt{(1-z^2)\tilde{P}_6(z)}},$$

where the sixth order polynomials read:

$$P_6(u) = -f(u) + hu^2 - \lambda, \qquad \tilde{P}_6(v) = -g(v) - hv^2 + \lambda.$$

The change of integration variable: $z^2 = \bar{z}$, renders both integrals to identical canonical form:

$$\int \frac{d\bar{z}}{\sqrt{2\bar{z}(1-\bar{z})(a^4\bar{z}^3 - a^2(a^2+2)\bar{z}^2 + (1+2a^2-2h)\bar{z} + 2\lambda - 1)}}$$

i.e., they are hyperelliptic integrals of genus 2.

The inverse gnomonic projection leads us to the Liouville Type I separable system in S^2_+ characterized by the rational functions:

$$F(U) = \frac{(U^2-\bar{\sigma}^2)(1-2U^2)^2}{2(1-U^2)^2}, \quad G(V) = \frac{(\bar{\sigma}^2-V^2)(1-2V^2)^2}{2(1-V^2)^2}$$

where, in this non-dimensional setting, we have identified the parameters in the form: $a = \frac{\bar{\sigma}}{\sigma}$.

The corresponding potential in terms of $(X, Y, Z) \in S^2_+$ is

$$\mathcal{U}(X, Y, Z) = \frac{1}{2\sigma^2}\left(\frac{1-\bar{\sigma}^2 X^2}{Z^4} - \frac{1+3\sigma^2}{Z^2}\right)$$

that is singular in the equator. Thus in this case, even if we extend $\mathcal{U}(X, Y, Z)$ to the whole S^2 sphere, the orbits cannot cross the equator, and unbounded planar orbits are mapped into spherical trajectories that approach the equator asymptotically.

6 Summary and Further Comments

In this report we have analyzed separable classical Hamiltonian systems in a unified way. We have focused in systems of two degrees of freedom for which the configuration space is either an S^2 sphere or the Euclidean plane $\mathbb{R}^2$. In the first case that we denote as Liouville Type I in S^2, we have selected those systems for which the Hamilton–Jacobi equation is separable in sphero-conical coordinates. In the planar case the separability of the HJ equation is demanded in Euler elliptic coordinates, thus restricting ourselves to Liouville Type I systems in $\mathbb{R}^2$.

The main contribution in this essay is the construction of a bridge between Liouville Type I systems, respectively, in S^2 and $\mathbb{R}^2$. The path is traced following the gnomonic projection from both the North and South hemispheres to the plane. The idea is inspired by the connection between the two Keplerian center problem, respectively, in S^2 and $\mathbb{R}^2$ established in [7–9]. We provide a geometric structure to the Borisov–Mamaev map which allows to extend the idea to any Liouville Type I system. As particular cases we construct the bridge between the Neumann problem and its partner in the plane, besides reconstructing the Borisov–Mamaev map between the Killing problem, two Keplerian centers in S^2, and the Euler problem, two Keplerian centers in $\mathbb{R}^2$, in this geometric setting. Moreover, we also consider a distinguished Liouville Type I system in $\mathbb{R}^2$, the Garnier system, and its mapping back in S^2 by using the inverse of the gnomonic projection. A remarkable fact emerges: the two center problems in S^2 and $\mathbb{R}^2$ exhibit separable potentials in terms of either trigonometric or polynomial functions but identical, up to a constant, strengths: in both manifolds Keplerian potentials arise.

Having established this equivalence for Liouville Type I systems, the idea of extending this relation to other types of Liouville systems in the plane and the sphere is very promising. The results of this work can be also extended to the quantum framework. It would be very interesting to analyze the relation between separable Schrödinger equations in S^2 and the corresponding projected equations in $\mathbb{R}^2$. Connecting paths between the classical and quantum worlds are provided by the WKB quantization procedure. Separable quantum systems in elliptic coordinates, for instance, [21, 22] or [23] would be analyzed in the spherical case.

Finally, it is adequate to remind that the search for solitary waves arising in $(1+1)$-dimensional relativistic scalar field theories is tantamount to solve an analogue mechanical system. In this framework, the application of the equivalence between separable systems in S^2 and $\mathbb{R}^2$ could be a fruitful source of information about the links between solitary waves in non-linear and linear sigma models, [24–26].

Acknowledgements The authors thank the Spanish Ministerio de Economía y Competitividad (MINECO) for financial support under grant MTM2014-57129-C2-1-P and the Junta de Castilla y León (Projects VA057U16, VA137G18, and BU229P18). We gratefully acknowledge the constructive comments on the paper offered by the anonymous referee.

References

1. J. Liouville, Mémoire sur l'integration des équations différentielles du mouvement d'un nombre quelconque de points matériels. J. Math. Pures et Appl. **14**, 257–299 (1849)
2. G. Morera, Sulla separazione delle variabili nelle equazioni del moto di un punto materiale su una superficie Atti. Sci. di Torino **16**, 276–295 (1881)

3. A.M. Perelomov, *Integrable Systems of Classical Mechanics and Lie Algebras* (Birkhäuser, Basel 1990)
4. M.A. Gonzalez Leon, J. Mateos Guilarte, M. de la Torre Mayado, Orbits in the problem of two fixed centers on the sphere. Regul. Chaotic Dyn. **22**, 520–542 (2017)
5. H.W. Killing, Die Mechanik in den nicht-euklidischen Raumformen. J. Reine Angew. Math. **98**(1), 1–48 (1885)
6. V.V. Kozlov, A.O. Harin, Kepler's problem in constant curvature spaces. Celest. Mech. Dyn. Astr. **54**(4), 393–399 (1992)
7. A.V. Borisov, I.S. Mamaev, Relations between integrable systems in plane and curved spaces. Celest. Mech. Dyn. Astr. **99**(4), 253–260 (2007)
8. A. Albouy, The underlying geometry of the fixed centers problems, in *Topological Methods, Variational Methods and Their Applications*, ed. by H. Brezis, K.C. Chang, S.J. Li, P. Rabinowitz (World Scientific, Singapore, 2003), pp. 11–21
9. A. Albouy, T. Stuchi, Generalizing the classical fixed-centres problem in a non-Hamiltonian way. J. Phys. A **37**, 9109–9123 (2004)
10. P. Appell, De l'homographie en mécanique. Am. J. Math. **12**(1), 103–114 (1890)
11. P. Appell, Sur les lois de forces centrales faisant décrire à leur point d'application une conique quelles que soient les conditions initiales. Am. J. Math. **13**(2), 153–158 (1891)
12. P.W. Higgs, Dynamical symmetry in a spherical geometry I. J. Phys. A: Math. Gen. **12**(3), 309–323 (1979). H.I. Leemon, Dynamical symmetry in a spherical geometry II. J. Phys. A: Math. Gen. **12**(4), 489–501 (1979)
13. A. Albouy, Projective dynamics and classical gravitation. Regul. Chaotic Dyn. **13**(6), 525–542 (2008)
14. A. Albouy, There is a projective dynamics. Eur. Math. Soc. Newsl. **89**, 37–43 (2013)
15. A. Albouy, Projective dynamics and first integrals. Regul. Chaotic Dyn. **20**(3), 247–276 (2015)
16. A.V. Borisov, I.S. Mamaev, I.A. Bizyaev, The spatial problem of 2 bodies on a sphere. Reduction and stochasticity. Regul. Chaotic Dyn. **21**(5), 556–580 (2016)
17. C. Neumann, De problemate quodam mechanico, quod ad primam integralium ultraellipticorum classem revocatur. J. Rein Angew. Math. **56**, 46–63 (1859)
18. B.A. Dubrovine, Theta functions and non-linear equations. Russ. Math. Surv. **36**(2), 11–92 (1981)
19. V.Z. Enolski, E. Hackmann, V. Kagramanova, J. Kung, C. Lämerzahl, Inversion of hyperelliptic integrals of arbitrary genus with applications to particle motion in general relativity. J. Geom. Phys. **61**, 899–921 (2011)
20. R. Garnier, Sur une classe de systems differentielles abelians deduits de la theorie de equations lineaires. Rend. Circ. Mat. Palermo **43**, 155–191 (1919)
21. O. Babelon, M. Talon, Separation of variables for the classical and quantum Neumann model. Nucl. Phys. B **379**, 321–339 (1992)
22. M. Gadella, J. Negro, L.M. Nieto, G.P. Pronko, Two charged particles in the plane under a constant perpendicular magnetic field. Int. J. Theor. Phys. **50**, 2019–2028 (2011)
23. M.A. Gonzalez Leon, J. Mateos Guilarte, M. de la Torre Mayado, Elementary solutions of the quantum planar two center problem. Europhys. Lett. **114**, 30007 (2016)
24. A. Alonso Izquierdo, M.A. Gonzalez Leon, J. Mateos Guilarte, Kinks in a non-linear massive sigma model. Phys. Rev. Lett. **101**, 131602 (2008). BPS and non BPS kinks in a massive non-linear S^2-sigma model. Phys. Rev. D **79**, 125003 (2009)
25. A. Alonso Izquierdo, M.A. Gonzalez Leon, J. Mateos Guilarte, M. de la Torre Mayado, On domain walls in a Ginzburg-Landau non-linear S2-sigma model. J. High Energy Phys. **2010**(8), 1–29 (2010)
26. A. Alonso Izquierdo, A. Balseyro Sebastian, M.A. Gonzalez Leon, Domain walls in a non-linear $\mathbb{S}^2$-sigma model with homogeneous quartic polynomial potential. J. High Energy Phys. **2018**(11), 1–23 (2018)

Construction of Partial Differential Equations with Conditional Symmetries

Decio Levi, Miguel A. Rodríguez, and Zora Thomova

Dedicated to our colleague and friend, Véronique Hussin, on her 60th anniversary.

Abstract Nonlinear PDEs having *given* conditional symmetries are constructed. They are obtained starting from the invariants of the *conditional symmetry* generator and imposing the extra condition given by the characteristic of the symmetry. Series of examples of new equations, constructed starting from the conditional symmetries of Boussinesq, are presented and discussed thoroughly to show and clarify the methodology introduced.

Keywords Lie symmetries · Partial differential equations · Conditional symmetries · Point transformations · Boussinesq equation

1 Introduction

Our capability of solving complicated physical problems described by mathematical formulas (say equations) is based on the existence of symmetries, i.e., transformations which leave the equations invariant. Towards the end of the nineteenth century, Sophus Lie introduced the notion of Lie group of symmetries in order to study the solutions of differential equations. He showed the following main property: if an

D. Levi (✉)
INFN, Sezione Roma Tre, Roma, Italy
e-mail: levi@roma.infn.it; decio.levi@roma3.infn.it

M. A. Rodríguez
Dept. de Física Teórica, Universidad Complutense de Madrid, Madrid, Spain
e-mail: rodrigue@ucm.es

Z. Thomova
SUNY Polytechnic Institute, Utica, NY, USA
e-mail: zora.thomova@sunpoly.edu

Ş. Kuru et al. (eds.), *Integrability, Supersymmetry and Coherent States*, CRM Series in Mathematical Physics, https://doi.org/10.1007/978-3-030-20087-9_17

equation is invariant under a one-parameter Lie group of point transformations, then we can reduce the equation and possibly construct an invariant solution. This observation unified and extended the available integration techniques such as separation of variables or integrating factors.

A partial differential equation (PDE) $\mathcal{E} = 0$ in the independent variables x_i and dependent variables u_j is invariant under a continuous group of Lie point transformations if the corresponding infinitesimal symmetry generator $\hat{X}$ satisfies

$$\text{pr}\hat{X}\mathcal{E}\Big|_{\mathcal{E}=0} = 0, \tag{1}$$

$$\hat{X} = \sum_i \xi_i \partial_{x_i} + \sum_j \phi_j \partial_{u_j}, \tag{2}$$

where ξ_i and ϕ_j are functions of the independent and dependent variables appearing in $\mathcal{E} = 0$. By the symbol pr we mean the prolongation of the infinitesimal generator to all derivatives appearing in $\mathcal{E} = 0$. See, for example, the references [6, 26] for this construction.

Given an infinitesimal generator of a symmetry $\hat{X}$ a function $\mathcal{I}$ is an invariant if it is such that

$$\text{pr}\hat{X}\mathcal{I} = 0. \tag{3}$$

Equation (3) is a first order PDE which can be solved on the characteristic and provide the set of invariants and differential invariants $\mathcal{I}_k$, $k = 0, 1, \cdots$ depending on x_i, u_j, and their partial derivatives up to arbitrary order. Then the PDEs invariant with respect to the infinitesimal generator (2) can be written as

$$\mathcal{E} = \mathcal{E}(\{\mathcal{I}_k\}) = 0, \quad k = 0, 1, \cdots. \tag{4}$$

Lie method is a well-established technique to search for exact solutions of differential or difference equations of any type, integrable or non-integrable, linear or nonlinear. However, many equations may have no symmetries and there is no simple algorithm to prove the existence of symmetries other than looking for them. Moreover, the obtained solutions do not always fulfill the conditions imposed by the physical requests (boundary conditions, asymptotic behavior, etc.). So one looks for extension or modification of the construction which could overcome some of these problems. One looks for more *symmetries*,

- not always expressed in local form in terms of the dependent variable of the differential equations,
- not satisfying all the properties of a Lie group but just providing solutions.

In the first class are the potential symmetries introduced by Bluman [4] and Bluman et al. [7], the nonlocal symmetries by Krasil'shchik and Vinogradov [21], Vinogradov and Krasil'shchik [36], Hernández-Heredero and Reyes [19], Reyes

[29, 30], Muriel and Romero [24], Catalano Ferraioli [10], and Levi et al. [23], while in the second one are the conditional symmetries [5, 15, 16, 22, 25].

In this paper we will be interested in showing how one can construct equations having given conditional symmetries.

In Sect. 2 we will provide the theory behind the construction of the conditional symmetries clarifying in this way the difference between symmetries and conditional symmetries. Then in Sect. 3 we will verify the proposed construction in the case of a few of the conditional symmetries of the Boussinesq equation (9) and, in correspondence with these conditional symmetries, construct new conditionally invariant equations which may have physical significance. Section 4 is devoted to the summary of the result, some concluding remarks, and prospects of future works.

2 What Is a Conditional Symmetry?

For simplicity of the presentation we will consider PDEs in only one dependent variable. The extension to systems is straightforward.

Bluman and Cole [5] introduced the method for finding conditional symmetries, the *non-classical method*. The non-classical method adds an auxiliary first order equation build up in terms of the coefficients of the infinitesimal generator $\hat{X}$ to $\mathcal{E} = 0$, namely

$$\mathcal{C} = \mathcal{C}(x_i, u, u_{x_i}) = \sum_i \xi_i u_{x_i} - \phi = 0. \tag{5}$$

Equation (5) is the infinitesimal symmetry generator (2) written in characteristic form [26] set equal to zero. Equation (5) is as yet unspecified and it will be determined together with the vector field $\hat{X}$, as it involves the same functions ξ_i and ϕ.

It is easy to prove that (5) is invariant under the first prolongation of (2)

$$\mathrm{pr}\hat{X}\mathcal{C} = -\left(\sum_i \xi_{i,u} u_{x_i} - \phi_u\right)\mathcal{C} = 0, \tag{6}$$

without imposing any conditions on the functions ξ_i and ϕ. Consequently, to get the conditional symmetries of $\mathcal{E} = 0$ we need just to apply the following invariance condition:

$$\mathrm{pr}\hat{X}\mathcal{E}\Big|_{\substack{\mathcal{E}=0\\ \mathcal{C}=0}} = 0. \tag{7}$$

Equation (7) gives nonlinear determining equations for ξ_i and ϕ which provide at the same time classical and non-classical symmetries.

As $\mathcal{C} = 0$ appears in (7) as a condition imposed on the determining equations the resulting symmetries are called *conditional symmetries*.

There are several works devoted to using the non-classical method to construct solutions of PDEs that are different from the ones obtained by the classical method using the Lie point symmetries, for example [2, 14, 17, 18, 20, 27, 31]. Moreover, we can find programs [3] to compute them and algorithms [11] for showing the existence of nontrivial non-classical symmetries for given PDEs. In the case of integrable equations let us mention the works of Sergyeyev [33, 34] where he considered the classification of all $(1+1)$-dimensional evolution systems that admit a generalized (Lie–Bäcklund) vector field as a generalized conditional symmetry.

In this paper we want to look at the conditional symmetries from a different perspective. Given an infinitesimal group generator characterized by a vector field $\hat{X}$ with specific values of the functions ξ_i and ϕ, we want to construct equations $\mathcal{E} = 0$ which have this symmetry as a *conditional symmetry* and not as a Lie point symmetry. Taking into account that an equation invariant under a given symmetry is written in terms of its invariants (4), a PDE invariant under a conditional symmetry will be given by

$$\mathcal{E}(\{\mathcal{I}_j\})\Big|_{\{\mathcal{C}=0\}} = 0, \quad j = 0, 1, \dots . \tag{8}$$

The constraint $\{\mathcal{C} = 0\}$ in (8) is to be interpreted as the differential equation (5) and its consequences (see Sect. 3 for details when constructing explicit examples).

3 A Few Examples from the Conditional Symmetries of the Boussinesq Equation

The Boussinesq equation

$$u_{yy} + uu_{xx} + (u_x)^2 + u_{xxxx} = 0 \tag{9}$$

was introduced in 1871 by Boussinesq to describe the propagation of long waves in shallow water [8, 9] and it is of considerable physical and mathematical interest. It also arises in several other physical applications including one-dimensional nonlinear lattice waves [35, 37], vibrations in a nonlinear string [38], and ion sound waves in a plasma [32]. It has two independent variables $x_1 = x$ and $x_2 = y$, consequently $i = 1, 2$ in (2). For simplicity we will call $\xi_1 = \xi$ and $\xi_2 = \eta$. If η in (2) is different from zero, the resulting determining equations for conditional symmetries do not fix it and we can always put $\eta = 1$. If $\eta = 0$ and $\xi \neq 0$, we can put $\xi = 1$ for the same reason. Consequently the condition is the same if we consider $\hat{X}$ or $f(x, y, u)\hat{X}$, although the invariants in the two cases are different.

The conditional symmetries of the Boussinesq equation for $\eta \neq 0$ were obtained in [22], and in [13] by non-group techniques. The case $\eta = 0$ has been considered later and can be found in [12]. In [12, 22] we find the following generators of the conditional symmetries for (9):

$$\hat{X}_1 = \partial_y + y\partial_x - 2y\partial_u \tag{10}$$

$$\hat{X}_2 = \partial_y - \frac{x}{y}\partial_x + \left(\frac{2}{y}u + \frac{6}{y^3}x^2\right)\partial_u \tag{11}$$

$$\hat{X}_3 = \partial_y + \left(-\frac{x}{y} + y^4\right)\partial_x + \left(\frac{2}{y}u + \frac{6}{y^3}x^2 - 2y^2x - 4y^7\right)\partial_u \tag{12}$$

$$\hat{X}_4 = \partial_y + \left(\frac{x}{2y} + y\right)\partial_x - \frac{1}{y}(u + 2x + 4y^2)\partial_u \tag{13}$$

$$\hat{X}_5 = \partial_y + \frac{\wp'}{2\wp}(x + c_1 W)\partial_x - \left[\frac{\wp'}{\wp}u + 3\wp' x^2 + c_1\left(\frac{1}{2\wp} + 6\wp' W\right)\right. \tag{14}$$

$$\left. + c_1^2 W\left(\frac{1}{2\wp} + 3\wp' W\right)\right]\partial_u, \quad W = \int_0^y \frac{\wp(s)}{\wp'(s)^2}ds,$$

$$\hat{X}_6 = \partial_x + \left(\frac{2}{x}u + \frac{48}{x^3}\right)\partial_u, \tag{15}$$

$$\hat{X}_7 = \partial_x + \left[-2x\wp + c_2\wp\int^y \frac{ds}{\wp(s)^2}\right]\partial_u, \tag{16}$$

where $\wp$ is a special case of the Weierstrass P function, $\wp(y, g_2, g_3)$ [1] with $g_2 = 0$ satisfying the differential equations $\wp'^2 = 4\wp^3 - g_3$, and g_3, c_1, c_2 are arbitrary constants.

The generators $\hat{X}_1, \cdots, \hat{X}_5$ were obtained assuming $\eta = 1$, thus are defined in (10)–(14) up to an arbitrary function $\eta(x, y, u)$, while $\hat{X}_6$ and $\hat{X}_7$ were obtained assuming $\eta = 0$ and $\xi = 1$, thus are defined in (15) and (16) up to an arbitrary function $\xi(x, y, u)$.

In the following we will not go through all the generators presented above but we will just consider as particular examples, $\hat{X}_1$, $\hat{X}_2$ and $\hat{X}_6$.

3.1 Conditional Invariant Equations Associated to $\hat{X}_1$

For the infinitesimal generator $\hat{X}_1$ and its prolongation up to fourth order we obtain the following invariants:

$$\begin{aligned} &\mathcal{I}_0 = -2x + y^2, \quad \mathcal{I}_1 = 2x + u, \quad \mathcal{I}_2 = u_x, \quad \mathcal{I}_3 = 2y + yu_x + u_y, \quad \mathcal{I}_4 = u_{xx}, \\ &\mathcal{I}_5 = yu_{xx} + u_{xy}, \quad \mathcal{I}_6 = u_{yy} + 2yu_{xy} + 2(y^2 - x)u_{xx}, \quad \mathcal{I}_7 = u_{xxx} \\ &\quad \ldots, \mathcal{I}_{11} = u_{xxxx}. \end{aligned} \tag{17}$$

The condition is given by $\mathcal{I}_3 = 0$, i.e., $\mathcal{C} = 2y + yu_x + u_y = 0$. As a verification of the correctness of the procedure proposed in the previous section we construct the Boussinesq equation in terms of the invariants (17). It is

$$\begin{aligned} \mathcal{I}_1\mathcal{I}_4 + \mathcal{I}_6 + \mathcal{I}_{11} + \mathcal{I}_2^2\Big|_{\mathcal{C}_x=0} &= u_{yy} + 2y(u_{xy} + yu_{xx}) + uu_{xx} + u_{xxxx} \\ + (u_x)^2\Big|_{u_{xy}+yu_{xx}=0} &= u_{yy} + uu_{xx} + u_{xxxx} + (u_x)^2 = 0. \end{aligned} \tag{18}$$

Now, we want to construct a new nonlinear PDE which has as a conditional symmetry the infinitesimal generator $\hat{X}_1$. Let us consider the following combination of the invariants $\mathcal{I}_1, \mathcal{I}_4, \mathcal{I}_6$, and $\mathcal{I}_7$ given in (17):

$$\mathcal{I}_1\mathcal{I}_4 + \mathcal{I}_6 + \mathcal{I}_7\Big|_{\mathcal{C}_x=0} = u_{yy} + 2y(u_{xy} + yu_{xx}) + uu_{xx} + u_{xxx}\Big|_{u_{xy}+yu_{xx}=0} = 0.$$

So we have

$$u_{xxx} + uu_{xx} + u_{yy} = 0, \tag{19}$$

a nonlinear dispersive Laplace equation.

The point symmetries of (19) are

$$\hat{H}_1 = \partial_x, \quad \hat{H}_2 = \partial_y, \quad \hat{H}_3 = 2x\partial_x + 3y\partial_y - 2u\partial_u. \tag{20}$$

Equation (19) has a conditional symmetry ($\hat{K}_1$) with $\eta = 1$ (which is the chosen conditional symmetry to construct the equation) and three conditional symmetries ($\hat{L}_1$, $\hat{L}_2$, and $\hat{L}_3$) with $\eta = 0$, $\xi = 1$ (after removing the Lie point symmetries previously computed (20) and using these points symmetries to remove inessential parameters):

$$\hat{K}_1 = \partial_y + y\partial_x - 2y\partial_u \tag{21}$$

$$\hat{L}_1 = \partial_x + \frac{2xu}{x^2 + c_1}\partial_u \tag{22}$$

$$\hat{L}_2 = \partial_x - \wp(y; 0, g_3)\left[6x - c_2\int_0^y \frac{1}{\wp(s; 0, g_3)^2}\,ds\right]\partial_u \tag{23}$$

$$\hat{L}_3 = \partial_x + \left[\frac{u}{x} - \wp(y; 0, g_3)\left(3x + \frac{c_3}{x}\int_0^y \frac{1}{\wp(s; 0, g_3)^2}\,ds\right)\right]\partial_u, \tag{24}$$

where $c_i, i = 1, 2, 3$ and g_3 are arbitrary constants.

3.2 *Conditional Invariant Equations Associated to $\hat{X}_2$*

For the infinitesimal generator $\hat{X}_2$ and its prolongation up to derivatives of fourth order we obtain the following invariants:

$$\begin{aligned}
\mathcal{I}_0 =& xy, \quad \mathcal{I}_1 = x^2 u + \frac{x^4}{y^2}, \quad \mathcal{I}_2 = x^3 u_x + 2\frac{x^4}{y^2}, \\
\mathcal{I}_3 =& \frac{x^2}{y}(y u_y - x u_x - 2u - 6\frac{x^2}{y^2}), \quad \mathcal{I}_4 = x^4 u_{xx} + 2\frac{x^4}{y^2}, \\
\mathcal{I}_5 =& \frac{x^3}{y}\left(y u_{xy} - x u_{xx} - 3u_x - 12\frac{x}{y^2}\right), \\
\mathcal{I}_6 =& \frac{x^2}{y}\left(y u_{yy} - 2x u_{xy} + \frac{x^2}{y} u_{xx} - 4u_y + 6\frac{x}{y} u_x + 6\frac{u}{y} + 42\frac{x^2}{y^3}\right), \\
\mathcal{I}_7 =& x^5 u_{xxx}, \ldots, \ \mathcal{I}_{11} = x^6 u_{xxxx}.
\end{aligned} \tag{25}$$

The condition is given by $\mathcal{I}_3 = 0$ which in this case we choose to be

$$\mathcal{C} = y u_y - x u_x - 2u - 6\frac{x^2}{y^2} = 0. \tag{26}$$

We can easily check that the Boussinesq equation can be constructed in terms of the invariants (25). It is

$$\mathcal{I}_1 \mathcal{I}_4 + \mathcal{I}_{11} + (\mathcal{I}_2)^2 \Big|_{\{\mathcal{C}=0\}} = x^6 \left[u_{yy} + u u_{xx} + u_{xxxx} + (u_x)^2\right]. \tag{27}$$

A new nonlinear dispersive Laplace equation can be written by considering the invariants $\mathcal{I}_0, \mathcal{I}_1, \mathcal{I}_2, \mathcal{I}_4$, and $\mathcal{I}_7$, i.e.,

$$\mathcal{I}_1 \mathcal{I}_4 + \mathcal{I}_0 \mathcal{I}_7 + (\mathcal{I}_2)^2 \Big|_{\{\mathcal{C}=0\}} = x^6 [u_{yy} + u u_{xx} + (u_x)^2 + y u_{xxx}] = 0. \tag{28}$$

The equation

$$u_{yy} + u u_{xx} + (u_x)^2 + y u_{xxx} = 0 \tag{29}$$

has point symmetries

$$\hat{H}_1 = \partial_x, \quad \hat{H}_2 = x\partial_x + y\partial_y, \tag{30}$$

and conditional symmetries with $\eta = 1:$

$$\hat{K}_1 = \partial_y + \left(c_1 y^4 - \frac{x}{y} \right) \partial_x + \left(\frac{2u}{y} - \frac{2\left(c_1 y^5 - x\right)\left(2c_1 y^5 + 3x\right)}{y^3} \right) \partial_u,$$

$$\hat{K}_2 = \partial_y + c_2 y^2 \partial_x + \left(\frac{u}{y} - 3c_2^2 y^3 \right) \partial_u, \tag{31}$$

where $\hat{X}_2$ is recovered from $\hat{K}_1$ by setting $c_1 = 0$. $\hat{K}_2$ is a genuine new conditional symmetry of (29) for any value of c_2. Also $\hat{K}_1$ with $c_1 \neq 1$ is a new conditional symmetry of (29).

When $\eta = 0, \ \xi = 1$ we have

$$\hat{L} = \partial_x + [f(x)u + g(x, y)]\partial_u, \tag{32}$$

where $f(x)$ satisfy the ODE:

$$f'' + 5ff' + 2f^3 = 0, \tag{33}$$

while the overdetermined system of equations for $g(x, y)$, depending explicitly of $f(x)$, is given in the Appendix in (45) and (46). Equation (33) can be solved completely but the solution we find by the use of Lie point symmetries is implicit, i.e., we have x as an integral function of $f(x)$. So we cannot use it to solve the overdetermined system (45) and (46). We are able to present, however, three explicit solutions for $f(x)$

$$\chi_1 = 0, \quad \chi_2 = \frac{2}{x}, \quad \chi_3 = \frac{1}{2x}. \tag{34}$$

Solving (45) and (46) with $f(x)$ given by χ_1 we get

$$\hat{L}_1 = \partial_x + \wp(y + c_1; 0, g_3) \left(-2x + c_2 \int_0^y \frac{1}{\wp(s + c_1; 0, g_3)^2} ds \right) \partial_u, \tag{35}$$

where c_1, c_2 are integration constants, while for χ_2 we get $g(x, y) = 0$ and then

$$\hat{L}_2 = \partial_x + \frac{2u}{x} \partial_u. \tag{36}$$

Finally, for χ_3 the overdetermined system, solved using the differential Grobner bases [28], is incompatible.

3.3 Conditional Invariant Equations Associated to $\hat{X}_6$

For the infinitesimal generator $\hat{X}_6$ and its prolongation up to third order we obtain the following invariants:

$$\begin{aligned}\mathcal{I}_0 =& y, \quad \mathcal{I}_1 = \frac{1}{x^4}\Big(12 + x^2 u\Big), \quad \mathcal{I}_2 = \frac{u_y}{x^2}, \\ \mathcal{I}_3 =& \frac{1}{x^5}\Big(x^3 u_x - 2x^2 u - 48\Big), \quad \mathcal{I}_4 = \frac{u_{yy}}{x^2}, \dots, \\ \mathcal{I}_{10} =& \frac{1}{x^7}\Big(- 1440 - 24x^2 u + 18x^3 u_x + x^5 u_{xxx} - 6x^4 u_{xx}\Big).\end{aligned} \tag{37}$$

The condition is given by $\mathcal{I}_3 = 0$ which in this case we choose to be

$$\mathcal{C} = u_x - \frac{2u}{x} - \frac{48}{x^3} = 0. \tag{38}$$

As in the previous case, the Boussinesq equation can be written in terms of the invariants (37):

$$6(\mathcal{I}_1)^2 + \mathcal{I}_4\Big|_{\{\mathcal{C}=0\}} = \frac{1}{x^2}\Big[u_{yy} + uu_{xx} + u_{xxxx} + (u_x)^2\Big]. \tag{39}$$

In correspondence with $\hat{X}_6$ we may also have

$$6(\mathcal{I}_1)^2 + \mathcal{I}_2\Big|_{\{\mathcal{C}=0\}} = \frac{1}{x^2}[u_y + uu_{xx} + u_{xxxx} + (u_x)^2] = 0. \tag{40}$$

The equation

$$u_y + uu_{xx} + u_{xxxx} + (u_x)^2 = 0 \tag{41}$$

has the point symmetries

$$\hat{H}_1 = \partial_x, \quad \hat{H}_2 = \partial_y, \quad \hat{H}_3 = x\partial_x + 4y\partial_y - 2u\partial_u. \tag{42}$$

The only conditional symmetry of (37) with $\eta = 1$ turns out to be, after multiplication by $4(y + c)$, a linear combination of the Lie point symmetries (42). When $\eta = 0, \; \xi = 1$ we have

$$\hat{L} = \partial_x + \phi(x, y, u)\partial_u, \tag{43}$$

where the function $\phi(x, y, u)$ is defined by just one PDE which we present in the Appendix in (47). For any entire function $\phi(x, y, u)$ in u (47) has, apart from the solution $\hat{X}_6$, the following conditional symmetries:

$$\hat{L}_1 = \partial_x + \frac{2u}{x}\partial_u, \quad \hat{L}_2 = \partial_x + \frac{x}{3y}\partial_u. \tag{44}$$

4 Conclusions

In this article we presented a construction of nonlinear PDEs having *given* conditional symmetries. They are obtained starting from the invariants of the symmetry and imposing the extra condition obtained by equating to zero the characteristic and its differential consequences. Starting from the conditional symmetries of the Boussinesq equation we reconstructed the Boussinesq equation itself as well as other nonlinear equations (19), (29), and (41). The obtained equations have Lie point symmetries and conditional symmetries.

An important point not touched in this work but on which we are presently working is understanding a priori when we can construct a conditionally invariant equation. Moreover work is also in progress on solving by symmetry reduction the obtained conditionally invariant equations and on the construction of conditional symmetry preserving discretizations of the Boussinesq equation.

Acknowledgements DL has been supported by INFN IS-CSN4 *Mathematical Methods of Nonlinear Physics*. DL thanks ZT and the SUNY Polytechnic Institute for their warm hospitality at Utica when this work was started. DL thanks the Departamento de Física Téorica of the Complutense University in Madrid for its hospitality. MAR was supported by the Spanish MINECO under project FIS 2015-63966-P. D. Nedza, summer student of ZT, contributed to the verification of some of the computations.

Appendix: Determining Equations for $\hat{X}_2$ and $\hat{X}_6$

For completeness we present here the determining equation of the conditional symmetries when $\xi = 1$ and $\eta = 0$ for which we have been able just to present some particular solutions

$$g_{xx} + 3fg_x + (5f' + 4f^2)g + 4f^4y + 7f'f^2y - 2f'f = 0, \tag{45}$$

$$g_{yy} + \left(5yf^2 - 5yf' + 3g\right)g_x + 2\left(8yf^2 + 10yf' + g\right)fg$$
$$+ 2y^2f\left(13f^4 - 20f'^2 + 16f^2f'\right) = 0. \tag{46}$$

$$\phi_{uuuu}\phi^4 + (4\phi_{uu}^2 + 6\phi_u\phi_{uuu} + 4\phi_{xuuu})\phi^3 + (7\phi_{uu}\phi_u^2 + 2(6\phi_{xuu} + 1)\phi_u$$
$$+ \phi_{uu}(12\phi_{xu} + u) + 6(\phi_{uuu}\phi_x + \phi_{xxuu}))\phi^2 + (8\phi_{xu}^2 + 2(2\phi_u^2 + u)\phi_{xu}$$
$$+ \phi_x(10\phi_u\phi_{uu} + 12\phi_{xuu} + 3) + 4\phi_{uu}\phi_{xx} + 6\phi_u\phi_{xxu} + 4\phi_{xxxu})\phi$$
$$+ 3\phi_{uu}\phi_x^2 + 4\phi_u\phi_x\phi_{xu} + u\phi_{xx} + 4\phi_{xu}\phi_{xx} + 6\phi_x\phi_{xxu} + \phi_{xxxx} + \phi_y = 0. \tag{47}$$

References

1. M. Abramowitz, I.A. Stegun, *Handbook of Mathematical Functions* (Dover, New York, 1972)
2. D.J. Arrigo, B.P. Ashley, S.J. Bloomberg, T.W. Deatherage, Nonclassical symmetries of a nonlinear diffusion–convection/wave equation and equivalents systems. Symmetry **8**, 140 (2016)
3. N. Bîlă, J. Niesen, On a new procedure for finding nonclassical symmetries. J. Symb. Comput. **38**, 1523–1533 (2004)
4. G.W. Bluman, Use and construction of potential symmetries. Math. Comput. Model. **18**, 1–14 (1993)
5. G.W. Bluman, J.D. Cole, The general similarity solutions of the heat equation. J. Math. Mech. **18**, 1025–1042 (1969)
6. G.W. Bluman, S. Kumei, *Symmetries of Differential Equations* (Springer, New York, 2002)
7. G.W. Bluman, S. Kumei, G.J. Reid, New classes of symmetries for partial differential equations. J. Math. Phys. **29**, 806–811 (1988)
8. J. Boussinesq, Théorie de l'intumescence liquide appelée onde solitaire ou de translation se propageant dans un canal rectangulaire. Comptes Rendus **72**, 755–759 (1871)
9. J. Boussinesq, Théorie des ondes et des remous qui se propagent le long d'un canal rectangulaire horizontal, en communiquant au liquide contenu dans ce canal des vitesses sensiblement pareilles de la surface au fond. J. Math. Pures Appl. **7**, 55–108 (1872)
10. D. Catalano Ferraioli, Nonlocal aspects of λ-symmetries and ODEs reduction. J. Phys. A: Math. Theor. **40**, 5479–5489 (2007)
11. T. Chaolu, G. Bluman, An algorithmic method for showing existence of nontrivial nonclassical symmetries of partial differential equations without solving determining equations. J. Math. Anal. Appl. **411**, 281–296 (2014)
12. P.A. Clarkson, Nonclassical symmetry reductions of the Boussinesq equation. Chaos Solitons Fractals **5**, 2261–2301 (1995)
13. P.A. Clarkson, M.D. Kruskal, New similarity reductions of the Boussinesq equation. J. Math. Phys. **30**, 2201–2213 (1989)
14. P.A. Clarkson, E.L. Mansfield, Symmetry reductions and exact solutions of shallow water wave equations. Acta Appl. Math. **39**, 245–276 (1995)
15. W.I. Fushchich, Conditional symmetry of the equations of nonlinear mathematical physics. Ukr. Math. J. **43**, 1350–1364 (1991)
16. W.I. Fushchich, R.Z. Zhdanov, Conditional symmetry and reduction of partial differential equations. Ukr. Math. J. **44**, 875–886 (1993)
17. R.K. Gupta, M. Singh, Nonclassical symmetries and similarity solutions of variable coefficient coupled KdV system using compatibility method. Nonlinear Dyn. **87**, 1543–1552 (2017)
18. M.S. Hashemi, M.C. Nucci, Nonclassical symmetries for a class of reaction-diffusion equations: the method of heir-equations J. Nonlinear Math. Phys. **20**, 44–60 (2013)
19. R. Hernández–Heredero, E.G. Reyes, Nonlocal symmetries and a Darboux transformation for the Camassa-Holm equation. J. Phys. A: Math. Theor. **42**, 182002 (2009)

20. L. Ji, C.Z. Qu, S. Shen, Conditional Lie-Backlund symmetry of evolution system and application for reaction-diffusion system. Stud. Appl. Math. **133**, 118–149 (2014)
21. I.S. Krasil'shchik, A.M. Vinogradov, Nonlocal trends in the geometry of differential equations: symmetries, conservation laws, and Bäcklund transformations. Symmetries of partial differential equations, Part I. Acta Appl. Math. **15**, 161–209 (1989)
22. D. Levi, P. Winternitz, Nonclassical symmetry reduction: example of the Boussinesq equation. J. Phys. A: Math. Gen. **22**, 2915–2924 (1989)
23. D. Levi, M.C. Nucci, M.A. Rodríguez, λ symmetries for the reduction of continuous and discrete equations. Acta Appl. Math. **122**, 311–321 (2012)
24. C. Muriel, J.L. Romero, New methods of reduction for ordinary differential equations. IMA J. Appl. Math. **66**, 111–125 (2001)
25. M.C. Nucci, P.A. Clarkson, The nonclassical method is more general than the direct method for symmetry reductions. An example of the Fitzhugh-Nagumo equation. Phys. Lett. A **164**, 49–56 (1992)
26. P.J. Olver, *Applications of Lie Groups to Differential Equations* (Springer, New York, 1993)
27. R.O. Popovych, N.M. Ivanova, O.O. Vaneeva, Potential nonclassical symmetries and solutions of fast diffusion equation. Phys. Lett. A **362**, 166–173 (2007)
28. G.J. Reid, A.D. Wittkopf, A. Boulton, Reduction of systems of nonlinear partial differential equations to simplified involutive forms. Eur. J. Appl. Math. **7**, 604–635 (1996)
29. E.G. Reyes, Nonlocal symmetries and the Kaup-Kupershmidt equation. J. Math. Phys. **46**, 073507 (2005)
30. E.G. Reyes, On nonlocal symmetries of some shallow water equations. J. Phys. A: Math. Theor. **40**, 4467–4476 (2007)
31. P.M.M. Rocha, F.C. Khannab, T.M. Rocha Filhoa, A.E. Santana, Non-classical symmetries and invariant solutions of non-linear Dirac equations. Commun. Nonlinear Sci. Num. Simul. **26**, 201–210 (2015)
32. A.C. Scott, in *Bäcklund Transformations*, ed. by R. M. Miura. Lecture Notes in Mathematics, vol. 515 (Springer, Berlin, 1975), pp. 80–105
33. A. Sergyeyev, Constructing conditionally integrable evolution systems in (1+1) dimensions: a generalization of invariant modules approach. J. Phys. A: Math. Gen. **35**, 7653–7660 (2002)
34. A. Sergyeyev, On the classification of conditionally integrable evolution systems in (1+1) dimensions. J. Math. Sci. **136**, 4392–4400 (2006)
35. M. Toda, Studies of a nonlinear lattice. Phys. Rep. **18**, 1–125 (1975)
36. A.M. Vinogradov, I.S. Krasil'shchik, A method of calculating higher symmetries of nonlinear evolutionary equations, and nonlocal symmetries (Russian). Dokl. Akad. Nauk SSSR **253**(6), 1289–1293 (1980)
37. N.J. Zabusky, A synergetic approach to problems of nonlinear dispersive wave propagation and interaction, in *Nonlinear Partial Differential Equations*, ed. by W.F. Ames (Academic, New York, 1967), pp. 233–258
38. V.E. Zakharov, On stochastization of one-dimensional chains of nonlinear oscillations. Sov. Phys. JETP **38**, 108–110 (1974)

An Integro-Differential Equation of the Fractional Form: Cauchy Problem and Solution

Fernando Olivar-Romero and Oscar Rosas-Ortiz

To prof. Veronique Hussin on his 60th birthday.

Abstract We solve the Cauchy problem defined by the fractional partial differential equation $[\partial_{tt} - \kappa\mathbb{D}]u = 0$, with $\mathbb{D}$ the pseudo-differential Riesz operator of first order, and the initial conditions $u(x, 0) = \mu(\sqrt{\pi}x_0)^{-1}e^{-(x/x_0)^2}$, $u_t(x, 0) = 0$. The solution of the Cauchy problem resulting from the substitution of the Gaussian pulse $u(x, 0)$ by the Dirac delta distribution $\varphi(x) = \mu\delta(x)$ is obtained as corollary.

Keywords Fractional partial differential equations · Fox H-functions · Dirac delta distribution · Pseudo-differential Riesz operator · Complementary equation

1 Introduction

Linear partial differential equations of second order are useful in physics to model phenomena like wave propagation, heat diffusion, and transport processes [1–3]. In analogy to conics of analytic geometry, the wave equation is hyperbolic, while the heat and transport equations are parabolic. In a recent work [4] we have reported a fractional formulation that permits the study of such equations in unified form. Additionally, we have introduced an integro-differential version of the parabolic equation $u_{tt} - \kappa u_x = 0$ (hereafter called *complementary equation*) that is solvable in analytic form. That is, in [4] we have solved the Cauchy problem for $u_{tt} - \kappa\mathbb{D}u = 0$ with zero initial velocity and the Dirac delta pulse $\varphi(x) = \mu\delta(x)$ as initial condition. The symbol $\mathbb{D}$ stands for the pseudo-differential Riesz operator [5] (for contemporary notions on the matter, see, e.g., [6]). In the present work we provide

F. Olivar-Romero (✉) · O. Rosas-Ortiz
Physics Department, Cinvestav, México City, Mexico
e-mail: folivar@fis.cinvestav.mx; orosas@fis.cinvestav.mx

Ş. Kuru et al. (eds.), *Integrability, Supersymmetry and Coherent States*, CRM Series in Mathematical Physics, https://doi.org/10.1007/978-3-030-20087-9_18

the solutions for the Cauchy problem with zero initial velocity and the Gaussian distribution $u(x,0) = \mu(\sqrt{\pi}x_0)^{-1}e^{-(x/x_0)^2}$ as initial disturbance.

The manuscript is structured as follows: In Sect. 2 we give the solution of the Cauchy problem for the modified complementary equation when the Gaussian distribution is considered as initial condition with zero initial velocity. We recover the results reported in [4] as a byproduct. In Sect. 3 we analyze the results. Some final conclusions are given in Sect. 4.

2 Statement of the Problem and Solution

The main results of this contribution are summarized in the following Proposition and Corollary.

Proposition 1 *The Cauchy problem defined for the integro-differential equation*

$$\frac{\partial^2}{\partial t^2}u(x,t) + \frac{\kappa}{\pi}\frac{\partial}{\partial x}\int_{\mathbb{R}}\frac{u(y,t)}{x-y}dy = 0, \quad \kappa > 0, \tag{1}$$

with the initial conditions

$$u(x,0) = \frac{\mu}{x_0\sqrt{\pi}}e^{-(x/x_0)^2}, \quad u_t(x,0) = 0, \quad x_0 \geq 0, \tag{2}$$

is solved by the function

$$u(x,t) = \frac{\mu}{\kappa t^2}\sum_{k=0}^{\infty}\frac{(-1)^k}{k!}\left(\frac{x_0}{2\kappa t^2}\right)^{2k}\theta_k(x,t), \tag{3}$$

where θ_k is the following Fox H-function:

$$\theta_k(x,t) = H^{2,1}_{3,3}\left[\frac{|x|}{\kappa t^2}\left|\begin{matrix}(-2k,1), (\frac{1}{2},\frac{1}{2}), (-1-4k,2)\\ (0,1), (-2k,1), (\frac{1}{2},\frac{1}{2})\end{matrix}\right.\right]. \tag{4}$$

Proof First note that Eq. (1) is indeed the fractional partial differential equation

$$\left[\frac{\partial^2}{\partial t^2} - \kappa\mathbb{D}\right]u(x,t) = 0, \tag{5}$$

with $\mathbb{D}$ the pseudo-differential Riesz operator [5, 6]. We may consider a generalized version of the latter equation [4, 7], defined as

$$[D^\alpha - v^2_{\alpha,\beta}\mathbb{D}^\beta]u(x,t) = 0, \quad 1 \leq \alpha \leq 2, \quad 1 \leq \beta \leq 2, \tag{6}$$

where the fractional time-derivative D^{α} is taken in the sense of Caputo [8] (see also [6]), and $\mathbb{D}^{\beta}$ is the Riesz operator of order β. In [4] we had already solved Eq. (6) for the initial conditions (2). The solution is written as the series

$$u(x,t) = \frac{\mu}{\beta t^{\frac{\alpha}{\beta}} v_{\alpha,\beta}^{2/\beta}} \sum_{k=0}^{\infty} \frac{(-1)^k}{k!} \left(\frac{x_0}{2t^{\frac{\alpha}{\beta}} v_{\alpha,\beta}^{2/\beta}} \right)^{2k} \Theta_k(x,t;\alpha,\beta), \tag{7}$$

with

$$\Theta_k(x,t;\alpha,\beta) = H_{3,3}^{2,1}\left[\frac{|x|}{t^{\frac{\alpha}{\beta}} v_{\alpha,\beta}^{2/\beta}} \middle| \begin{matrix} \left(\frac{\beta-(1+2k)}{\beta}, \frac{1}{\beta}\right), \left(\frac{1}{2}, \frac{1}{2}\right), \left(\frac{\beta-\alpha(1+2k)}{\beta}, \frac{\alpha}{\beta}\right) \\ (0,1), \left(\frac{\beta-(1+2k)}{\beta}, \frac{1}{\beta}\right), \left(\frac{1}{2}, \frac{1}{2}\right) \end{matrix} \right]. \tag{8}$$

Here

$$H_{p,q}^{m,n}\left[x \middle| \begin{matrix} (a_1,\alpha_1), \ldots, (a_p,\alpha_p) \\ (b_1,\beta_1), \ldots, (a_p,\beta_p) \end{matrix} \right] \tag{9}$$

$$= \frac{1}{2\pi i} \int_L \frac{\prod_{j=1}^{m} \Gamma(b_j + \beta_j z) \prod_{i=1}^{n} \Gamma(1 - a_i - \alpha_i z) x^{-z} dz}{\prod_{i=n+1}^{p} \Gamma(a_i + \alpha_i z) \prod_{j=m+1}^{q} \Gamma(1 - b_j - \beta_j z)}$$

is the Fox H-function [9, 10] for which the labels m, n, p, and q are integers such that $0 \le m \le q$, and $0 \le n \le p$. Besides $a_i, b_j \in \mathbb{C}$ and $\alpha_i, \beta_j \in (0, \infty)$.

The solution to the Cauchy problem (1)–(2) is obtained by evaluating Eqs. (7) and (8) at the point $(\alpha, \beta) = (2, 1)$, with $\kappa = v_{2,1}^2$ and $\Theta_k(x,t;2,1) = \theta_k(x,t)$ ◇.

Corollary 1 *If the Gaussian profile of the initial condition $u(x,0)$ of Proposition 1 is substituted by the Dirac delta distribution $\varphi(x) = \mu\delta(x)$, then the solution is given by*

$$u^{(\delta)}(x,t) = \left(\frac{\mu}{\kappa t^2}\right) H_{3,3}^{2,1}\left[\frac{|x|}{\kappa t^2} \middle| \begin{matrix} (0,1), \left(\frac{1}{2}, \frac{1}{2}\right), (-1,2) \\ (0,1), (0,1), \left(\frac{1}{2}, \frac{1}{2}\right) \end{matrix} \right]. \tag{10}$$

Proof The delta pulse $\varphi(x)$ is recovered from the Gaussian distribution $u(x,0)$ at the limit $x_0 \to 0$. The proof is simple by noticing that, with the exception of the term with $k = 0$, the coefficients of (3) become zero at such a limit. Therefore, $u(x,t) \to u^{(\delta)}(x,t)$ as $x_0 \to 0$ ◇.

3 Analysis of Results

The behavior of the solutions $u(x,t)$ defined in (3)–(4) is shown in the panel of Fig. 1 for $\mu = \kappa = 1$. From top to bottom, the rows correspond to $x_0 = 1, \sqrt{0.5}, \sqrt{0.1}$. From left to right, the columns refer to $t = 0.1, 1.7, 5, 6.5$. An interesting profile of these functions is the emergence of zeros as time goes pass. They arise in pairs, superposed at $x = 0$, and then propagate in opposite directions (symmetrically with respect to $x = 0$). As the function $u(x,t)$ is initially a nonnegative pulse, the zeros are indeed nodes that propagate, together with the maxima and minima of the disturbance, in wavelike form. For fixed values of μ and κ, the times at which we find new pair of nodes depend on the width of the initial Gaussian distribution. That is, they arise at shorter times for smaller values of x_0. We are interested in studying the behavior of such nodes as the disturbance $u(x,t)$ propagates.

First, we use Theorems 1.2 and 1.4 of Ref. [9] to rewrite θ_k as the absolutely convergent series [7]

$$\theta_k(x,t) = \frac{4^k}{\sqrt{\pi}} \left(\frac{\kappa t^2}{|x|}\right)^{1+2k} \sum_{\ell=0}^{\infty} \frac{(-1)^\ell \Gamma(\frac{1}{2}+k+\frac{\ell}{2})}{\Gamma(1+4k+2\ell)\Gamma(-k-\frac{\ell}{2})} \left(\frac{\kappa t^2}{|x|}\right)^\ell, \tag{11}$$

where $x \neq 0$. The divergences of $\Gamma(-k-\frac{\ell}{2})$ eliminate the terms with even values of ℓ in the above expression, then

$$u(x,t) = \frac{\mu}{\sqrt{\pi}|x|} \sum_{k,n=0}^{\infty} \frac{(-1)^{k+1}}{k!} \left(\frac{x_0}{x}\right)^{2k} \left(\frac{2\kappa t^2}{|x|}\right)^{2n+1} \lambda(n,k), \tag{12}$$

with

$$\lambda(n,k) = \frac{\Gamma(1+n+k)}{\Gamma(3+4n+4k)\Gamma\left(-\frac{1}{2}-n-k\right)}. \tag{13}$$

The latter formulae give us information about the nodes of the disturbance generated by the initial Gaussian-like perturbation defined in (2). Of course, as the point $x = 0$ has been omitted, the following description does not automatically hold for $|x| \leq \epsilon$ as ϵ approaches to zero.

The straightforward calculation shows that (12) can be rewritten in the form

$$u(x,t) = \frac{\mu}{\pi}\left(\frac{\kappa t^2}{x^2}\right) \sum_{s=0}^{\infty} \frac{\Gamma(2s+2)}{\Gamma(4s+3)} \left(\frac{\kappa t^2}{|x|}\right)^{2s} \Lambda_s(x_0,t), \quad x \neq 0, \tag{14}$$

where $\Lambda_s(x_0, t)$ is the polynomial of $x_0 t^{-2}$ given by

$$\Lambda_s(x_0, t) = \sum_{n+k=s} \frac{(-1)^n}{\Gamma(k+1)} \left(\frac{x_0}{2\kappa t^2}\right)^{2k}. \tag{15}$$

Now, let us analyze the series (14) in terms of $\xi = \frac{\kappa t^2}{|x|} > 0$. For $\xi << 1$ we may consider the power with $s = 0$ only. We have

$$u(x, t) \approx \frac{\mu\kappa}{2\pi} \left(\frac{t}{x}\right)^2, \quad x \neq 0. \tag{16}$$

The latter means that, no matter the value of x_0, the solution is free of zeros at short times. To illustrate the phenomenon, Fig. 1a, e and i show the behavior of $u(x, t)$ for the indicated values of x_0 at $t = 0.1$.

At slightly larger times we may hold only the powers with $s = 0$ and $s = 1$. Thus, dropping the terms with $s \geq 2$ we arrive at the expression

$$u(x, t) \approx \frac{\mu\kappa}{\pi} \left(\frac{t}{x}\right)^2 \left[\frac{1}{2} + \frac{\Gamma(4)}{\Gamma(7)} \left(\frac{\kappa t^2}{|x|}\right) \Lambda_1(x_0, t)\right], \quad x \neq 0, \tag{17}$$

where

$$\Lambda_1(x_0, t) = -1 + \left(\frac{x_0}{2\kappa t^2}\right)^2. \tag{18}$$

Given $x_0 \geq 0$, the values of t such that $\Lambda_1 < 0$ permit the presence of a pair of zeros in function (17). Before such values, the function $u(x, t)$ exhibits a global minimum that is positive and goes to zero as t increases. Then the minimum becomes equal to zero (the time at which the first pair of nodes is created, both superposed at $x = 0$), and finally it takes negative values (the nodes start to propagate in opposite directions with respect to $x = 0$). The second column (from left to right) of Fig. 1 shows the situation in which the minimum of $u(x, t)$ is negative for three different values of x_0. Although the three graphics are evaluated at $t = 1.7$, notice that the minimum is as deep as x_0 is short. The latter shows that the positions of the nodes at a given time depend on x_0.

At larger times, the value of x_0 determines the number of zeros as well as their distribution. For example, in the third and fourth columns (from left to right) of Fig. 1 we appreciate that the number of nodes increases as x_0 decreases at a given time. In general, such number increases as $\xi \to \infty$. Then, the time t at which a new pair of nodes arises is shorter for smaller values of x_0. Remarkably, at the limit $x_0 \to 0$, for the polynomial (15) we have

$$\lim_{x_0 \to 0} \Lambda_s(x_0, t) = (-1)^s. \tag{19}$$

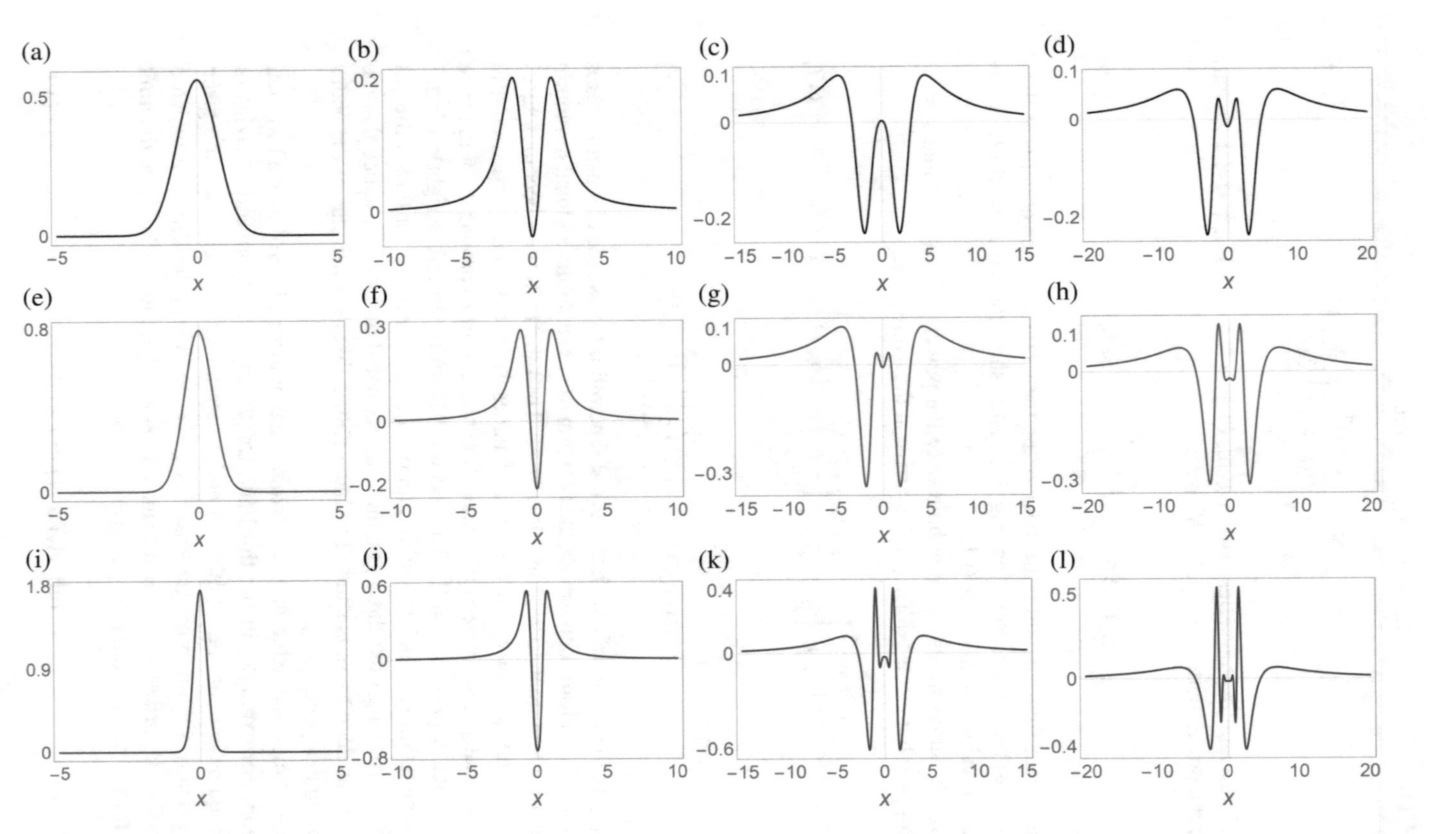

Fig. 1 Time-evolution of the function $u(x, t)$ defined in (3)–(4) with $\mu = \kappa = 1$. The rows (characterized by the indicated values of x_0) show the emerging of zeros in $u(x, t)$ as t increases. The columns exhibit the behavior of $u(x, t)$ as $x_0 \to 0$ (from top to bottom) at the indicated times. (**a**) $x_0 = 1, t = 0.1$. (**b**) $x_0 = 1, t = 1.7$. (**c**) $x_0 = 1, t = 5$. (**d**) $x_0 = 1, t = 6.5$. (**e**) $x_0 = \sqrt{0.5}, t = 0.1$. (**f**) $x_0 = \sqrt{0.5}, t = 1.7$. (**g**) $x_0 = \sqrt{0.5}, t = 5$. (**h**) $x_0 = \sqrt{0.5}, t = 6.5$. (**i**) $x_0 = \sqrt{0.1}, t = 0.1$. (**j**) $x_0 = \sqrt{0.1}, t = 1.7$. (**k**) $x_0 = \sqrt{0.1}, t = 5$. (**l**) $x_0 = \sqrt{0.1}, t = 6.5$

From (14) and (19) one has

$$\lim_{x_0 \to 0} u(x,t) = \frac{\mu}{\pi}\left(\frac{\kappa t^2}{x^2}\right)\sum_{s=0}^{\infty}(-1)^s \frac{\Gamma(2s+2)}{\Gamma(4s+3)}\left(\frac{\kappa t^2}{|x|}\right)^{2s}, \quad x \neq 0, \tag{20}$$

which corresponds to the series expansion of $u^{(\delta)}(x,t)$ reported in [4].

4 Concluding Remarks

We have shown that the (modified) complementary equation $u_{tt} - \kappa \mathbb{D}u = 0$ can be solved in analytic form by considering the Cauchy problem for zero initial velocity and the Gaussian distribution as initial disturbance. The solutions exhibit nodes that arise in pairs at different times and propagate from $x = 0$ in wavelike form. The number of zeros in a given time-interval increases as the width of the distribution is reduced. At the very limit in which the width becomes equal to zero we recover the solutions to the Cauchy problem with the initial disturbance as a Dirac delta pulse. The possible physical applications of the modified complementary equation represent an open problem, which we shall face elsewhere.

Acknowledgements Financial support from Ministerio de Economía y Competitividad (Spain) grant MTM2014-57129-C2-1-P, Consejería de Educación, Junta de Castilla y León (Spain) grants VA057U16, and Consejo Nacional de Ciencia y Tecnología (México) project A1-S-24569 is acknowledged.

References

1. A.N. Tikhonov, A.A. Samarskii, *Equations of Mathematical Physics* (Pergamon Press, New York, 1963)
2. D.G. Duffy, *Green's Functions with Applications*, 2nd edn. (CRC Press, Boca Raton, 2015)
3. D. Borthwick, *Introduction to Partial Differential Equations* (Springer, Basel, 2018)
4. F. Olivar-Romero, O. Rosas-Ortiz, Transition from the wave equation to either the heat or the transport equations through fractional differential expressions. Symmetry-Basel **10**, 524 (2018)
5. M. Riesz, L'integrale de Riemann-Liouville et le probléme de Cauchy. Acta Math. **81**, 1–222 (1949)
6. S. Umarov, *Introduction to Fractional and Pseudo-Differential Equations with Singular Symbols* (Springer, Basel, 2015)
7. R. Gorenflo, A. Iskenderov, Y. Luchko, Mapping between solutions of fractional diffusion wave equations. Fractional Calculus Appl. Anal. **3**, 75–86 (2000)
8. M. Caputo, Linear models of dissipation whose Q is almost frequency independent, part II. Geophys. J. R. Astr. Soc. **13**, 529–539 (1967)
9. A.A. Kilbas, *H-Transforms: Theory and Applications* (CRC Press, Boca Raton, 2004)
10. A.M. Mathai, R.K. Saxena, H.J. Haubold, *The H-Function: Theory and Applications* (Springer, New York, 2010)

Quasi-Integrability and Some Aspects of $SU(3)$ Toda Field Theory

Wojtek Zakrzewski

Abstract In this talk I discuss results of our recent paper in which we have studied various properties of solitonic solutions of deformed $SU(3)$ Toda field theories in $(1+1)$ dimensions. The aim of that work was to check whether the results of scattering of such solitons supported our ideas on quasi-integrability. The deformations we considered preserved the symmetries which for other models were sufficient to guarantee their quasi-integrability. The results of our simulations had indeed led to the expected results, thus broadening the class of models which are quasi-integrable.

The simulations had also found some interesting properties of the scattering (like the interesting dependence of the interaction between solitons being related to the sign of the deformation). In this talk we also present some recently found understanding of this behaviour based on the collective coordinate approximation to the description of the scattering of such solitons.

Keywords Solitons · Nonlinear · Integrability · Quasi-integrability · Toda field theories · Collective coordinates · Evolution of solitons · Deformations of Lagrangians

1 Introduction

Field theories possessing soliton-like solutions have been studied a lot, both analytically and numerically. In particular, many papers have been written describing various aspects of the scattering of solitons in various $(d+1)$ models [1]. They included pure S^2 σ models and various 'baby Skyrme' models. In addition, there are

The talk is based on the work performed in collaboration with L.A. Ferreira and P. Klimas.

W. Zakrzewski (✉)
Durham University, Durham, UK
e-mail: w.j.zakrzewski@durham.ac.uk

Ş. Kuru et al. (eds.), *Integrability, Supersymmetry and Coherent States*, CRM Series in Mathematical Physics, https://doi.org/10.1007/978-3-030-20087-9_19

also papers which described the scattering of solitons in a modified chiral model—the so-called Ward model. Of these models this Ward model [2] is integrable, while the other models are not; surprisingly, some properties of such scatterings were found to be quite similar. This led Sutcliffe (P.M. Sutcliffe, private communication) to look at a model which is a superposition of the Ward model and the 'baby Skyrme' model, and again the properties of the scatterings were not very different. By such properties we mean very few waves of radiation being emitted during the scattering (i.e. the scattering being 'almost' elastic). In fact, this is quite close to what is seen in 'nature'. At CERN, at high energy scattering of protons a significant percent of the total cross section (just over 20%) is elastic. In the Skyrme model [3], which describes such processes quite well, radiation is represented by the emission of pions and so, given the energies available for the production of pions—this suggests the possibility that physics equations that describe 'nature' maybe be quite close to being integrable, and so in some sense 'almost' integrable or 'quasi-integrable'.

Of course, these are all very loose ideas but they have led L.A. Ferreira and myself to attempt to formulate more precisely this concept of 'quasi-integrability' [4]. As soon as one start thinking about this one realises that this is like opening a 'can of worms'. Mathematically (in particular for systems of finite numbers of degrees of freedom) the integrability of any model is reasonably well defined. The integrable systems have large numbers of conserved quantities (for finite dimensional system their number equals the number of the degrees of freedom of such systems). For field theories all of this is much more complicated and I do not think there is an agreed definition what integrability really means. And quasi-integrability is even more complicated. Quasi-conserved quantities? For all field configurations or only for some of them? And how small effects of what type of perturbations of integrable systems should they be (to define this 'closeness' to being integrable)?

The approach we have formulated in our early papers [4–6] was based on studying various deformed integrable systems which possessed soliton-like solutions. The studied models included various deformed modifications of the Sine-Gordon and NLS models. The deformations were such that the models still possessed exact one-soliton solutions and then we considered the modifications of the quantities which for vanishing deformations corresponded to the conserved quantities. We have found that in many of such models these modifications were very small and so these quantities were quasi-conserved. By studying such models further we have managed to connect the quasi-conservation of these quantities to some symmetry properties of the field configurations describing two-soliton fields in such models.

Our first studies were restricted to models of one field in $(1 + 1)$ dimensions and the discussions were relatively simple. Next we considered more complicated systems, involving more solitons (like the various modifications of the KdV equation) [7] or models with more fields (still in $(1 + 1)$ dimensions). In this talk I want to describe our recent study of one of the models of this second class—and here our work involved the study of some deformations of the $SU(3)$ Toda model. So my talk is based on a paper published in the Journal of High Energy Physics [8] but I also want to mention some unpublished results.

2 Toda Model

The $SU(3)$ model is described by the Lagrangian [9]

$$L = \frac{1}{2}[(\partial_\mu \phi_1)^2 + (\partial_\mu \phi_2)^2 - \partial_\mu \phi_1 \, \partial^\mu \phi_2] \\ + 2[e^{i(2\phi_1 - \phi_2)} + e^{i(2\phi_2 - \phi_1)} + e^{-i(\phi_1 + \phi_2)} - 3]. \tag{1}$$

The equations of motion of this model take the form:

$$\partial_+ \partial_- \phi_1 = -i[e^{i(2\phi_1 - \phi_2)} - e^{-i(\phi_1 + \phi_2)}], \\ \partial_+ \partial_- \phi_2 = -i[e^{i(2\phi_2 - \phi_1)} - e^{-i(\phi_1 + \phi_2)}], \tag{2}$$

where $\partial_\pm = \frac{1}{2}(\partial_x \pm \partial_t)$ and where we have set $c = 1$.

Note that the fields of this model are complex and so the energy does not have to be real. This is potentially a serious problem (at least from the point of view of any 'physical' applications) but this has not stopped many mathematicians studying it at length. This is partly due to its properties. Apart from the fact that the model can be formulated in a more 'mathematical' form (using group theory, etc.), it is fully integrable and many of its solutions are well known. Moreover, although the fields are complex its soliton solutions have real energy.

So what are the solutions of this model? In fact, the model possesses two one-soliton solutions. The difference lies in their complex phase rotating differently. (real and imag. parts of ϕ_i). The fields of the solitons are defined in terms of Hirota's τ functions. They are given by

$$\phi_a = i \ln \frac{\tau_a}{\tau_0}, \qquad a = 1, 2. \tag{3}$$

Then, the one-soliton solution of the first type is given by

$$\begin{pmatrix} \tau_0 \\ \tau_1 \\ \tau_2 \end{pmatrix} = \begin{pmatrix} 1 \\ 1 \\ 1 \end{pmatrix} + \begin{pmatrix} 1 \\ \omega \\ \omega^2 \end{pmatrix} e^{\Gamma(z)}, \tag{4}$$

and the one-soliton solution of the second type is given by

$$\begin{pmatrix} \tau_0 \\ \tau_1 \\ \tau_2 \end{pmatrix} = \begin{pmatrix} 1 \\ 1 \\ 1 \end{pmatrix} + \begin{pmatrix} 1 \\ \omega^2 \\ \omega \end{pmatrix} e^{\Gamma(z)} \tag{5}$$

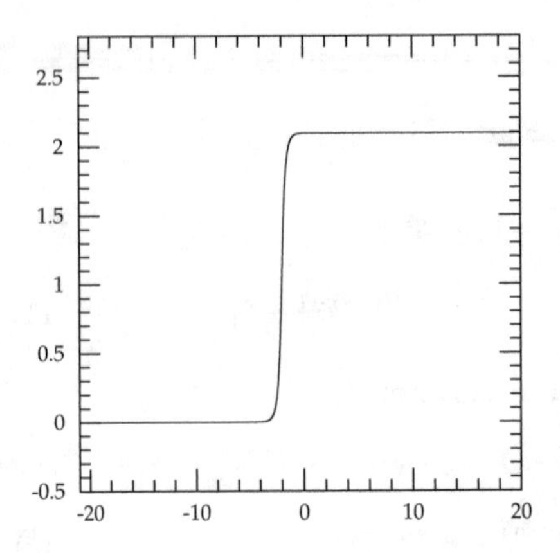

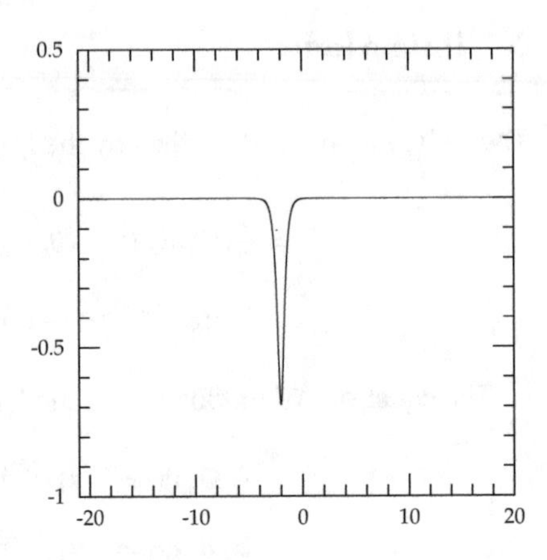

Fig. 1 The real and imaginary parts of ϕ_1

with $\Gamma(z)$ given by

$$\Gamma(z_k) = \sqrt{s}\left(z_k x_+ + \frac{x_-}{z_k}\right) + \xi_k = 2\sqrt{3}\eta_k \frac{(x - v_k t - x_0^k)}{\sqrt{1 - v_k^2}} \tag{6}$$

and where ω is a cubic root of unity, different from unity itself. So we take

$$\omega = e^{i\,2\pi/3}, \qquad 1 + \omega + \omega^2 = 0. \tag{7}$$

The real and imaginary parts of ϕ_1 are shown in Fig. 1; the field ϕ_2 is similar (its real part is the same but its imaginary part is opposite of that of ϕ_1).

Note that the real parts of these fields look very similar to the fields of the Sine-Gordon solitons.

The model possesses also three classes of two-soliton solutions. They are obtained by combining the expressions for one solitons:

The species-11 two-soliton solution is given by

$$\begin{pmatrix} \tau_0 \\ \tau_1 \\ \tau_2 \end{pmatrix} = \begin{pmatrix} 1 \\ 1 \\ 1 \end{pmatrix} + \begin{pmatrix} 1 \\ \omega \\ \omega^2 \end{pmatrix} e^{\Gamma(z_1)} + \begin{pmatrix} 1 \\ \omega \\ \omega^2 \end{pmatrix} e^{\Gamma(z_2)} + \begin{pmatrix} 1 \\ \omega^2 \\ \omega \end{pmatrix} e^{\Gamma(z_1)+\Gamma(z_2)+\Delta_{11}}. \tag{8}$$

The species-22 two-soliton solution takes the form:

$$\begin{pmatrix} \tau_0 \\ \tau_1 \\ \tau_2 \end{pmatrix} = \begin{pmatrix} 1 \\ 1 \\ 1 \end{pmatrix} + \begin{pmatrix} 1 \\ \omega^2 \\ \omega \end{pmatrix} e^{\Gamma(z_1)} + \begin{pmatrix} 1 \\ \omega^2 \\ \omega \end{pmatrix} e^{\Gamma(z_2)} + \begin{pmatrix} 1 \\ \omega \\ \omega^2 \end{pmatrix} e^{\Gamma(z_1)+\Gamma(z_2)+\Delta_{11}}. \tag{9}$$

The mixed (i.e. species-12) two-soliton solution is given by

$$\begin{pmatrix}\tau_0\\ \tau_1\\ \tau_2\end{pmatrix}=\begin{pmatrix}1\\ 1\\ 1\end{pmatrix}+\begin{pmatrix}1\\ \omega\\ \omega^2\end{pmatrix}e^{\Gamma(z_1)}+\begin{pmatrix}1\\ \omega^2\\ \omega\end{pmatrix}e^{\Gamma(z_2)}+\begin{pmatrix}1\\ 1\\ 1\end{pmatrix}e^{\Gamma(z_1)+\Gamma(z_2)+\Delta_{12}}, \quad (10)$$

where $\Gamma(z_k)$ are given as before and the quantities Δ_{11} and Δ_{12} are given by

$$e^{\Delta_{11}}=\begin{cases}\dfrac{4\sinh^2\left(\frac{\alpha_2-\alpha_1}{2}\right)}{4\cosh^2\left(\frac{\alpha_2-\alpha_1}{2}\right)-1} & \text{if} \quad \eta_1\eta_2=1\\[2ex] \dfrac{4\cosh^2\left(\frac{\alpha_2-\alpha_1}{2}\right)}{4\sinh^2\left(\frac{\alpha_2-\alpha_1}{2}\right)+1} & \text{if} \quad \eta_1\eta_2=-1\end{cases} \quad (11)$$

and

$$e^{\Delta_{12}}=\begin{cases}\dfrac{4\sinh^2\left(\frac{\alpha_2-\alpha_1}{2}\right)+1}{4\cosh^2\left(\frac{\alpha_2-\alpha_1}{2}\right)} & \text{if} \quad \eta_1\eta_2=1\\[2ex] \dfrac{4\cosh^2\left(\frac{\alpha_2-\alpha_1}{2}\right)-1}{4\sinh^2\left(\frac{\alpha_2-\alpha_1}{2}\right)} & \text{if} \quad \eta_1\eta_2=-1\end{cases}. \quad (12)$$

In these expressions α_a, $a=1,2$, are the rapidities of the solitons related to the velocities by $v_a=\tanh\alpha_a$. Note that the two-soliton solutions satisfy the conditions $\tau_1=\tau_2^\star$ and $\tau_0^\star=\tau_0$ and so the Hamiltonian for these solutions is real.

2.1 Some Properties of All These Solutions

Note that the energy and its density of all these solutions is real. Moreover, one-soliton solutions can be static, and, from the two-soliton solutions, the mixed ones can be static too. Furthermore, all solutions are stable, and the energies of static one-soliton solutions are the same and the energy of the static two solitons (the mixed case) is exactly twice the energy of one soliton. This suggests to us that the model may possess a BPS-like property [10]; however, so far we have not managed to demonstrate it.

The fields of two solitons possess various symmetries (like $\phi_1^\star=-\phi_2$, etc.) which guarantee the existence of the solutions in the first place and are responsible for some of their properties. Moreover, these symmetries are also responsible for the integrability of the model and, in this, the model resembles the Sine-Gordon model.

To use this model to test our ideas on quasi-integrability, like we have done this for the Sine-Gordon model, we have to deform it. So in the next section we discuss our deformation of this model.

3 Deformed Model

Of course, there are various possible deformations of the model. For our purposes the deformations of the potential $V(\phi_1, \phi_2)$ are the most convenient and so we considered various changes of

$$2[e^{i(2\phi_1-\phi_2)} + e^{i(2\phi_2-\phi_1)} + e^{-i(\phi_1+\phi_2)} - 3] \quad (13)$$

and have finally settled on [8] its change to

$$2[e^{i(2\phi_1-(1+\epsilon)\phi_2)} + e^{i(2\phi_2-(1+\epsilon)\phi_1)} + e^{-i(1-\epsilon)(\phi_1+\phi_2)} - 3], \quad (14)$$

in which ϵ is the parameter of the deformation.

However, the changes introduced by this deformation are not insignificant. Not only the equations of motion change appropriately but also one needs to change the boundary conditions (so that each term in V vanishes).

So what are the properties of our deformation and can we use it to study the quasi-integrability properties of our deformed model?

In the various cases of the deformed Sine-Gordon model we used the conserved quantities of the original (undeformed) Sine-Gordon model as test quantities which helped us to understand better the quasi-integrability. In the deformed model these quantities were no longer conserved (they constituted the so-called 'anomalies'). These anomalies could have any form but when the deformed model still possessed some important symmetries these anomalies were small and non-zero only when the interacting solitons were close together. Moreover, their total contribution vanished asymptotically—so that conserved were still conserved asymptotically (i.e. at large times before and after the interaction of solitons). We have also noticed that this asymptotical conservation always held if the original field configuration possessed some specific properties (symmetries with respect to a reflection around some point in space time (x, t)). Our previous studies involved models with one field; in this work we have studied models involving more fields (here still only two).

Our paper [8] describes all this in detail, here we have no space to present the details of our procedure of calculation of these anomalies and of relating them to the study of the above-mentioned symmetry. Interested reader can find all needed details in [8]. Here we recall the main results of our work and present new results which provide explanations to some intriguing features of the previous results.

As shown in [8] our deformation preserves many symmetries of the original Toda system and we have performed the calculation of the anomalies in a similar way to the Sine-Gordon case. In [8] we showed that the lowest anomaly density is proportional to:

$$
\begin{aligned}
&-[6(\partial_-\phi_1)^2-(3-\epsilon)(\partial_-\phi_2)^2-2(3+\epsilon)\partial_-\phi_1\partial_-\phi_2]E_1\\
&\quad+[(3-\epsilon)(\partial_-\phi_1)^2-6(\partial_-\phi_2)^2+2(3+\epsilon)\partial_-\phi_1\partial_-\phi_2]E_2\\
&\quad\quad+3(1-\epsilon)[(\partial_-\phi_1)^2-(\partial_-\phi_2)^2]E_0,
\end{aligned}
\tag{15}
$$

in which E_i stand for the three terms of the modified potential.

To go further we had to perform the simulations, but for them we needed initial conditions and we do not even have expressions for one-soliton fields of the deformed model in an analytic form.

So we have had to construct the relevant initial conditions numerically. We did this in two steps. First we constructed one-soliton fields (with appropriately modified boundary conditions). This was done by replacing ∂_t^2 by ∂_t in the equations of motions, i.e. running the simulations of the relevant dissipative system. This took a while but gradually the fields settled to the one-soliton solutions of the deformed equation. We repeated this procedure for various values of ϵ and so obtained the numerical expressions of the deformed one-soliton fields for various values of the deformation parameter. Then the two-soliton fields were constructed from one-soliton fields by superpositions and reflections, etc. (aping the unmodified Toda case). And then we performed our numerical simulations of our main equations for the case of two solitons of the mixed case. As we mentioned before, this case describes the solitons which can be static when $\epsilon = 0$. For the other two-soliton system we have repulsion like in the simple Sine-Gordon system.

4 Results: Static Cases

First of all we have looked at the case involving two solitons placed too far apart (see the figure below); this has effectively tested our program and we were satisfied as, as required, there was no motion. In Fig. 2 we present plots of the energy density at $t = 0$ and at $t = 1000$. They clearly show that the solitons have not moved.

Then we started the solitons much closer together. In Fig. 3 we present the results observed in three simulations ($\epsilon = 0$, where we do not expect any motion, and for two other values (one positive and one negative) of ϵ).

The plots are very representative of what we have seen in all our simulations. First of all the two-soliton configurations for $\epsilon = 0$ appear to be static, e.g. there is no motion, while for $\epsilon > 0$ we have always observed repulsion and for $\epsilon < 0$ we saw attraction followed by an interesting behaviour (involving a reflection). In the static $\epsilon = 0$ case, in fact, we do see infinitesimally small attraction—but this can be shown to be a numerical artifact (due to the too simple approximation of the second derivatives in the equation of motion). We checked this by including more

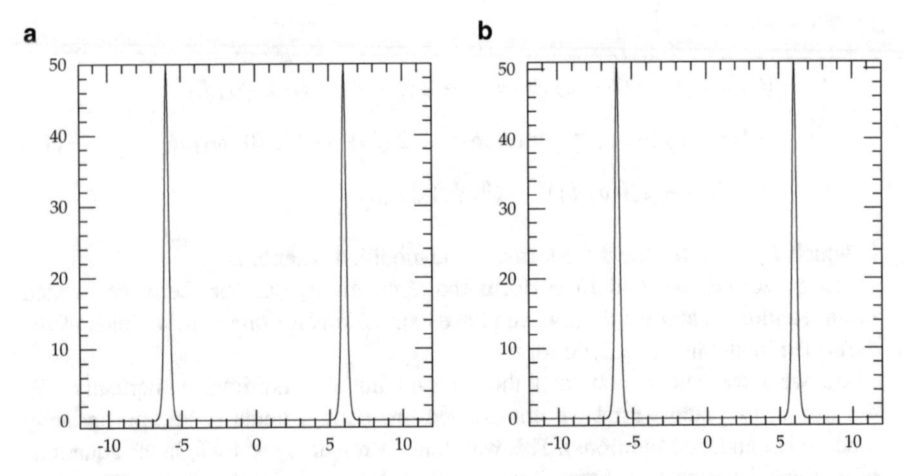

Fig. 2 Plots of energy densities see for values of t (**a**) $t = 0$, (**b**) $t = 1000$ seen in a static simulation $\epsilon = 0.01$

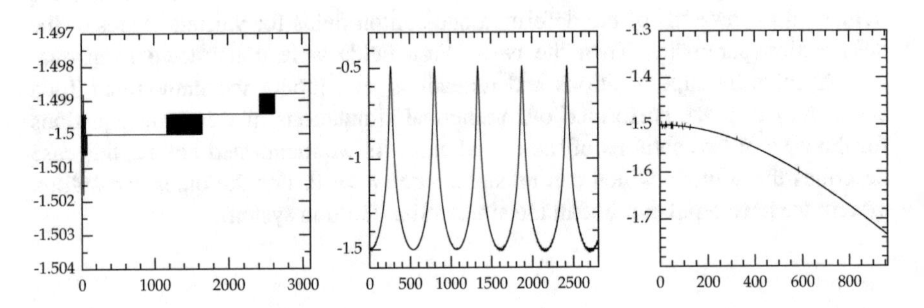

Fig. 3 Plots of trajectories for $\epsilon = 0$, $\epsilon = -0.01$ and $\epsilon = 0.005$

terms in the derivatives and attraction was smaller. However, including more terms requires more computer time so all our results mentioned here were obtained using the simplest approximation to the second order derivatives. The other cases are clear cut and are genuine. No small corrections to the numerical approximations of the derivatives would change them.

We have also obtained plots for the time evolution of the anomaly for the simulations with $\epsilon \neq 0$. In the next two figures we present plots (or the real and imaginary parts) of the anomaly for $\epsilon = -0.01$ (Fig. 4) and in the following one for $\epsilon = 0.001$. These anomalies are exceptionally small suggesting quasi-integrability if not full integrability. However, staring at these and some other similar results, discussed in more detail in [8], we claim that we have only quasi-integrability.

The observed phenomena persist for more general values of $\epsilon \neq 0$.

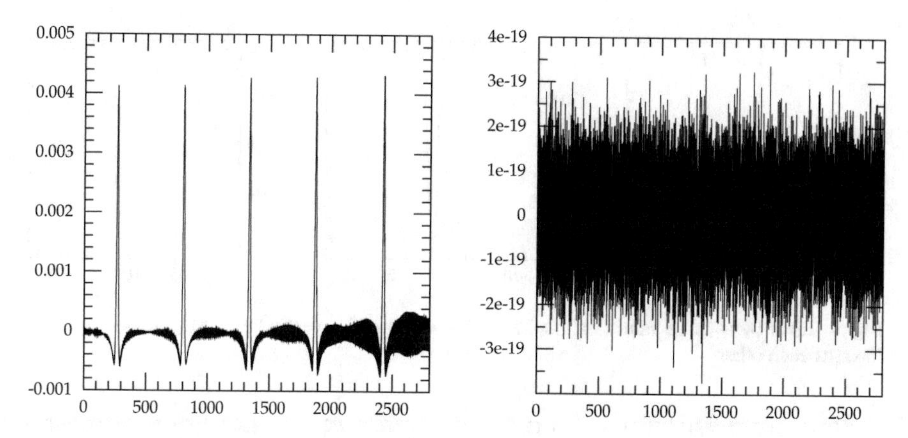

Fig. 4 Time dependence of real and imaginary parts of the anomaly seen in the simulation for $\epsilon = -0.01$

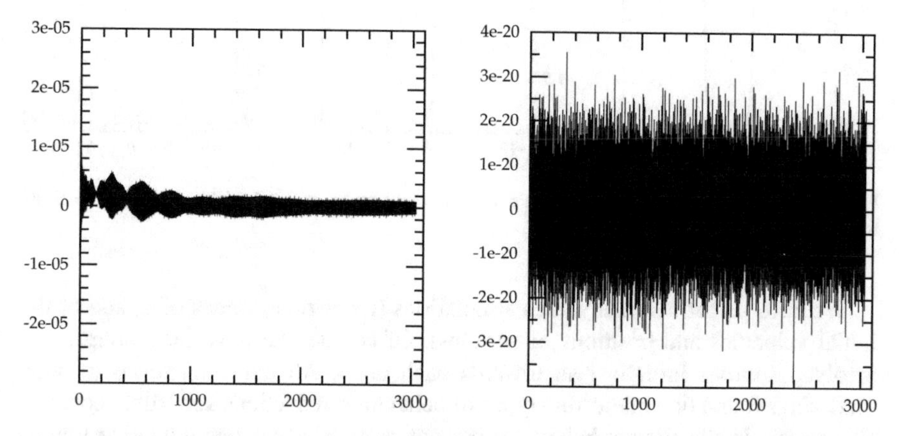

Fig. 5 Time dependence of real and imaginary parts of the anomaly seen in the simulation for $\epsilon = 0.001$

5 Non-static Cases

We have also performed many simulations for various values of velocities of initial solitons (and for various values of ϵ). This involved taking non-static initial conditions; we constructed such solutions taking $\frac{d\phi_a}{dt}$ proportional to $\frac{d\phi_a}{dx}$, treating each soliton separately. Moreover, we started our simulations with solitons further (i.e. at positions where their interactions were negligible). As usual we compared this procedure with the exact non-static solitons for the undeformed case and we did not see any difference. Hence we are very confident of the validity of our results (Fig. 5).

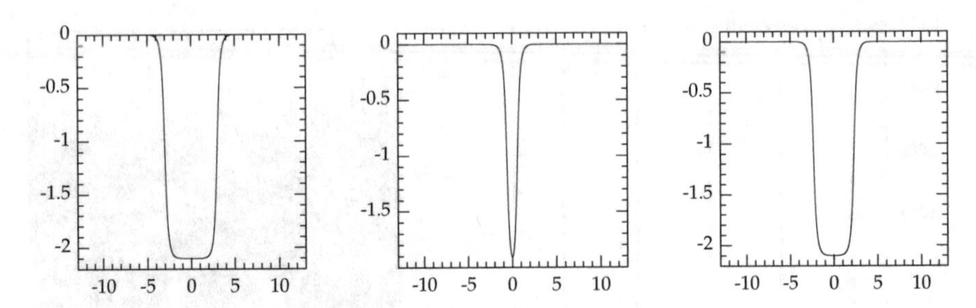

Fig. 6 Field ϕ_1 at three values of time: $t = 0$, $t = 40$ and $t = 80$. Initial velocities $v = 0.06$, towards each other

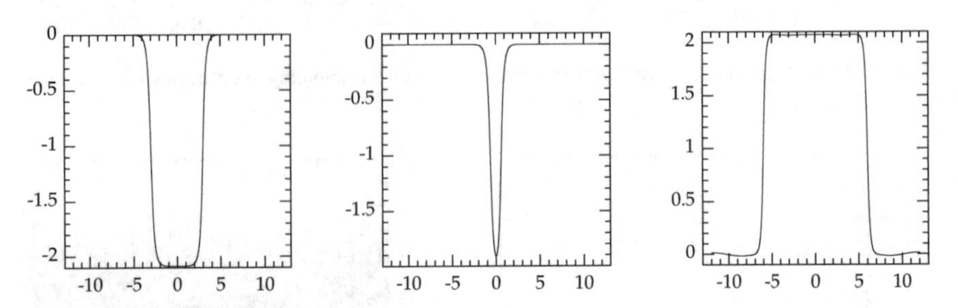

Fig. 7 Field ϕ_1 at three values of time: $t = 0$, $t = 40$ and $t = 80$. Initial velocities $v = 0.06$, towards each other

We have performed many such simulations (for various values of ϵ, and of the initial velocities and positions of solitons). Of course, the most interesting cases involved solitons initially sent towards each other. All our simulations of such scatterings found two interesting types of behaviour: reflections and 'flips' between ϕ_1 and ϕ_2. In the figures below we present plots of some representative figures showing both types of scattering.

In Fig. 6 we show the field ϕ_1 at three values of time, which together demonstrate the reflection (case with $\epsilon = 0.001$). The field ϕ_2 performed the corresponding reflective evolution.

The next figure presents the plots of ϕ_1 seen in a case which involved 'flip'. Again the velocity was $v = 0.06$ and the simulation was performed for $\epsilon = 0.01$ (Fig. 7).

In Fig. 8 we present more plots of a typical 'flipped' case. This case corresponds to $\epsilon = -0.001$ with the starting velocities being $v = 0.1$. The three plots in this figure show the trajectory of the field ϕ_1, its shape at the 'flipped' point (in fact at $t = 58.5$) and the energy density of the system at that value of time.

Looking at the picture of the trajectory in detail we note that the 'trajectory' of the 'flipped' soliton is a bit different (as if the solitons moved with a slightly modified velocity). We have also looked at the anomalies seen in our simulations. They were

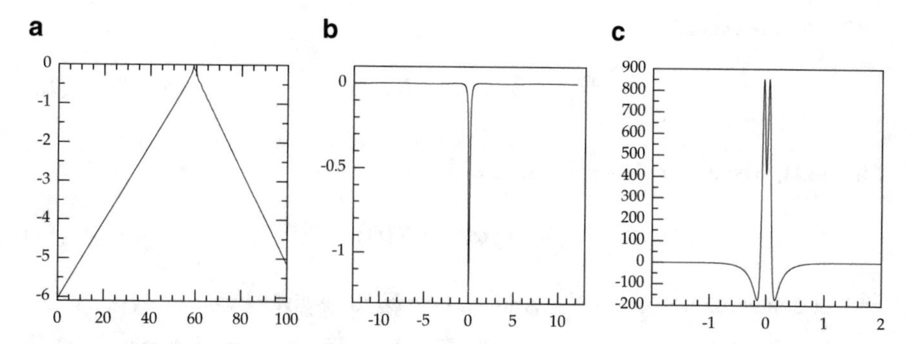

Fig. 8 (**a**) Trajectory, (**b**) field ϕ_1 and (**c**) energy density

always all very small and we do not include any plots of them here. Any interested reader can see them in [8].

Instead, we return the static case in which we saw an essentially different behaviour of the solitons (for different values of the sign of ϵ). We have tried to understand it, since [8] was published, and in the next section we present our current understanding of the observed difference in behaviour. This explanation is based on performing a collective field approximation to the scattering in such cases.

6 Collective Coordinate Approximation

In the collective coordinate field approximation we replace each soliton field by a variable describing the position of a soliton, and then construct a Lagrangian describing the evolution of these variables. Of course, working in the centre of mass we can restrict ourselves to considering only one variable which we denote by $X(t)$. The question then is: what is the Lagrangian for $X(t)$?

There are various ways for obtaining such an approximation. In our work we follow the standard procedure as described in, e.g., [10]. This procedure involves approximating fields $f(x,t)$ by $f(x - X(t))$ and then substituting this approximation into the original Lagrangian and deriving the Lagrangian for $X(t)$ by integrating over x.

6.1 *The Case of* $\epsilon = 0$

So let us first consider the $\epsilon = 0$ case. We 'join together' two single solitons—one in the region $x < 0$, the other located in $x > 0$.

To do this we take

$$\phi_i = i \ln \frac{\tau_1}{\tau_0}, \tag{16}$$

for $x < 0$, where $\tau_0 = 1 + e^{\Gamma_1}$ and $\tau_1 = 1 + \omega e^{\Gamma_1}$ with

$$\Gamma_1 = \alpha(x + X(t)) \tag{17}$$

and, of course, $\omega = -\frac{1}{2} + i\,\frac{\sqrt{3}}{2}$. And we take ϕ_2 to be given by $\phi_2 = -(\phi_1)^\star$.

For $x \geq 0$ take $\phi_i = i \ln \frac{\tau_1'}{\tau_0'}$, where $\tau_0' = 1 + e^{\Gamma_2}$, and $\tau_1' = 1 + \omega e^{\Gamma_2}$. For Γ_2 we take $\Gamma_2 = \alpha(-x + X(t))$.

As we said above, $X(t)$ is the collective coordinate of one kink and α is a constant. Note that at $x = 0$ $\Gamma_1 = \Gamma_2 = \alpha X$ and so the fields are continuous.

To get equations of motion for $X(t)$ we integrate the Lagrangian density (1) over x from $-\infty$ to ∞. Interestingly, the spatial derivatives terms exactly cancel with the potential terms in (1) and only the time derivative term remains. This gives us:

$$L(X, \dot{X}) = (\dot{X})^2 f(X), \tag{18}$$

where $f(X)$ is given by

$$f(X) = \int_{-\infty}^{0} \frac{e^{3\Gamma_1}\, dx}{(1 - e^{\Gamma_1} + e^{2\Gamma_1})^2 (1 + e^{\Gamma_1})^2} \tag{19}$$

with $\Gamma_1(x, X) = 2\sqrt{3}\,[x + X(t)]$. Next we calculate $f(X)$ and find

$$f(X) = \int_{-\infty}^{0} \frac{e^{3\Gamma_1}\, dx}{(1 - e^{\Gamma_1} + e^{2\Gamma_1})^2 (1 + e^{\Gamma_1})^2} = \frac{1}{6\sqrt{3}}\,\frac{e^{6\sqrt{3}X}}{1 + e^{6\sqrt{3}X}}. \tag{20}$$

We note that we have two solutions for $X(t)$, either

$$X = 0 \qquad \text{or} \qquad X(t) = \frac{1}{12\sqrt{3}} \ln\left[\sinh \sqrt{L}|t - t_0|\right] \tag{21}$$

The first one is what we wanted to get. However, what is the meaning of the second one? This is unclear, and we have not understood it. It is likely to be an artifact of our collective coordinate approximation.

6.2 The $\epsilon \neq 0$ Case

Next we consider the $\epsilon \neq 0$ case. As the initial configurations for our simulations have only been calculated numerically we have to consider only small values of ϵ. To

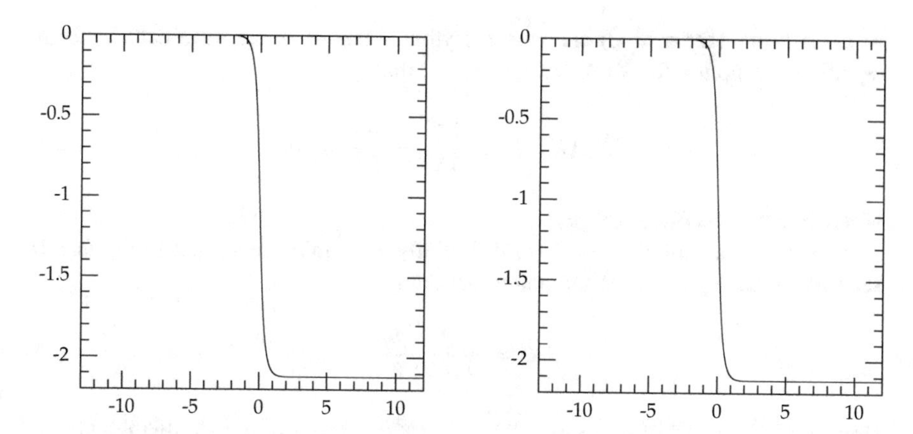

Fig. 9 ϕ_1 Fields of one kink for $\epsilon = 0.04$ at the beginning and at the end of diffusive simulation

get the correct boundary conditions it was convenient to introduce $\kappa = \frac{3}{3+\epsilon} \sim 1+2\alpha$ and to approximate all expressions by including only the first order corrections in α.

However, to get the numerical solutions of the soliton fields, for $\epsilon \neq 0$ we modified ϕ_1 and ϕ_2 (and their derivatives) by multiplying them by $\frac{3}{3+\epsilon}$ and then evolved them numerically (using a diffusive equation) to their correct values (i.e. we evolved them until the energy did not change by more than 0.01% and the fields seemed to be independent of the diffusive variable). When we compared the plots we have found that the change of the fields was extremely small; time original and the final fields looked essentially the same. In Fig. 9 we present the plots of the real parts of ϕ_1 fields of one solitons at the beginning and final times obtained in such a diffusion for $\epsilon = 0.04$. It is clear that these fields are essentially indistinguishable.

This suggested to us that, in our collective coordinate discussion we could take the original fields (i.e. the errors introduced by this approximation would be very small (essentially negligible)). This simplified significantly the calculations but, of course, introduced small (but hopefully almost negligible) errors. Then the detailed calculations gave us the following expression for our Lagrangian:

$$L = \kappa^2(\partial_t X)^2 g(X) \; - \kappa^2\, g(X)\beta + (g(X)\beta)^\kappa, \tag{22}$$

where $g(X)$, in this approximation, is given by

$$g(X) \sim f(X) = \frac{1}{6\sqrt{3}}\,\frac{e^{6\sqrt{3}X}}{1 + e^{6\sqrt{3}X}} \tag{23}$$

and β is a constant very close to 1.

This time the cancellation of the spatial derivative terms with the potential (which occurred for $\epsilon = 0$) does not take place and so the resultant Lagrangian has three terms. Of course, when $\kappa = 1$, the last two terms cancel. The non-cancellation of these terms (for $\epsilon \neq 0$) changes everything. So, when we have calculated the equations of motion for $X(t)$, we have found that

$$2\ddot{X}g(X) + \dot{X}^2\frac{\partial g}{\partial X} - \frac{\partial h}{\partial X} = 0, \tag{24}$$

where $h(X) = \alpha\, g\beta[2 - \ln(g\beta)]$.

So what is the solution for $X(t)$ if, initially, the kinks are started from rest? To see this we have put $\dot{X} = 0$ and noted that then

$$\ddot{X} = \frac{1}{2g(X)}\frac{\partial h}{\partial X}. \tag{25}$$

However, $\frac{\partial h}{\partial X} = \alpha\beta\frac{\partial g}{\partial X}[1 - \ln(g\beta)]$ and $g(X) < 1$ and so we see that the sign of $\frac{\partial h}{\partial X}$ agrees with the sign of α. This clearly explains our results as to the direction of the initial motion of solitons (i.e. for $\epsilon < 0$ we initially have the attraction, and ϵ we have repulsion). However, it is much harder to find an explanation, why for $\epsilon < 0$, the initial attraction gets reversed when the kinks come close together resulting in an effective oscillation. This would probably require more detailed evaluations of the integrals and lies beyond the scope of this work.

7 Conclusions

The modifications of the Toda models possess interesting properties which may help us in our study of quasi-integrability. We have obtained interesting results and we have demonstrated that the fields of our deformed models possess interesting properties. Their scattering properties depend on the deformations and we have showed that, at least initially, the motion of the solitons can be explained by ideas based on a collective field approximation.

We have also found that the anomaly is real (this is guaranteed initially by the symmetry but, it seems, that this symmetry is essentially preserved by the simulations).

Our anomaly (its real part) does change with time but it preserves our condition of 'returning to its original form', i.e. provides the support for quasi-integrability of the model. Thus we feel that this work has provided further arguments for the existence of quasi-integrability.

Our future plans involve looking at breather-like fields, etc. but at this meeting ... 'Many happy returns to Veronique Hussin'.

Acknowledgements This talk was delivered at the 6th International Workshop on New Challenges in Quantum Mechanics: Integrability and Supersymmetry, which was held in Valladolid, Spain, 27–30 June 2017. I would like to thank Prof. L.M. Nieto for inviting me to this very interesting and well organised meeting and for the hospitality in Valladolid.

References

1. N. Manton, P. Sutcliffe, *Topological Solitons* (Cambridge University Press, Cambridge, 2004), pp. 102–108
2. R. Ward, Nontrivial scattering of localized solitons in a $(2+1)$-dimensional integrable system. Phy. Lett. A 102–108, 203–208 (1995)
3. See e.g., Y.-L. Ma, M. Harada, Lecture notes on the Skyrme model (2016). arXiv:1604.04850
4. L.A. Ferreira, W.J. Zakrzewski, The concept of quasi-integrability: a concrete example. J. High Energy Phys. 05 **130**, 1–38 (2011)
5. L.A. Ferreira, W.J. Zakrzewski, Numerical and analytical tests of quasi-integrability in modified Sine-Gordon models. J. High Energy Phys. 01 **58**, 1–29 (2014)
6. L.A. Ferreira, G. Luchini, W.J. Zakrzewski, The concept of quasi-integrability for modified non-linear Schrodinger models. J. High Energy Phys. 09 **103**, 1–35 (2012)
7. F. ter Braak, L.A. Ferreira, W.J. Zakrzewski, Quasi-integrability of deformations of the KdV equation (2017). arXiv:1710.00918 [hep-th]
8. L.A. Ferreira, P. Klimas, W.J. Zakrzewski, Quasi-integrable deformations of the SU(3) Affine Toda theory. J. High Energy Phys. 05 **65**, 1–51 (2016)
9. H. Aratyn, C.P. Constantinidis, L.A. Ferreira, J.F. Gomes, A.H. Zimerman, Hirota's solitons in the affine and the conformal affine Toda models. Nucl. Phys. B **406**, 727–770 (1993)
10. See e.g., the relevant part of the book in Ref. 2

On Some Aspects of Unitary Evolution Generated by Non-Hermitian Hamiltonians

A Unitary Way Towards Quantum Collapse

Miloslav Znojil

Abstract The possibility of nontrivial quantum-catastrophic effects caused by the mere growth of the imaginary component of a non-Hermitian but $\mathcal{PT}$-symmetric *ad hoc* local-interaction potential $V(x)$ is revealed and demonstrated. Via a replacement of coordinate $x \in \mathbb{R}$ by a non-equidistant discrete lattice x_n with $n = 0, 1, \ldots, N+1$ the model is made exactly solvable at all N. By construction, the energy spectrum shrinks with the growth of the imaginary strength. The boundary of the unitarity of the model is reached in a certain strong non-Hermiticity limit. The loss-of-stability instant is identified with the Kato's exceptional point of order N at which the model exhibits a complete N-state degeneracy. This phase-transition effect is accessible as a result of a unitary-evolution process in an amended physical Hilbert space.

Keywords Quantum systems · Unitary evolution · Three-Hilbert-space representation of states · Non-Hermitian observables · Quantum phase transitions · Quantum catastrophes · Exactly solvable model

1 Introduction

The challenge of using non-Hermitian Hamiltonians in Schrödinger equation dates back to the very early days of quantum mechanics. Typically, these Hamiltonians were usually used in the study of resonances (cf., e.g., [1]). Still, in the context of stable bound states, the very recent enormous growth of interest in the concept (cf.,

Work supported by the GAČR Grant No. 16-22945S.

M. Znojil (✉)
NPI ASCR, Řež, Czech Republic
e-mail: znojil@ujf.cas.cz

Ş. Kuru et al. (eds.), *Integrability, Supersymmetry and Coherent States*, CRM Series in Mathematical Physics, https://doi.org/10.1007/978-3-030-20087-9_20

e.g., reviews [2, 3]) may be perceived as evoked by the single short 1998 letter by Bender with Boettcher [4]. These two authors managed to persuade the international physics community that the study of non-Hermitian Hamiltonians $H \neq H^\dagger$ with real spectra may be well motivated even if one studies a unitarily evolving quantum system.

Already the early concrete applications of the abstract theoretical idea proved surprisingly productive. In the context of quantum field theory, for example, the use of non-Hermitian toy-model interactions helped to clarify certain non-perturbative aspects of the spontaneous symmetry breaking [5]. In a certain parallel to the conventional Hermitian models, unfortunately, even the innovation of the mathematical frame did not lead to a discovery of any experimentally detectable mechanism of the spontaneous breakdown of supersymmetry [6]. The parallel with Hermitian theory only survived in a successful extension of the well-known non-relativistic supersymmetric quantum mechanics to its versions using non-Hermitian Hamiltonians (cf., e.g., [7]).

In the literature one can find many other, immediately visible paradoxes connected with the development of the new formulation of quantum theory. Let us, for example, mention here the rather unpleasant absence of a unified name of the formalism. Thus, Bender [2] calls the formalism $\mathcal{PT}$-symmetric quantum mechanics. Reason: he and his coauthors strongly recommend the work with non-Hermitian Hamiltonians $H \neq H^\dagger$ exhibiting an additional property $H\mathcal{PT} = \mathcal{PT}H$. Here, $\mathcal{P}$ denotes parity, while $\mathcal{T}$ is time reversal. In a later review [3], in contrast, Mostafazadeh called (more or less the same) formalism a pseudo-Hermitian representation of quantum mechanics. Reason: as long as the operator $\mathcal{T}$ of time reversal is antilinear, the above-mentioned $\mathcal{PT}$-symmetry of H is to be more naturally interpreted as its pseudo-Hermiticity.

An additional source of terminological misunderstandings has been, in parallel, found in the fact that the non-Hermitian Hamiltonians with real spectra (though not necessarily with the pseudo-Hermiticity property) were already constructed and used in the past. Indeed, in the 1956 study of ferromagnetism by Dyson [8], the Hamiltonians in question have been called "quasi-Hermitian." Subsequently, in the 1970s and 1980s the idea found a number of successful practical applications in the physics of heavy nuclei (these developments—called interacting boson model—were partially reviewed, in 1992, by Scholtz et al. [9]).

The productivity of the concept, under any name, attracted attention of a broader scientific community. Thus, mathematicians claimed that one should certainly speak, more precisely, about the so-called Krein-space self-adjointness of H [10]. In parallel, the experimental physicists working in classical optic often translate the $\mathcal{PT}$-symmetry of the medium (with a complex refraction index) as a balance between the gain (i.e., sources) and loss (i.e., sinks), simulated by the contemporary highly advanced nanotechnologies [11].

In our present paper we will only marginally mention such a broadened physical perspective. Our attention will be concentrated upon the comparatively narrow field of the unitary quantum theory in which the Dyson's (i.e., father's founder's) 60 years old ideas found their new and surprising applications recently, ranging from the relativistic quantum mechanics up to the phenomenological quantum cosmology.

The essence and consequences of these ideas will be explained via certain technically (though not phenomenologically) simplified illustrative examples. Our text will be organized as a compact introduction in the theory (Sect. 2) followed by an outline of models living on discrete lattices (Sect. 3). In Sect. 4 we then pick up a family of solvable models in which we will be able to simulate the unitary evolution leading, in a finite time, to an ultimate quantum energy-complexification catastrophe.

The latter result may be perceived as a phenomenological climax of our present message. In Appendix we shall complement it by a concise summary of some of its mathematical aspects.

2 Unitary $\mathcal{PT}$-Symmetric Quantum Theory *In Nuce*

For the conventional quantum systems $\mathcal{S}$ which are assigned self-adjoint one-parametric families of Hamiltonians $\mathfrak{h}(z) = \mathfrak{h}^\dagger(z)$ [with a real parameter $z \in (z_{\min}, z_{\max})$ which can vary, slowly, in time] the evolution described by Schrödinger equation

$$\mathrm{i}\frac{d}{dt}\,|\psi(t)\succ = \mathfrak{h}(z)\,|\psi(t)\succ\,, \qquad |\psi(t)\succ\, \in \mathcal{H}^{(\text{conventional})} \tag{1}$$

is unitary [12]. This is a traditional formulation of the theory which may prove, in some situations, over-restrictive. For example, without an additional, *ad hoc* symmetry of the Hamiltonian such a traditional formalism does not admit several less usual but still utterly elementary evolution processes involving, e.g., an unavoided crossing, or a merger and complexification (i.e., the loss of observability) of a pair (or, in general, of an N-plet) of energy levels. The reason was formulated by Kato [13]. He explained that the mergers of the energy levels can only occur at the so-called exceptional points (EP) which are, for a generic self-adjoint Hamiltonian, not real, i.e., which lie out of the range of an experimental realization.

2.1 Non-Hermitian Hamiltonians Enter the Scene

Whenever one needs to describe the process of an energy merger (marking, e.g., the quantum phase transition [14] or a "catastrophic" evolution scenario [15]), it is necessary to employ the so-called pseudo-Hermitian representation (PHR) of quantum mechanics [3] admitting non-Hermitian Hamiltonians $H(z) \neq H^\dagger(z)$ for which the EP values of z can be real, $z^{(EP)} \in \mathbb{R}$.

Naturally, the search for such PHR models of quantum dynamics is of imminent interest [2, 16]. For their description one only has to replace the conventional Schrödinger Eq. (1) by its "false space" [17] generalization

$$\mathrm{i}\frac{d}{dt}\,|\psi(t)\rangle = H(z)\,|\psi(t)\rangle\,, \qquad |\psi(t)\rangle \in \mathcal{H}^{(\text{auxiliary})} \tag{2}$$

in which the uppercase Hamiltonian H is manifestly non-Hermitian. One of the most popular illustrative examples of the applicability of the non-Hermitian evolution Eq. (2) using ordinary differential Hamiltonians with real spectrum is due to Bender and Boettcher [4] who studied the models

$$H(z) = -\frac{d^2}{dx^2} + V(x,z) \neq H^\dagger(z) \tag{3}$$

where the potential represents the imaginary cubic oscillator $V(x,0) = \mathrm{i}x^3$ or the family of its generalizations

$$V(x,z) = (\mathrm{i}x)^z\, V(x,0)\,. \tag{4}$$

In the special case of $z \in (-1,1)$ these authors worked in $\mathcal{H}^{(\text{auxiliary})} = L^2(\mathbb{R})$. They conjectured that the model could be compatible with the unitarity-of-evolution hypothesis (cf. [2, 3] and also several updated reviews in [18]).

2.2 The Emergence of Difficulties

A few years later it was revealed that the Bender's and Boettcher's concrete model cannot be made compatible with the Stone theorem so that, *eo ipso*, Hamiltonian (3) + (4) cannot be interpreted as a generator of unitary evolution of a quantum system (see the detailed explanations as given, e.g., in Refs. [19, 20]). Similar mathematical objections proved also to apply to a number of other non-Hermitian local-interaction models [21]. Hence, the search for an update of a widely accepted benchmark illustrations of the pseudo-Hermitian representation theory has been reopened [22].

The search is not yet completed. Besides the attempts aimed at the rather complicated mathematics behind the unbounded operators (3) and reviewed by Antoine and Trapani [23], a more pragmatic strategy has been developed in Ref. [9]. Scholtz et al. described and recommended there a bounded-operator version of the formalism. Along these lines we also proposed, in our recent paper [24], a replacement of the differential-operator models (3) by their various discrete, bounded-operator descendants: These Hamiltonians may be found recalled in Sect. 3.

One of the shortcomings of the illustrative models of Ref. [24] was their numerical nature. In our present text we will get rid of such a weakness. We will introduce a family of amended discrete-operator benchmark Hamiltonians which remain closely related to their paradigmatic ordinary differential predecessors (3). Our project will involve also the following, phenomenologically oriented aims.

1. Via Eq. (2) we intend to simulate the quantum unitary-evolution process in which the Hamiltonian is a one-parametric $N \times N$ matrix $H^{(N)}(z)$. This toy-model matrix will be Hermitian at $z = 0$ and non-Hermitian but $\mathcal{PT}$-symmetric at $z \in (0, 1)$ [for the definition of the concept of (discrete) $\mathcal{PT}$-symmetry see Sect. 4.4].
2. An exact linear-algebraic solvability of our model will be achieved via its discrete-operator construction based on a replacement of the real axis of coordinates $x \in \mathbb{R}$ in (3) by a finite lattice of grid points x_n, $n = 0, 1, \ldots, N + 1$.
3. We shall guarantee that the spectra $\{E_n(z)\}$ of our matrices $H^{(N)}(z)$ will be real. The task will prove facilitated by an interplay between the tridiagonality and $\mathcal{PT}$-symmetry of $H^{(N)}(z)$.
4. We shall find a special subfamily of our matrices $H^{(N)}(z)$ for which the spectrum will prove obtainable, at all of the relevant parameters $z \in (0, 1)$ as well as at all of the quantum numbers $n = 0, 1, \ldots, N - 1$, in closed form (cf. Sect. 4).

In the latter, exactly solvable special case, all of the energies as well as all of the wave functions will be required to degenerate at the real exceptional point $z = z^{(EPN)} = 1$ of the N-th order

$$\lim_{z\to 1} E_n(z) = E_0(1)\,, \qquad \lim_{z\to 1} |\psi_n(z)\rangle = |\psi_0(1)\rangle\,, \qquad n = 0, 1, \ldots, N - 1 \tag{5}$$

(cf. the rigorous mathematical definition of the exceptional points in [13]). Parameter $z \in (0, 1)$ will be treated as a measure of the degree of non-Hermiticity of our discrete local potential $V(x) = V(x, z)$. At the maximal non-Hermiticity boundary with $z = 1$ the specific EPN degeneracy (5) will be rendered possible by the use of a non-equidistant lattice, with the self-consistently fine-tuned separation distances $h_n = x_n - x_{n-1}$, $n = 1, 2, \ldots, N + 1$ at any $N = 2, 3, \ldots$. An independent, inseparable part of the solvability feature of our benchmark model will be shown to lie in the user-friendly nature of the so-called Hilbert-space metric Θ which defines the correct physical inner products. Interested readers may find a brief outline of these technical details in Appendix.

3 Simplified Mathematics and the Discrete-Coordinate Models

According to the recent study [21] the spectrum generated by $V \sim ix^3$ could be unstable and "a very small perturbation …can create …eigenvalues very far from the spectrum of the unperturbed operator H." The danger is real. Its possible emergence was already predicted by Dieudonné [25]. The authors of the quantum-physics-oriented review [9] decided to restrict, therefore, the scope of the PHR approach to bounded Hamiltonians, $H \in \mathbb{B}(\mathcal{H})$. In this way they managed to stay on the mathematically safe side. The ill-behaved unbounded-operator models were excluded.

In many subsequent publications which accepted this philosophy, multiple realistic calculations as well as conceptual considerations were even based on the mere finite-dimensional benchmark matrix models [14, 26, 27]. The same strategy will be also accepted in our present paper.

3.1 Discrete Imaginary Cubic Oscillators

In Ref. [24] we followed and tested the consequences of the latter tendency. For the model of Eq. (4) which manifestly violates the boundedness condition $H \in \mathbb{B}(\mathcal{H})$ a remedy has been proposed. It consisted in a discretization of the continuous coordinate $x \in \mathbb{R}$ yielding the lattice of points $x_n = \text{const} + h\,n$ with $n = 0, 1, \ldots, N+1$, and with a constant grid-mesh size $h > 0$.

The use of the discrete coordinates led to the replacement of the continuous-coordinate kinetic-energy term by its difference-operator analog

$$-\frac{d^2}{dx^2}u(x_n) \;\to\; \frac{1}{h^2}\left[-u(x_{n-1}) + 2\,u(x_n) - u(x_{n+1})\right]. \tag{6}$$

We fixed $\delta = 0$, premultiplied Hamiltonian (4) by h^2, abbreviated $h^5/8 = a$ and arrived at the imaginary cubic discrete-oscillator sequence of the N by N matrix models

$$H^{(2)}_{(IC)}(a) = \begin{pmatrix} 2-ia & -1 \\ -1 & 2+ia \end{pmatrix}, \quad H^{(3)}_{(IC)}(a) = \begin{pmatrix} 2-8ia & -1 & 0 \\ -1 & 2 & -1 \\ 0 & -1 & 2+8ia \end{pmatrix},$$

$$H^{(4)}_{(IC)}(a) = \begin{pmatrix} 2-27ia & -1 & 0 & 0 \\ -1 & 2-ia & -1 & 0 \\ 0 & -1 & 2+ia & -1 \\ 0 & 0 & -1 & 2+27ia \end{pmatrix}, \ldots. \tag{7}$$

In comparison with the formal $N \to \infty$ limit (4) we emphasized the technical merits of the study of family (7) at a few not too large N. This offered an immediate theoretical insight in several interesting physical phenomena and, in particular, an access to a transparent exemplification of the two-level mergers of the energy levels.

3.2 General Power-Law Potentials

In the limit $N \to \infty$, dynamical models (7) yield the traditional PHR Hamiltonian (4) at $\delta = 0$. From this perspective the methodical role of the Bender's and Boettcher's variable exponent δ has been transferred to the variable mesh-size a. The original EP singularities [4, 28] appeared paralleled by the dimension-dependent EP values $a^{(N)}_{(EP)}$ (cf. their sample in Table Nr. 1 in Ref. [24]).

Feeling encouraged by such a success we decided to introduce, in [24], another variable parameter. We made its value γ intimately connected with the exponent δ in Eq. (4). Our two-parametric toy-Hamiltonian construction resulted in another sequence of power-law matrices

$$H^{(2)}_{(PL)}(a,\gamma) = \begin{pmatrix} 2-ia & -1 \\ -1 & 2+ia \end{pmatrix},$$

$$H^{(3)}_{(PL)}(a,\gamma) = \begin{pmatrix} 2-2^{\gamma}ia & -1 & 0 \\ -1 & 2 & -1 \\ 0 & -1 & 2+2^{\gamma}ia \end{pmatrix}, \tag{8}$$

$$H^{(4)}_{(PL)}(a,\gamma) = \begin{pmatrix} 2-3^{\gamma}ia & -1 & 0 & 0 \\ -1 & 2-ia & -1 & 0 \\ 0 & -1 & 2+ia & -1 \\ 0 & 0 & -1 & 2+3^{\gamma}ia \end{pmatrix}, \ldots . \tag{9}$$

Such a $\gamma \neq 3$ generalization of Eq. (7) enhanced the flexibility of the predictions. Serendipitously, it also led to the emergence of a number of topologically non-equivalent spectral loci [24].

4 Physics of Unitary Evolution Towards a Collapse

A less satisfactory aspect of the above-outlined results lies in a serious methodical problem which emerged in connection with the necessity of the construction of the hermitizing inner products in the correct, physical Hilbert spaces. Only the purely numerical techniques were found to be able to deal with the problem, i.e., in the language of physics, with the necessity of the guarantee of the positive definiteness of the inner product.

4.1 Non-uniform Lattices with Site-Dependent Spacings

On an equidistant lattice with a fixed spacing $h > 0$ the discrete Laplacian is defined by Eq. (6). On the more general one-dimensional lattices with the site-dependent spacings $h_n > 0$ an analogous formula holds, just with the modified, site-dependent coefficients [29]

$$-\frac{d^2}{dx^2}u(x_n) \rightarrow -a_n u(x_{n-1}) + 2\,b_n u(x_n) - c_n u(x_{n+1})\,, \tag{10}$$

$$a_n = \frac{2}{h_n(h_n + h_{n+1})}\,, \quad b_n = \frac{1}{h_n\, h_{n+1}}\,, \quad c_n = \frac{2}{h_{n+1}(h_n + h_{n+1})}\,. \tag{11}$$

The old formula is correctly reproduced in the equidistant limit $h_n \rightarrow h$. Along these lines the original discrete Laplacian $\triangle$ becomes generalized

$$\begin{bmatrix} 0 & -1 & 0 & \dots & 0 \\ -1 & 0 & \ddots & \ddots & \vdots \\ 0 & -1 & \ddots & -1 & 0 \\ \vdots & \ddots & \ddots & 0 & -1 \\ 0 & \dots & 0 & -1 & 0 \end{bmatrix} \rightarrow \begin{bmatrix} 2b_1 & -c_1 & 0 & \dots & 0 \\ -a_2 & 2b_2 & -c_2 & \ddots & \vdots \\ 0 & -a_3 & 2b_3 & \ddots & 0 \\ \vdots & \ddots & \ddots & \ddots & -c_{N-1} \\ 0 & \dots & 0 & -a_N & 2b_N \end{bmatrix}. \tag{12}$$

Once we add a local potential $V(x_n)$ we arrive at the ultimate discrete version of ansatz (3)

$$H = \begin{bmatrix} 2b_1 + V(x_1) & -c_1 & 0 & \dots & 0 \\ -a_2 & 2b_2 + V(x_2) & -c_2 & \ddots & \vdots \\ 0 & -a_3 & 2b_3 + V(x_3) & \ddots & 0 \\ \vdots & \ddots & \ddots & \ddots & -c_{N-1} \\ 0 & \dots & 0 & -a_N & 2b_N + V(x_N) \end{bmatrix}. \tag{13}$$

The variability of the sequence of spacings h_n may be reinterpreted as the emergence of a new model-building freedom. The tridiagonal discrete-oscillator Hamiltonians can now contain more or less arbitrary *real* off-diagonal matrix elements. Naturally, such a freedom may be expected to become highly welcome in phenomenological context.

4.2 Simplification Requirement: $\mathcal{PT}$-Symmetry

A family of user-friendly PHR quantum models $H = \Delta + V(x)$ can be built using the generalized kinetic-energy component (12). We are also allowed to pick up such a (non-Hermitian) discrete local interaction $V(x)$ which would guarantee the stable, unitary evolution of the system, i.e., the strict reality of the whole energy spectrum. That's why the $\mathcal{PT}$-symmetry of H is often required. Indeed, such a property of H proves methodically useful because in the complex plane of energies, it guarantees that even if not real, all of the formal complex energy eigenvalues are very specific in always occurring in complex conjugate pairs. Hence, under the apparently redundant $\mathcal{PT}$-symmetry the loss of the reality is much more easily traced and detected.

This observation makes the concept of $\mathcal{PT}$-symmetry useful and productive. For the matrix Hamiltonians, in particular, the antilinear operator $\mathcal{T}$ can be visualized as causing Hermitian conjugation (i.e., transposition plus complex conjugation). In parallel, the linear operator of parity acquires the matrix form

$$\mathcal{P} = \begin{bmatrix} & & & & 1 \\ & & & 1 & \\ & & \iddots & & \\ & 1 & & & \\ 1 & & & & \end{bmatrix}. \tag{14}$$

Having selected, for the sake of definiteness, the value of the exponent $\gamma = 1$ in our guiding example (8), we then obtain a $\mathcal{PT}$-symmetric non-uniform-grid more-parametric generalizations

$$H^{(2)}_{(\mathcal{PT})}(f, g, a) = \begin{pmatrix} f - ia & g \\ g & f + ia \end{pmatrix},$$

$$H^{(3)}_{(\mathcal{PT})}(\mathbf{f}^{(2)}, g, a) = \begin{pmatrix} f_1 - 2ia & g & 0 \\ g & f_2 & g \\ 0 & g & f_1 + 2ia \end{pmatrix}, \tag{15}$$

$$H^{(4)}_{(\mathcal{PT})}(\mathbf{f}^{(2)}, \mathbf{g}^{(2)}, a) = \begin{pmatrix} f_1 - 3ia & g_1 & 0 & 0 \\ g_1 & f_2 - ia & g_2 & 0 \\ 0 & g_2 & f_2 + ia & g_1 \\ 0 & 0 & g_1 & f_1 + 3ia \end{pmatrix}, \ldots \tag{16}$$

of the respective two-parametric Hamiltonian matrices in Eq. (8).

4.3 Construction of the Collapse

The higher$-N$ elements of our $\mathcal{PT}$-symmetric family (15) of the candidates for a solvable Hamiltonian contain so many free parameters that one cannot really hope that the determination of the energies would become non-numerical at $N > 4$. At the same time, one can try to restrict such a redundant freedom and obtain, in this manner, a non-numerical form of the energy levels. In what follows we shall see that it is possible to achieve even more: The variability of the parameters will enable us to fine-tune their values in a way aimed at making our EP values of some parameters real, i.e., accessible to an experimental simulation.

In the light of the latter comment our present model-building strategy will be based on a step-by-step search for the most efficient simplifications. Naturally, such a search has to start from $H^{(N)}_{(\mathcal{PT})}(\mathbf{f}^{(J)}, \mathbf{g}^{(K)}, a)$ at $N = 2$ and $J = K = 1$. Thus, from the available elementary formula for the energy spectrum $E_\pm = f \pm \sqrt{g^2 - a^2}$ one deduces that the role of f, meaning just a shift of the spectrum, is trivial so that we may put this parameter equal to zero. Similarly, the value of g just fixes the scale so that we may set it equal to one.

The naive strategy of such a type fails immediately at $N = 3$, $J = 2$, and $K = 1$. Indeed, the Cardano formulae for the energies are found complicated and far from being transparent even with $f_2 = 0$ or with $f_1 = 0$. In contrast, the simultaneous choice of $f_1 = f_2 = 0$ leads to the satisfactory elementary formulae $E_0 = 0$ and $E_\pm = \sqrt{2\,g^2 - 4\,a^2}$. The tests made at $N = 4$ and $N = 5$ gave similar answers. On these grounds we arrived at our following ultimate multiparametric ansatz for $H^{(N)}(\mathbf{g}^{(K)}, z)$:

$$\left[\begin{array}{cccccc}
-i\,(N-1)z & g_1 & 0 & 0 & \ldots & 0\\
g_1 & -i\,(N-3)z & g_2 & 0 & \ddots & \vdots\\
0 & g_2 & -i\,(N-5)z & \ddots & \ddots & 0\\
0 & 0 & g_3 & \ddots & g_2 & 0\\
\vdots & \ddots & \ddots & \ddots & i\,(N-3)z & g_1\\
0 & \ldots & 0 & 0 & g_1 & i\,(N-1)z
\end{array}\right]. \tag{17}$$

It coincides with the sequence of models (15) under the trivial choice of $\mathbf{f}^{(J)} = 0$, i.e., it still varies with the K-plet of optional real parameters $\mathbf{g}^{(K)}$ such that $N = 2K$ or $N = 2K + 1$.

Theorem 1 *For the special choice of parameters*

$$g_k = \pm\sqrt{k(N-k)} \tag{18}$$

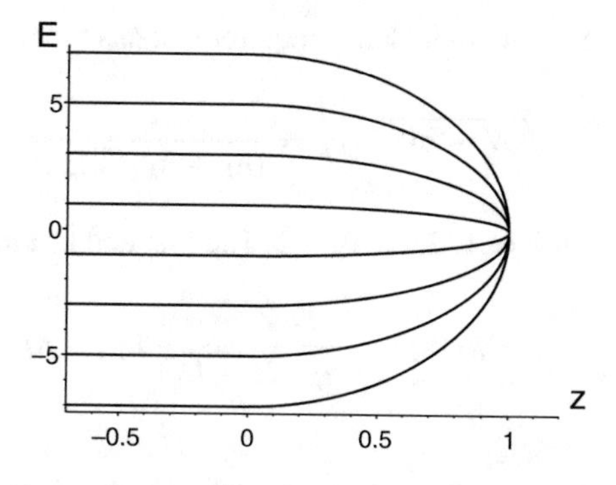

Fig. 1 Time-dependent energies (19) for the choice of $z = z(t) = \max(0, t)$ and $N = 8$. The EP8 collapse is encountered at $t = 1$

in the complex symmetric Hamiltonian (17), the set of the energy eigenvalues has the following elementary form:

$$E_k(z) = (N - 2k + 1)\sqrt{1 - z^2}\,, \qquad k = 1, 2, \ldots, N\,. \tag{19}$$

Proof Due to the tridiagonality of the Hamiltonian one can recall the recurrent construction recipe of Ref. [30]. Once we introduce the reduced variable $s = \sqrt{E/(1 - z^2)}$ we find that at any dimension N the secular polynomial $S(s)$ has just integer coefficients and that it factorizes into elementary factors, with the roots given by Eq. (19).

Figure 1 offers an example of the N-level collapse. We choose there $N = 8$ and a specific time dependence of $z = z(t) = \max(0, t)$. The picture shows that the spectrum remains constant before zero time $t = 0$, then it shrinks and, ultimately, degenerates at $t = t^{(EPN)} = 1$. It becomes purely imaginary and unobservable later on, at $t > 1$.

4.4 Multiparametric Hamiltonians

Let us now assume that our toy-model Hamiltonian (13) is exactly solvable, i.e., that it coincides, up to an inessential overall multiplicative factor, with the solvable matrix (17) and (18), i.e., the matrix $H^{(N)}(z)$ is

$$\left[\begin{array}{cccccc}
-i\,(N-1)z & -\sqrt{N-1} & 0 & 0 & \ldots & 0 \\
-\sqrt{N-1} & -i\,(N-3)z & -\sqrt{2(N-2)} & 0 & \ddots & \vdots \\
0 & -\sqrt{2(N-2)} & -i\,(N-5)z & \ddots & \ddots & 0 \\
0 & 0 & -\sqrt{3(N-3)} & \ddots & -\sqrt{2(N-2)} & 0 \\
\vdots & \ddots & \ddots & \ddots & i\,(N-3)z & -\sqrt{N-1} \\
0 & \ldots & 0 & 0 & -\sqrt{N-1} & i\,(N-1)z
\end{array}\right]. \tag{20}$$

Such a coincidence requirement has the form of algebraic set

$$\sqrt{(N-k)k} = c_k = \frac{2}{(h_k + h_{k+1})h_{k+1}} = a_{k+1} = \frac{2}{h_{k+1}(h_{k+1} + h_{k+2})}, \tag{21}$$

for $k = 1, 2, \ldots N - 1$. The first and last items enable us to eliminate h_1 and h_{N+1},

$$h_1 = h_1(h_2) = \frac{2}{h_2\sqrt{N-1}} - h_2\,, \quad h_{N+1} = h_{N+1}(h_N) = \frac{2}{h_N\sqrt{N-1}} - h_N\,. \tag{22}$$

Thus, at $N = 2$ we are left with the single variable mesh-size $h_2 = h$ and its two functions (22). At any larger $N > 2$ we may simplify relations (22) by evaluating the ratios c_k/a_k. This yields the $N - 2$ recurrences

$$h_{k+2}\sqrt{(N-k-1)(k+1)} = h_{k+1}\sqrt{(N-k)k}\,, \quad k = 1, 2, \ldots, N-2, \tag{23}$$

which interrelate the remaining $N - 1$ spacings $h_2, h_3, \ldots, h_N$. One quantity is always left as an independent mesh-size variable. It is $h = h_2 = h_3$ at $N = 3$, etc.

Such a determination of the non-uniform spacings $h_k = h_k(h)$ may be followed by the subsequent evaluation of components $2b_k$ of the diagonal matrix elements $2b_k + V(x_k)$ in Eq. (13). Once we want our matrix to coincide with the exactly solvable model of Eq. (20), we must make these matrix elements equal to the purely imaginary quantities $-\mathrm{i}(N+1-2k)z$. In this manner, our exactly solvable potential will be defined by the relations

$$\mathrm{Re}\, V(x_k) = -2b_k(h)\,, \quad \mathrm{Im}\, V(x_k) = -\mathrm{i}(N+1-2k)z\,. \tag{24}$$

The resulting potential is fine-tuned to guarantee the degeneracy (5). Its (non-vanishing!) real part remains fixed and only its imaginary part varies with the real coupling-strength parameter z (cf. [31]). Our construction of the local potential defined on the non-equidistant discrete lattice of grid points x_k and supporting the existence of the (in principle, experimentally accessible) real N-tuple exceptional point is completed.

5 Conclusions

In our present paper we described an answer to the truly challenging problem of the description of the physics of unitary evolution towards a collapse using a *local* interaction potential. Our successful construction was rendered possible by a patient trial-and-error search for a maximally user-friendly toy-model. Besides a suitable modification of the more or less routine discretization strategy, the main merit of our approach can be seen in a strictly non-numerical tractability of our N by

N Hamiltonians. We managed to construct an amended family of matrix models containing a sufficient number of variable parameters while still keeping the bound-state energies themselves in a closed, non-numerical form.

On the basis of our present results we believe that many of the above-discussed weaknesses of the currently available non-Hermitian local-interaction models will be further suppressed along the indicated methodical lines. We are persuaded that, first of all, the purely physical credit of the finite-dimensional Hamiltonians will be further enhanced, mainly due to our demonstration that even at the higher dimensions N, the form of the spectrum of the toy models need not necessarily be purely numerical. Secondly, the future analyses of toy models will certainly be encouraged by our present constructive demonstration that the linear-algebraic construction of the physical Hilbert space may often decisively be facilitated by the finiteness of their matrix dimensions $N < \infty$.

Appendix: The Metric as a Degree of Model-Building Freedom

The specification of quantum system $\mathcal{S}$ requires not only the knowledge of its Hamiltonian $H^{(N)}(z)$ [i.e., at any preselected dimension N and parameter z, the knowledge of matrix (20) in our case] but also a constructive access to the correct probabilistic interpretation of experiments. In other words, having solved the time-dependent Schrödinger equation (2) we still need to replace our manifestly unphysical working Hilbert space $\mathcal{H}^{(\text{auxiliary})}$ by the correct physical Hilbert space, i.e., we must modify the inner product accordingly [3].

The Abstract Theory Revisited

In the context of quantum theory of many-body systems it was Freeman Dyson [8] who conjectured that in some cases, an enormous simplification of the variational determination of the bound-state spectra could be achieved via a suitable *non-unitary* similarity transformation of the given realistic Hamiltonians

$$\mathfrak{h} \to H = \Omega^{-1}\mathfrak{h}\Omega\,, \quad \Omega^\dagger\Omega \neq I. \tag{25}$$

The trick proved particularly efficient in nuclear physics [9]. An amendment of the calculations has been achieved via a judicious choice of the operators Ω converting, e.g., the strongly correlated pairs of nucleons into weakly interacting effective bosons.

In spite of the initial success, the trial-and-error nature of the Dyson-inspired recipes and the fairly high formal mathematical costs of the replacement of the self-adjoint "realistic" operator $\mathfrak{h} = \mathfrak{h}^\dagger$ by its manifestly non-Hermitian, quasi-

Hermitian [9] alternative

$$H = \Theta^{-1} H^{\dagger} \Theta \neq H^{\dagger} \,, \qquad \Theta = \Omega^{\dagger} \Omega \tag{26}$$

have been found, beyond the domain of nuclear physics, strongly discouraging (cf., e.g., [25]).

Undoubtedly, the idea itself is sound. In the context of abstract quantum theory its appeal has been rediscovered by Bender with Boettcher [4]. In effect, these authors just inverted the arrow in Eq. (25). They conjectured that one might start a model-building process *directly* from Eq. (26), i.e., *directly* from a suitable trial-and-error choice of a sufficiently simple non-Hermitian Hamiltonian with real spectrum. Their conjecture was illustrated by the family of perturbed imaginary cubic oscillator Hamiltonians

$$H_\epsilon = -\frac{d^2}{dx^2} + V_\epsilon(x) \neq H_\epsilon^{\dagger} \,, \quad V_\epsilon(x) = \mathrm{i}x^3(\mathrm{i}x)^\epsilon \,, \quad x \in (-\infty, \infty), \quad \epsilon \in (-1, 1). \tag{27}$$

Technical details may be found in reviews [2, 3, 9, 18] in which several formulations of the "inverted" stationary version of the quantum model-building strategy may be found.

The Unitarity of Evolution Reestablished

It is worth adding that strictly speaking, the latter strategies are not always equivalent (cf. also further comments in [22, 32]). For our present purposes we may distinguish between the older, more restrictive "quasi-Hermitian" formulation of quantum mechanics (QHQM) of Ref. [9], and the "$\mathcal{PT}$-symmetric" version of quantum mechanics (PTQM, [2]).

The key difference between the latter two pictures of quantum reality lies in the strictly required *non-admissibility* of the unbounded Hamiltonians in the QHQM framework of Ref. [9]. This requirement is by far not only formal, and it also makes the QHQM theory mathematically better understood. In contrast, the process of the rigorous mathematical foundation of the extended, phenomenologically more ambitious PTQM theory (admitting the unbounded Hamiltonians as sampled by Eq. (27)) is still unfinished (cf., e.g., the concise progress reports [23, 33]). Hence, also the toy models with the local but not real potentials are far from being widely accepted as fully understood and consistent at present (cf., e.g., [20, 21]).

One is forced to conclude that the ordinary differential (but, unfortunately, unbounded) benchmark model (27) of a $\mathcal{PT}$-symmetric quantum system (where $\mathcal{P}$ means parity, while symbol $\mathcal{T}$ denotes the time reversal [4]) is far from satisfactory. At the same time, its strength may be seen in its methodical impact as well as in its simplicity and intuitive appeal. For all of these reasons one is forced to search for alternative $\mathcal{PT}$-symmetric quantum models which share the merits while not suffering of the inconsistencies.

Needless to add that the unitarity of the quantum evolution can be reestablished for many non-Hermitian models with real spectra. One just has to return to the standard quantum theory in QHQM formulation. The details of the implementation of the idea may vary. Thus, Bender [2] works with an auxiliary nonlinear requirement $H\mathcal{PT} = \mathcal{PT}H$ called "$\mathcal{PT}$-symmetry of H." In a slightly more general setting Mostafazadeh [3] makes use of the same relation written in the equivalent form $H^\dagger\mathcal{P} = \mathcal{P}H$, and he calls it "$\mathcal{P}$-pseudo-Hermiticity of H." Still, both of these authors respect the Stone theorem. This means that both of them introduce the correct physical Hilbert-space metric Θ and both of them use it in the postulate

$$H = \Theta^{-1} H^\dagger \Theta := H^\ddagger. \tag{28}$$

of the so-called *quasi-Hermiticity* property of the acceptable Hamiltonians. Rewritten in the form

$$H^\dagger \Theta = \Theta\, H \tag{29}$$

the equation can be interpreted as a linear-algebraic system which defines, for a given Hamiltonian matrix H with real spectrum, the N-parametric family of all of the eligible matrices of metric Θ. For the tridiagonal input matrices H, the solution is particularly straightforward because the algorithm can be given a recurrent form implying that the solutions exist at any input H [34].

References

1. N. Moiseyev, *Non-Hermitian Quantum Mechanics* (Cambridge University Press, Cambridge, 2011), pp. 1–394
2. C.M. Bender, Making sense of non-Hermitian Hamiltonians. Rep. Prog. Phys. **70**, 947–1018 (2007)
3. A. Mostafazadeh, Pseudo-Hermitian representation of quantum mechanics. Int. J. Geom. Meth. Mod. Phys. **7**, 1191–1306 (2010)
4. C.M. Bender, S. Boettcher, Real spectra in non-Hermitian Hamiltonians having PT symmetry. Phys. Rev. Lett. **80**, 5243–5246 (1998)
5. C.M. Bender, K.A. Milton, Nonperturbative calculation of symmetry breaking in quantum field theory. Phys. Rev. D **55**, 3255–3259 (1997)
6. C.M. Bender, K.A. Milton, Model of supersymmetric quantum field theory with broken parity symmetry. Phys. Rev. D **57**, 3595–3608 (1998)
7. M. Znojil, Non-Hermitian SUSY and singular PT-symmetrized oscillators. J. Phys. A Math. Gen. **35**, 2341–2352 (2002)
8. F.J. Dyson, General theory of spin-wave interactions. Phys. Rev. **102**, 1217 (1956)
9. F.G. Scholtz, H.B. Geyer, F.J.W. Hahne, Quasi-Hermitian operators in quantum mechanics and the variational principle. Ann. Phys. (NY) **213**, 74–101 (1992)
10. H. Langer, C. Tretter, A Krein space approach to PT symmetry. Czech. J. Phys. **54**, 1113–1120 (2004)

11. R. El-Ganainy, K.G. Makris, M. Khajavikhan, et al., Non-Hermitian physics and PT symmetry. Nat. Phys. **14**, 11 (2018)
12. M.H. Stone, On one-parameter unitary groups in Hilbert space. Ann. Math. **33**, 643–648 (1932)
13. T. Kato, *Perturbation Theory for Linear Operators* (Springer, Berlin, 1966), pp. 1–592
14. M. Znojil, Hermitian-to-quasi-Hermitian quantum phase transitions. Phys. Rev. A **97**, 042117 (2018)
15. M. Znojil, Quantum catastrophes: a case study. J. Phys. A Math. Theor. **45**, 444036 (2012); G. Lévai, F. Růžička, M. Znojil, Three solvable matrix models of a quantum catastrophe. Int. J. Theor. Phys. **53**, 2875 (2014)
16. V.V. Konotop, J.-K. Yang, D.A. Zezyulin, Rev. Mod. Phys. **88**, 035002 (2016)
17. M. Znojil, Time-dependent version of cryptohermitian quantum theory. Phys. Rev. D **78**, 085003 (2008); M. Znojil, Three-Hilbert space formulation of quantum theory. SIGMA **5**, 001 (2009) (e-print overlay: arXiv:0901.0700)
18. F. Bagarello, J.-P. Gazeau, F.H. Szafraniec, M. Znojil (eds.), *Non-Selfadjoint Operators in Quantum Physics: Mathematical Aspects* (Wiley, Hoboken, 2015), pp. 1–407
19. L.N. Trefethen, M. Embree, *Spectra and Pseudospectra* (Princeton University Press, Princeton, 2005)
20. P. Siegl, D. Krejčiřík, On the metric operator for the imaginary cubic oscillator. Phys. Rev. D **86**, 121702(R) (2012)
21. D. Krejčiřík, P. Siegl, M. Tater, J. Viola, Pseudospectra in non-Hermitian quantum mechanics. J. Math. Phys. **56**, 103513 (2015)
22. M. Znojil, in *Non-Selfadjoint Operators in Quantum Physics: Mathematical Aspects*, ed. by F. Bagarello, J.-P. Gazeau, F.H. Szafraniec, M. Znojil (Wiley, Hoboken, 2015), pp. 7–58
23. J.-P. Antoine, C. Trapani, in *Non-Selfadjoint Operators in Quantum Physics: Mathematical Aspects*, ed. by F. Bagarello, J.-P. Gazeau, F.H. Szafraniec, M. Znojil (Wiley, Hoboken, 2015), pp. 345–402
24. M. Znojil, N-site-lattice analogues of $V(x) = ix^3$. Ann. Phys. (NY) **327**, 893–913 (2012)
25. J. Dieudonné, Quasi-Hermitian operators, in *Proceedings of the International Symposium on Linear Spaces* (Pergamon, Oxford, 1961), pp. 115–122
26. M. Znojil, Solvable quantum lattices with nonlocal non-Hermitian endpoint interactions. Ann. Phys. (NY) **361**, 226–246 (2015)
27. M. Znojil, H.B. Geyer, Phys. Lett. B **640**, 52–56 (2006); M. Znojil, Gegenbauer-solvable quantum chain model. Phys. Rev. A **82**, 052113 (2010); M. Znojil, I. Semorádová, F. Růžička, H. Moulla, I. Leghrib, Problem of the coexistence of several non-Hermitian observables in PT-symmetric quantum mechanics. Phys. Rev. A **95**, 042122 (2017); M. Znojil, Bound states emerging from below the continuum in a solvable PT-symmetric discrete Schrodinger equation. Phys. Rev. A **96**, 012127 (2017)
28. Z. Ahmed, S. Kumar, D. Sharma, Ann. Phys. (NY) **383**, 635 (2017)
29. N. Sukumar, J.E. Bolander, Numerical computation of discrete differential operators on non-uniform grids. Comput. Model. Eng. Sci. **4**, 691–706 (2003), eq. (27)
30. M. Znojil, Maximal couplings in PT-symmetric chain models with the real spectrum of energies. J. Phys. A Math. Theor. **40**, 4863–4875 (2007); M. Znojil, Tridiagonal PT-symmetric N by N Hamiltonians and a fine-tuning of their observability domains in the strongly non-Hermitian regime. J. Phys. A Math. Theor. **40**, 13131–13148 (2007)
31. S. Longhi, PT-symmetric mode-locking. Optics Lett. **41**, 4518–4521 (2016); C.-F. Huang, J.-L. Zeng, Opt. Laser Technol. **88**, 104 (2017)
32. M. Znojil, Admissible perturbations and false instabilities in PT-symmetric quantum systems. Phys. Rev. A **97**, 032114 (2018)
33. F. Bagarello, M. Znojil, Nonlinear pseudo-bosons versus hidden Hermiticity. II: the case of unbounded operators. J. Phys. A Math. Theor. **45**, 115311 (2012)
34. M. Znojil, Quantum inner-product metrics via recurrent solution of Dieudonne equation. J. Phys. A Math. Theor. **45**, 085302 (2012); F. Růžička, Hilbert space inner product for PT-symmetric Su-Schrieffer-Heeger models. Int. J. Theor. Phys. **54**, 4154–4163 (2015)

Index

A
Abstract quantum theory, 423–424
Accidental degeneracy, 105
Adapted Susskind–Glogower coherent states, 89–90
Algebraic Jacobi functions (AJF), 269
algebra representations, 273–276
ladder operators, 276–278
representation of *su*(2, 2), 278–280
structure, 270–273
Algebraic special functions (ASF), 268
Ambient space, 6
AN class
definition, 73–74
displacement operator, 97–98
as Eigenstates, 96–97
quantization, 95–98
Anisotropic oscillators
caged, 14
Euclidean plane, 12–15
harmonic, 121
on S^2 and H^2, 16–22
Askey scheme, 268

B
Baby Skyrme model, 395, 396
Bardeen–Cooper–Schrieffer (BCS) approximation, 256
Barut–Girardello CS, 88
BCS gap equation, 263–266
Beltrami coordinates, 9–11
Berggren representation, 256, 261
Bertrand's theorem, 105
Bogolyubov transformation, 258
Borisov–Mamaev map, 372
Boussinesq equation, conditional symmetry of
$\hat{X}_1$ infinitesimal generator, 379–380
$\hat{X}_2$ infinitesimal generator, 381–382
$\hat{X}_6$ infinitesimal generator, 383–384
Breit–Wigner distribution, 259

C
Canonical coherent states, *see* Klauder–Glauber–Sudarshan states
Canonical transformation, 258
Cardano formulae, 420
Cartan subalgebra, 273
Cauchy problem, 387–393
Chazy class
in Cartesian coordinates, 106
Painlevé property and, 110
Classical property of states, 189
Clebsch–Gordan decompositions, 352
Coherent states (CS)
adapted Susskind–Glogower, 89–90
AN class, 73–74
AN quantization, 95–98
bare essentials of, 214
Barut–Girardello, 88
canonical, 200
even and odd, 212–213
field correlations, 198
in Fock space, 70–71
generalized, 214
generalized oscillator algebras, 215–216
Glauber–Sudarshan (*see* Glauber–Sudarshan coherent states)

Ş. Kuru et al. (eds.), *Integrability, Supersymmetry and Coherent States*, CRM Series in Mathematical Physics, https://doi.org/10.1007/978-3-030-20087-9

Coherent states (CS) (*cont.*)
Glauber–Sudarshan P-and Fock–Bargmann representations, 201–202
H and $\widetilde{H}$, 50–52
harmonic oscillator, 75
Hermite-Glauber, 329–330
Hermite-$su(1, 1)$, 330–333
Hermite-$su(2)$, 333–337
holomorphic Hermite polynomials, PHIN CS with, 72–73
hydrogen atom wave packets, 208–210
Klauder–Glauber–Sudarshan states, 199–201
Mandel parameter, 199
non-linear, 80–83, 191
oscillator wave packets, 202–204
Perelomov, 86–88
PHIN class, 71–72
photon counting, 92–94
position-dependent mass systems and quantum-classical analogies, 216–217
quantized radiation fields, 196–197
quantum-family portrait, 206–207
quantum-pictures controversy, 208–210
Schrödinger's wave packets of minimum uncertainty, 204–205
self-interference, single photons, 195–196
single-mode, 197–198
spin CS as optical CS, 83–86
standard, 75
SUSY QM, 38
from symmetric deformed binomial distributions, 90–92
time-dependent oscillator wave packets, 205–206
Weyl–Heisenberg, 79–80
Conditional symmetry
of Boussinesq equation, 378–384
non-classical method, 377–378
Confluent SUSY technique
infinite square-well potential with moving barrier, 295–297
time dependent, 294
Continuum complex-energy representation
Berggren representation, 261
pole approximation, 262
Continuum real-energy representation
BCS solution, 260
Bogolyubov transformation, 258
canonical transformation, 258
many-body problem, 256
particle number operator, 258, 259
residual interaction, 257
Continuum single-particle level density (CSPLD), 256, 258, 262
Coupling constant metamorphosis mechanism, 181
$\mathbb{C}P^{N-1}$ *sigma models*
action integral for, 342, 343
description, 342
projector formalism and solitons, 342–350
spectral problem, 347
spin matrices, 354–355
Curvature
constant, 2
deformation approach, 22
dependent formalism, 4
Gaussian, 2
parameter, 2
trigonometric functions, 6

D
Darboux–Crum transformations, 177
Deformed binomial distributions (DFB)
coherent states, 90–92
on complex plane, 92
spin coherent states, 92
Deformed Toda model, 400–401
Dirac delta distribution, 387, 389
Dirichlet boundary value problem, 232
Dirichlet Laplacian, 232
Discrete imaginary cubic oscillators, 416
Dual shape invariance, 149–151

E
Einstein–Maxwell equations, 163, 164
Ermakov equation, 326
Euler–Lagrange (EL) equations, 343, 345, 350–354, 356
Euler–Poincaré characters, 349
Even and odd coherent states, 212–213
Exceptional orthogonal polynomials (EOP), 39, 134
Exotic supersymmetry
Darboux–Crum transformations, 177
Lax–Novikov integral, 169
nature of generators, 180
nonlinear $N = 4$, 166–172
structure of, 170
unbroken and broken phases, 171
Extended supersymmetry
arbitrary vector-potentials, 138–139
discrete transformations, 137–138

F
Factorization method, 40, 64, 274, 278
  development of, 38
  Infeld–Hull, 38
Fiducial state, 211
Field correlations, 198
Field theories, 395, 396
  See also SU (3) Toda model
Fock–Bargmann functions, 202
Fock space
  coherent states, 70–71
  DFB coherent states, 92
  finite-dimensional subspace, 83
Fourier coefficients, 202
Fourth order superintegrable systems, 110–120
Fox H-function, 389
Fractional partial differential equation, 388
Free particle Hamiltonian model, 309–310

G
Gamow state, 259, 261
Gamow vectors, 315, 316
Garnier system, 370–371
Gaussian error-curve, 204
Generalized Weierstrass formula for immersion (GWFI), 342, 345
Generic matrix superpotentials, 143
Geodesic dynamics
  geodesic parallel and polar coordinates, 8–9
  on sphere, 3
Geodesic parallel and polar coordinates, 8–9
Glauber's approach, 189
  first-order correlation function, 194
Glauber–Sudarshan coherent states
  adjective coherent, 76–79
  definition, 74–75
  properties, 74–75
  Weyl–Heisenberg CS with Laguerre polynomials, 79–80
Glauber–Sudarshan P-and Fock–Bargmann representations, 201–202
Glauber–Sudarshan P-representation, 191
Graphene
  Dirac electrons in, 39
  motion of electrons in, 61
Grassmannian manifolds, 356
Green operator, 310

H
Hamilton–Jacobi equation, 105
Hartree–Fock–Bogoliubov framework, 256
Heisenberg's theory, 207
Hénon–Heiles system
  curved counterpart, 26
  on Euclidean plane, 24–26
  integrable, 22–26
  KdV hierarchy, 3
  on S^2 and H^2, 26–29
Hermite functions, 269
Hermite–Gaussian modes, 324
  in homogeneous medium, 325
  ladder operators, 326–329
  stationary, 325–326
Hermite–Glauber coherent states, 329–330
Hermite-*su*(1, 1) coherent states, 330–333
Hermite-*su*(2) coherent states, 333–337
Hidden symmetry
  definition, 163
  nature of generators, 180
  See also Nonlinear supersymmetry
Higgs oscillator, 19
Higher order integrals, 120
  See also Superintegrable systems
Hilbert-space metric, 415
Holomorphic Hermite polynomials
  PHIN class, 72–73
Hydrogen atom wave packets, 208–210
Hyperboloid
  geodesic dynamics on, 4–9
  plane, 2

I
Immersion functions, 351–354
Incoherent, defined, 188
Infeld–Hull factorization method, 38
Infinite square-well potential with moving barrier, 288–290
  confluent SUSY partner of, 295–297
Integrable perturbations, 24
Intertwining operator, 286
Irreducible representations (IR), of *su*(2, 2), 280
Isotropic oscillator, 14
  harmonic, 120

J
Jacobi polynomials, 84, 85, 269–271

K
Kadomtsev–Petviashvili equation, 106
Kepler–Coulomb system, 105
Kerr-NUT-(A)dS solutions, 163

Killing system, 369–370
Killing–Yano tensors, 164
Klauder–Glauber–Sudarshan states, 199–201
Korteweg-de Vries (KdV) equation, 106, 164
Kronig–Penney model, 310

L
Ladder operators, for Hermite–Gaussian modes, 326–329
Lagrangian expression, 407
Laguerre polynomials, 79–80
Laplace–Runge–Lenz vector, 105
Lax–Novikov integrals, 164, 169
Lie algebras, 268, 269, 274, 275, 278
Lie group of symmetries, 375
Lie point transformations, 376
Liouville integrable, 104
Liouville type I dynamical systems, in sphere and plane, 359
  Euler elliptic coordinates, 365
  Garnier system, inverse gnomonic projection of, 370–371
  gnomonic projections, 360–363
  Killing system, 369–370
  Neumann system, 367–369
  sphero-conical coordinates, 363, 364
  trajectory isomorphism, 366–367
Lippmann–Schwinger formula, 310

M
Mandel parameter, 199
Many-body systems, 255
Matrix superpotentials
  of dimension 2 X 2, 144–147
  of dimension 3 X 3, 144–147
  dual shape invariance, 149–151
  generic matrix superpotentials, 143
  Pron'ko–Stroganov problem, 142–143
  scalar superpotentials, 144
  shape invariant QM systems with, 148–149
Meijer G-function, 54
Morse potential, 305
Multiparametric Hamiltonians, 421–422

N
Neumann system, 367–369
Non-Abelian integrability, 106
Non-classical property of states, 189
Non-conserving particle number pairing solution
  continuum complex-energy representation, 261–262
  continuum real-energy representation, 256–260
Non-Hermitian Hamiltonians, 411–414
Non-linear coherent states, 191
  definition, 80
  deformed Poissonian, 81
  q deformations of integers, 82–83
Non-linear supersymmetry
  characteristics, 164
  exotic nonlinear $N = 4$ supersymmetry, 166–172
  invisible *PT*-symmetric zero-gap systems, 172–174
  PT-symmetric zero-gap Calogero systems, 174–177
  and quantum anomaly, 164–166
  rationally extended harmonic oscillator and conformal mechanics systems, 177–181
Non-uniform lattices with site-dependent spacings, 418
N-th order integral, 108

O
One dimensional Salpeter Hamiltonian with *N* deltas, 317–320
One dimensional Schrödinger Hamiltonian with *N* Dirac delta interactions
  bound states calculation, 313–315
  Green function, 313
  Lippmann–Schwinger equation, 311, 312
  resonances and Gamow states, 315–317
  scattering states calculation, 312–313
Optical coherence
  classification, 187
  definition, 187
Orthogonal polynomials, 124
Oscillator potentials
  harmonic, 46
  superintegrable, 16
Oscillator wave packets, 202–204

P
Painlevé equations
  IV transcendents, 56–58
  SUSY QM and, 54–61
  V transcendents, 58–61
Painlevé transcendents
  anisotropic oscillator, 110
  ODEs with, 115–120
  orthogonal polynomials, 124

Painlevé property and, 106
Schrödinger equation with, 123–124
Superintegrable systems, 110
Parasupersymmetry, 134
Paraxial beams, 324
Partial coherence, 189
Partial differential equations (PDE), 376
See also Conditional symmetry
Perelomov coherent states, 86–88, 331
Peyrard–Bishop–Dauxois (PBD) model, for DNA molecule, 302
PHIN class
holomorphic Hermite polynomials, 72–73
normalisation condition, 71
orthonormality condition, 72
Photon counting
coherent states, 92–94
count rate, 93
Pochhammer symbol, 270
Point transformation, 286–288
Polynomial Heisenberg algebras (PHA), 38, 54, 56, 191
Position dependent mass
function, 302
rotationally invariant systems, 155–156
system dependent, two parameters, 159–160
two strategies in construction of exact solutions, 156–159
Power-law matrices, 417
Pron'ko–Stroganov problem, 142–143
Pseudo-differential Riesz operator, 387, 388
Pseudo-Hermitian representation (PHR) of quantum mechanics, 413, 414
PT-symmetric version of quantum mechanics (PTQM), 424
PT-symmetry, 419
quantum mechanics, 412

Q

Quadratic constants, 15
Quadratic refractive index optical media
Hermite–Glauber coherent states, 329–330
Hermite-$su(1, 1)$ coherent states, 330–333
Hermite-$su(2)$ coherent states, 333–337
Quadratures, 74
Quantum catastrophic evolution scenario, 413
Quantum-family portrait, 206–207
Quantum infinite square-well potential, 286–287
Quantum optics
coherent states in (*see* Coherent states (CS))
photon counting, 92
role in, 70
Quantum phase transition, 413
Quantum-pictures controversy, 208–210
Quasi-Hermitian formulation of quantum mechanics (QHQM), 424, 425
Quasi-Hermiticity property, 425
Quasi-integrability, 396
See also $SU(3)$ Toda model

R

Ramani–Dorizzi–Grammaticos (RDG) series, 3, 23, 25
Reflectionless and finite-gap
periodic systems, 172
quantum systems, 164
Repulsive oscillator, 46
Resonance condition, 116
Resonant (Gamow) state, 259, 261
Riemann zeta (ζ) function
basic formulas for, 246–252
spectral trace formula applications, 233–245
Rigged Hilbert spaces, 315
Rosochatius–Winternitz potential, 15

S

Salpeter one dimensional Hamiltonian with N deltas, 317–320
Schrödinger equation (SE)
exact solutions, 134
generic approach and energy values, 151–153
ground state solutions, 153–154
isospectrality, 155
with Painlevé potential, 123–124
Schrödinger-like equation, 302, 304, 305
Schrödinger one dimensional Hamiltonian with N Dirac delta interactions
bound states calculation, 313–315
Green function, 313
Lippmann–Schwinger equation, 311, 312
resonances and Gamow states, 315–317
scattering states calculation, 312–313
Schrödinger–Pauli equations
arbitrary vector-potentials, extended SUSY, 138–139
extended SUSY, 137–138
QM systems with exact SUSY, 136–137
Schrödinger's formulation, 207
Schrödinger's wave packets of minimum uncertainty, 204–205

Second order superintegrable systems, 106–108
Seed solutions, 41
Self-interference, single photons, 195–196
Shape invariant systems
  position dependent mass, 155–160
  QM systems with matrix superpotentials, 148–149
  Schrödinger equations (*see* Schrödinger equation (SE))
Single-mode coherent states, 197–198
Single photons, self-interference, 195–196
Smorodinsky–Winternitz system, 15
Solitons
  collective coordinate field approximation, 405–408
  non-static cases, 403–405
  scattering of, 395, 396
  static cases, 401–403
Spatial attention, 192
Spectral design, quantum mechanics, 63
Spectral zeta functions, 232
Sphere
  ambient space coordinates, 5–8
  geodesic dynamics on, 3
Spin CS as optical CS
  Barut–Girardello CS, 88
  detection probability distribution, 83
  Jacobi polynomials, 84, 85
  optical phase space, 84
  Perelomov CS, 86–88
Spin matrices, 354–355
Squeezed states
  coherent states, 213
  definition, 190
  even and odd coherent states, 212–213
  expression, 217
  generalized oscillator algebras, 215–216
  group approach and, 210–211
  position-dependent mass systems and quantum-classical analogies, 216–217
  two-mode, 213–214
  vacuum, 212
Standard coherent states, 75
Stationary Hermite–Gaussian modes, 325–326
Stefan problems, 285
su(2, 2) AJF representation, 278–280
Sub-Poissonian statistics, 191
Supercharges, 137
Superintegrability (SI), 135
Superintegrable systems, 3, 16
  Cartesian coordinates, 112–114
  classification of, 120
  equations of motion, 110
  fourth order, 110–120
  four three-parameter families, 108
  Kepler-Coulomb system, 105
  Painlevé transcendents, 110
  second order, 106–108
  theory of, 106
Superpartner, 141
Superposition, coherent states, 190
Supersymmetric quantum mechanics (SUSY QM)
  applications, 61–63
  birth of, 37–39
  coherent states of H and $\widetilde{H}$, 50–51
  confluent second-order, 44–45
  confluent Transformations, 42–45
  construction and wave functions, 124–127
  definition, 40
  and exactly solvable potentials, 45–48
  first-order, 47
  H, algebraic structure, 49
  $\widetilde{H}$, algebraic structure, 50
  harmonic oscillator potential, 46–48, 52–54, 60–61
  higher-order, 40
  non-confluent, 45
  and Painlevé equations, 54–61
  second-order, 47–48
  standard transformations, 41–42
  1-SUSY, 286
Supersymmetry (SUSY)
  arbitrary vector-potentials, extended, 138–139
  definition, 133
  extended, 137–138
  matrix superpotential, 135
  QM systems with exact, 136–137
  quantum mechanics, 134
  technique, 286 (*see also* Confluent SUSY technique)
su(1, 1) representations, 276–278
1-SUSY technique, 286
SU (3) Toda model
  deformation of, 400–401
  equations of motion, 397
  soliton solutions, 397–399
Sym-Tafel (ST) formulas, 347

T
Time-dependent oscillator wave packets, 205–206
Time dependent Schrödinger equation, 286
Time dependent supersymmetric quantum mechanics, 291–292

Trace class operators, 233
Tridiagonal discrete-oscillator Hamiltonians, 418
Trigonometric Pöschl-Teller potential with moving barrier, 292–293
2D orthogonal Cayley–Klein geometries, 5
Two-mode squeezed states, 213–214

U
Unitary evolution, physics of
  construction of collapse, 420–421
  multiparametric Hamiltonians, 421–422
  non-uniform lattices with site-dependent spacings, 418
  PT-symmetry, 419
Unitary PT-symmetric quantum theory, 414–415
  non-Hermitian Hamiltonians, 413–414
  pseudo-Hermitian representation, 413, 414

V
Vacuum squeezed states, 212
Variational method, 303–304
  exponential position-dependence, mass and potential with, 304–305
  ground state solution, 302
  mass with quadratic position-dependence, 306–307
  potential with quadratic and inverse quadratic position-dependence, 306–307
von Roos ambiguity parameters, 302

W
Ward model, 396
Wave packets
  hydrogen atom, 208–210
  minimum, 203
  oscillator, 202–204
  time-dependent oscillator, 205–206
Weierstrass coordinates, 6
Weyl–Heisenberg coherent states, Laguerre polynomials, 79–80
Wigner d_j rotation matrices, 269, 272
Willmore functionals, 349

Printed in the United States
By Bookmasters